45
Topics in Heterocyclic Chemistry

Aims and Scope

The series Topics in Heterocyclic Chemistry presents critical reviews on present and future trends in the research of heterocyclic compounds. Overall the scope is to cover topics dealing with all areas within heterocyclic chemistry, both experimental and theoretical, of interest to the general heterocyclic chemistry community.

The series consists of topic related volumes edited by renowned editors with contributions of experts in the field.

More information about this series at http://www.springer.com/series/7081

Tamás Patonay • Krisztina Kónya
Editors

Synthesis and Modification of Heterocycles by Metal-Catalyzed Cross-coupling Reactions

With contributions by

Á. Balázs · G. Dormán · J. Gerencsér · M. Koley · A. Kotschy · M.D. Mihovilovic · Z. Novák · L. Rycek · M. Schnürch · R.S.G.R. Seixas · A.M.S. Silva · V.L.M. Silva · L. Wimmer

Springer

Editors
Tamás Patonay
Department of Organic Chemistry
University of Debrecen
Debrecen
Hungary

Krisztina Kónya
Department of Organic Chemistry
University of Debrecen
Debrecen
Hungary

ISSN 1861-9282 ISSN 1861-9290 (electronic)
Topics in Heterocyclic Chemistry
ISBN 978-3-319-32608-5 ISBN 978-3-319-32610-8 (eBook)
DOI 10.1007/978-3-319-32610-8

Library of Congress Control Number: 2016938783

Printed on acid-free paper

This Springer imprint is published by Springer Nature
The registered company is Springer International Publishing AG Switzerland

Dedicated to the memory of Professor Tamás Patonay

Preface

This volume is focused on new and innovative metal-catalysed reactions that lead to the formation of a carbon–carbon or a carbon–nitrogen bond. In the first chapter entitled 'Metal-Catalyzed Cross-Coupling Reactions in the Decoration of Pyridines', Moumita Koley, Michael Schnürch and Marko D. Mihovilovic outline cross-coupling reactions of pyridine derivatives applying the approaches that involve pyridine as organometal or as (pseudo)halide species. A lot of research is dedicated to cross-coupling reactions in aqueous medium that is of particular importance for environmental reasons. This chapter brings all types of cross-coupling methods covering mainly the literature from the past decade.

The chapter 'Metal-Catalyzed Cross-Coupling Reactions in the Decoration of Pyrimidine, Pyridazine, and Pyrazine' by Laurin Wimmer, Lukas Rycek, Moumita Koley and Michael Schnürch provides us with the comprehensive review on the cross-coupling chemistry of electron-deficient N-heterocycles containing two nitrogen atoms. The focus lies on new developments in the field, e.g. regarding new catalytic systems, and covers literature from 2008 until late 2013.

The third chapter entitled 'Metal-Catalysed Cross-Coupling Reactions in the Synthesis and Transformations of Quinolones and Acridones' by Raquel S. G. R. Seixas, Vera L. M. Silva and Artur M. S. Silva is an overview of the applications of cross-coupling reactions in the synthesis and transformations of quinolones and acridones. These compounds are widely recognized by their diverse bioactivity being useful structural moieties for drug candidates. Furthermore, they hold significant interest due to their host–guest chemistry; applications in chemical, biochemical and environmental analyses; and utility in synthetic methods' development.

The chapter 'Synthesis and Transformations of Oxygen Heterocycles' by Zoltán Novák and András Kotschy deals with the recent developments in the transition metal-catalysed synthesis and transformations of such oxygen-containing heteroaromatic systems where the oxygen is part of a five-membered ring.

The last chapter 'Transition Metal-Catalyzed Coupling Reactions in Library Synthesis' authored by János Gerencsér, Árpád Balázs and György Dormán describes the recent progress in utilizing carbon–halogen bonds for a variety of transition metal-catalysed reactions, including Suzuki–Miyaura, Sonogashira,

Buchwald–Hartwig, Stille, Negishi and Heck couplings. Furthermore, several C–H activation/arylation protocols are reported for a direct derivatization of the heterocyclic cores. Cross-coupling reactions are also implemented into domino and cascade reactions that enable the construction of novel ring systems.

Let me take this opportunity to express my sincere thanks to the authors for their outstanding contributions as well as to Elizabeth Hawkins, Judith Hinterberg and other people from Springer that are involved in the project for their patience and for their dedicated support in completing this volume.

The volume was edited by Professor Tamás Patonay. He was extremely motivated in accepting this challenge and immediately invited several distinguished authors to contribute the chapters of their expertise. Unfortunately, he passed away on June 23, 2015.

Professor Patonay was born in Debrecen, Hungary, on September 26, 1951. After finishing his undergraduate education at the University of Debrecen, he continued his studies at the same institution to reach the M.Sc. degree (1976) and Ph.D. in chemistry (1980). Furthermore, he was promoted to the D.Sc. in chemistry at the Hungarian Academy of Sciences in 2002. He was employed as a research associate at the Research Group for Antibiotics of HAS, Debrecen, from 1976 till 1982. Then, he joined the Department of Organic Chemistry, University of Debrecen, as a senior lecturer, where he became an associate professor in 1990 and was promoted to professor of organic chemistry in 2004. He served as a head of the department from 2008 onwards. He spent some time abroad. Namely, he was a visiting research associate with Professor R. V. Hoffman at the Department of Chemistry, New Mexico State University, Las Cruces, USA (1993–1994), and with Professor L. Hevesi at the Departement de Chimie, Facultes Universitaire Notre-Dame de la Paix, Namur, Belgium (1996). He also developed very intensive collaborations with several other laboratories in Europe and in the USA.

Although his research interest was very broad, let me mention only some of his favourite topics: metal-catalysed C–C and C–N cross-coupling reactions of oxygen heterocycles; synthesis of oxygen- and sulphur-containing heterocycles; the chemistry of 2-azido-, 2-halo- and 2-sulphonyloxy ketones; an application of new oxidizing and reducing agents for heterocyclic compounds; asymmetric oxidations and reductions; and highly efficient synthetic methods involving microwave-assisted and parallel syntheses. He published about 125 papers in the peer-reviewed international journals, 17 review articles and book chapters, as well as 3 patents. As an invited speaker, he delivered a number of lectures at international conferences, at universities and at research institutes. He also served as a guest editor of the special issue published in 2009 by the *Current Organic Chemistry* entitled 'Nitrogen

Heterocycles and Their Acyclic Precursors: New Routes and Methods'. His successful and fruitful career was recognized by several honours: Higher Education Medal (1977), Outstanding Work (1986), FUNDP Scholarship (1996), Széchenyi Professor Scholarship (1998), Novicardin Prize (2006) and Academy Award (2015).

His sense of humour and direct style of communication were well known among his students. His extremely practical way of thinking in solving different problems was also legendary as he always reached the most elegant solutions to the research problems. While he carried out an extensive work in organic chemistry, he also served as an outstanding supervisor of students to reach their degrees. He always supported their experimental work giving them the opportunities to enjoy the beauty of the research experiences. It happened very often that even his former students visited him to get support or a friendly advice. Whenever possible, he spent the holidays discovering the wilderness of nature. He particularly loved the mountains, and he was an excellent teammate not only during the hiking trips but also on other occasions. I met Tamás for the first time about two decades ago at the University of Ljubljana when he delivered an outstanding lecture devoted to his research work. He came back to Slovenia several times and I had also a pleasure of visiting him twice at his department. We also met each other at several conferences around the world. I have had always the privilege to discuss with him various aspects of organic chemistry and numerous other topics as well.

The organic chemistry community has lost an excellent scientist and a man who was able to accept challenges and solve them with optimism and positive approach. He is greatly missed not only by his family but also by numerous colleagues and friends around the world. He will be always remembered but never replaced.

Ljubljana, Slovenia
February 2016

Slovenko Polanc

Contents

Top Heterocycl Chem (2016) 45: 1–60
DOI: 10.1007/7081_2015_141

Published online: 18 April 2015

Metal-Catalyzed Cross-Coupling Reactions in the Decoration of Pyridines

Moumita Koley, Michael Schnürch, and Marko D. Mihovilovic

Abstract Cross-coupling reactions involving pyridine derivatives are discussed for both approaches involving pyridine as organometal or as (pseudo)halide species. All types of cross-coupling methods are included, and mainly literature from the past decade is covered. Older landmark contributions are included as well whenever it is necessary to communicate key concepts. This chapter is organized according to the coupling type and the role of the pyridine derivative, either as (pseudo)halide or as organometal species.

Keywords (Pseudo)halide · Boronic acid · Cross-coupling · Metal organyl · Palladium

Contents

M. Koley, M. Schnürch, and M.D. Mihovilovic (✉)
Institute of Applied Synthetic Chemistry, Vienna University of Technology, Getreidemarkt 9/163, 1060 Vienna, Austria
e-mail: marko.mihovilovic@tuwien.ac.at

Abbreviations

9-BBN	3-Borabicyclo[3.3.1]nonane
acac	Acetylacetonate
BMIM	1-Butyl-3-methylimidazolium
BrettPhos	2-(Dicyclohexylphosphino)3,6-dimethoxy-2′,4′,6′-tri-*i*-propyl-1,1′-biphenyl
cBRIDP	Di-*tert*-butyl(2,2-diphenyl-1-methyl-1-cyclopropyl)phosphine
CM-phos	2-[2-(Dicyclohexylphosphino)phenyl]-1-methyl-1*H*-indole
cod	Cyclooctadiene
C-Phos	2-(2-Dicyclohexylphosphanylphenyl)-*N1*,*N1*,*N3*,*N3*-tetramethylbenzene-1,3-diamine
CuMeSal	Copper(I) 3-methylsalicilate
CuTC	Copper(I) thiophene-2-carboxylate
Cy	Cyclohexyl
CyPF-*t*Bu	(*R*)-1-[(S_P)-2-(Dicyclohexylphosphino)ferrocenyl]ethyldi-*tert*-butylphosphine
DABCO	Diazabicyclo[2.2.2]octane
DavePhos	2-Dicyclohexylphosphino-2′-(*N*,*N*-dimethylamino)biphenyl
dba	Dibenzylideneacetone
DIPEA	*N*,*N*-Diisopropylethylamine
DMF	*N*,*N*-Dimethylformamide
DPE-Phos	Bis[(2-diphenylphosphino)phenyl] ether
dppb	1,4-Bis(diphenylphosphino)butane
dppf	1,1′-Bis(diphenylphosphino)ferrocene
dppp	1,3-Bis(diphenylphosphino)propane
DSC	Dye-sensitized solar cell
D-*t*-BPF	1,1′-Bis(di-*t*-butylphosphino)ferrocene
EDG	Electron donating group
EWG	Electron withdrawing group
HASPO	Heteroatom-substituted secondary phosphine oxide
IMes HCl	1,3-Bis(2,4,6-trimethylphenyl)imidazolium chloride
IPr	Bis(2,6-diisopropylphenyl)imidazol-2-ylidene

IPrNi(allyl)Cl	Allyl[1,3-bis(2,6-diisopropylphenyl)imidazol-2-ylidene] chloropalladium(II)
JosiPhos	(*R*)-1-[(S_P)-2-(Dicyclohexylphosphino)ferrocenyl]ethyldi-*tert*-butylphosphine
MIDA	*N*-Methyliminodiacetic acid
MTBE	Methyl *t*-butyl ether
NHC	*N*-Heterocyclic carbene
NMP	*N*-Methyl-2-pyrrolidone
OLED	Organic light-emitting diode
PEG	Polyethylene glycol
PEPPSI-IPent	Dichloro[1,3-bis(2,6-Di-3-pentylphenyl)imidazol-2-ylidene] (3-chloropyridyl)palladium(II)
PEPPSI-IPr	[1,3-Bis(2,6-diisopropylphenyl)imidazol-2-ylidene] (3-chloropyridyl)palladium(II) dichloride
PMB	*p*-Methoxybenzyl
PTS	Polyoxyethanyl-α-tocopheryl sebacate
PyBroP	Bromo-tris-pyrrolidino-phosphonium hexafluorophosphate
rt	Room temperature
RuPhos	2-Dicyclohexylphosphino-2′,6′-diisopropoxybiphenyl
Salen	2,2′-Ethylenebis(nitrilomethylidene)diphenol
SDPP	Silicadiphenylphosphinite
SES-NH_2	2-(Trimethylsilyl)ethanesulfonyl
S_NAr	Substitution nucleophilic aromatic
SPhos	2-Dicyclohexylphosphino-2′,6′-dimethoxybiphenyl
SPO	Secondary phosphine oxide
TBAB	Tetrabutylammonium bromide
TBAF	Tetrabutylammonium fluoride
TBS	Tert-butyldimethylsilyl
Tedicyp	*cis,cis,cis*-1,2,3,4-tetrakis(diphenylphosphinomethyl)-cyclopentane
t-BuJosiPhos	(*R*)-1-[(S_P)-2-(Diphenylphosphino)ferrocenyl]ethyldi-*t*-butylphosphine
THF	Tetrahydrofuran
TMEDA	*N,N,N′,N′*-Tetramethylethylenediamine
tmp	Tetramethylpiperidide
TON	Turnover number
Xantphos	4,5-Bis(diphenylphosphino)-9,9-dimethylxanthene
XPhos	2-Dicyclohexylphosphino-2′,4′,6′-triisopropylbiphenyl

1 Introduction

Nitrogen-containing heterocycles are compounds of high relevance in many areas of chemistry and in everyday life. Among them, pyridine can be regarded as a privileged structure. Highly functionalized pyridines, including aryl and heteroaryl

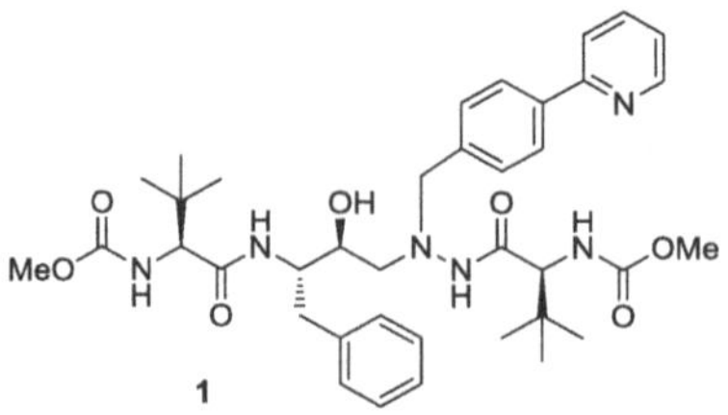

1

Atazanavir

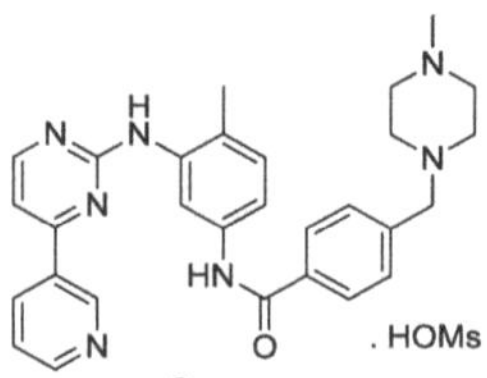

2

Imatinib Mesylate

Fig. 1 Pyridine containing drugs on the market

substituted derivatives, are among the most prevalent heterocycles encountered in natural products, pharmaceutical and agrochemical industry, as well as in material science [1–4]. According to a recent MDL Drug Data Report, the most common heterocycle in pharmaceutically active compounds is pyridine. Examples of drugs based on pyridine are atazanavir (Reyataz) (**1**) [5] and imatinib mesylate (Gleevec) (**2**) [6] (Fig. 1) which are prescribed against human immunodeficiency virus (HIV) and chronic myelogenous leukemia, respectively.

As pyridines are the most common heterocyclic motif found in pharmaceutically active compounds, preparative methods of pyridine derivatives remain an essential research topic in organic synthesis. Besides constructing the pyridine core via classical cyclization strategies, decoration of pyridine building blocks via different types of reactions is applied frequently. As an electron poor heterocycle, nucleophilic substitutions can be carried out in a relatively facile fashion. Via this method, a C-heteroatom bond is formed between a pyridine C (most commonly in position 2 or 4) and the heteroatom of a nucleophile. For introducing a C-residue, other types of reactions are better suited, one example being metal-catalyzed cross-coupling reactions [7]. The success story of these reactions started in the 1970s and peaked in 2010 by awarding the Nobel Prize in chemistry to Akira Suzuki, Ei-ichi Negishi, and Richard Heck. Cross-coupling reactions have been applied on heterocyclic systems and, naturally, also on pyridine with a substantial number of examples being reported.

Within this chapter, the various types of cross-coupling reactions applied on pyridine will be discussed. The chapter is organized according to the coupling type and the role of the pyridine derivative, either as (pseudo)halide or as organometal species. We believe that this will be convenient for readers in order to most effectively find an answer to a specific question. Regarding the timeframe covered, focus was put on the past decade. Older landmark contributions are included as well whenever it is necessary to communicate key concepts.

2 Pyridine in Suzuki–Miyaura Reactions

The palladium-catalyzed cross-coupling reaction between different types of organoboron compounds and various electrophiles such as halides and triflates was initially described by Suzuki and Miyaura in 1979 [8]. It is now generally referred to as the Suzuki–Miyaura reaction (often only the Suzuki reaction). Since then, continuous advances have been made with respect to reaction scope, including coupling partners, catalyst loading, temperature profile, etc. It is nowadays also possible to couple sterically demanding substrates, and even asymmetric variants have been reported. Suzuki–Miyaura coupling is now considered as one of the most powerful, versatile, and general methodologies for the construction of C–C bonds via metal catalysis. Among the attractive features of the Suzuki–Miyaura reaction are the wide availability, stability (to air and moisture), and low toxicity of boronic acids, as well as the facile removal of the boron-containing side products of the coupling process.

2.1 Pyridine as (Pseudo)halide in Suzuki–Miyaura Reactions

Many pyridine halides are commercially available or easily accessible, and hence, they have found widespread application in cross-coupling chemistry. Pyridine is considered as a privileged structure, and hence, it is not surprising that pyridine halides have been applied in Suzuki reactions frequently. Within this subchapter, new developments regarding functional group tolerance, coupling of less reactive (pseudo)halides (mainly chlorides), and regioselective or one-pot procedures are particularly highlighted.

2.1.1 Recent Developments in Suzuki–Miyaura Coupling of Pyridine Bromides

Santelli and coworkers reported an efficient catalyst system for the coupling of bromopyridine derivatives giving high yields of the coupling products [9, 10]. As the phosphine ligands used in Suzuki–Miyaura coupling have important influence on stability and rate of the catalytic reaction, *cis,cis,cis*-1,2,3,4-tetrakis(diphenylphosphinomethyl)-cyclopentane (Tedicyp) (**3**) [11–13] was tested in Suzuki–Miyaura reactions as novel phosphine ligand. The ligand contains a cyclopentane ring to which four diphenylphosphinoalkyl groups are stereospecifically bound to the same face. In combination with $[PdCl(C_3H_5)]_2$, a catalyst system is formed which proved to be very efficient in Suzuki–Miyaura reactions, especially for heterocyclic substrates, including a number of bromopyridines (Scheme 1).

Extremely low catalyst loadings could be applied. 3-Bromopyridine as best performing pyridine halide required a substrate to catalyst ratio of 10,000,000

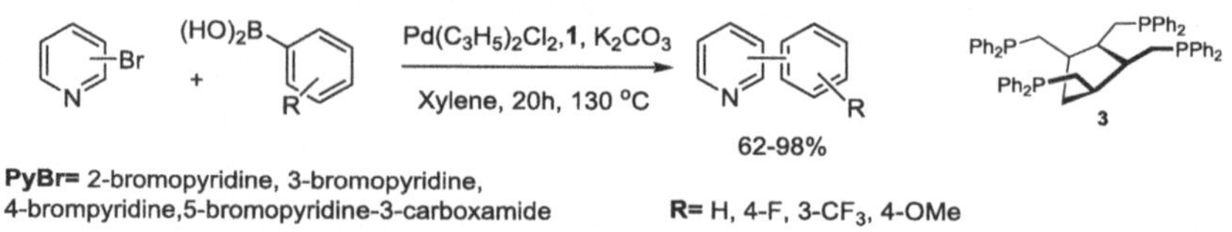

Scheme 1 Tedicyp as ligand in Suzuki-Miyaura couplings

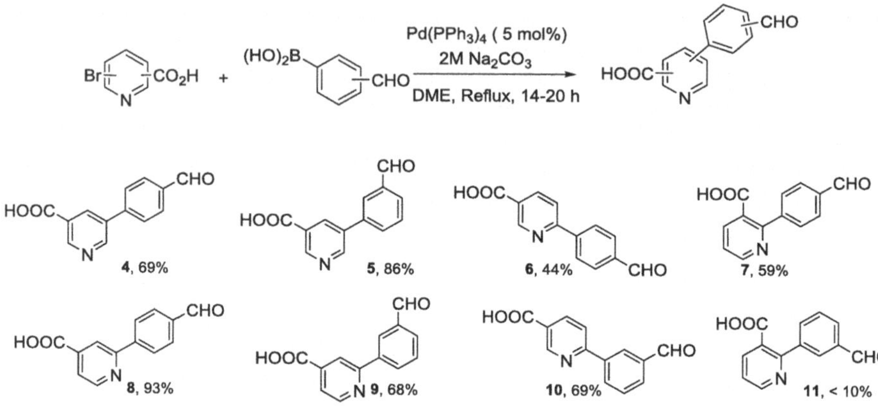

Scheme 2 Coupling between formylphenylboronic acids and bromobyridine carboxylic acids

only. This is rather counterintuitive since due to the electronegativity of the nitrogen atom, the 2- and 4-positions of halopyridines are expected to be more reactive towards oxidative addition to Pd(0). The highest TON was observed for 3-bromopyridine (2,500,000) followed by 4-bromopyridine (810,000) and 2-bromopyridine (62,000).

Suzuki–Miyaura coupling reactions of unprotected bromopyridine carboxylic acids with formylphenylboronic acids were reported by Schaub and coworkers [14]. $Pd(PPh_3)_4$ was applied as catalyst in a mixture of an aqueous Na_2CO_3 solution and 1,2-dimethoxyethane. The coupling products were synthesized in multi-gram scale. Conversions and yields of the coupling reactions depended on electronic effects (substitution pattern of the bromopyridine carboxylic acids). *Meta*-bromo-substituted-carboxylic acids gave the highest yields of coupling products with 4- and 3-formylphenylboronic acids (Scheme 2, examples **4**, **5**, **8**, **9**) and also 2-bromopyridine-4-carboxylic acid gave reasonable good yields (Scheme 2, examples **6** and **10**). 2-Bromopyridine-3-carboxylic acid gave a reasonable yield of 59% of **7** with 4-formylphenylboronic acid, but was inefficient (<10%, **11**) with 3-formylphenylboronic acid, most likely due to the increased steric hindrance. 5-Bromopyridine-2-carboxylic acid gave a similarly poor result with all three boronic acids due to unfavorable electronic effects. The sterically demanding 2-formylphenylboronic acid gave a good yield only with 5-bromopyridine-3-carboxylic acid (75%, not shown), and all other bromides gave low yields (Scheme 2).

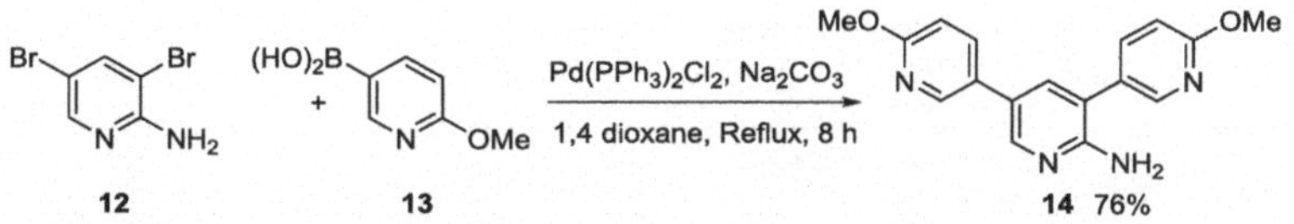

Scheme 3 Suzuki-Miyaura coupling in presence of a free amine functionality

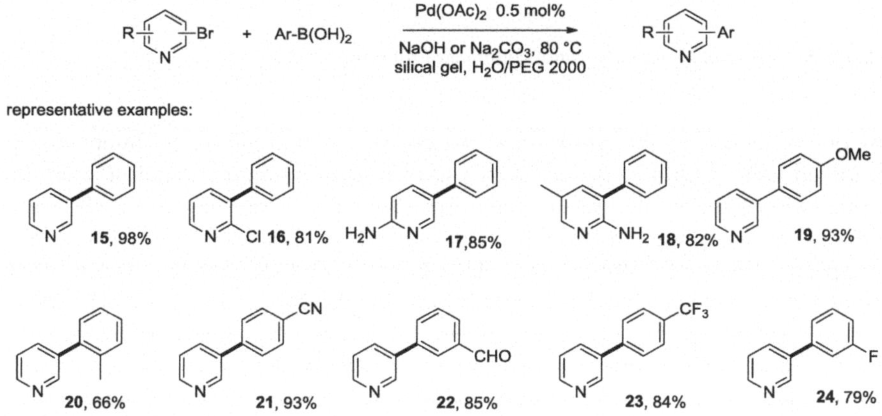

Scheme 4 Suzuki-Miyaura coupling in a PEG/water mixture

Bryce and coworkers reported a systematic study of a Suzuki–Miyaura coupling method which tolerates free primary amine functionality [15]. Within standard conditions, Na_2CO_3 (aqueous 1 M) was used as base and $Pd(PPh_3)_2Cl_2$ was employed as catalyst (5 mol%) in 1,4-dioxane at reflux for 8 h. In this work, pyridine was applied as halide and as boronic acid component as well. Both, pyridine bromides and chlorides were tested with generally good results. Besides the amine, other functional groups were tolerated, as well, such as NO_2, CF_3, OMe, or acetamide. One remarkable example is depicted in Scheme 3: in this case, both coupling partners **12** and **13** were pyridines, and double coupling gave a good yield of 76% of **14** (Scheme 3).

One convenient feature of the Suzuki reaction is the possibility to carry out reactions in aqueous medium. Zhang disclosed such a protocol under phosphine-free conditions (Scheme 4) [16]. A mixture of H_2O and PEG (3:3.5) was used as solvent. Interestingly, silica gel had a beneficial effect on the transformation. A series of pyridine bromides (and other heteroaryl bromides) could be coupled with good yields of arylated pyridine (**15–24**) in general.

A site-selective one-pot double Suzuki–Miyaura cross-coupling on dihalogenated pyridines was reported by Handy and coworkers (Scheme 5) [17], using 2,5-dibromopyridine **26** as the most suitable substrate. Naturally, the first substitution occurs at the more reactive 2-position. A series of 2,5-bisarylated pyridines (in some cases hepteneboronic acid was used as one coupling partner) were synthesized successfully with yields around 40% in most cases with no

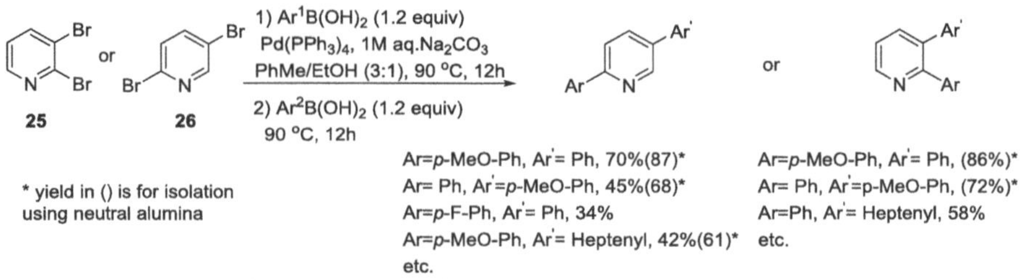

Scheme 5 Sequential Suzuki-Miyaura coupling of dibromopyridines

obvious influence of the substitution pattern of the boronic acids. Interestingly, improved yields could be observed by use of neutral alumina instead of silica for purification of the crude reaction mixtures (typically ~20% increase). The same protocol can be applied to 2,3-dibromopyridine **25** with similar effectiveness, and as expected, the first coupling took place in position 2 (Scheme 5). The reaction conditions can be considered as standard for a Suzuki–Miyaura reaction. The use of a ternary solvent mixture consisting of EtOH, toluene, and water (derived from the Na_2CO_3 solution) is noteworthy, since this is critical for the success of the protocol.

Using the same reaction conditions and 2-bromo-3-iodopyridine as substrate, the one-pot protocol led to initial cross-coupling in position 3 based on the enhanced reactivity of the C–I bond in position 3 overcoming the electronic effect of the pyridine ring which would actually favor cross-coupling in position 2 (not shown). However, no yield was reported for this specific transformation, and hence, the efficiency and utility cannot be assessed.

Rault and coworkers published an interesting study on the cross-coupling behavior of dihalopyridines bearing two different halogen atoms with halopyridylboronic acids, whereas the carbon–halogen bond on the boronic acid coupling partner had a lower reactivity than the carbon–halogen bonds in the dihalopyridines [18]. They found the heteroarylation reaction to occur at the more reactive carbon–halogen bond of the halopyridine substrates irrespective of its position on the ring. A few examples are displayed in Scheme 6. When typical reaction conditions were used (dioxane, aqueous Na_2CO_3, $Pd(PPh_3)_4$, reflux), the reaction of 2-bromo-5-chloropyridine **27** with 2-fluoro-3-pyridylboronic acid **28** gave 5-chloro-2′-fluoro-2,3′-bipyridine **29** in 42% yield (Scheme 6, Eq. (a)). Moreover, 2,6′-dichloro-3,3′-bipyridine **32** was synthesized in 73% yield from 2-chloro-5-iodopyridine **30** and boronic acid **31** (Scheme 6, Eq. (b)) and the Suzuki–Miyaura reaction between **33** and boronic acid **34** gave 6-bromo-6′-chloro-3,3′-bipyridine **35** in 74% yield (Scheme 6, Eq. (c)). When 5-bromo-2-iodopyridine **36** was coupled with compound **37** 5,6-dibromo-2,3′ bipyridine **38** was obtained in 77% yield. This shows that bromine in the presence of iodine remains inert. The procedure has its limitations when all halides are of identical nature (e.g., all bromides). In such cases, low yields and mixtures of products were obtained.

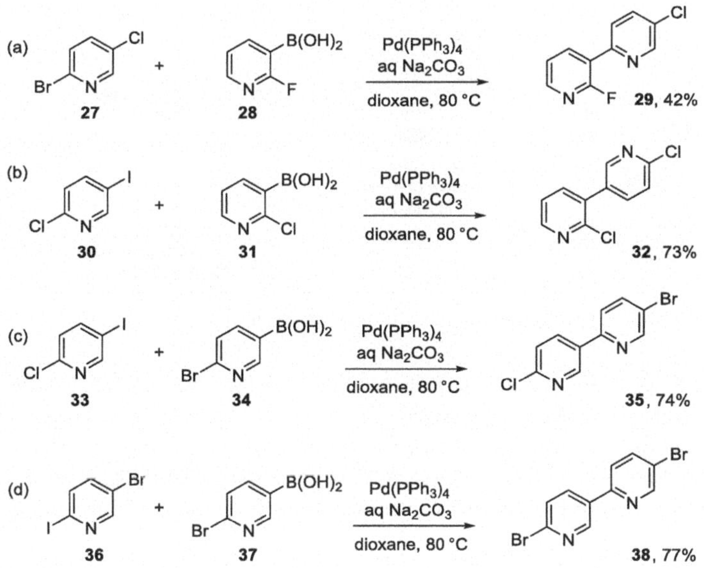

Scheme 6 Coupling of dihalopyridines with halopyridineboronic acids

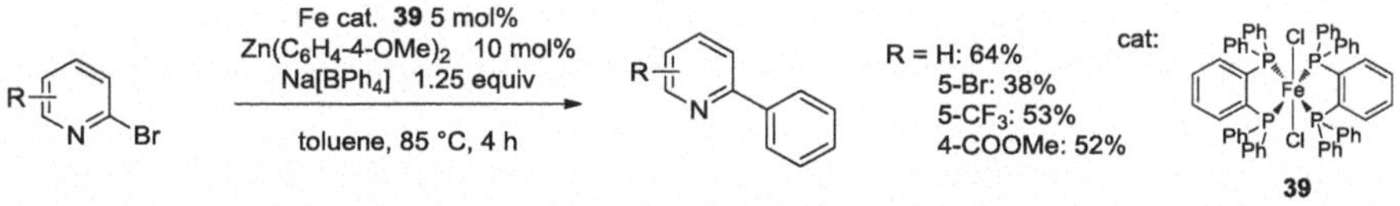

Scheme 7 Iron catalyzed Suzuki-Miyaura coupling of bromopyridines

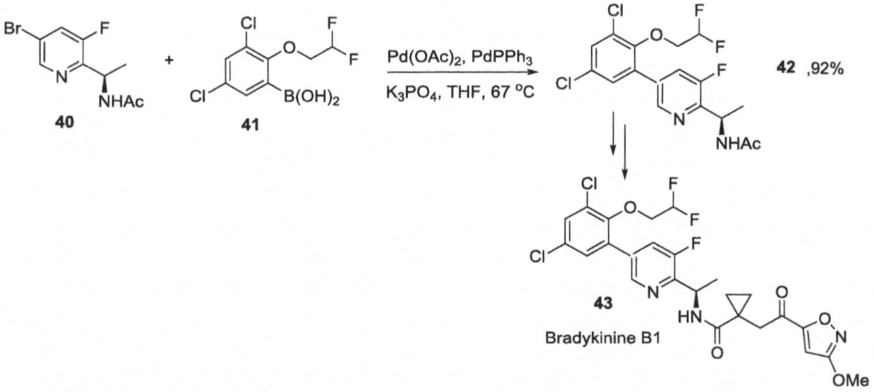

Scheme 8 Suzuki-Miyaura coupling as key step in the synthesis of Bradykinine B1

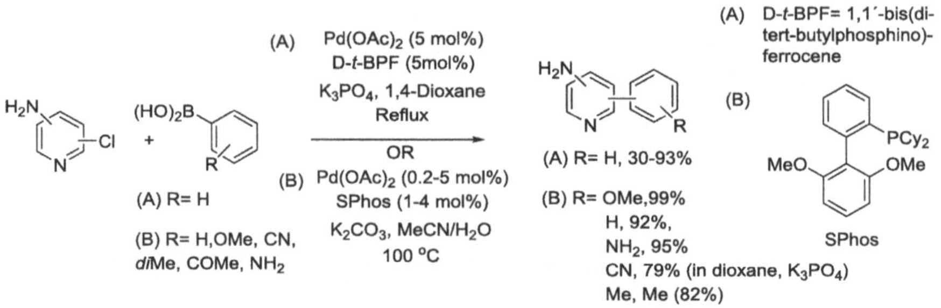

Scheme 9 Suzuki-Miyaura coupling of chloropyridineamines

Catalytic systems based on non-precious metals are of increasing interest. Bedford and coworkers disclosed such an example with their iron–zinc co-catalyzed Suzuki reaction. Also, the organometal coupling partner is unusual since sodium tetraphenylborate gave the best results. Several bromopyridines were subjected to the coupling conditions; however, only moderate yields were obtained (Scheme 7).

Many interesting applications of Suzuki–Miyaura coupling can be encountered in crucial steps of natural products synthesis. The compound responsible for the physiological processes accompanying inflammation and chronic pain is bradykinin B1 (**43**). The synthesis of (**43**) involves a Suzuki–Miyaura coupling to generate intermediate (**42**) where arylboronic acid (**41**) and bromopyridine amide (**40**) are coupled using $Pd(OAc)_2$ and PPh_3 in a 1:2 ratio with 2 M K_3PO_4 in THF (Scheme 8). This intermediate (**42**) can then be converted to bradykinin B1 in two more synthetic steps, (Scheme 8) [19].

2.1.2 Recent Developments in Suzuki–Miyaura Coupling of Pyridine Chlorides

(Hetero)aryl chlorides are even more attractive coupling partners since there are more examples commercially available, often at a cheaper price, and additionally, the atom efficiency is significantly higher when chlorides are used instead of bromides or iodides. Hence, significant efforts are undertaken to develop general protocols for the coupling of (hetero)aryl chlorides including pyridine chlorides.

Chloro-aminopyridines are considered as difficult substrates in Pd-catalyzed cross-coupling processes since the free amino group is often not well tolerated. Itoh reported the Suzuki–Miyaura reaction of 2-chloropyridine-3-amine, 6-chloropyridine-3-amine, 6-chloropyridine-2-amine, and 5-chloropyridine-2-amine with phenylboronic acid in good yields (68–93%). 2-Chloropyridine-4-amine was less successful and gave only 30% yield (Scheme 9, conditions A) [20]. The key to successful coupling is the use of the sterically demanding ligand 1,1′-bis(di-*tert*-butylphosphino)-ferrocene (D-*t*-BPF). The choice of ligand proved to be crucial also in other cases as demonstrated by Buchwald and coworkers, who

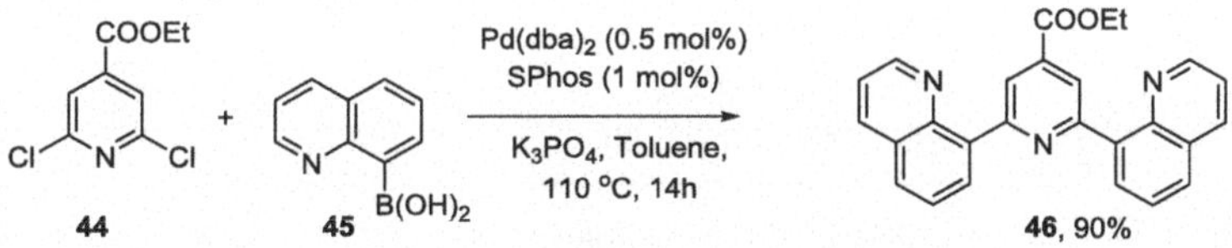

Scheme 10 Twofold Suzuki-Miyaura coupling of ethyl 2,6-dichloroisonicotinate

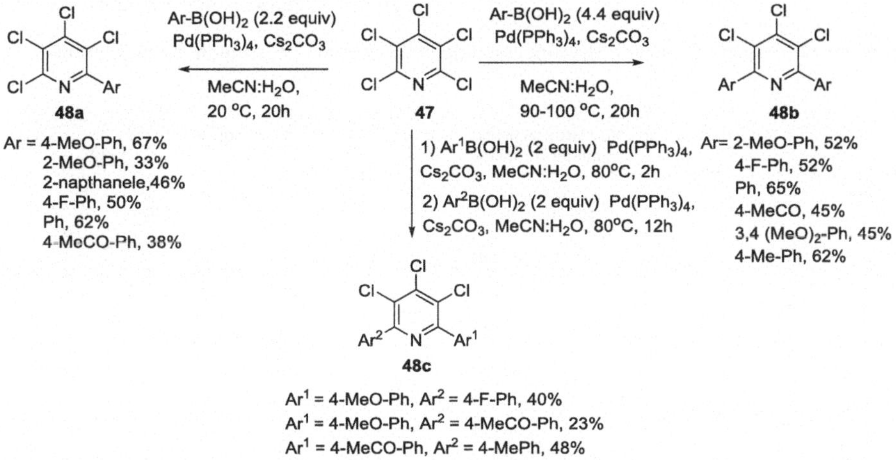

Scheme 11 Pentachloropyridine in Suzuki-Miyaura couplings

had used the highly active dialkylbiphenylphosphino ligand SPhos (Scheme 9, conditions B) which then again did not require protection of the amino group [21].

SPhos is also effective in the preparation of 2,6-di(quinolin-8-yl)-pyridine compounds, an excellent tridentate ligand for Ru(II), providing complexes with micro-second luminescent lifetimes. Johansson et al. reported the synthesis of this compound class by the double Suzuki–Miyaura coupling of a heteroarylboronic acid and halopyridine derivatives [22]. As shown in Scheme 10, an ester group of **44** was unaffected while coupling with **45** under the applied reaction conditions (Scheme 10).

Site-selective arylation reactions of pentachloropyridine (**47**) with arylboronic acids were reported by Langer and coworkers (Scheme 11) [23]. Optimization of the selective mono-arylation in position 2 revealed that simple $Pd(PPh_3)_4$ outperformed more advanced catalytic systems such as $Pd(OAc)_2$/XPhos, significantly (Scheme 11, **48a**). Using acetonitrile/water as solvent mixture and Cs_2CO_3 as base, the coupling can be achieved at room temperature while only 2.2 equiv. of boronic acid are required. 4-Methoxyphenylboronic acid gave the highest yield of 67% which is even better than for phenylboronic acid (62%). Sterically demanding boronic acids are ineffective indicated by lower yields as expected (2-naphtyl 46%, 2-MeO-phenyl 33%). Also, electron withdrawing groups are less tolerated (4-F-phenyl 50%, 4-AcO-phenyl 38%) in the reaction. Doubling the amount of boronic

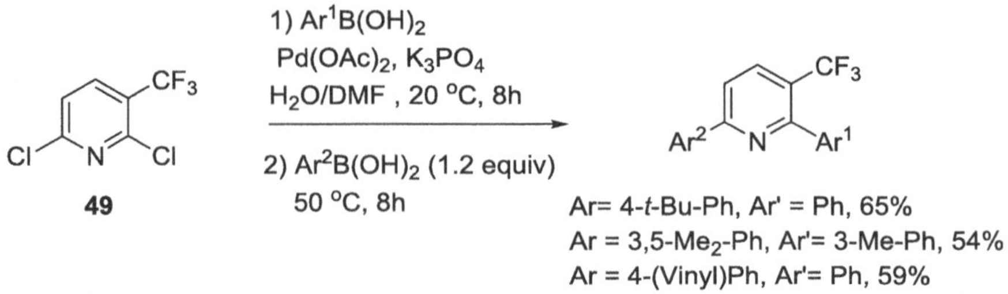

Scheme 12 Sequential Suzuki-Miyaura coupling of 2,6-dichloro-3-(trifluoromethyl)pyridine

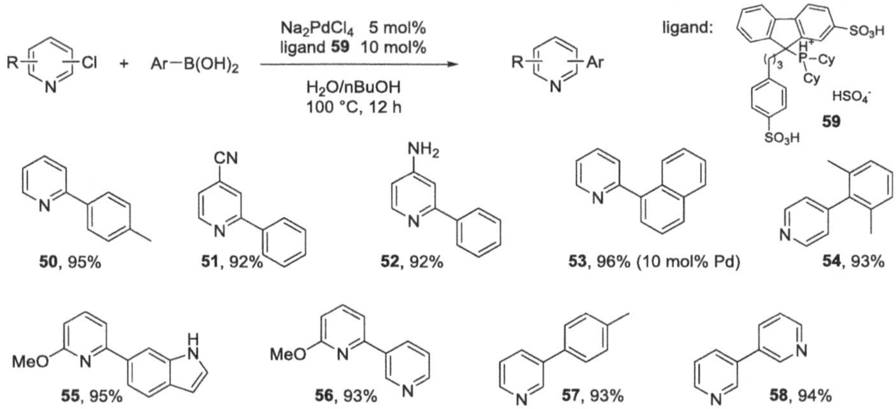

Scheme 13 A water soluble ligand in Suzuki-Miyaura coupling of pyridine chlorides

acid to 4.4 equiv. led to bis-arylation in positions 2 and 6 with moderate yields, ranging between 45% and 64% (Scheme 11, **48b**). A one-pot reaction with consecutive introduction of two different boronic acids can be performed successfully as well, and three examples were reported (Scheme 11, **48c**). In the best performing example, 48% yield was achieved, corresponding to approximately 70% yield for each individual cross-coupling step.

An example of Suzuki–Miyaura coupling on pyridine derivatives containing a CF_3 group is displayed in Scheme 12. This is an important result as trifluoromethylated pyridines [24], along with other fluorinated and trifluoromethylated heterocycles [25], are of special relevance in medicinal and agricultural chemistry due to the metabolic stability of the C–F bond. In general, palladium-catalyzed cross-coupling reactions of polyhalogenated substrates usually proceed regioselectively in favor of the sterically less hindered and electronically more deficient position. In the case of pyridine halide **49**, the Suzuki–Miyaura reaction proceeded regioselectively in favor of the electronically more deficient position 2, which is sterically more hindered than position 6, due to the proximity of the CF_3 substituent (Scheme 12) [26].

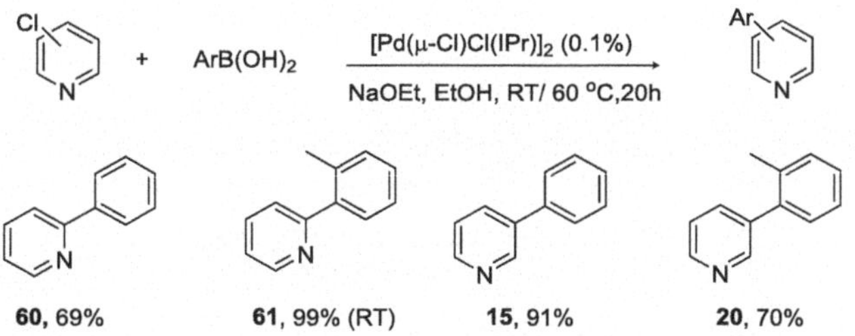

Scheme 14 NHC-assisted Suzuki-Miyaura coupling of pyridine chlorides

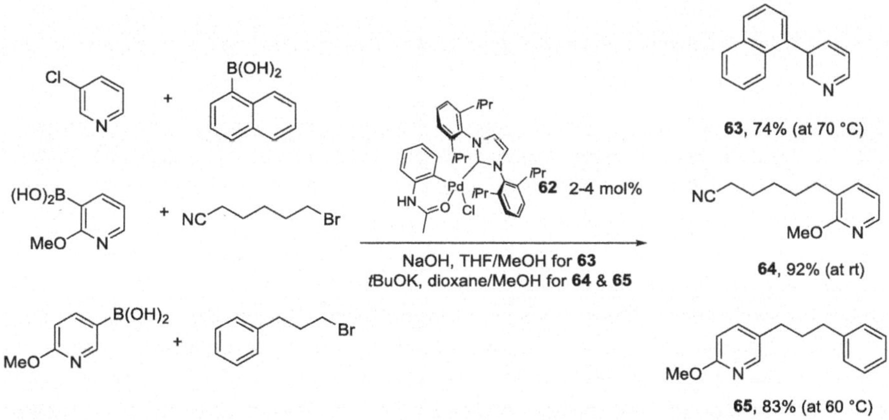

Scheme 15 A Pd-NHC catalyst in Suzuki-Miyaura coupling of pyridine derivatives

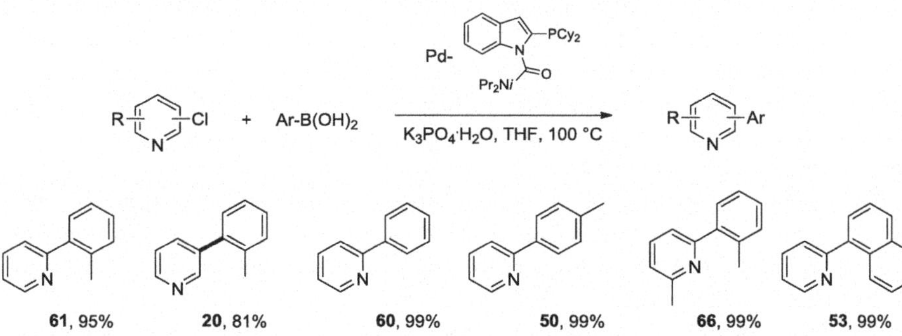

Scheme 16 An indolyl-based ligand in the coupling of pyridine chlorides

Plenio and coworkers designed water-soluble phosphine ligands (e.g., **59**) which gave excellent results in the coupling of pyridine and quinolone chlorides [27]. A water/n-BuOH mixture was used as reaction medium and Na_2PdCl_4 served as palladium source. In all reported cases involving pyridine and quinolone, yields >90% were achieved (Scheme 13, e.g., **50–58**)! Both aromatic and heteroaromatic

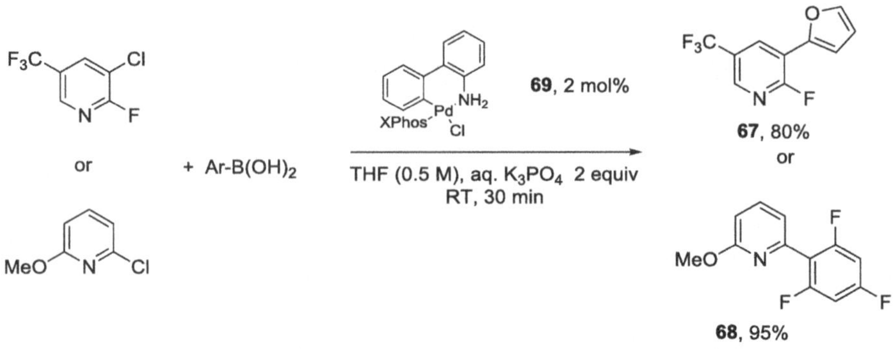

Scheme 17 Suzuki-Miyaura coupling towards flurinated building blocks

boronic acids could be applied leading to valuable bi-heteroaryls (e.g., asymmetrically substituted bipyridines). Functional group tolerance was excellent as well since NH_2, CF_3, OMe, CN, and CHO were well accepted. Representative examples are given in Scheme 13.

N-Heterocyclic carbenes (NHC) as different class of ligands were applied to Suzuki–Miyaura transformations as well. In recent years, well-defined mono-NHC palladium complexes have been developed which promote Suzuki–Miyaura reactions with low catalyst loading and under mild reaction conditions. One such NHC–Pd complex, the air- and moisture-stable $[Pd(\mu\text{-}Cl)Cl(IPr)]_2$ (IPr=bis (2,6-diisopropylphenyl)imidazol-2-ylidene), has been reported as an efficient precatalyst for coupling of 2- and 3-chloropyridines at room temperature with very low catalyst loading of only 0.1 mol% (Scheme 14, e.g., **15**, **20**, **60,** and **61**) [28].

Kantchev and Ying reported an example in which 3-chloropyridine and 2-chloro-3-methoxypyridine were coupled with boronic acids, successfully [29]. Additionally, the same catalyst **62** was able to promote coupling of pyridineboronic acid (esters) as well, in two case (**64** and **65**) even with an alkyl bromide which is a rare example of a sp^2–sp^3 coupling of pyridine under Suzuki–Miyaura conditions (Scheme 15).

Kwong and coworkers applied an indolyl-based ligand in the coupling of several pyridine chlorides with boronic acids [30]. Remarkably low catalyst loadings (0.01–0.067 mol% Pd) were sufficient to achieve almost quantitative product yields in most cases (Scheme 16).

Fluorine is a recurring substituent in commercialized and developmental drugs. Coupling methods which tolerate multiply fluorinated reaction partners are hence of great importance. One such example was reported by the group of Buchwald in 2010 [31]. They described new Pd-precatalysts **69** which were able to give the desired polyfluorinated coupling products (**67** and **68**). Two different pyridine chlorides were coupled as well in both cases with excellent yield (Scheme 17).

Most Suzuki couplings are dedicated to the coupling of two arene systems. However, also alkenylboronic acid derivatives are useful coupling partners. In their efforts towards pyrrole-fused pyridines, Hoelder and coworkers applied such

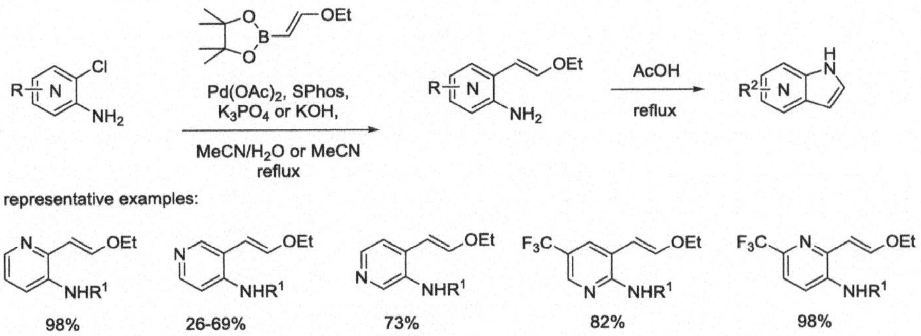

Scheme 18 Alkenylboronic esters in Suzuki-Miyaura couplings of pyridine chlorides

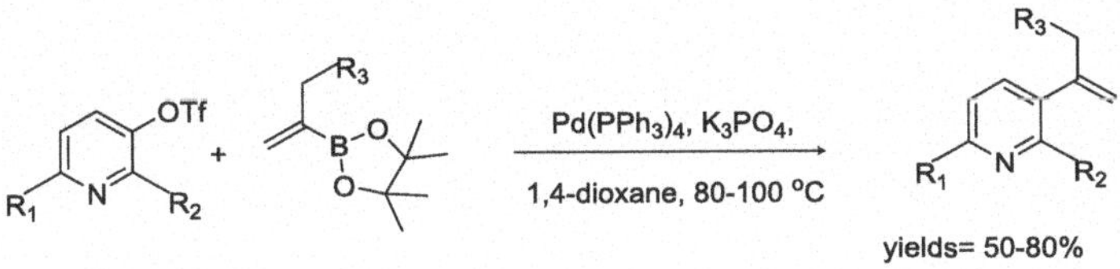

Scheme 19 Alkenylboronic esters in Suzuki-Miyaura couplings of pyridine triflates

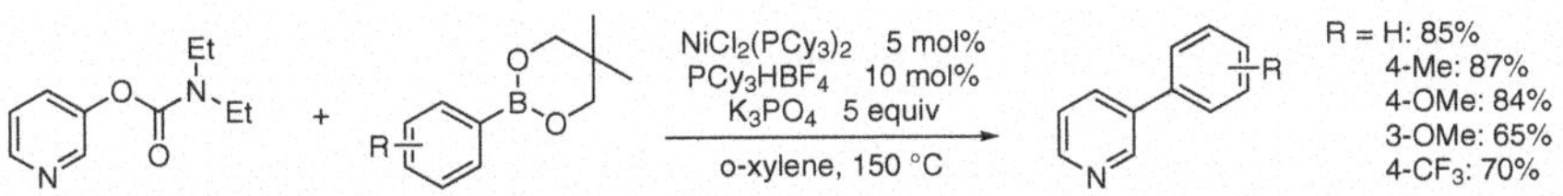

Scheme 20 *N,N*-Dietylcarbamate as leaving group in Ni-catalyzed Suzuki-Miyaura coupling

a reaction on different halogenated pyridine amines (Scheme 18) [32]. The catalytic system utilized ($Pd(OAc)_2$/SPhos) can be considered as well established by now. Yields in the alkenylation step were usually good to excellent. The same is true for the final cyclization.

2.1.3 Other Leaving Groups in Suzuki–Miyaura Reactions

Due to the better availability of 3-hydroxypyridines as compared to the corresponding halides, this substance class would be an interesting alternative in cross-coupling reactions. The hydroxyl group can be converted to triflates easily [33], which is an excellent leaving group in Suzuki–Miyaura reactions. In this regard, 3-alkenyl-substituted pyridines can be synthesized via Suzuki–Miyaura coupling as described by Vyvyan and coworkers using 3-pyridyl triflates with alkenylpinacolboronates as coupling partners (Scheme 19). The reactions proceed in moderate to good yields using simple $Pd(PPh_3)_4$ (10 mol%) as catalyst with K_3PO_4 (3 equiv.) as base in 1,4-dioxane [34]. The method can be applied to the

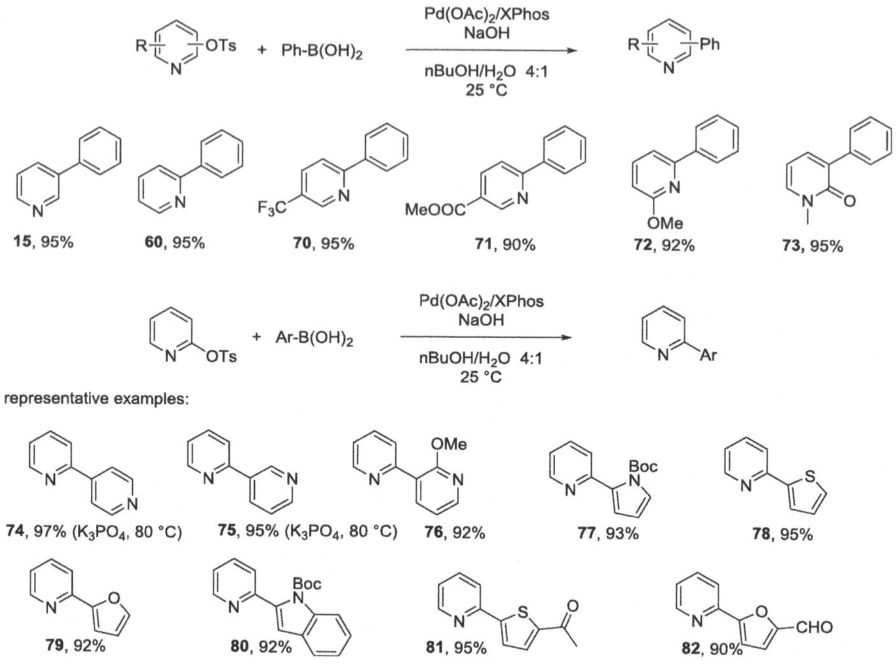

Scheme 21 Pyridine tosylates in Suzuki-Miyaura couplings

synthesis of a potential precursor for the synthesis of the natural product cananodine.

The group of Snieckus pioneered directed ortho metalation methods in which the directing group is used as leaving group (i.e., the pseudohalide) in cross-coupling reactions [35]. When employing *N,N*-diethyl *O*-carbamate as directing group in position 3 of pyridine, coupling with several boronic acid esters worked well, independent of the electronic nature of the organometal compound (Scheme 20). Noteworthy, nickel catalysis proved to be efficient which is often neglected in cross-coupling chemistry.

Tosylates are another potential leaving group accessible from readily available hydroxyl pyridines. Inamoto and Doi reported a nickel-catalyzed protocol in which pyridine-3-tosylate was coupled with phenylboronic acid (69%), 4-MeO-phenylboronic acid (86%), and 4-CN-phenylboronic acid (83%) with good results (not shown) [36].

A pyridine-bridged bis-benzimidazolylidene pincer nickel(II) complex was very effective in Suzuki coupling of various (hetero)aryl halides and pseudohalides. Regarding pyridine, 3-bromopyridine, 3-chloropyridine, as well as 3-pyridyl tosylate and 3-pyridyl mesylate were coupled with phenylboronic acid. The latter two gave 3-phenyl pyridine in 95% and 91%, respectively (not shown) [37].

Mesylates are rarely applied as leaving groups in cross-coupling chemistry since their leaving group ability is less pronounced as compared to other (pseudo)halides.

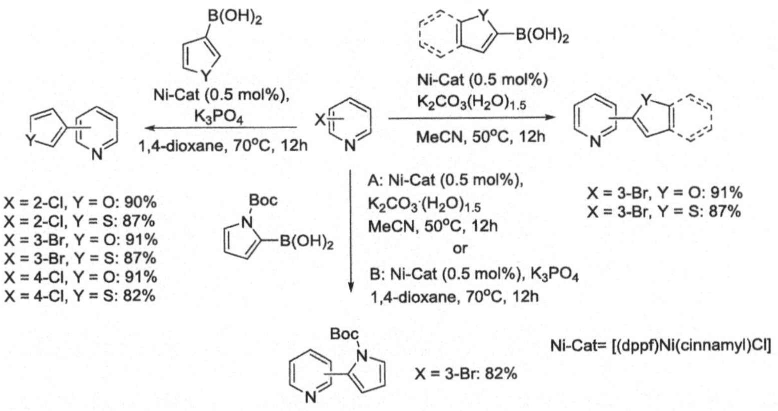

Scheme 22 Pyridine halides in Suzuki-Miyaura couplings with heterocyclic boronic acids

Again, Kwong and coworkers established a general protocol for Suzuki–Miyaura coupling of potassium aryltrifluoroborates and aryl mesylates [38]. In one example, they coupled 3-pyridyl mesylate with thiophene-3-potassium trifluoroborate which gave 66% of the coupling product (not shown). A catalytic system consisting of $Pd(OAc)_2$ and the indole-based phosphine ligand CM-Phos proved to be efficient.

A comprehensive study of pyridine chlorides but also pyridine tosylates was reported by Zhou and coworkers [39]. Yields >90% were reported within short reaction time (5 min–1 h) no matter whether pyridine chlorides or tosylates were applied. In this study, various pyridine tosylates were coupled with phenylboronic acid (Scheme 21, products **15**, **60**, **70**–**73**). Additionally, 2-pyridyl tosylate was coupled with a number of heterocyclic boronic acids, always with excellent yields (Scheme 21, products **74**–**82**). A simple catalytic system of $Pd(OAc)_2$ and XPhos proved to be ideal, and in most cases, room temperature was sufficient.

2.2 *Nickel-Catalyzed Suzuki–Miyaura Coupling*

Reduction of cost associated with metal-catalyzed reactions is important for industrial application. In this regard, replacement of palladium by the cheaper and more abundant metal nickel is an interesting option. Hartwig and coworkers reported the well-defined nickel catalyst dppf-ligated cinnamylnickel(II) chloride (dppf=1,1′-bis(diphenylphosphanyl)-ferrocene) [40]. This catalyst allows coupling of pyridine halides along with pyrimidine, pyrazine, and quinoline halides with five-membered heteroarylboronic acids (Scheme 22). These coupling reactions can be carried out with as low as 0.5% catalyst loading, and very high yields (37 examples, 81–97% yield) were obtained with no additional ligand necessary. Additionally, 50°C was sufficient to promote the reaction. Scheme 22 shows coupling

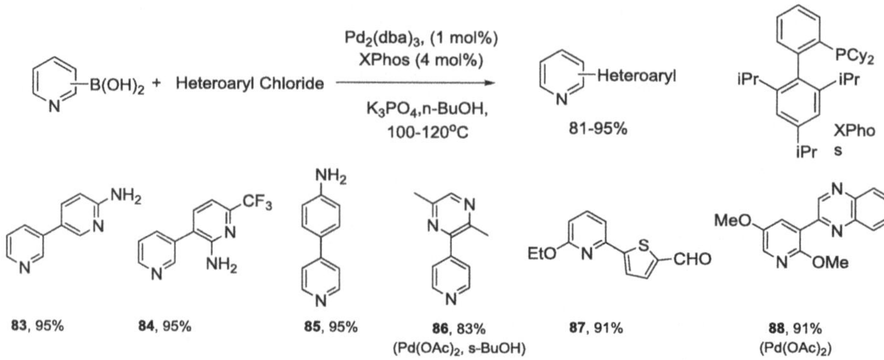

Scheme 23 Pyridineboronic acids in Suzuki-Miyaura couplings with heteroaryl chlorides

examples of simple pyridine halides. However, also substituted pyridine halides were applied (MeO, Me, CF_3, CHO, acetyl, CN) with similar efficiency.

2.3 *Pyridine as Boronic Acid Derivatives*

Pyridine-derived boronic acids have proven to be a particularly difficult class of substrate for the Suzuki–Miyaura coupling due to a relatively low stability (especially 2-pyridylboronic acid) and the formation of trimeric anhydrides which display low reactivity.

Buchwald and coworkers reported a successful utilization of pyridine boronic acids in Suzuki–Miyaura coupling. A $Pd(OAc)_2$/Xphos catalyst system provided good yields for the reaction of pyridine boronic acids with activated heteroaryl chlorides (Scheme 23, products **83–88**), although in selected cases Sphos is more effective as ligand [21].

Fu and coworkers have reported a versatile coupling method for a variety of heteroarylboronic acids, including 3-pyridineboronic acid, with a range of chlorobenzenes, including ortho-substituted and electronically deactivated substrates [41]. Using bulky trialkylphosphines, including $P(^tBu)_3$ and PCy_3, as effective ligands along with Pd as catalyst, good yields were obtained [42]. PCy_3 is relatively cheap compared to other ligands such as $P(^tBu)_3$, aryldialkylphosphines, and carbenes and shows high reactivity in Suzuki–Miyaura coupling. After some optimization efforts, the authors could establish Pd/PCy_3/K_3PO_4/dioxane/H_2O as effective catalytic system to promote cross-coupling of 3-pyridylboronic acids with different halides, especially aryl/heteroaryl chlorides in excellent yield. This catalyst system is also effective for Suzuki–Miyaura cross-couplings towards bipyridyls providing these products (**74, 75, 89–94**) generally in good yield (Scheme 24). Free NH_2 and OH groups were well tolerated as well as MeO or benzoyl.

This Pd/PCy_3/K_3PO_4/dioxane/H_2O-based method is effective not only for Suzuki–Miyaura cross-couplings of heteroarylboronic acids but also for boronate

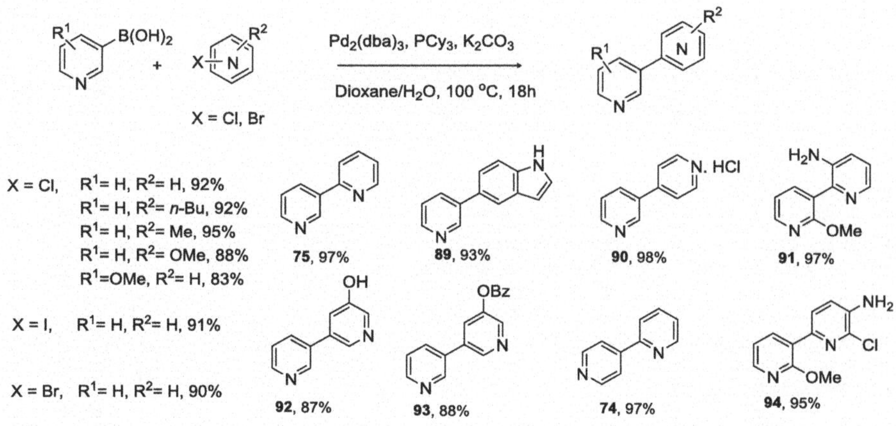

Scheme 24 Pyridineboronic acids in Suzuki-Miyaura couplings with pyridine halides

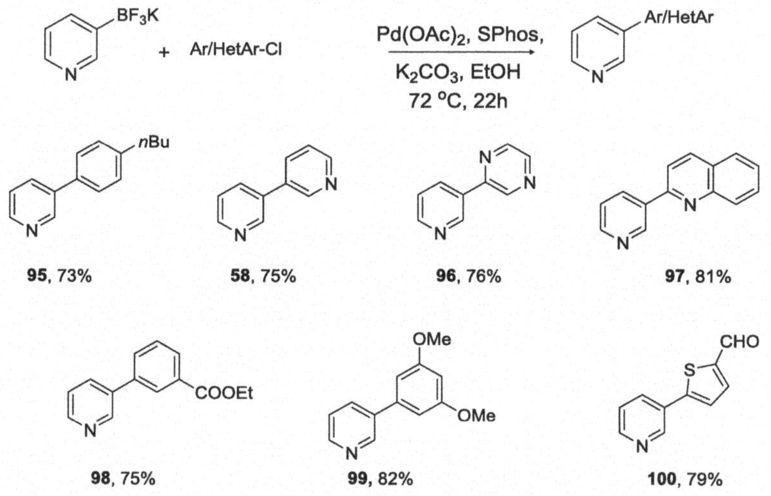

Scheme 25 Potassium pyridine-3-trifluoroborate in Suzuki-Miyaura couplings

esters and trifluoroborates as well, as demonstrated in the two examples (not shown).

The drawbacks associated with the use of boronic acids can be overcome by using boronate esters and trifluoroborates. A coupling method of (hetero)aryl potassium trifluoroborates was reported by Buchwald and coworkers including potassium 3-pyridyl trifluoroborate. Different aryl and heteroaryl chlorides including 3-pyridyl chloride can be applied as coupling partners, giving rise to 3,3′-bipyridine (**58**) (75% yield). Several other (hetero)aryl potassium trifluoroborates were used as coupling partners as well in generally good yield (Scheme 25,

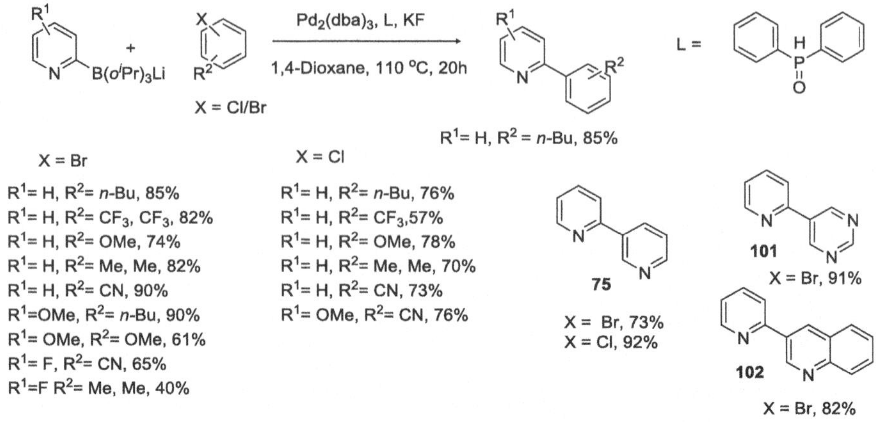

Scheme 26 Lithium triisopropyl 2-pyridylborates in Suzuki-Miyaura couplings

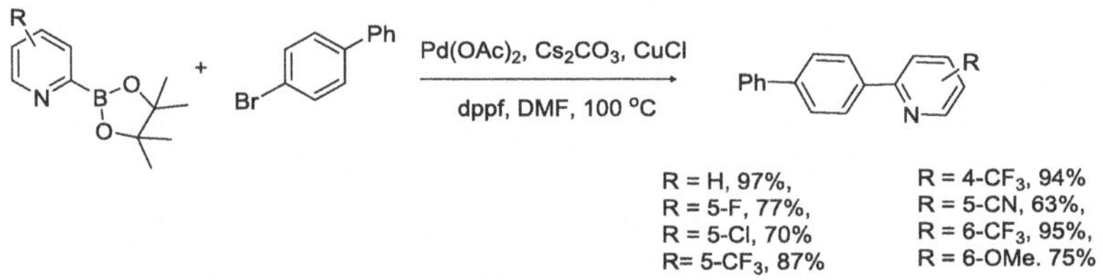

Scheme 27 2-Pyridyl pinacolboronates in Suzuki-Miyaura couplings

products **95–100**) [43]. $Pd(OAc)_2$, SPhos, K_2CO_3 in ethanol was applied as catalytic system. A similar study was published in 2012 by the group of Wu and Zou [44].

There are certain issues attributed for difficulties in Suzuki–Miyaura coupling of 2-pyridyl nucleophiles with aryl halides: (1) slow transmetalation rate of electron-deficient heteroaryl boron species and (2) rapid decomposition of reagents via protodeboronation. An efficient method to overcome these difficulties is by using lithium triisopropyl 2-pyridylborates as reported by Buchwald and coworkers. The borates required in Suzuki–Miyaura coupling can be obtained readily in one step from the corresponding 2-bromo- or 2-iodopyridine derivatives. Suzuki–Miyaura coupling reaction of these borates with different aryl/hetaryl halides is a general method for the synthesis of 2-substituted pyridine derivatives such as **75**, **101,** and **102** (Scheme 26) [45]. A similar method was reported by Ackermann and coworkers later on [46].

Another approach to overcome instability of 2-pyridineboronic acid is its conversion into the corresponding boronate. Generally, the use of pinacolboronates is attractive due to their commercial availability and stability towards air and moisture. Deng and coworkers have reported a protocol for coupling of 2-pyridyl pinacolboronate as nucleophile with aryl bromide (Scheme 27) [47]. Yields were generally good, and electron donating and electron withdrawing substituents were

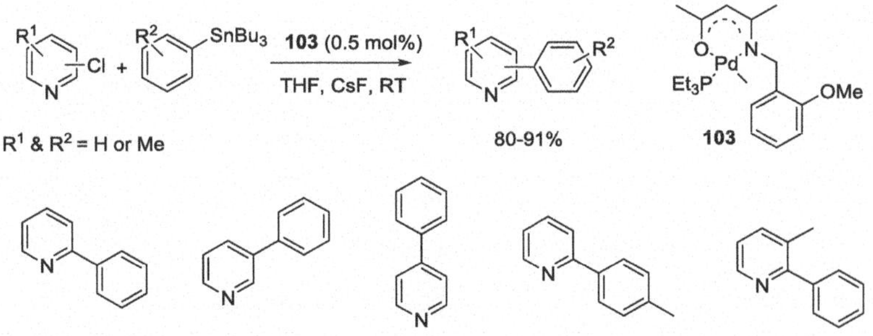

Scheme 28 Pyridine chlorides in the Stille coupling with arylstannanes

well tolerated. CuCl was required as additive, and the authors suggested that in its presence, the pyridine boronic ester might be transmetalated to a cuprate species.

3 Stille–Migita Coupling

The palladium-catalyzed cross-coupling of aryl or vinyl (pseudo)halides with organostannanes is known as the Stille–Migita coupling reaction [48, 49] more often referred to as Stille coupling, only. This is another widely used method for C–C bond formation offering the advantage that air and moisture stable organotin reagents can be used, along with an excellent functional group tolerance. In contrast to Suzuki, Kumada, and Heck reactions, which operate under basic conditions, usually neutral conditions are employed for Stille couplings (although in some cases base may be required). The main drawback associated with the Stille cross-coupling procedure is toxicity of the tin compounds which make them problematic for industrial applications. Stille coupling, using aryl bromide and iodide, is already well established, but use of aryl chloride as coupling partner is not well developed [50, 51]. Examples of recent applications of Stille couplings of pyridine halides are discussed in this chapter. Simple applications of pyridine in Stille reactions under well-established reaction conditions are not discussed unless the substrates or products are of special interest or complexity. As will be demonstrated, most new developments are based on the establishment of new ligands which are able to create a more reactive metal species.

3.1 Pyridine as (Pseudo)halide in Stille–Migita Reactions

A Stille coupling at room temperature is achieved using a phosphanyl–βketoiminate palladium complex (**103**). Chloropyridines bearing a chloride in 2-, 3-, or 4-position can be coupled with tributylphenylstannane in high yields of 80–91% within 4–6 h, with low catalyst loading of 0.5% (Scheme 28). Although the coupling process of

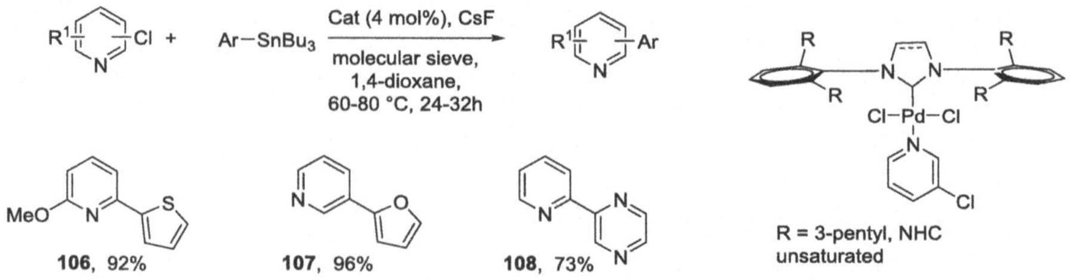

Scheme 29 Pyridine chlorides in the Stille coupling with heteroarylstannanes

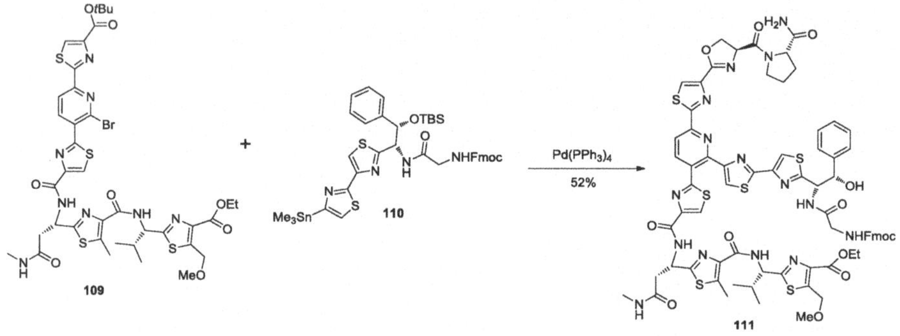

Scheme 30 Stille coupling of highly complex precursors

sterically hindered 2-chloro-3-methylpyridine is slower compared to other substrates, the desired coupling product **105** can be obtained with 81% isolated yield by prolonging the reaction time to 8 h [52].

The group of Organ reported an extremely efficient catalytic system for the coupling of heterocyclic halides with heterocyclic stannanes under Stille conditions. In three examples, pyridine chlorides were used: 2-chloro-6-methoxypyridine reacted with 2-(tributylstannyl)thiophene in 92% yield (**106**), whereas 3-chloropyridine coupled with 2-(tributylstannyl)oxazole and 2-chloropyridine with 2-(tributylstannyl)pyrazine in 96% (**107**) and 73% (**108**), respectively (Scheme 29) [53]. The Pd-PEPPSI-iPent catalyst proved to be most efficient. The reaction temperature was elevated but below the boiling point of the reaction solvent. One drawback is the relatively long reaction time.

The group of Bach applied Stille coupling of a very complex pyridine halide in a late stage of the synthesis of the thiazolylpeptide antibiotic GE2270 A [54]. This compound consists of a complex macrocycle which contains five thiazole and one pyridine ring. In an early approach towards the total synthesis of the target, the Stille reaction was used to connect two advanced fragments which set the stage for the macrocyclization. Pyridine bromide **109** was coupled with thiazole stannane **110** using simple $Pd(PPh_3)_4$ as catalyst in 52% yield (Scheme 30). This yield is actually quite good considering the complexity of the structures and the presence of several amide moieties.

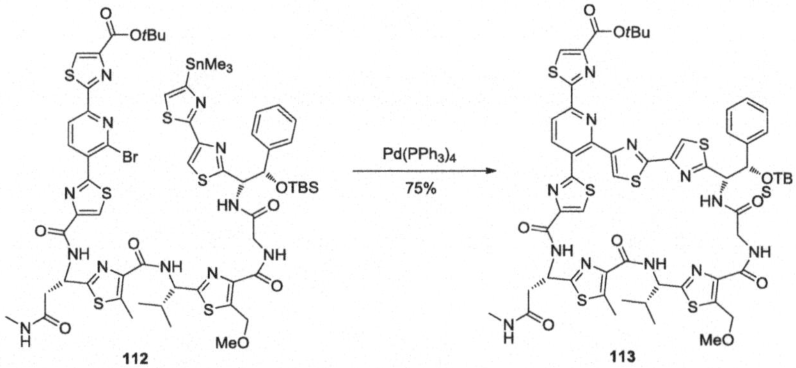

Scheme 31 Mocrocyclization via intramolecular Stille coupling

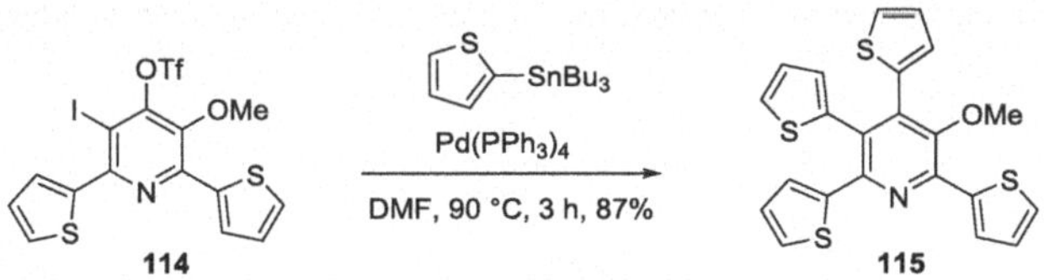

Scheme 32 Per(2-thienyl)pyridines via Stille cross-coupling

In the second variant, the Stille reaction was used specifically for the macrocyclization step of **112** to **113** (Scheme 31). Applying the same conditions, the yield was even better with 75% which makes up for the long reaction time of 3 days. In order to favor intramolecular Stille coupling, the reaction had to be carried out in high dilution (0.001 M).

In subsequent work, the structurally closely related natural products amythiamicin C and D were synthesized applying the same strategy [55]. The group of Alvarez applied Stille coupling reactions as well in their synthetic efforts towards natural products structurally related to amythiamicin. However, they applied the method at different stages in the synthesis, i.e., on smaller precursors [56].

Reissig and coworkers reported the synthesis of several per(2-thienyl)pyridines since they were interested in their photophysical properties [57]. Two out of five thienyl residues were introduced via Stille coupling in one-pot giving a good yield of 87% of product **115** starting from **114** (Scheme 32).

An unusual example of a Stille coupling reaction was reported by Legoupy and coworkers [58]. They used ionic liquid supported organotin reagents for the decoration of various (hetero)aryl bromides or alkenylbromides. Reaction optimization was carried out using 3-bromopyridine as coupling partner. It turned out that homo-coupling of the arylation reagent was a major problem which had to be overcome. In the end, they identified a relatively simple catalytic system since $Pd(OAc)_2$ without any ligand gave the best results. Suppression of homo-coupling was

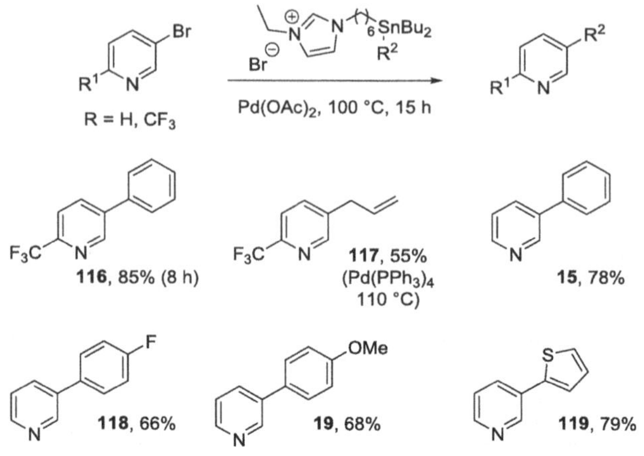

Scheme 33 Ionic liquid supproted tin reagents in the Stille coupling with pyridine bromides

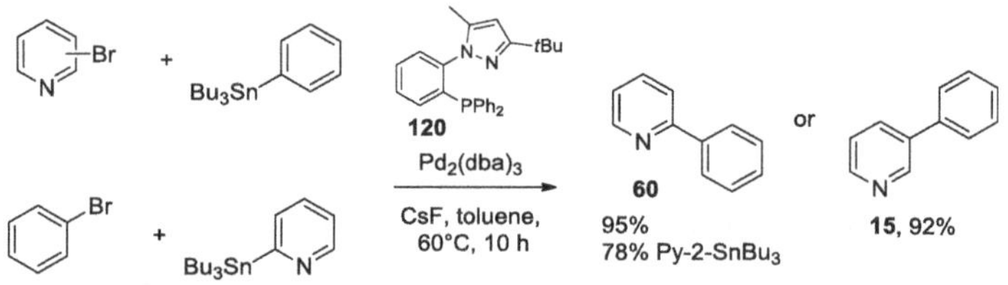

Scheme 34 A pyrazole-tethered phodphine ligand enables milder Stille coupling

achieved largely by optimizing the reaction temperature; 100°C proved to be optimal with higher or lower temperature giving larger quantities of homo-coupling products. The examples involving pyridine halides are compiled in Scheme 33. As can be seen, products **15**, **19**, and **116–119** were obtained in good yields.

In order to obtain even milder and more generally applicable catalytic systems, efforts towards new ligands for coupling reactions are undertaken, continuously. Sarkar and coworkers developed pyrazole-tethered phosphine ligand **120** for Pd (0) which could be applied successfully in Stille, Hiyama, and Kumada coupling reactions [59]. Examples employing pyridine as halide and organometal species were reported, as well. The reaction proceeds at 60°C in toluene in excellent yields for the studied pyridine examples **15** and **60**. However, the reaction time of 10 h is rather long (Scheme 34).

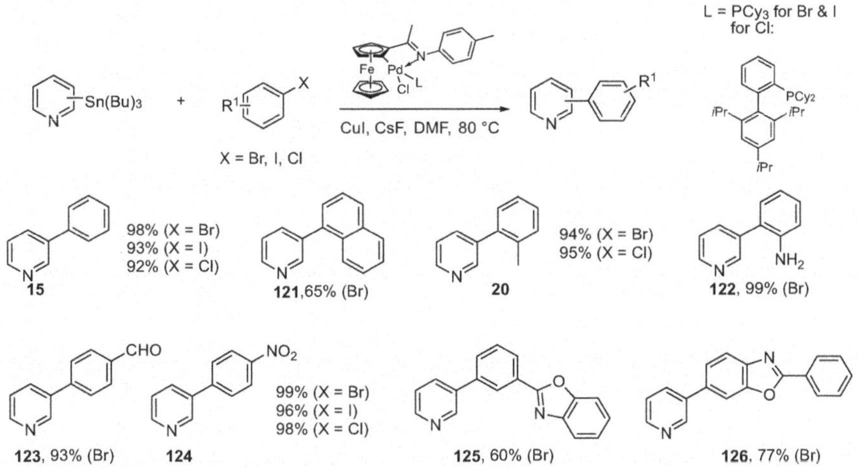

Scheme 35 A ferrocene-palladium species for Stille coupling of pyridinstannanes

3.2 *Pyridine as Organometal Species in Stille–Migita Reactions*

In 2008, Hocek and coworkers exploited the Stille protocol in the synthesis of certain 4- and 3-substituted benzene and aniline C-ribonucleosides (not shown) [60]. They used 3- or 4-bromophenyl C-nucleoside intermediates as substrates carrying TBS protecting groups on the hydroxyl functions. In two examples, 2-Bu_3Sn-pyridine was used as coupling partner. Reaction conditions can be considered as conventional ($Pd(PPh_3)_2Cl_2$, DMF, 100°C), and good yields were obtained.

A ferrocene-based palladacyclic system was reported to promote the Stille coupling between 3-alkylstannylpyridines and aryl halides (Cl, Br, I) (Scheme 35) [61]. In contrast to many other examples of Stille reactions, a base had to be applied and CsF turned out to be most suitable. Additionally, CuI was required as additive (but its role not further commented), and reactions were carried out in DMF as solvent. The requirement for base and additive can be understood as drawback; however, the reaction is quite robust giving yields often >90% independent of the halide species (see examples **15**, **20**, **121–126**). Sterically demanding halides are well tolerated as well as functional groups such as NH_2, MeO, Me, CHO, COOR, CF_3, NO_2, COMe, and OH. Selected examples are depicted in Scheme 35.

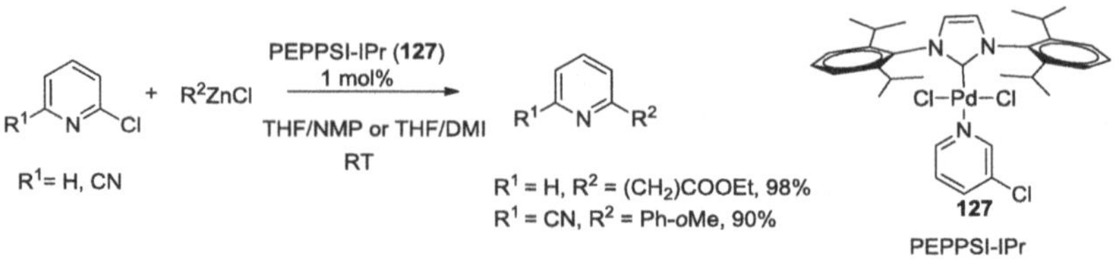

Scheme 36 Room temperature Negishi coupling of pyridine chlorides

4 Negishi Coupling

Pd-catalyzed cross-coupling of organozinc reagents with organohalides or triflates is known as the Negishi reaction. Reported by Negishi in 1977 for the first time, this was the first reaction that allowed the preparation of unsymmetrical biaryls in reasonably good yields [62–64]. The advantages associated with Negishi reactions are employment of mild reaction conditions and broad functional group compatibility, including ketones, esters, amines, and nitriles. The organozinc reagents can be generated and used in situ by transmetalation of Grignard or organolithium reagents with $ZnCl_2$ [65]. New developments in the field of application of pyridine in Negishi reactions are reported in this chapter independent of the role of pyridine (halide or organometal). Simple applications of pyridine in Negishi reactions under well-established reaction conditions are not discussed.

4.1 Pyridine as (Pseudo)halide

A widely applicable Negishi cross-coupling protocol must fulfill several criteria, such as ease of handling of reagents or broad substrate scope. In this regard, an *N*-heterocyclic carbene (NHC)-based precatalyst (**127**) (PEPPSI-IPr) is reported to be useful in Negishi coupling with a broad substrate scope which includes pyridines as well. Utility of precatalyst **127** was demonstrated in various types of cross-coupling reactions in recent years. It can be synthesized easily, is air stable and highly active as well as structurally defined, and can surpass the phosphine-ligated Negishi processes in terms of activity and usefulness. Both alkyl and arylzinc reagents can be coupled with pyridine-2-chlorides (Scheme 36) using this active catalyst system at room temperature [66]. The so termed Pd–PEPPSI complexes have been very successful in Negishi reactions in recent years, and the field was reviewed previously [67].

In 2013, Tu and coworkers disclosed another catalyst type structurally closely related to the PEPPSI systems [68]. Their Pd–NHC complexes **128** were designed to have strong σ-donor and π-acceptor properties. Using these catalysts, Negishi couplings could be carried out at room temperature with low catalyst loading (0.25 mol%) within 30 min (Scheme 37). Cyclopentylzinc bromide reacted well with aryl bromides and even chlorides including pyridine derivatives (towards **129**

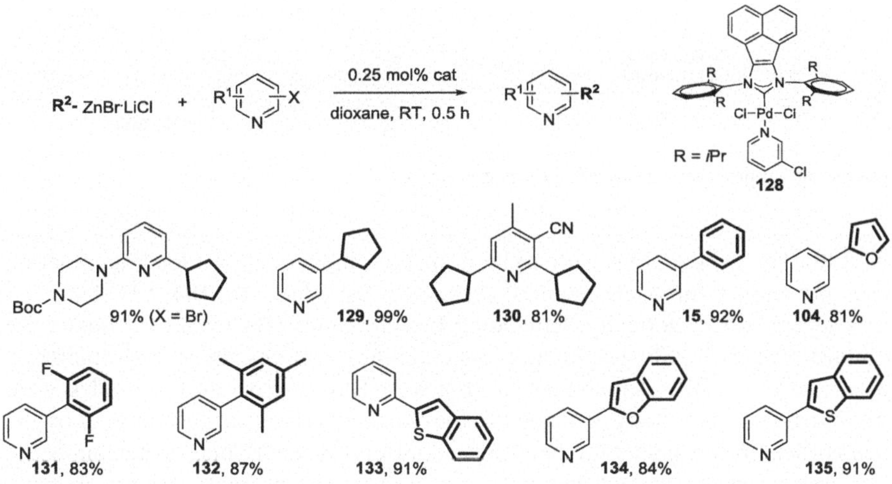

Scheme 37 Pyridine chlorides were used as coupling partners unless indicated differently

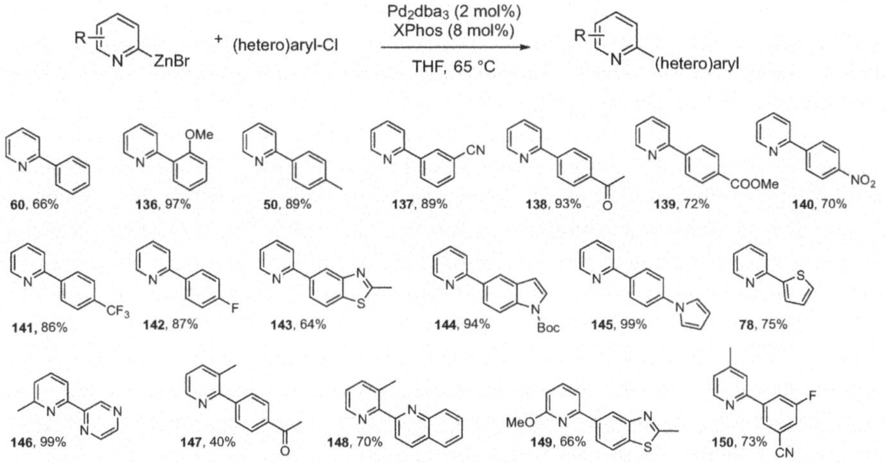

Scheme 38 Negishi coupling of pyridine zinc bromides with (hetero)aryl chlorides

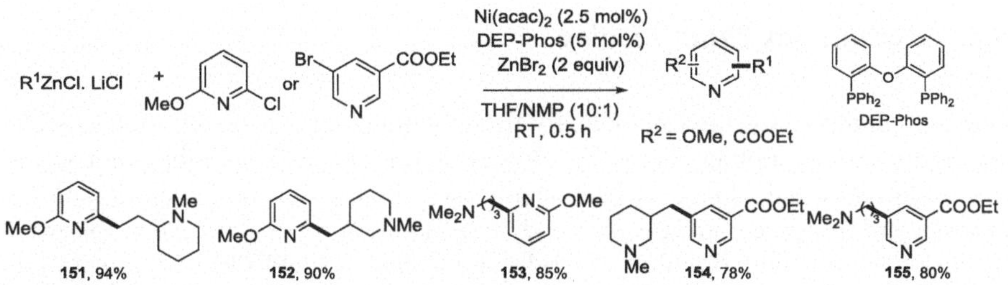

Scheme 39 Nickel catalyzed Negishi coupling at room temperature

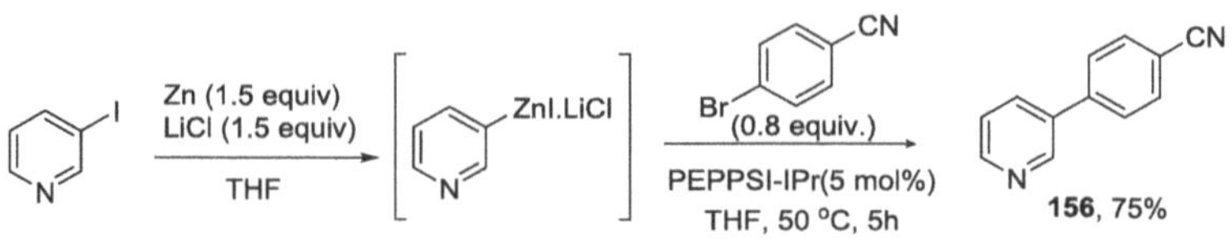

Scheme 40 Improving stability of pyridinezincates - 1

and **130**). For the coupling with arylzinc species, higher catalyst loading, reaction time, and temperature were required (Scheme 37, products **15**, **104**, **131–135**).

In recent years, sterically demanding biaryl-ligands (Buchwald ligands) were demonstrated as highly useful in cross-coupling reactions, most importantly in cases when sterically demanding substrates or less reactive aryl chlorides were employed. One such study dedicated to pyridinezinc bromides was reported by the group of Yin [69]. Under their conditions, a series of aryl chlorides were coupled to pyridinylzinc species, and typically good to excellent yields were obtained (Scheme 38, products **50**, **60**, **78**, **136–150**). Functional groups (NO_2, CF_3, OMe, CN, Ac, COOMe, Boc) were well tolerated as well as ortho substitution of the halide species (Scheme 38). Naturally, the group of Buchwald contributed to the field as well. In their recent report of Negishi cross-couplings on structurally quite diverse systems, few examples employing pyridine in one role or the other were reported as well (not shown) [70].

The aminoalkyl moiety is one of the most frequently occurring structural motives in biologically active molecules. In this regard, Ni-catalyzed Negishi coupling for aminoalkylation of heteroarenes including pyridine is a valuable method for decorating a pyridine scaffold (Scheme 39) [71, 72]. The alkylzinc species required as nucleophile for Negishi cross-coupling can be prepared easily from the corresponding chlorides via a Grignard reagent and transmetalation with $ZnCl_2$. The optimal ratio of Ni-catalyst to ligand is 1:2. Both bromo- and chloropyridine derivatives can undergo cross-coupling with aminoalkylzinc reagents at room temperature, for example, towards compounds (**151–153**). An ester functionality on pyridine is reported to be tolerated also under the reaction conditions (Scheme 39, products **154** and **155**).

4.2 *Pyridine as Zinc Species*

Due to high reactivity, environmentally friendly organozinc reagents are a suitable choice for cross-coupling chemistry. However, zinc reagents are labile and sensitive to air and water. This drawback can be overcome by in situ generation of the zinc reagent from the corresponding iodide by addition of zinc and LiCl, followed by cross-coupling with bromide, chloride, and triflate using PEPPSI-IPr as catalyst. This method is a one-pot protocol, and handling of sensitive zinc reagents can be

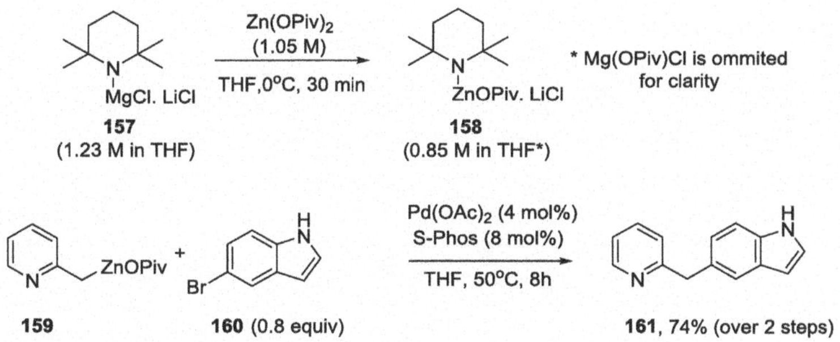

Scheme 41 Improving stability of pyridinezincates - 2

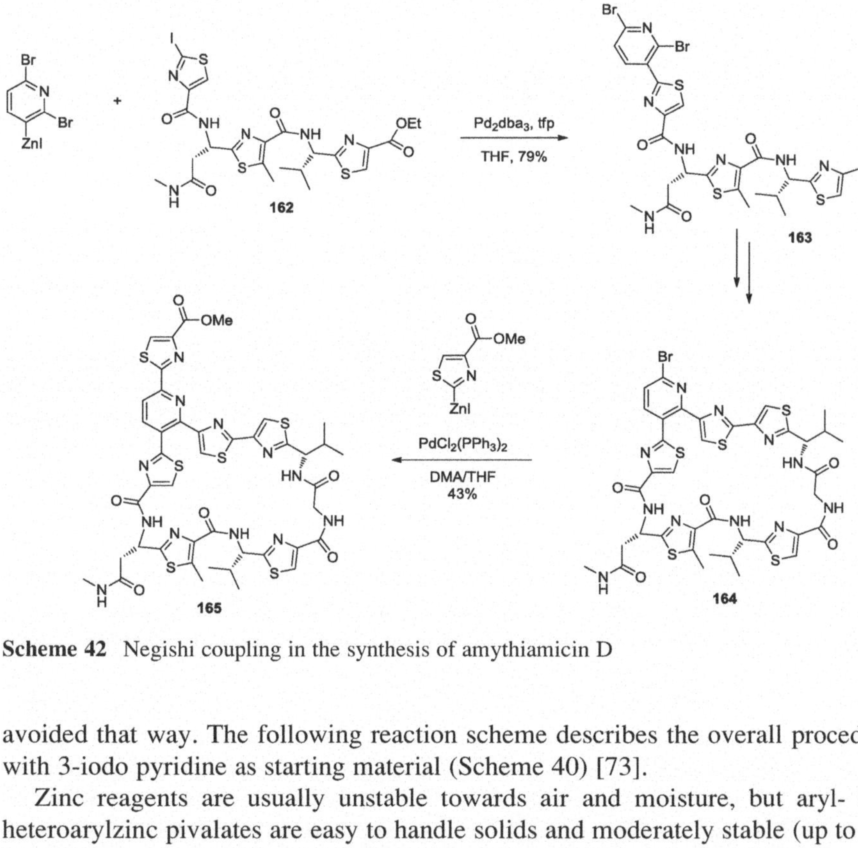

Scheme 42 Negishi coupling in the synthesis of amythiamicin D

avoided that way. The following reaction scheme describes the overall procedure with 3-iodo pyridine as starting material (Scheme 40) [73].

Zinc reagents are usually unstable towards air and moisture, but aryl- and heteroarylzinc pivalates are easy to handle solids and moderately stable (up to 4 h in air) even when handled in air. Moreover, zinc pivalates can be prepared easily either via Mg insertion in the presence of $Zn(OPiv)_2$ 2LiCl [74] or directed metalation using TMP-MgCl LiCl (TMP = 2,2,6,6-tetramethylpiperidyl) and subsequent transmetalation to $Zn(OPiv)_2$ [75]. Still, both methods are incompatible with the presence of electron withdrawing functional groups such as nitro, aldehyde, etc. In this regard, the milder reagent TMP-ZnOPiv LiCl proved to be

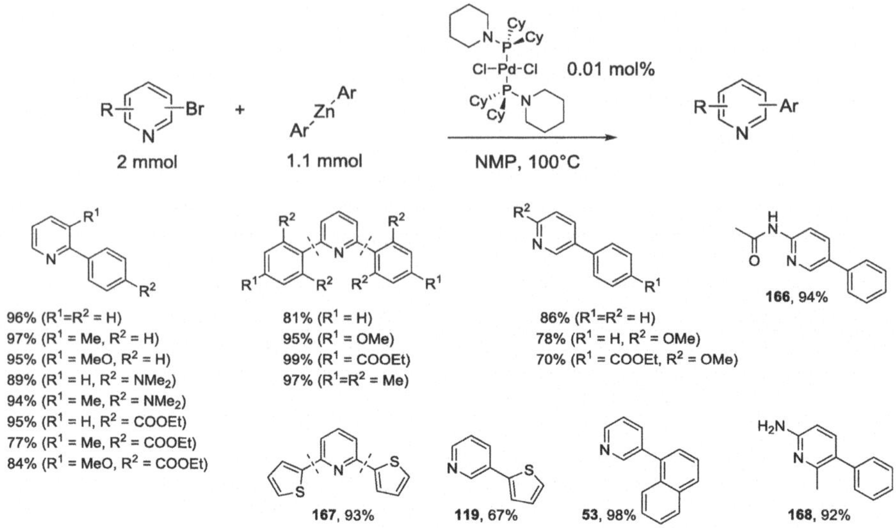

Scheme 43 Diarylzinc reagents in Negishi coupings with pyridine bromides

successful for the metalation at the benzylic position of five- and six-membered heterocycles and is tolerated by sensitive functionalities such as nitro and aldehyde as well. TMP-ZnOPiv LiCl (**158**) is prepared by the addition of $Zn(OPiv)_2$ (1.05 equiv., 0°C) to a solution of TMP-$MgCl_3$ LiCl (**157**, 1.23 M in THF) followed by dilution with dry THF until a clear solution is achieved (final concentration: 0.85 M; Scheme 26). For example, 2-picoline can be zincated smoothly using this reagent after 1 h at 25°C to provide the corresponding zinc pivalate **159** which can then be used effectively in a Negishi cross-coupling with 5-bromoindole **160** using Pd $(OAc)_2$ and SPhos as the catalytic system. The desired product **161** can be obtained in 74% yield over two steps (Scheme 41) [76].

The total synthesis of the thiopeptides amythiamicin C and D proved to be a fruitful area for cross-coupling chemistry. The group of Bach reported a total synthesis in which they carried out three cross-coupling reactions on the pyridine scaffold [55], one Stille reaction (see Sect. 3.1) and two Negishi reactions. In the case of amythiamicin D (**165**), pyridine served as the zinc organyl in the first Negishi reaction and was coupled to a complex thiazole-peptide building block **162** to intermediate **163**. The second Negishi coupling represented the last step in the synthesis of amythiamicin D where the remaining bromine on pyridine of substrate **164** was coupled to a 2-thiazolyl-zinc species (Scheme 42). In both reactions, the coupling conditions can be considered as standard, but the complexity of the target product is remarkable. Removal of triphenylphosphine oxide proved to be tricky in the last step which lowered the yield of **165** to 43%. In the synthesis of amythiamicin C, the order of events was slightly different, but overall the same coupling reactions were applied.

Frech reported a new Pd-catalyst with extremely high activity in Negishi reactions [77]. The catalyst is readily prepared from commercially available $[Pd(cod)(Cl)_2]$ (cod=cyclooctadiene) with 1-(dicyclohexylphosphanyl)piperidine. The authors demonstrated a wide substrate scope and excellent functional group tolerance at low catalyst loading (0.01 mol%), and often the reactions were completed within minutes (Scheme 43). Among the many examples, several coupling reactions of pyridine bromides with diarylzinc reagents were reported (Scheme 43, e.g., **166–168**). Also in these cases, good yields were obtained demonstrating the power of this new catalytic system.

The group of Twieg showed that a Negishi reaction can be carried out in the presence of a stannyl group employing simple $Pd(PPh_3)_4$ as catalyst [78]. Alkyl- and arylzinc chlorides were used as coupling partners, and it was demonstrated that the Bu_3Sn group can be further converted in a Stille coupling reaction (not shown).

5 Mizoroki–Heck Reaction on Pyridine

Palladium-catalyzed replacement of vinylic hydrogen by aryl, alkenyl, or benzyl moieties, independently discovered by Mizoroki and Heck in the early 1970s, is known as the Mizoroki–Heck (MH) coupling reaction, often referred to as the Heck reaction only [79, 80]. This reaction has special value for industrial and academic research as mild conditions are sufficient to activate the olefin. This procedure is now broadly defined as the Pd(0)-mediated cross-coupling of an aryl or vinyl halide or triflate with an alkene, and many catalytic systems were reported to catalyze this cross-coupling reaction [81–83].

Reactive aryl halides (Br or I) and activated alkenes are used frequently in cross-coupling processes. Aryl chlorides are definitely more attractive substrates due to their more facile availability. However, due to the relatively strong C–Cl bond, oxidative addition is difficult to achieve. So far, considerable advances in terms of catalytic systems used in MH coupling have been reported, but poor regioselectivity is one of the major disadvantages in synthetic utility. Regioselectivity of this reaction is defined by the resulting substitution pattern on the olefins. As a general rule, electron poor alkenes selectively provide products of β-substitution, whereas linear β- and branched α-substituted products are formed with electron-rich alkenes [84–87]. To achieve selective α-substitution with electron-rich olefins, a cationic palladium (II) complex need to be generated in situ under the reaction conditions, to promote an ionic pathway [88, 89]. So far, several methods have been successfully developed to circumvent the problem of regioselectivity, but concomitant disadvantages are the use of expensive substrates such as triflates or toxic additives (halide scavengers like silver and thallium) or applications limited to particular olefins [90, 91].

A general and efficient method for coupling of pyridine bromides with vinyl ethers and hydroxyalkyl vinyl ethers in ethylene glycol was reported (Scheme 44) [92]. In this green solvent, highly regioselective heteroarylation of electron-rich

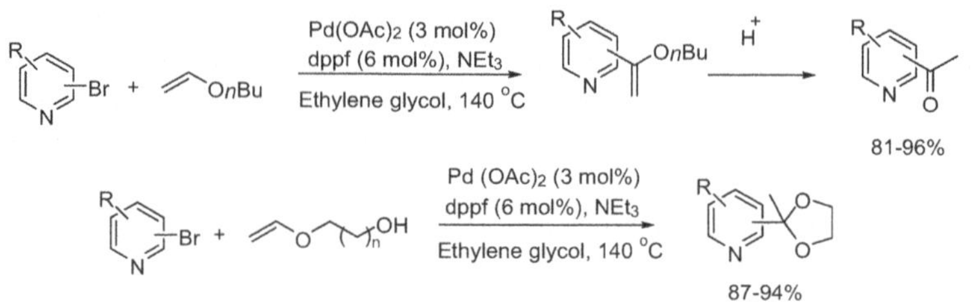

Scheme 44 Coupling of vinyl ethers with pyridine bromides

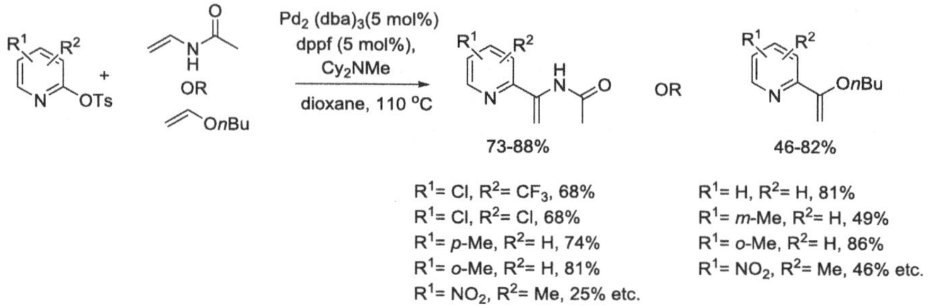

Scheme 45 Coupling of vinyl ethers and enamides with pyridine tosylates

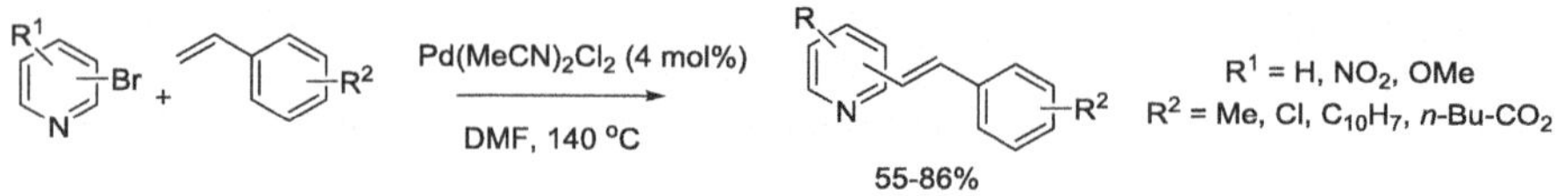

Scheme 46 Ligandless Heck coupling of pyridine bromides

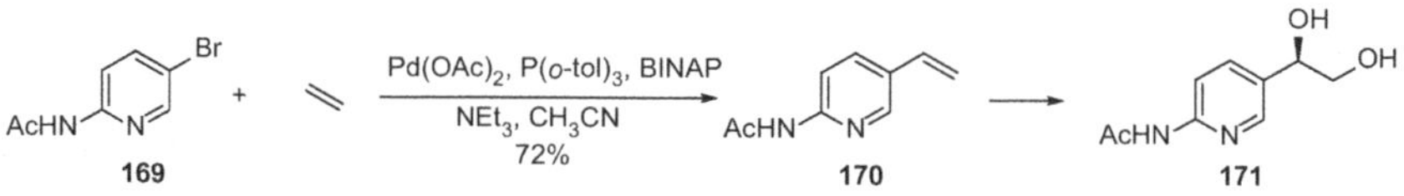

Scheme 47 Heck coupling of 169 with ethylene

olefins was achieved with pyridyl bromides. The advantages of this method are that there is no necessity for using expensive triflates, halide scavengers, or ionic liquids (which often help to promote an ionic pathway). Ethylene glycol supposedly facilitates the formation of the ionic palladium intermediate, which promotes the ionic route and preferentially affords the branched products.

Heck coupling of pyridine tosylates with electron-rich enamides and vinyl ethers was reported by the group of Skrystrup [93]. Excellent regioselectivity for the desired products and no loss in reactivity of the applied olefin are observed (Scheme 45). 2-Pyridyl tosylate derivatives used in this protocol can be synthesized easily from the corresponding commercially available 2-hydroxy pyridines by

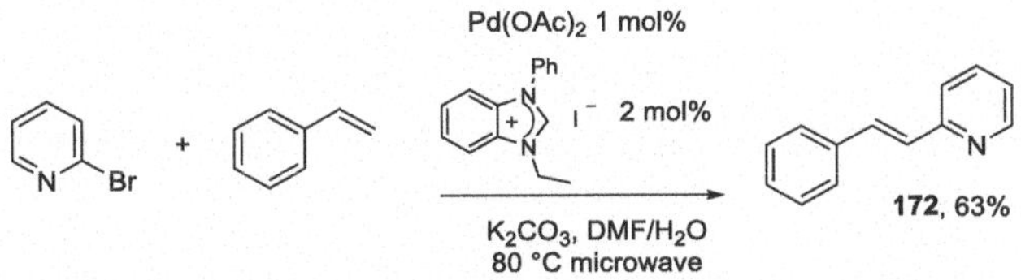

Scheme 48 Microwave assisted Heck coupling of styrene with 2-bromopyridine

simple treatment with tosylchloride (TsCl) and a base. Coupling of 2-pyridyl tosylates with *N*-vinyl acetamide using $Pd(dba)_2$ as catalyst precursor and dppf (1,1′-bis(diisopropylphosphino) ferrocene) as ligand provides the desired product with good yields of 68–79% within 22 h (Scheme 45). In the case of electron withdrawing nitro-substituted pyridines, the yield drops to 25%. In the case of bistosylates, double MH coupling is observed with *N*-vinyl acetamide. Couplings of butyl vinyl ether with 2-pyridyl tosylates are also successful under the same reaction conditions giving 49–85% yield of the desired products. Electron withdrawing groups are well tolerated as well in this case. Usually, a slightly lower yield was observed with substituents in the 3-position of the 2-pyridyl tosylates.

Pyridine bromides and substituted pyridine bromides (Scheme 46) are reported to undergo Heck reaction with styrene derivatives under ligand and base-free conditions. This makes the transformations more convenient and economic, although a high temperature of 140°C is mandatory to obtain the maximum yield of the products [94]. MH coupling of styrene with 3-bromopyridine gives higher yields than 2-bromopyridine, which is a bit counterintuitive.

The MH reaction is not only widely investigated for small-scale applications in laboratory, but also some industrial applications are reported [95]. For example, 2-acetamido-5-vinylpyridine **171**, a key intermediate for drug candidates at Pfizer, can be synthesized by Sharpless dihydroxylation of **170**, which is obtained effectively in multi-kilogram scale by MH coupling of **169** with ethylene using rac-BINAP as ligand in combination with $Pd(OAc)_2$ and $P(o\text{-tolyl})_3$ (Scheme 47) [96].

Küçükbay and coworkers investigated Suzuki–Miyaura and MH reactions in the presence of *N*-phenylbenzimidazolium salts [97]. Their catalytic system proved to be very reactive in the case of aryl bromides and aryl iodides. Unfortunately, coupling of styrene with 2-bromopyridine gave a significantly lower yield. Still, around 60% of product **172** could be obtained within 10 min of reaction time (Scheme 48).

6 Sonogashira Coupling

The palladium (or copper)-catalyzed coupling of aryl or vinyl halides with terminal acetylenes is known as Sonogashira coupling reaction which was first published in 1975 by Sonogashira, Tohda, and Hagihara [98]. This cross-coupling reaction is

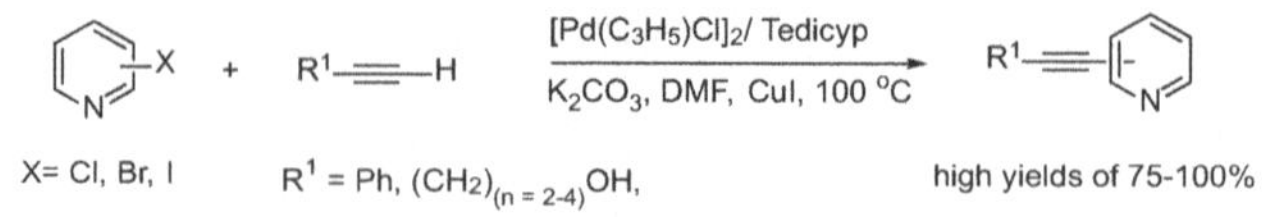

Scheme 49 Generally applicable Sonogashira coupling of pyridine halides

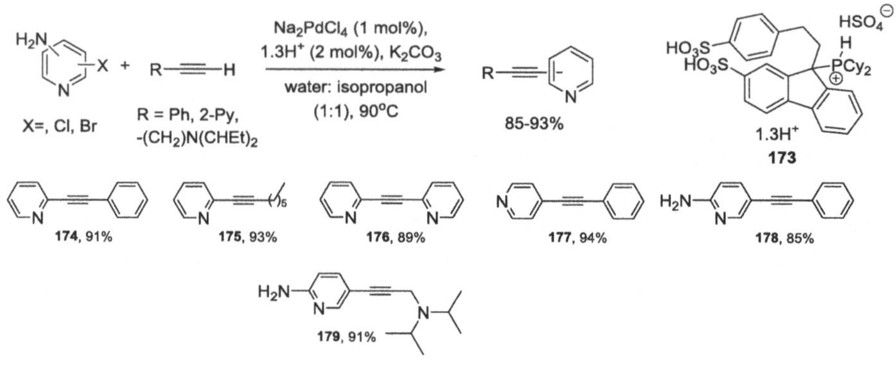

Scheme 50 Sonogashira coupling in aqueous media

one of the most synthetically significant routes to achieve sp^2–sp C–C bond formation. Application of Sonogashira coupling includes natural product synthesis, synthesis of heterocyclic and biologically active compounds, conjugated polymers, molecular electronics, etc. [99]. In this section, new developments in Sonogashira coupling of recent years are reported. Only reactions are reported in which pyridine acts as halide coupling partner.

A general protocol for coupling of pyridyl halides with terminal alkynes (Scheme 49) with low catalyst loading was reported by Santelli and coworkers [100, 101]. This group employed a stable and efficient palladium catalyst based on a phosphine ligand, *cis,cis,cis*-1,2,3,4-tetrakis(diphenylphosphinomethyl)-cyclopentane or TedicyP [11]. As general trend, pyridyl iodides are more reactive than bromides, whereas chlorides showed the slowest kinetic profile. Due to the electronegativity of the nitrogen atom, the 2 and 4 positions of halopyridines are expected to be most susceptible to the oxidative addition to palladium. However, with this reported catalytic system, in the case of pyridyl bromides, the position of halide has generally a minor effect. In contrast, in the case of chloropyridines, reactions of 2- and 4-chloropyridines are faster than with 3-chloropyridine. This indicates that the oxidative addition of chloropyridines to palladium is the rate-limiting step.

On a more environmentally friendly note, Sonogashira coupling of pyridyl halides with terminal alkynes was reported in aqueous media without use of any copper cocatalyst [102]. 2-Chloropyridine and 5-bromopyridine-2-amine were coupled with phenyl, 2-pyridyl acetylene, or oct-1-yne in good yields (Scheme 50, products **174–179**). Interestingly, there is no need for protection of amine in the

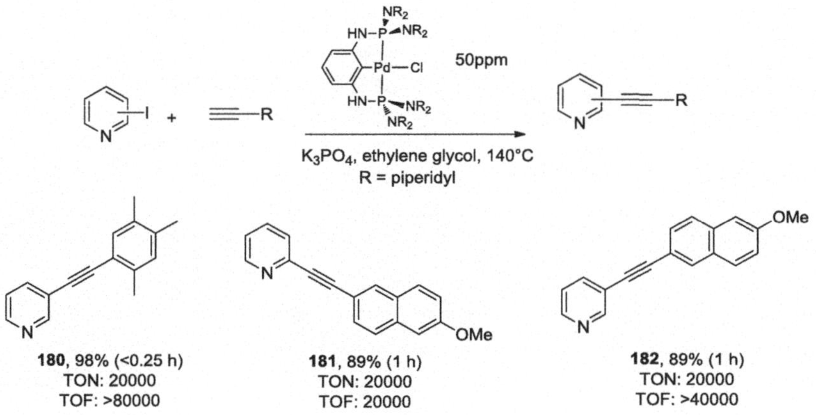

Scheme 51 Low catalyst loading Sonogashira coupling of pyridine iodides

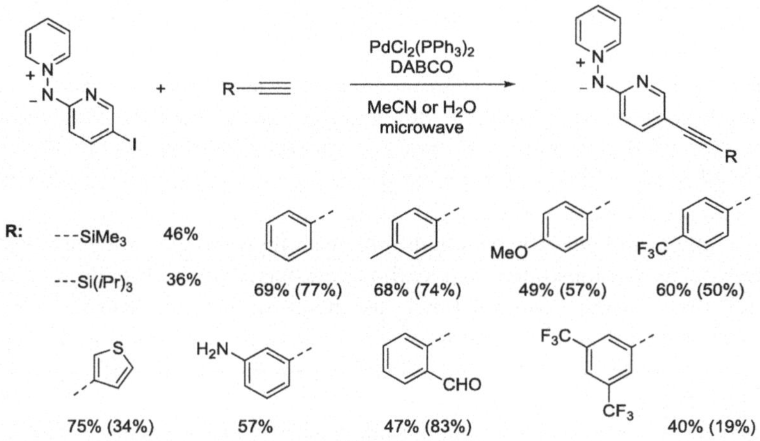

Scheme 52 Yields in brackets correspond to the reaction in water

case of 2-amino-5-bromopyridine, which gives almost quantitative yield upon coupling with phenylacetylene **178**. The active Pd-catalyst used in this transformation is generated in situ from Na_2PdCl_4, and two equivalents of triply protonated ligand $1.3H^+$ (**173**), in the presence of 5 equiv. of K_2CO_3 (Scheme 50). Ligand **173** is commercially available as well.

Another copper-free example was reported by Frech and Bolliger who disclosed amino-phosphine-based pincer complexes of palladium which promote the Sonogashira coupling of aryl bromides and aryl iodides in the absence of cocatalyst at extremely low catalyst loadings [103]. The examples including pyridine halides required only 50–100 ppm of catalyst and were completed between 0.25 and 6 h. TOF of up to >80,000 and TON up to 20,000 were measured. Conversions were complete in almost all cases. Eleven examples were reported involving pyridine

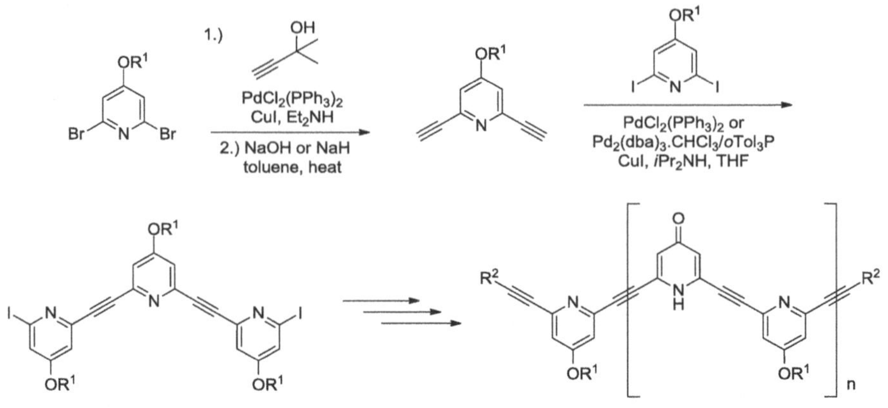

Scheme 53 Sonogashira coupling in the synthesis of pyridine–pyridone alternate oligomers

halides, but in only three cases (**180–182**), a yield was reported (although then excellent; Scheme 51).

Another copper-free method came from the laboratory of Alvarez-Builla [104]. In this report, pyridinium *N*-heteroarylaminides were used as substrates, and it was found that a simple catalytic system ($PdCl_2(PPh_3)_2$, DABCO, acetonitrile) proved to be sufficient to promote the desired coupling process (Scheme 52) even though yields were mediocre (30–75%).

Other copper (and amine)-free conditions were reported over the past years. For example, it was demonstrated that carboxylate-based ionic liquids are suitable media for that purpose. Simple $PdCl_2$ as metal source and KOAc as base at 100–120°C proved to be sufficient to give a good yield of 85–88% in the coupling of 3-bromopyridine and phenylacetylene depending on the IL applied (not shown) [105].

Li and coworkers investigated the tetradentate phosphine ligand *N,N,N′,N′*-tetra(diphenylphosphinomethyl)pyridine-2,6-diamine in Cu-free Sonogashira couplings in combination with $[Pd(\eta^3\text{-}C_3H_5)Cl]_2$ as metal source [106]. As coupling partners, 2- and 3-bromopyridine were used as well as 3-chloropyridine in the reaction with phenylacetylene. In the case of bromides, 96% yield was obtained after 3 h, whereas chloride gave only 12% after 20 h.

Palladium-free Sonogashira reactions have been reported as well. Bolm and coworkers investigated such a variant [107]. Initially, they tested the coupling between iodobenzene and phenylacetylene in the presence of various copper sources and an excess of DMEDA as ligand. Interestingly, all Cu sources gave the desired product in high yields after 22 h. When applied to 2-iodo and 3-iodopyridine as halide coupling partner, the corresponding products were obtained in good yields of 86% and 84%, respectively.

The Sonogashira reaction is well established on pyridine systems and has been applied to the synthesis of compounds with interesting properties in recent years. Abe and Inouye exploited the Sonogashira methodology in the synthesis of

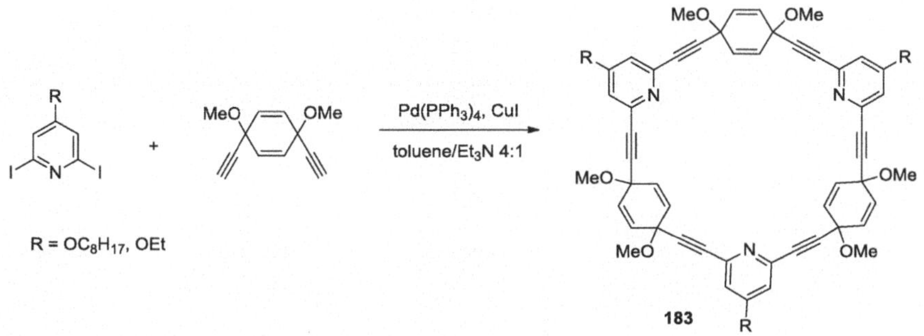

Scheme 54 Macrocycle synthesis via Sonogashira coupling

pyridine–pyridone alternate oligomers (Scheme 53) [108]. The target compounds displayed self-dimerization and glucoside recognition properties.

Ohe reported the synthesis of an interesting compound class applying standard Sonogashira coupling conditions. Pyridine-containing strained cyclynes were synthesized from 2,6-diiodopyridines and *cis*-3,6-diethynyl-3,6-dimethoxycyclohexa-1,4-diene as coupling partner (Scheme 54, product **183**) [109]. Not surprisingly, the yield in the macrocyclization was low (3–10%). Further transformation of these intermediates led to a product with strong fluorescence around 500 nm with a good fluorescence quantum yield.

7 Hiyama Coupling

Coupling of organosilanes with organohalides and triflates as coupling partners to generate biaryls and alkyl and vinyl aromatics is known as Hiyama coupling and has emerged to be a powerful synthetic method in organic chemistry. Compared to organoborons (Suzuki–Miyaura reaction) and organostannanes (Stille reaction), organosilanes have several advantages of being lower in toxicity, ease of handling, and comparatively high stability [110–114]. Previously, the application of organosilane derivatives as coupling partners was not attractive due to the weak polarization of the carbon–silicon bond. This problem has been solved through the generation of a penta-coordinate silicon intermediate in situ, using a fluoride activator [115] such as tetra-*n*-butylammonium fluoride (TBAF) or an inorganic base [116–120]. Subsequently, other efficient partners such as organo(trialkoxy) silanes [121, 122], organosilanolates [123, 124], and organohalosilanes [125, 126] have been utilized in Hiyama cross-coupling also referred to as Hiyama–Denmark coupling.

A variety of palladium catalysts have been studied extensively for Hiyama coupling. Aryl iodides and bromides have been predominantly used as the coupling partners of organosilanes. However, the use of aryl chlorides is highly desirable, due to ready availability and cost effectiveness. So far, only few examples of the

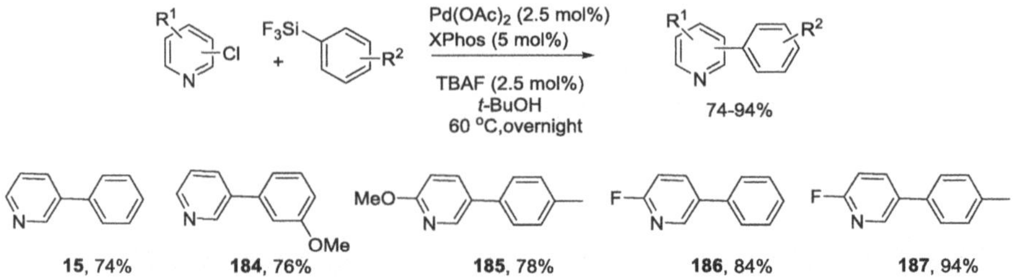

Scheme 55 Pd-catalyzed Hiyama coupling of pyridine chlorides

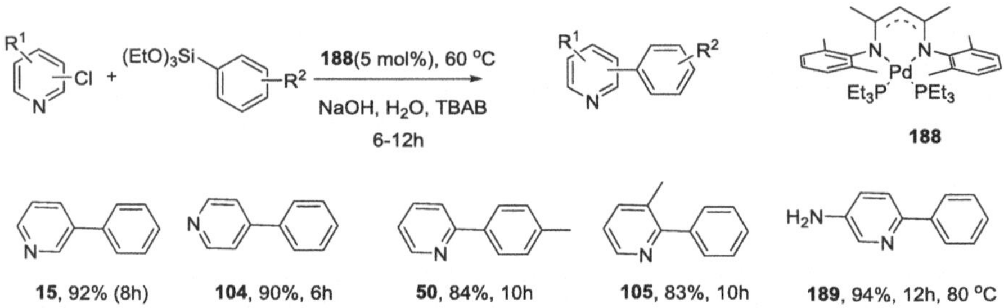

Scheme 56 Hiyama-Denmark coupling of pyridine chlorides

Hiyama coupling with aryl chlorides are reported, and they often proceed under microwave heating or high catalyst loading [120, 127, 128].

An effective Hiyama coupling of aryltrifluorosilanes with pyridyl chlorides (Scheme 55) has been reported by Molander and Iannazo [129]. Aryltrifluorosilanes are interesting as coupling partners in Hiyama coupling because of their stability to heat, air, and moisture, as well as their easy handling. Moreover, synthesis of these triflurosilanes is straightforward and can be achieved easily from commercially available trichlorosilanes. The substrate scope of pyridyl chloride includes electron donating (as in **185**) and withdrawing (as in **186** and **187**) substrates which make this method more attractive.

β-Diketiminatophosphane palladium complex **188** is reported to be a highly active catalyst for a fluoride-free Hiyama coupling of un-activated aryl and heteroaryl chlorides including chloropyridines and also substituted chloropyridines in aqueous media. 2-, 3-, and 4-chloro-substituted pyridines are known to undergo coupling with arylsilanes, and very basic and electron-rich 5-amino-2-chloropyridine is also found to participate in coupling to give the desired products **15**, **50**, **104**, **105**, and **189** in high yield (Scheme 56) [130]. This is remarkable since catalyst deactivation has often been an issue for coupling of electron-rich substrates, owing to the presence of free amino group, which allows it to bind to the Pd and thus deactivate the catalyst.

Halogenated 2-trimethylsilylpyridines have been reported to undergo Hiyama coupling with aryl and heteroaryl halides at room temperature as reported by Gros and coworkers (Scheme 57) [131]. The electron withdrawing effect of chlorine and

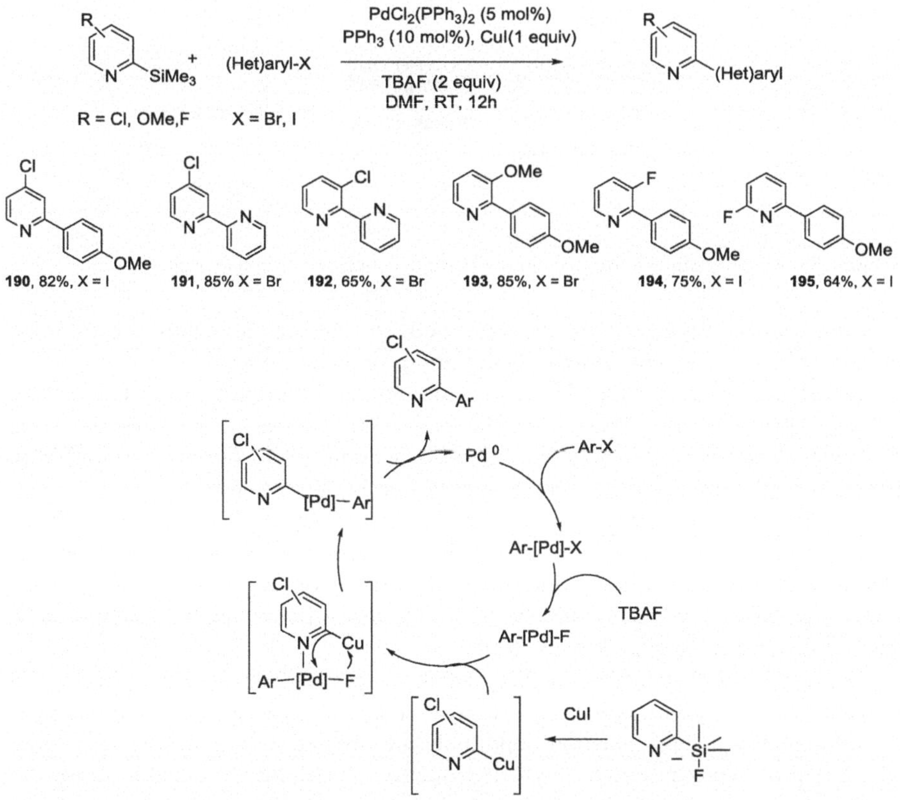

Scheme 57 Room temperature Hiyama coupling of (trimethylsilyl)pyridines

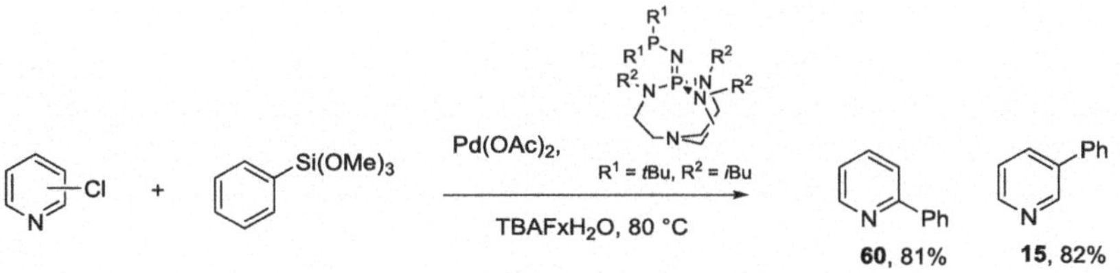

Scheme 58 Hiyama-Denmark coupling of 2- and 3-chloropyridine

other electron withdrawing substituent such as fluorine on the pyridine ring is believed to increase polarization of the C–Si bond, thus favoring the formation of the intermediate ate complex by reaction with fluoride ions, from TBAF (see products **190–192**, **194** and **195**). Interestingly, the authors observed that a methoxy group in 3 position was also tolerated (**193**), and they attributed an electron withdrawing effect to this group as well even though it is generally considered to be electron donating. Complexation of palladium by the pyridine nitrogen in transmetallation step (assisted by fluorine and copper) is an important factor in

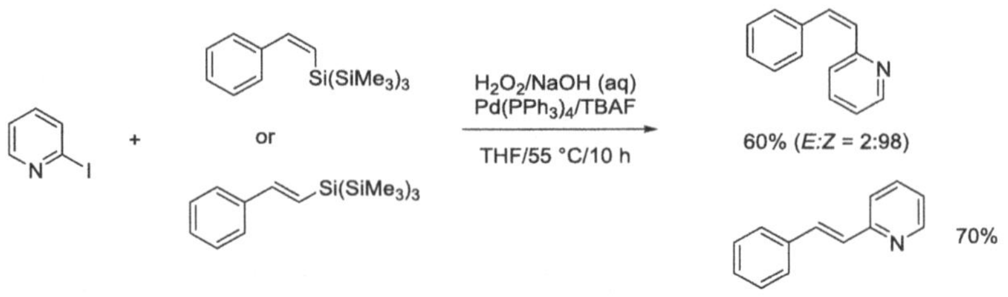

Scheme 59 Tris(trimethylsilyl)silanes in the Hiyama coupling of 2-iodopyridine

this reaction as suggested by the proposed mechanism, and thus the pyridine nitrogen must be α- to the trimethylsilyl group in the substrate.

Sarkar and coworkers developed pyrazole-tethered phosphine ligands for Pd(0) which were already discussed in the Stille section. One case of a Hiyama reaction was disclosed as well in that contribution. Coupling of 3-bromopyridine with trimethoxy(phenyl)silane gave 3-phenylpyridine in 88% yield [59].

Another class of ligands was applied by the group of Verkade [132]. They synthesized electron-rich amino-phosphines which promoted the Hiyama–Denmark coupling of aryl bromides but also of aryl chlorides. 2- and 3-chloropyridine gave 81% **60** and 82% **15**, respectively, in the reaction with trimethoxy(phenyl)silane (Scheme 58).

The trend in Hiyama-type coupling reactions clearly goes into the direction of fluoride-free conditions as realized in Hiyama–Denmark coupling. There, silanols deprotonated under the reaction conditions are applied as organosilicon species. The group of Wnuk reported vinyl tris(trimethylsilyl)silanes as suitable substrates for Hiyama coupling in the absence of fluoride anions [133]. In this case, a reactive silanol or siloxane species is generated due to the presence of aq. H_2O_2 and NaOH. *Z*-Tris(trimethylsilyl)silanes and *E*-tris(trimethylsilyl)silanes were reported as substrates (Scheme 59). In the first case, coupling with 2-iodopyridine gave 60% yield with an *E*:*Z* ratio of 2:98, the best of all investigated halides. In the case of *E*-substrates, the *E* product was formed exclusively in all cases investigated.

8 Kumada–Corriu–Tamao Coupling

The success story of modern cross-coupling chemistry started with the discovery of the nickel-catalyzed reactions of Grignard reagents with alkenyl and aryl halides, independently described by Kumada and Corriu in 1972. This transformation is now generally referred to as Kumada coupling, less frequently as Kumada–Corriu–Tamao reaction [134–136].

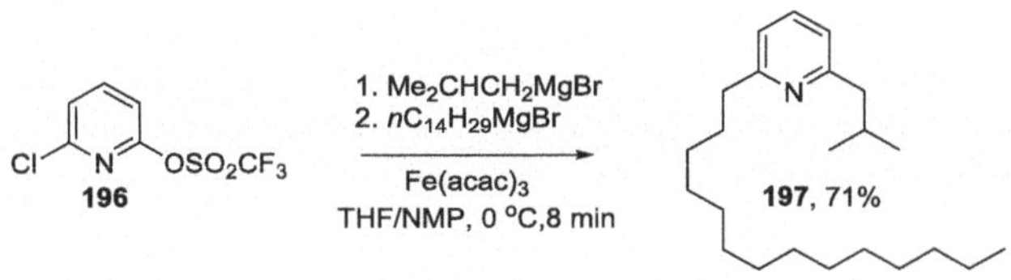

Scheme 60 Iron catalyzed Kumada coupling with simple $Fe(acac)_3$

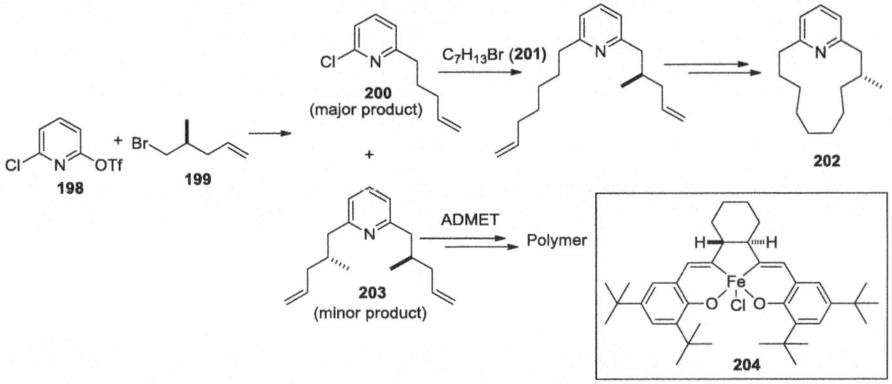

Scheme 61 Iron complex 204 in Kumada coupling of 198

8.1 *Pyridine as (Pseudo)halide*

Fürstner and coworkers described an iron-catalyzed cross-coupling process of alkylmagnesium halides and aromatic electrophiles, including pyridyl chlorides and sulfonates. The latter compounds are distinctly more reactive: the first coupling in pyridine substrate **196** proceeded at the sulfonate site and only then the chloride reacted, giving the desired product **197** in 71% in a one-pot protocol [137]. Cheap and nonhygroscopic $Fe(acac)_3$ is used as alternative to replace the expensive palladium catalysts usually used for such coupling procedures. Additionally, this catalyst shows unprecedentedly high reaction rates even at or below room temperature. This Fe-catalyzed method does not require any addition of special ligand, making it very attractive for coupling of pyridyl chlorides with alkyl, alkenyl, and alkylmagnesium bromides, as cheap and environmentally benign method. Only for *sec*-alkyl Grignard reagents, the use of [Fe(salen)Cl] (**204**) is recommended [138].

The utility of this iron-catalyzed method is illustrated in the synthesis of the alkaloid (R)-(+)-muscopyridine by the same research group [139]. The enantiomerically pure Grignard reagent **199** is converted with the difunctional pyridine derivative **198**, and the iron–salen complex **204** (5 mol%) in THF/NMP at 0°C for 20 min. Compound **200** is formed selectively by reaction at the triflate site accompanied by small amounts of the dialkylation product **201**. Without

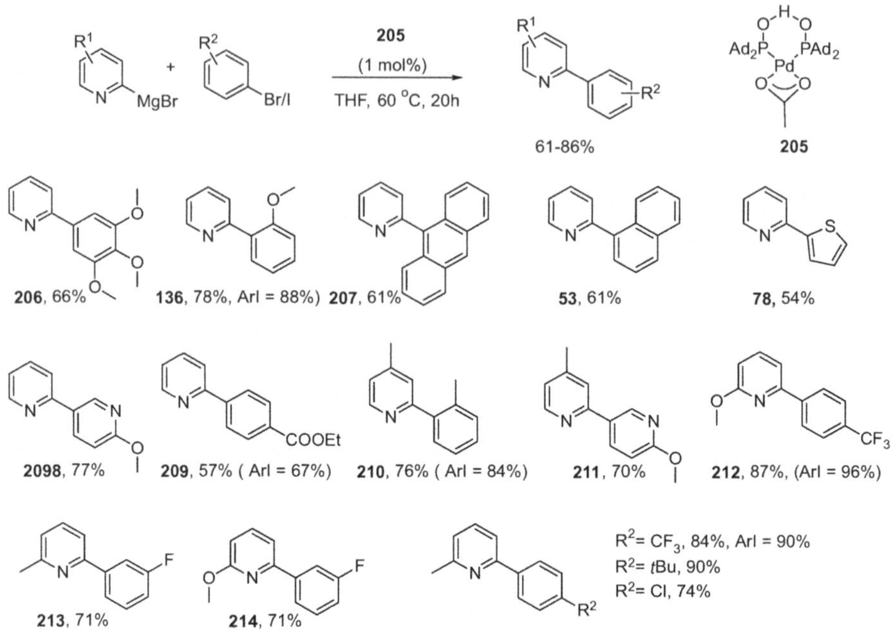

Scheme 62 Palladium catalyzed Kumada coupling of pyridine Grignards

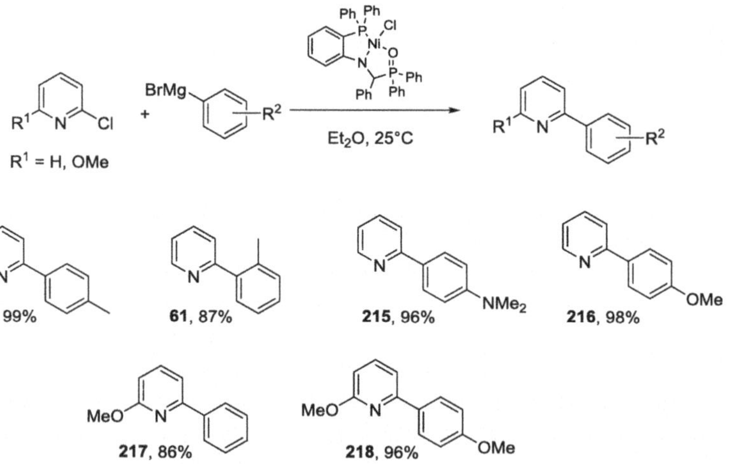

Scheme 63 Nickel catalyzed Kumada coupling of pyridine chlorides

intermediate purification to remove the dialkylated product, 6-heptenylmagnesium bromide **202** and additional catalyst **204** (5 mol%) are introduced and stirring can be continued for 30 min at 0°C. Subsequent ring-closing metathesis and reduction of the resulting double bond gives the natural product **203** (Scheme 61), while the

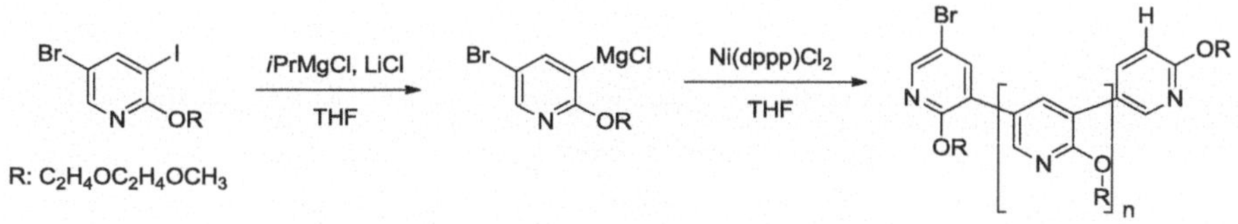

Scheme 64 Pyridine polymerization via Kumada cross-coupling

byproduct generated from the dialkylated intermediate can be transformed into a polymer by acyclic diene metathesis (ADMET) to facilitate easy separation.

8.2 Pyridine as Grignard Species

Generally applicable Kumada cross-coupling reactions of easily available 2-pyridyl organomagnesium reagents using $(1\text{-Ad})_2P(O)H$ as ligand were reported by Ackerman et al. [140]. The ligand forms the active catalyst (**205**) in the presence of $Pd_2(dba)_3$ in situ. The air- and moisture-stable secondary phosphine oxide $(1\text{-Ad})_2P(O)H$ turned out to be the ideal choice as pre-ligand. The generated catalyst is very active and gives good yields of desired products in the absence of any base. A variety of substituted 2-pyridyl nucleophiles and aryl halides (iodide and bromide) can be coupled using the reaction conditions described in Scheme 62. Generally, iodides give better yield than bromides. Interestingly, functional group tolerance is high as well for this cross-coupling process (see examples in Scheme 62).

Palladium is the predominant metal in cross-coupling chemistry. However, nickel is often an interesting alternative, and also in the original contribution on Kumada couplings, nickel was used as catalyst. Such catalysts can be extremely valuable as demonstrated by the group of Wang [141, 142]. They introduced amido pincer nickel complexes to promote biaryl formation of a series of halides including 2-chloropyridine. In all examples, excellent yields >90% were obtained at room temperature, even for sterically demanding *o*-tolyl Grignard coupling partners (Scheme 63, products **50**, **61**, **215–218**).

Yokozawa and coworkers reported two examples of pyridine polymerization via Kumada cross-coupling. Both 2,5-dihalopyridine [143] and a 3,5-dihalopyridine (Scheme 64) [144] served as substrates. Upon "mono-Grignardation," the polymerization proceeded under nickel (or Pd) catalysis. These are rare examples in which cross-coupling chemistry is exploited for a desired polymerization reaction.

As already mentioned in the Stille and Hiyama sections, pyrazole-tethered phosphine ligands are quite versatile in promoting various coupling reactions, in particular also Kumada coupling [59]. Sarkar disclosed indole-based bidentate compounds as alternative type of ligands for the nickel-catalyzed Kumada coupling of aryl chlorides including 2- and 3-chloropyridine [145]. Both halides were

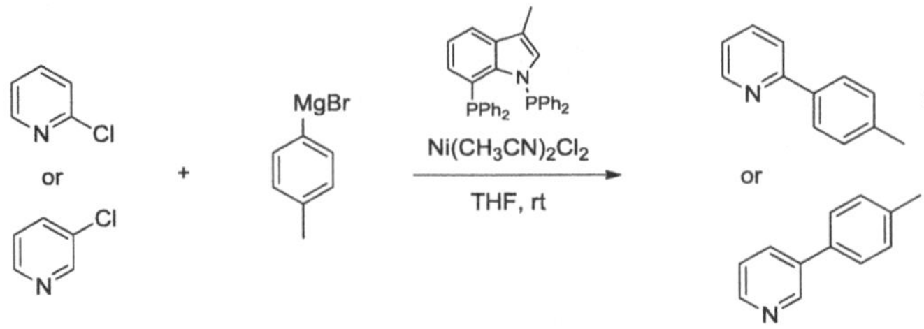

Scheme 65 Nickel catalyzed Kumada coupling of pyridine chlorides

coupled with 4-tolyl magnesium bromide at room temperature in almost quantitative yields (Scheme 65, products **50** and **57**).

NHC systems can be considered as privileged ligand class, and they were applied in nickel-catalyzed Kumada couplings, as well [146]. The activity of the metal complexes was excellent. Un-activated aryl chlorides could be reacted even at room temperature. The reaction of 2-chloropyridine with 4-tolyl magnesium bromide gave a quantitative yield, and also 2,6-dichloropyridine gave the bis-coupling products with phenylmagnesium bromide and 4-toluyl magnesium bromide in 81% and 88%, respectively (not shown).

9 Buchwald–Hartwig Amination Reactions on Pyridine

Palladium-catalyzed amination of aryl and heteroaryl (pseudo)halides is known as the Buchwald–Hartwig (B–H) reaction [147–150]. Since the discovery of this coupling process in 1995, significant progress has been made regarding improvement of substrate scope, lowering catalyst loading, application of highly active ligands for better yields, improved catalyst stability towards air and hydrolysis, or milder reaction conditions. Even after such progresses, a generalized procedure to couple any aryl/heteroaryl halides (or pseudohalides) with any nitrogen nucleophile is still lacking. Industrial applicability of Buchwald–Hartwig coupling is also limited, so far.

One possible side reaction in B–H coupling can be encountered due to the replacement of phosphine ligands in the active catalyst system by primary amine nucleophiles or by heteroaryl substrates through basic heterocyclic nitrogen. As a consequence, the lifetime of the catalyst is reduced. Therefore, a ligand which binds the palladium strongly to prevent ligand replacement but possesses the properties to promote oxidative addition and reductive elimination is ideally suited for Buchwald–Hartwig reactions. In this regard, commercially available, air- and moisture-stable CyPF-*t*Bu **219** reported by Hartwig is a versatile ligand. Its backbone is conformationally rigid due to the orientation of the methyl and ferrocenyl group, thus directing the phosphorous electron pair towards the metal center. A generalized and highly efficient method for coupling of aryl halides, including pyridyl chloride and bromides using $Pd(OAc)_2$ as catalyst along with CyPF-*t*Bu

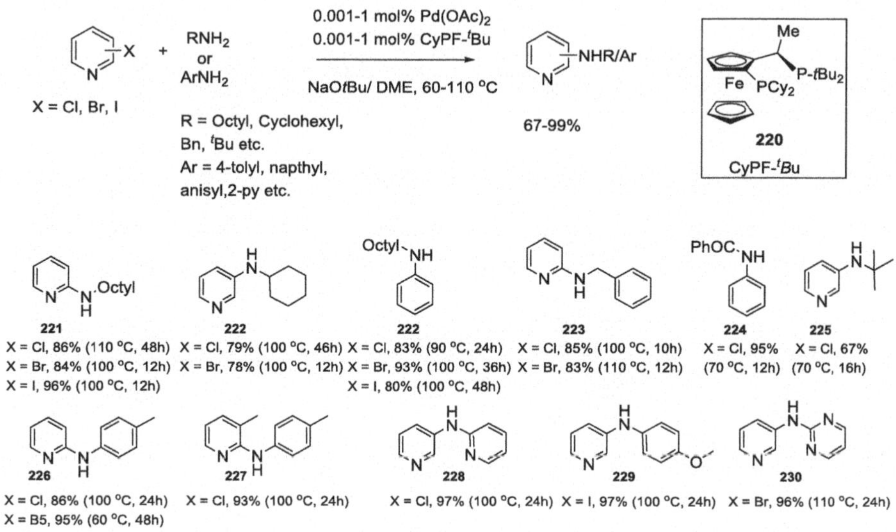

Scheme 66 Buchwald-Hartwig coupling of pyridine halides at low catalyst loading

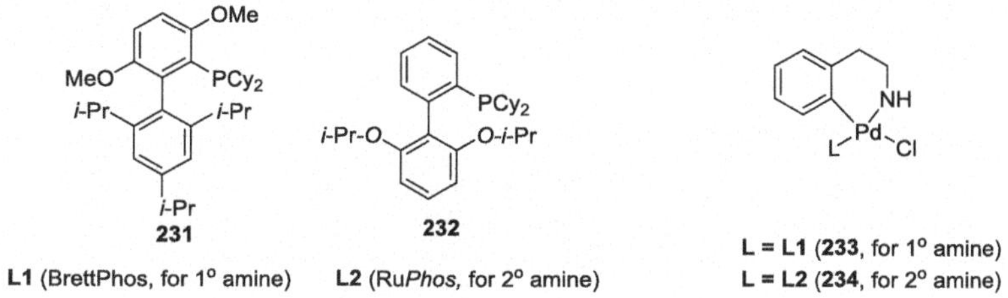

Scheme 67 Priviledged ligands in for Buchwald-Hartwig amination

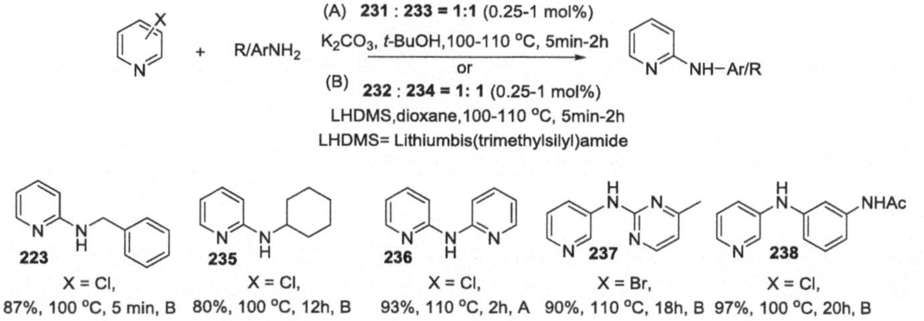

Scheme 68 Buchwald-Hartwig coupling of pyridine halides using BrettPhos and Ruphos ligands

as ligand (**220**), is described in Scheme 66 [151]. This reaction exhibits broad substrate scope with a variety of functional groups, including cyano, keto, free-carboxylate, amido, carbalkoxy, aromatic, and aliphatic hydroxyl and amino groups

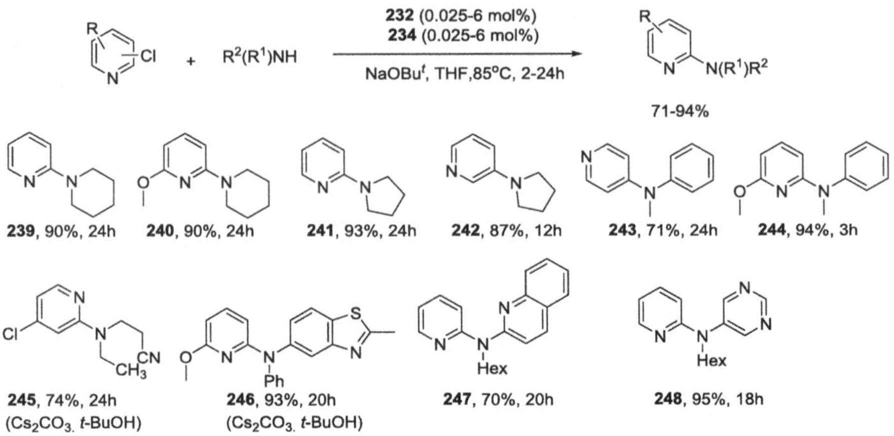

Scheme 69 Buchwald-Hartwig amination of pyridine chlorides

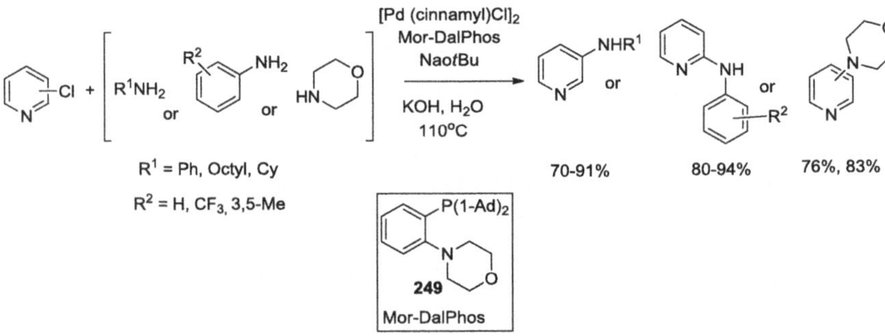

Scheme 70 Buchwald-Hartwig amination of pyridine chlorides in water

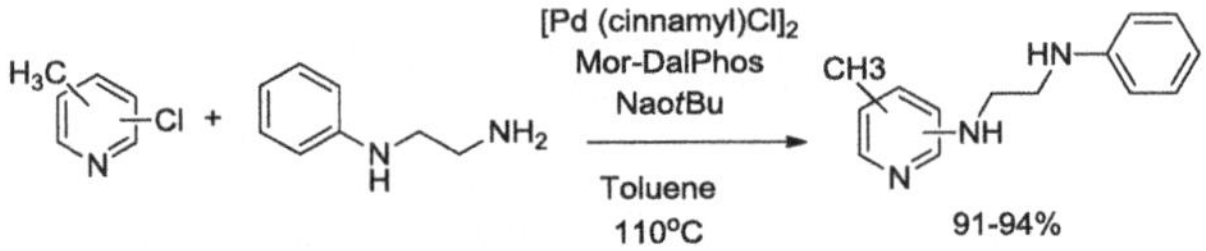

Scheme 71 Chemoselective Buchwald-Hartwig amination

employing very low catalyst loading (typically 0.001–1% with respect to both catalyst and ligand). Usually, high yields are obtained (Scheme 66, **221**–**230**). Secondary amines were also reported to couple under such conditions making this process more general.

Another generalized method for coupling of a wide variety of amines, including pyridyl amines, with aryl and heteroaryl halides including pyridine halides has been reported by Buchwald and coworkers using only the ligand systems BrettPhos (**231**), for primary amines, and RuPhos (**232**), for secondary amines (Scheme 67)

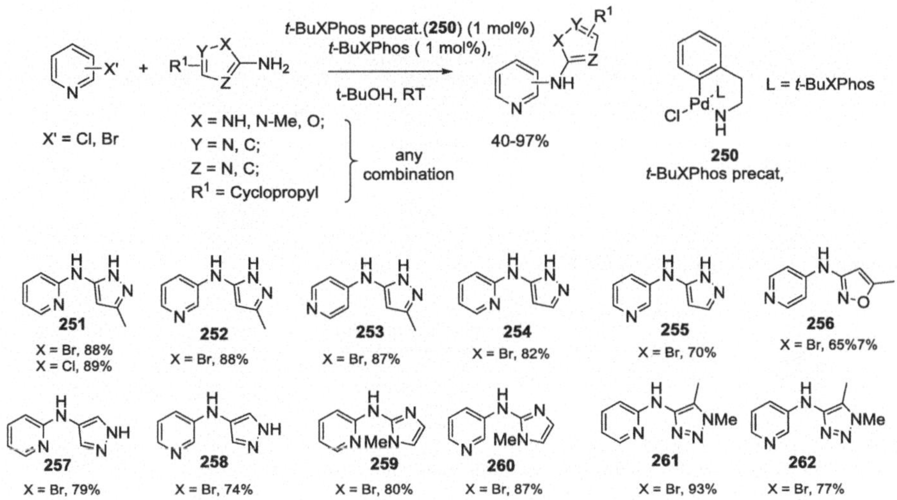

Scheme 72 Buchwald-Hartwig amination of pyridine halides with heterocyclic amines

[152]. Usually, a catalyst loading of 0.05–0.5 mol% is sufficient to accomplish the coupling, but in the case of pyridyl chlorides, a typical catalyst loading ranges from 0.25 to 1 mol%.

Coupling of 2-chloropyridine with primary amines (both aliphatic and aromatic) can be achieved using the BrettPhos ligand with Pd-catalyst system. Pyridine amines are usually considered as difficult substrate class due to lower nucleophilicity of the amine nitrogen and potential coordination to Pd with the pyridine nitrogen. The BrettPhos–Pd-catalyst system **233** proved to be useful in such cases. Representative examples of coupling are shown below (Scheme 68, products **223**, **235–238**).

Coupling of 2-chloropyridine with secondary amines requires lower catalyst loading (0.025–0.1 mol%), whereas 3-chloropyridines are more difficult substrates which require 1 mol% catalyst loading. Reaction conditions and examples (**239–248**) are given in Scheme 69.

The commercially available [Pd(cinnamyl)Cl]$_2$/Mor-DalPhos **249** catalyst system can be used for coupling of readily available and cheap pyridyl chlorides with primary or secondary amines in water, making the coupling of pyridyl chlorides more environmentally friendly. Details on reaction conditions along with substrates are given in Scheme 70 [153].

It is noteworthy that the [Pd(cinnamyl)Cl]$_2$/Mor-DalPhos system is effective in chemoselective amination with distinct preference for unhindered nucleophilic amine reaction partners (Scheme 71) [154].

Coupling of a wide range of five-membered heteroaryl amines with pyridyl chlorides and bromides at room temperature can be achieved, with a general set of reaction conditions using the *t*-BuXPhos precatalyst **250** (Scheme 72) [155]. Precatalyst **250** can be readily converted to the active monoligated

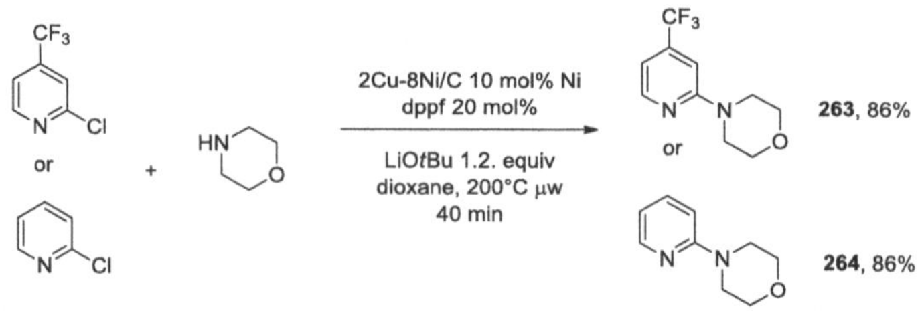

Scheme 73 Cu/Ni particles in the Buchwald-Hartwig amination

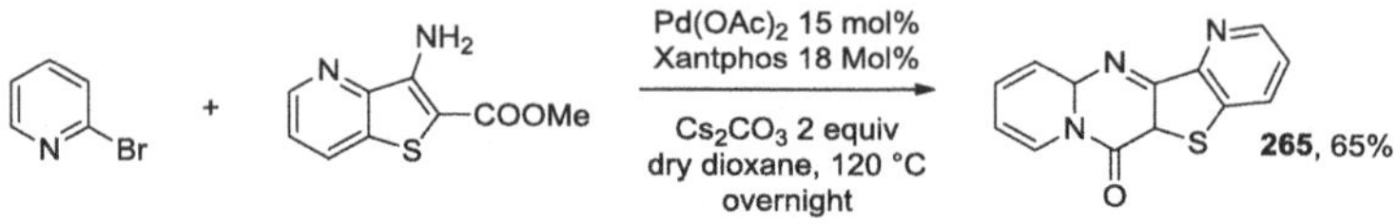

Scheme 74 Tandem C–N coupling - intramolecular cyclization

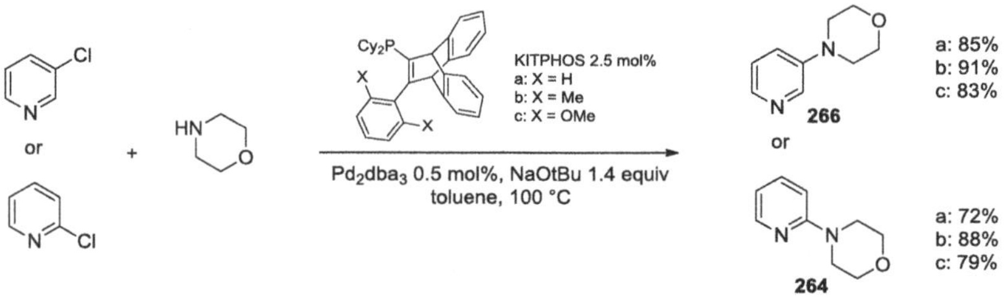

Scheme 75 KITPHOS ligands in Buchwald-Hartwig aminations

Pd(0) species within 2–3 min at or below room temperature. This allows subsequent room temperature C–N coupling at low catalyst loadings towards a series of products (Scheme 72, products **251–262**).

In the case of coupling of pyridyl halides and 1-*H*pyrazoles, coupling is completely selective for the exocyclic amine (**251–255**, **257**); no arylation of the N(H) heterocycle can be detected, so protection of –NH is not necessary, making the above transformation more attractive for industrial applications.

The group of Lipshutz developed a heterogeneous catalytic system composed of copper and nickel oxide particles which are supported on charcoal [156]. The additional ligand dppf had to be used in a relatively large amount of 20 mol%. The system proved to be an efficient catalyst for a number of metal-catalyzed reactions, including the Buchwald–Hartwig amination of two pyridine chlorides towards **263** and **264** (see Scheme 73). As can be seen, the reaction time was short, and good yields were obtained with morpholine as the amine coupling partner. Additional ligand and base were required as well. The advantage lies in the potential for recyclability of this heterogeneous catalyst.

Tandem C–N coupling and subsequent intramolecular cyclization was reported by Queiroz and coworkers [157]. A well-established catalytic system was employed

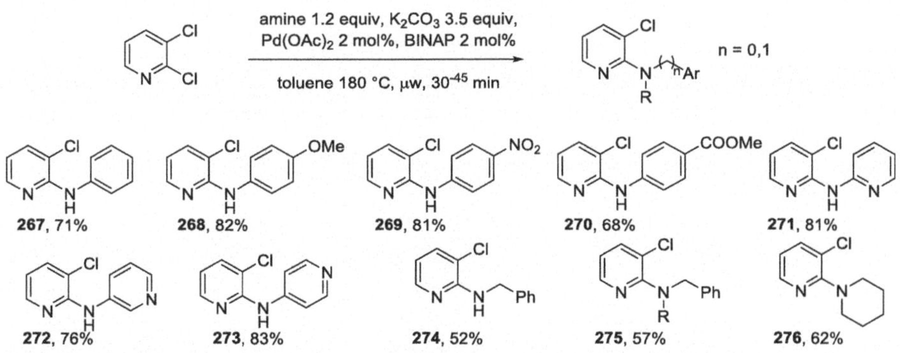

Scheme 76 Buchwald-Hartwig coupling of 2,3-dichloropyridine

for the C–N coupling, but high catalyst loading was required (Scheme 74). Overall, the reaction leads to interesting tetracyclic scaffolds (such as **265**) with good efficiency. An interesting feature of this transformation is the loss of aromatic character for the pyridine core.

The group of Doherty reported the synthesis of a ligand class called KITPHOS and their application in Buchwald–Hartwig reactions [158]. Both 2- and 3-chloropyridines were successfully coupled with morpholine (giving **264** and **266**), and depending on the applied phosphine ligand, up to 91% yield was obtained with low catalyst loading within 1–2 h reaction time (Scheme 75).

Schnürch, Mihovilovic, and coworkers disclosed a microwave-promoted Buchwald–Hartwig amination procedure for 2,3-dichloropyridine [159]. The amine was introduced selectively in 2 position (Scheme 76, products **267–276**). This protocol was an improvement to earlier reports by the group of Maes [160]. In this case, conventional heating was applied and 20 equiv. of K_2CO_3 was required to achieve high conversions. By applying microwave irradiation at higher reaction temperature, this amount could be reduced to 3.5 equiv.; moreover, reaction times could be shortened significantly (typically 30 min). Aniline derivatives, benzylic amines, and piperidine were applied as amine coupling partners. Additionally, 2-, 3-, and 4-aminopyridine were used as well, representing the highest yielding examples with coupling product yields at 81% (**271**), 76% (**272**), and 83% (**273**), respectively (Scheme 76). 3-Iodo-2-(methylthio)pyridine was employed as alternative substrate as well (not shown). Naturally, C–N coupling took place in position 3. The conditions differed in the applied base (Cs_2CO_3), and conventional heating was used instead of microwave irradiation. Again, aniline derivatives were used as coupling partners, and the remaining methylthio group was used as leaving group in a subsequent Liebeskind–Srogl coupling (see Sect. 10).

Miller and coworkers used Buchwald–Hartwig-type coupling reactions in their efforts to synthesize *N*-alkyl-*N*-(pyridin-2-yl)hydroxylamine scaffolds as selective antibacterial agents [161]. Reaction conditions applied were typical for such transformations, but the amine source was exceptional. Most frequently, primary or

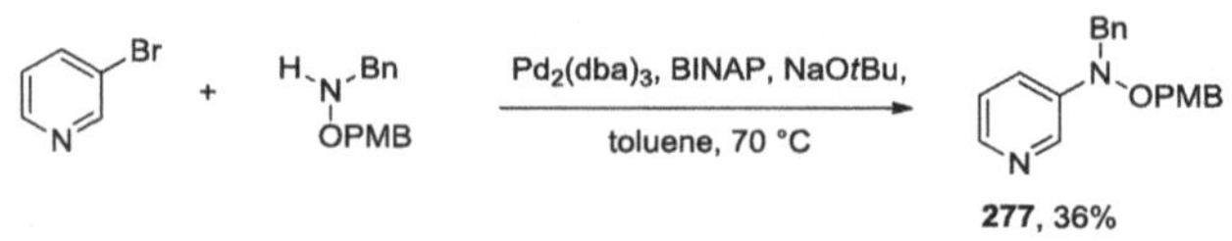

Scheme 77 PMB-protected benzylhydroxylamine in Buchwald-Hartwig coupling with 3-bromopyridine

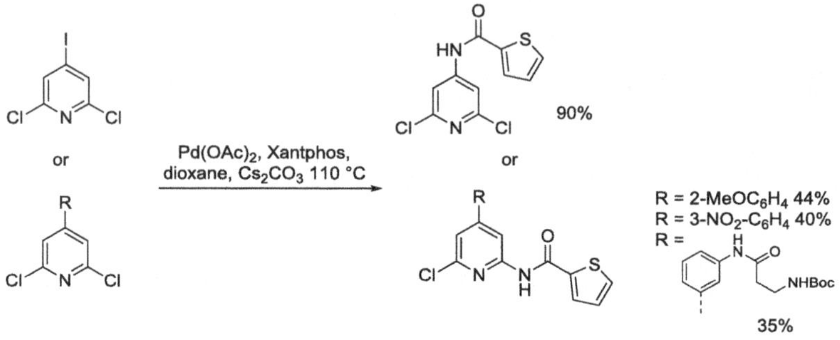

Scheme 78 Buchwald-Hartwig coupling with thiophene-2-amide

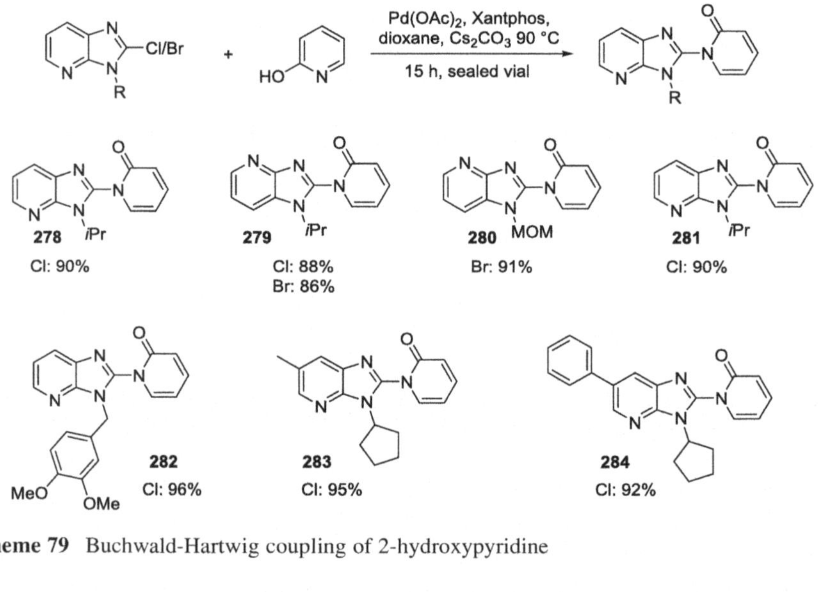

Scheme 79 Buchwald-Hartwig coupling of 2-hydroxypyridine

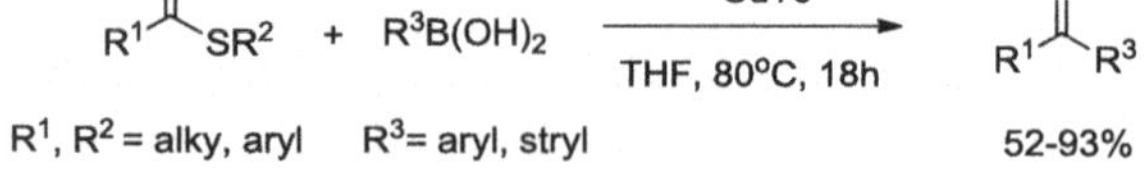

Scheme 80 The Liebeskind-Srogl reaction towards ketones

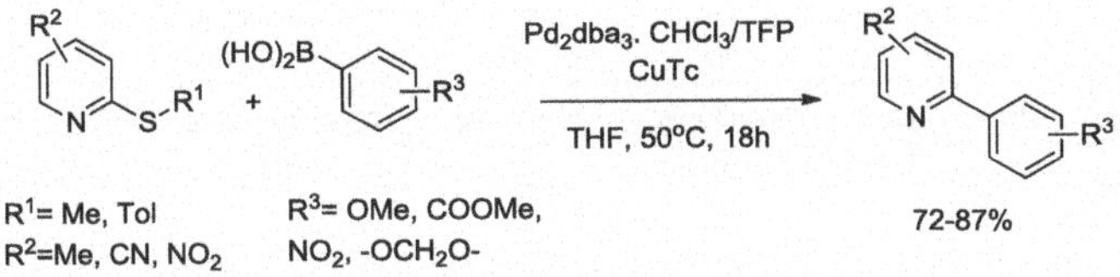

Scheme 81 Liebeskind-Srogl coupling of methylthio- or p-tolylthio-pyridine

secondary amines are used as coupling partners, but in this case, PMB-protected benzylhydroxylamine was applied. However, the yield was low with 36% of **277** (Scheme 77).

Also amides have been established as coupling partners for B–H type reactions. One such example was disseminated by Bihel and coworkers in 2013 [162]. 2,6-Dichloro-4-iodopyridine could be coupled with thiophene-2-amide in position 4 in 90% yield using $Pd(OAc)_2$ and Xantphos as catalytic system. If position 4 or positions 4 and 6 were already decorated, thiophene-2-amide or benzamide could be introduced in position 2 (Scheme 78). Besides Buchwald–Hartwig coupling, Sonogashira and Suzuki couplings were reported as well in this contribution. The three methods were exploited in sequential couplings towards 2,4,6-trisubstituted pyridines.

An unusual example came from the lab of Padusha [163]: In their synthesis of C2-substituted imidazo[4,5-*b*]pyridine, analogues 2-hydroxypyridine was coupled with various imidazo[4,5-*b*]pyridine halides leading to N-substitution of the pyridine nitrogen and formation of (imidazo[4,5-*b*]pyridin-2-yl)pyridin-2(1*H*)-ones **278–284**. Again, $Pd(OAc)_2$, Xantphos, and Cs_2CO_3 were used as catalytic system. A large number of examples were prepared giving yields >90% typically. Scheme 79 shows representative examples.

10 Liebeskind–Srogl Coupling

Liebeskind and Srogl reported in 2000 a palladium-catalyzed, Cu(I)-mediated, cross-coupling protocol for the synthesis of ketones from thioesters and boronic acids under neutral conditions (Scheme 80) [164]. A key feature of this coupling method is the requirement of stoichiometric amounts of a copper(I) carboxylate species, for example, copper(I) thiophene-2-carboxylate (CuTC), as a thiophilic metal cofactor.

Initially, the reaction scope was limited to thioesters as an organosulfur building block and a boronic acid as the organometallic reagent to access ketones. Since then, considerable advances have been made in terms of substrate scope and reaction conditions. A variety of thio-organic reagents, such as thioethers, thioamide, etc., are now known to undergo selective C–C cross-coupling, not only with boronic acids but also with other organometallic substrates, such as

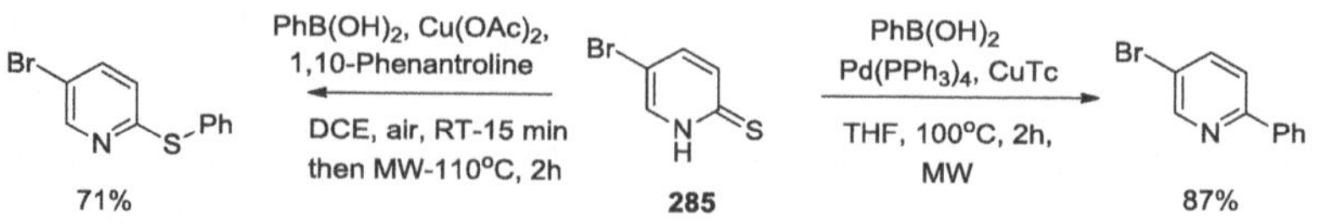

Scheme 82 Complementary couplings under Liebeskind-Srogl conditions

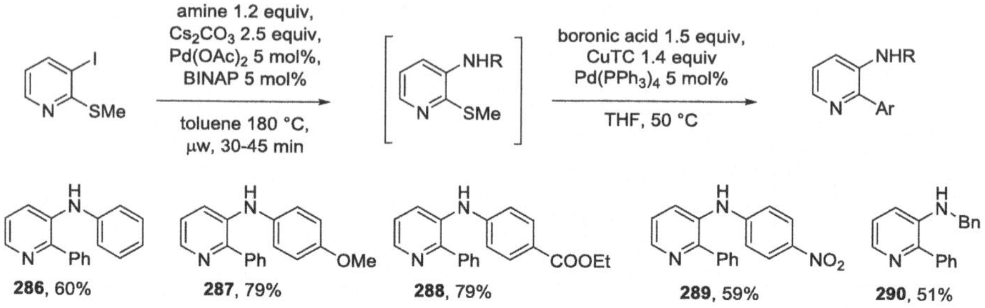

Scheme 83 A sequential Buchwald-Hartwig - Liebeskind-Srogl coupling sequence

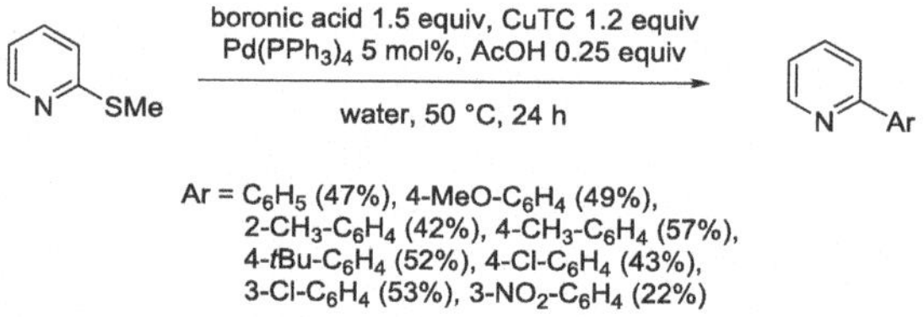

Scheme 84 Liebeskind-Srogl coupling in water

organostannes and organoindium reagents, with high yields and selectivity [165–167]. Heteroaromatic thioethers can be used as substrate in this Pd(0)-catalyzed, Cu (I)-mediated base-free desulfitative coupling to form biaryls.

2-Methylthio- and 2-tolylthio-pyridine derivatives (Scheme 81) are reported to undergo palladium-catalyzed copper(I) thiophene-2-carboxylate (CuTc)-mediated Liebeskind–Srogl(L–S) reaction as reported by Liebeskind and Srogl with the Pd_2dba_3/tris(2-furyl)phosphine (TFP) catalytic system in THF as solvent [168].

Five- and six-membered heterocycles, including pyridine, bearing a thioamide motif are also reported to undergo L–S coupling, but a high loading of CuTC (2–3 equiv.) is required [169, 170]. Using thioamide **285** as starting materials offers the advantage to tune reactivity from C–C coupling to C–S coupling by changing the catalyst system. Anaerobic Pd(0)/Cu(I) system favors C–C coupling, whereas aerobic Pd(II) system is directed towards C–S coupling as demonstrated for pyridine derivative **285** (Scheme 82).

Schnürch and coworkers exploited the orthogonality of the Liebeskind–Srogl coupling reaction vis-à-vis other methods such as Buchwald–Hartwig coupling [159]. Starting from 3-iodo-2-(methylthio)pyridine, first a Buchwald–Hartwig

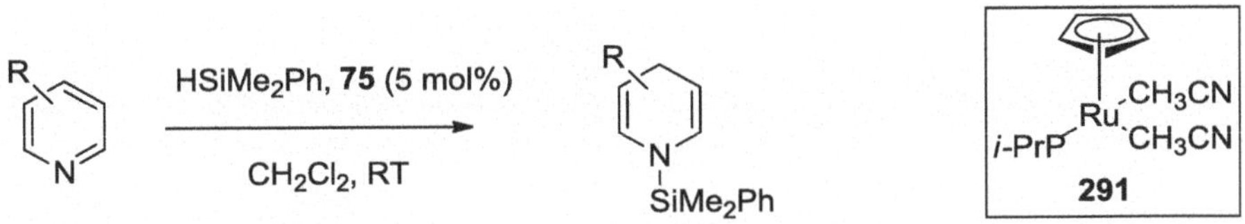

Scheme 85 Ruthenium catalyzed hydrosylation of pyridines

coupling was carried out. The reaction mixture was cooled to room temperature, diluted with DCM and filtered through a pad of Celite. After evaporation of DCM, anhydrous THF was added, followed by the boronic acid and the catalytic system for the Liebeskind–Srogl step. Via this facile protocol, a number of 2-arylated-3-aminated pyridines (**286–290**) could be prepared with reasonable good yields over two steps (Scheme 83).

Later, the same authors published the first Liebeskind–Srogl coupling in aqueous medium [171]. Using 2-(methylthio)pyridine as substrate yields of around 50% were obtained with a series of boronic acids. A strongly coordinating NO_2 substituent gave significantly lower yield (Scheme 84). Overall, reactions proceeded rarely beyond 50% conversion, and long reaction times were required in water as solvent. By addition of AcOH, the rate could be enhanced, but still 24 h was required. The authors hypothesized that the rate-enhancing effect originates from protonation of pyridine which increases the positive charge and speeds up the reaction.

11 Miscellaneous Reactions

11.1 Hydrosilylation of Pyridine

Few synthetic methods are available for the introduction of silyl groups in pyridine [172–175]. Silylated partially reduced pyridine derivatives are often used as selective reducing agents [176] and can be important as building blocks for organic synthesis [177].

A facile catalytic hydrosilylation method of pyridine is reported by Nikonov and coworkers using cationic ruthenium complex $[Cp(iPr_3P)Ru(NCCH_3)_2]^+$ (**291**) [178] (Scheme 85). The advantageous features of this method are that the transformation proceeds at room temperature and selectively generates 1,4-addition products; this behavior is of high significance, as conventional reduction methods usually give mixtures of 1,2- and 1,4-dehydropyridines [177].

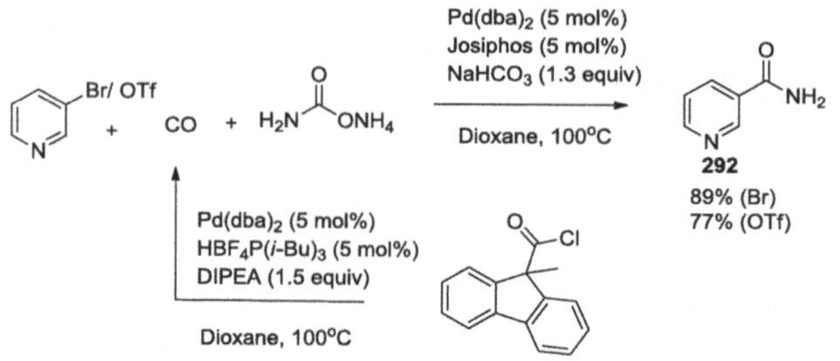

Scheme 86 Pd-catalyzed formation of nicotinamide

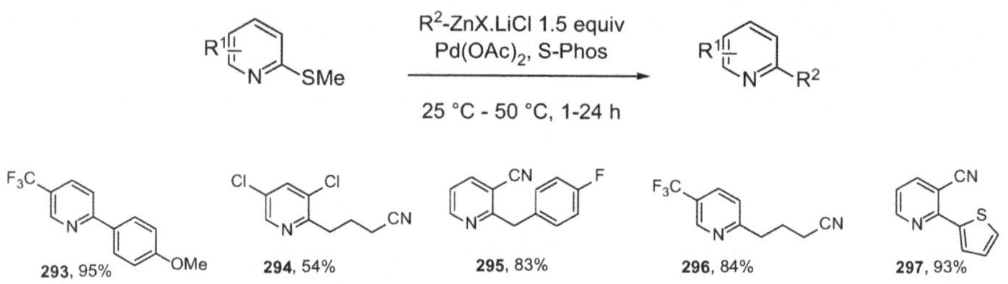

Scheme 87 Coupling of (methylthio)pyridine with organozinc reagents

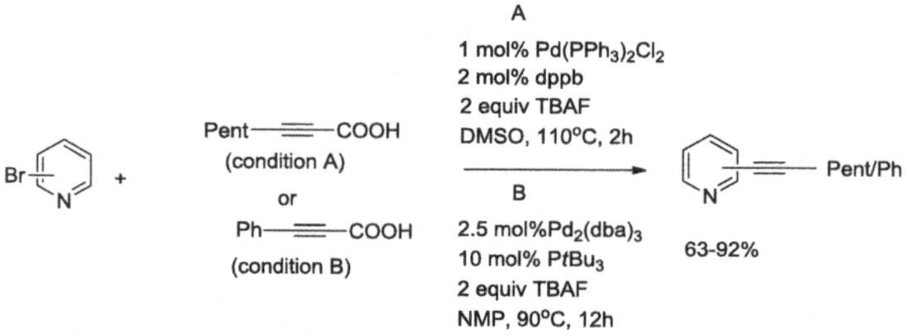

Scheme 88 Decarboxylative alkynylation of pyridine bromides

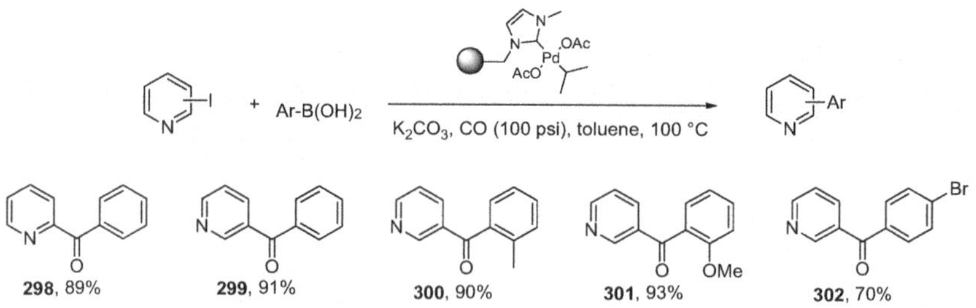

Scheme 89 Carbonylative Suzuki-Miyaura coupling towards ketones

11.2 Primary Amide Formation from Pyridine

A palladium-catalyzed method for the direct transformation of primary aryl amides from aryl bromides, including pyridyl bromides (or triflates), is reported by Skydstrap and coworkers (Scheme 86, product **292**) [179]. A two-chamber system is used for carrying out the reaction, where the solid precursors and the gaseous components, carbon monoxide, and ammonia are kept separately and allowed to react subsequently. Functional group tolerance of this methodology is also a noteworthy factor.

Knochel and coworkers reported a transformation which can be considered as a hybrid between a Liebeskind–Srogl and Negishi coupling reaction [180–183]. The methylthio group is employed as leaving group, whereas the organometal counterpart is a zinc organyl. A number of heterocyclic methylthio ethers were reacted including 2-(methylthio)-5-(trifluoromethyl)pyridine and 3-cyano-2-(methylthio) pyridine. In both cases, good to excellent yields of products (**293–297**) were obtained (Scheme 87).

A decarboxylative coupling of 2-octynoic- and phenyl propiolic acids with 2- and 3-bromopyridine was reported in 2009 by Moon and coworkers (Scheme 88) [184]. The reaction was catalyzed by a combination of either $Pd(PPh_3)_2Cl_2$ and dppb or $Pd_2(dba)_3$ and the bulky phosphine $P\mathit{t}Bu_3$ in the presence of tetrabutylammonium fluoride (TBAF), to yield the corresponding diarylated acetylene.

Bhanage and coworkers subjected 2- and 3-iodopyridine to carbonylative Suzuki coupling leading to the corresponding ketones (**298–302**) (Scheme 89) [185]. The reaction proved to be robust since steric bulk was well tolerated and also a bromo-substituted acid gave a good yield of 70%. A polymer-supported Pd–NHC complex served as catalyst amendable to recycling: It was demonstrated in the reaction of iodobenzene and phenylboronic acid that the catalyst can be used four times without a significant drop in yield and minor loss in activity as only observed within the fifth repetition. A CO pressure of 100 psi was applied representing relatively moderate reaction conditions.

12 Conclusion

The field of reaction optimization in cross-coupling chemistry is still a very active area. This is also reflected by the ongoing activities applying pyridine derivatives as coupling partners. Given the importance of the pyridine motif, this is not surprising, and we expected further research to be published on a regular basis. Recent trends focus on the development of milder and more environmentally benign reaction conditions. This goes in line with the general trend towards "green chemistry" in all chemical research fields.

One important aspect is the development of new ligands. We have seen in the recent years that ligand fine-tuning is a valuable tool to achieve the aforementioned goals in cross-coupling chemistry, with great discoveries to be expected in the future.

Another trend aims at the application of less reactive coupling partners such as chlorides or tosylates (derived from alcohols). This would lead to a broader range of potential coupling partners since much more chlorides and alcohols are commercially available as compared to bromides or iodides besides an often favorable cost issue.

Furthermore, a lot of research is dedicated to the reaction medium, most importantly, towards cross-coupling reactions in aqueous medium, again for environmental reasons. Also here significant advances have been made in recent years for several cross-coupling methods.

Finally, heterogeneous catalyst systems (metal nanoparticles or immobilized catalyst systems) have gained importance as well. They allow for catalyst recycling and would be compatible with continuous flow techniques, even though in this field research is scarce still.

References

1. Daly JW, Garraffo HM, Spande TF (1999) In: Pelletier WW (ed) Alkaloids: chemical and biological perspectives, vol 13. Elsevier, New York
2. Jones G (1996) In: Katritzky AR, Rees CW, Scriven EFV, McKillop A (eds) Comprehensive heterocyclic chemistry II, vol 5. Pergamon, Oxford, pp 167–243
3. Joule JA, Mills K (2000) Heterocyclic chemistry, 4th edn. Blackwell Science, Cambridge, pp 63–120
4. Michael JP (2005) Nat Prod Rep 22:627–646
5. Swainston Harrison T, Scott LJ (2005) Drugs 65:2309–2336
6. Deininger MWN, Druker BJ (2003) Pharmacol Rev 55:401–423
7. de Meijere A, Bräse S, Oestreich M (eds) (2014) Metal catalyzed cross-coupling reactions and more. Wiley, Weinheim
8. Miyaura N, Yamada K, Suzuki A (1979) Tetrahedron Lett 36:3437–3440
9. Feuerstein M, Laurenti D, Bougeant C, Doucet H, Santelli M (2001) Chem Commun 2001:325–326
10. Feuerstein M, Laurenti D, Bougeant C, Doucet H, Santelli M (2001) Tetrahedron Lett 42:5659–5662
11. Laurenti D, Feuerstein M, Pepe G, Doucet H, Santelli M (2001) J Org Chem 66:1633–1637
12. Feuerstein M, Laurenti D, Doucet H, Santelli M (2001) Chem Commun 2001:43–44
13. Feuerstein M, Laurenti D, Doucet H, Santelli M (2001) Tetrahedron Lett 42:2313–2316
14. Meier P, Legraverant S, Mueller S, Schaub J (2003) Synthesis 551–552
15. Thompson AE, Hughes G, Batsanov AS, Bryce MR, Parry PR, Tarbit B (2005) J Org Chem 70:388–390
16. Shi S, Zhang Y (2008) Green Chem 10:868–872
17. Handy ST, Wilson T, Muth A (2007) J Org Chem 72:8496–8500
18. Voisin-Chiret AS, Bouillon A, Burzicki G, llant MC, Legay R, El-Kashef H, Rault S (2009) Tetrahedron 65:607–612
19. O'Shea PD, Gauvreau D, Gosselin F, Hughes G, Nadeau C, Roy A, Shultz CS (2009) J Org Chem 74:4547–4553
20. Itoh T, Mase T (2005) Tetrahedron Lett 46:3573–3577
21. Billingsley K, Anderson KW, Buchwald SL (2006) Angew Chem Int Ed 45:3484–3488

22. Jager M, Eriksson L, Bergquist J, Johansson O (2007) J Org Chem 72:10227–10330
23. Ehlers P, Reimann S, Erfle S, Villinger A, Langer P (2010) Synlett 1528–1532
24. Xiong WN, Yang CG, Jiang B (2001) Bioorg Med Chem Lett 9:1773–1780
25. Ma JA, Cahard D (2004) Chem Rev 104:6119–6147
26. Ahmed S, Sharif M, Shoaib K, Reimann S, Iqbal J, Patonay T, Spannenberg A, Langer P (2013) Tetrahedron Lett 54:1669–1672
27. Fleckenstein CA, Plenio H (2008) Chem Eur J 14:4267–4279
28. Diebolt O, Braunstein P, Nolan SP, Cazin CSJ (2008) Chem Commun 3190–3192
29. Peh G-R, Kantchev EAB, Er J-C, Ying JY (2010) Chem Eur J 16:4010–4017
30. So CM, Yeung CC, Lau CP, Kwong FY (2008) J Org Chem 73:7803–7806
31. Kinzel T, Zhang Y, Buchwald SL (2010) J Am Chem Soc 132:14073–14075
32. Whelligan DK, Thomson DW, Taylor D, Hoelder S (2010) J Org Chem 75:11–15
33. Comins DL, Dehghani A, Foti CJ, Joseph SP (1998) Organic syntheses coll, vol 9. Wiley, New York
34. Vyvyan JR, Dell JA, Ligon TJ, Motanic KK, Wall HS (2010) Synthesis 3637–3644
35. Antoft-Fitch A, Blackburn T, Snieckus V (2009) J Am Chem Soc 131:17750–17752
36. Kuroda J-i, Inamoto K, Hiroya K, Doi T (2009) Eur J Org Chem 2251–2261
37. Tu T, Mao H, Herbert C, Xu M, Dotz KH (2010) Chem Commun 46:7796–7798
38. Chow WK, So CM, Lau CP, Kwong FY (2010) J Org Chem 75:5109–5112
39. Yang J, Liu S, Zheng J-F, Zhou J (2012) Eur J Org Chem 2012:6248–6259
40. Ge S, Hartwig JF (2012) Angew Chem Int Ed 57:12837–12841
41. Kudo N, Perseghini M, Fu GC (2006) Angew Chem Int Ed 118:1304–1306
42. Tsuji J (2004) Palladium reagents and catalysts. Wiley, New York
43. Barder T, Buchwald SL (2004) Org Lett 6:2649–2652
44. Ren W, Li J, Zou D, Wu Y, Wu Y (2012) Tetrahedron 68:1351–1358
45. Billingslay KS, Buchwald SL (2008) Angew Chem Int Ed 47:4695–4698
46. Ackermann L, Potukuchi HK (2009) Synlett 2852–2856
47. Deng JZ, Paone D, Ginnetti AT, Kurihara H, Dreher SD, Weissmann SA, Stauffer SR, Burgey CS (2009) Org Lett 11:345–347
48. Milstein D, Stille JK (1978) J Am Chem Soc 100:3636–3638
49. Kosugi M, Sasazawa K, Shimizu Y, Migita T (1977) Chem Lett 6:301–302
50. Su W, Urgaonkar S, McLaughlin PA, Verkade JG (2004) J Am Chem Soc 126:16433–16439
51. Naber JR, Buchwald SL (2008) Adv Synth Catal 350:957–961
52. Lee DH, Taher A, Ahn WS, Jin MJ (2010) Chem Commun 46:478–480
53. Dowlut M, Mallik D, Organ MG (2010) Chem Eur J 16:4279–4283
54. Delgado O, Muller HM, Bach T (2008) Chem Eur J 14:2322–2339
55. Ammer C, Bach T (2010) Chem Eur J 16:14083–14093
56. Just-Baringo X, Albericio F, Alvarez M (2013) Eur J Org Chem 6404–6419
57. Gholap SL, Hommes P, Neuthe K, Reissig HU (2013) Org Lett 15:318–321
58. Louaisil N, Pham PD, Boeda F, Faye D, Castanet A-S, Legoupy S (2011) Eur J Org Chem 2011:143–149
59. Pal A, Ghosh R, Adarsh NN, Sarkar A (2010) Tetrahedron 66:5451–5458
60. Štefko M, Pohl R, Klepetářová B, Hocek M (2008) Eur J Org Chem 2008:1689–1704
61. Ma G, Leng Y, Wu Y, Wu Y (2013) Tetrahedron 69:902–909
62. King AO, Okukado N, Negishi E (1977) J Chem Soc Chem Commun 1977:683–684
63. Negishi E, Wang G, Rao H, Xu Z (2010) J Org Chem 75:3151–3182
64. Phapale VB, Cardenas DJ (2009) Chem Soc Rev 38:1598–1607
65. Blaser HU, Pugin B, Spindler F (1996) In: Cornils B, Herrmann WA (eds) Applied homogeneous catalysis with organometallic compounds, vol 2. Wiley, Weinheim
66. Organ MG, Avola S, Dubovyk I, Hadei N, Kantchev AB, O'Brien CJ, Valentea C (2006) Chem Eur J 12:4749–4755
67. Valente C, Belowich ME, Hadei N, Organ MG (2010) Eur J Org Chem 2010:4343
68. Liu Z, Dong N, Xu M, Sun Z, Tu T (2013) J Org Chem 78:7436–7444

69. Luzung MR, Patel JS, Yin J (2010) J Org Chem 75:8330–8332
70. Yang Y, Oldenhuis NJ, Buchwald SL (2013) Angew Chem Int Ed 52:615–619
71. Melzig L, Gavryushin A, Knochel P (2007) Org Lett 9:5529–5532
72. Melzig L, Dannenwaldt T, Gavryushin A, Knochel P (2011) J Org Chem 76:8891–8906
73. Sase S, Jaric M, Metzger A, Malakhov V, Knochel P (2008) J Org Chem 73:7380–7382
74. Bernhardt S, Manolikakes G, Kunz T, Knochel P (2011) Angew Chem Int Ed 50:9205–9209
75. Stathakis CI, Bernhardt S, Quint V, Knochel P (2012) Angew Chem Int Ed 51:9428–9432
76. Stathakis CI, Manolikakes G, Knochel P (2013) Org Lett 15:1302–1305
77. Bolliger JL, Frech CM (2010) Chem Eur J 16:11072–11081
78. Getmanenko YA, Twieg RJ (2008) J Org Chem 73:830–839
79. Mizoroki T, Mori K, Ozaki A (1971) Bull Chem Soc Jpn 44:581–581
80. Heck RF, Nolley JP Jr (1972) J Org Chem 37:2320–2322
81. Heck RF (2006) Synlett 2006:2855–2860
82. Phan NTS, Van Der Sluys M, Jones CW (2006) Adv Synth Catal 348:609–679
83. Alonso F, Beletskaya IP, Yus M (2005) Tetrahedron 61:11771–11835
84. Ozawa F, Kubo A, Hayashi T (1991) J Am Chem Soc 113:1417–1419
85. Cabri W, Candiani I (1995) Acc Chem Res 28:2–7
86. Ludwig M, Stromberg S, Svensson M, Akermark B (1999) Organometallics 18:970–975
87. Larhed M, Hallberg A (2002) In: Negishi E (ed) Handbook of organopalladium chemistry for organic synthesis. Wiley, New York
88. Cabri W, Candiani I, Bedeschi A, Santi R (1992) J Org Chem 57:3558–3563
89. Mo J, Xu L, Xiao J (2005) J Am Chem Soc 127:751–760
90. Vallin KSA, Larhed M, Johansson K, Hallberg A (2000) J Org Chem 65:4537–4542
91. Sonesson C, Larhed M, Nyquist C, Hallberg A (1996) J Org Chem 61:4756–4763
92. Xu D, Liu Z, Tang W, Xu L, Hyder Z, Xiao J (2008) Tetrahedron Lett 49:6104–6107
93. Gogsig TM, Lindhardt AT, Dekhane M, Grouleff J, Skrydstrup T (2009) Chem Eur J 15:5950–5955
94. Kantam MK, Reddy PV, Srinivas P, Bhargava S (2011) Tetrahedron Lett 52:4490–4494
95. de Vries JG (1986) Can J Chem 79:1086–1092
96. Raggon JW, Snyder MW (2002) Org Process Res Dev 6:67–69
97. Yılmaz Ü, Küçükbay H, Deniz S, Şireci N (2013) Molecules 18:2501–2517
98. Sonogashira K, Tohda Y, Hagihara N (1975) Tetrahedron Lett 16:4467–4470
99. Chinchilla R, Najera C (2011) Chem Soc Rev 40:5084–5121
100. Feuerstein M, Doucet H, Santelli M (2005) Tetrahedron Lett 46:1717–1720
101. Feuerstein M, Doucet H, Santelli M (2006) J Mol Catal A 256:75–84
102. Fleckenstein CA, Plenio H (2008) Green Chem 10:563–570
103. Bolliger JL, Frech CM (2009) Adv Synth Catal 351:891–902
104. Córdoba M, Galájov M, Izquierdo ML, Alvarez-Builla J (2013) Tetrahedron 69:2484–2493
105. Iranpoor N, Firouzabadi H, Ahmadi Y (2012) Eur J Org Chem 2012:305–312
106. Zhou R, Wang W, Jiang Z, Fu H, Zheng X, Zhang C, Chen H, Li R (2014) Catal Sci Today 4:746–751
107. Zou L, Johansson AJ, Zuidema E, Bolm C (2013) Chem Eur J 19:8144–8152
108. Abe H, Machiguchi H, Matsumoto S, Inouye M (2008) J Org Chem 73:4650–4661
109. Miki K, Fujita M, Inoue Y, Senda Y, Kowada T, Ohe K (2010) J Org Chem 75:3537–3540
110. Nakao Y, Hiyama T (2011) Chem Soc Rev 40:4893–4901
111. Hiyama T (1998) In: Diederich F, Stang PJ (eds) Metal-catalyzed cross-coupling reactions. Wiley, Weinheim
112. Hiyama T (2002) J Organomet Chem 653:58–61
113. Denmark SE, Sweiss RF (2002) Acc Chem Res 35:835–846
114. Spivey AC, Gripton CJG, Hannah JP (2004) Curr Org Synth 1:211–226
115. Hatanaka Y, Hiyama T (1989) J Org Chem 54:268–270
116. Huang T, Li CT (2002) Tetrahedron Lett 43:403–405
117. Koike T, Mori A (2003) Synlett 1850–1852

118. Wolf C, Lerebours R (2004) Org Lett 6:1147–1150
119. Lerebours R, Wolf C (2005) Synthesis 2287–2292
120. Alacid E, Nàjera C (2006) Adv Synth Catal 348:945–952
121. Ackermann L, Born R (2005) Angew Chem Int Ed 44:2444–2447
122. So CM, Lee HW, Lau CP, Kwong FY (2009) Org Lett 11:317–320
123. Denmark SE, Smith RC, Chang WTT, Muhuhi JM (2009) J Am Chem Soc 131:3104–3118
124. Denmark SE, Werner NS (2010) J Am Chem Soc 132:3612–3620
125. Powell DA, Fu GC (2004) J Am Chem Soc 126:7788–7789
126. Strotman NA, Sommer S, Fu GC (2007) Angew Chem Int Ed 46:3556–3558
127. Clarke ML (2005) Adv Synth Catal 347:303–307
128. Ackermann L, Gshrei CJ, Althammer A, Riederer M (2006) Chem Commun 2006:1419–1421
129. Molander GA, Iannazo L (2011) J Org Chem 76:9182–9187
130. Lee DH, Jung JY, Jim MJ (2010) Chem Commun 46:9046–9048
131. Pierrat P, Gros P, Fort Y (2005) Org Lett 7:697–700
132. Raders SM, Kingston JV, Verkade JG (2010) J Org Chem 75:1744–1747
133. Wang Z, Pitteloud J-P, Montes L, Rapp M, Derane D, Wnuk SF (2008) Tetrahedron 64:5322–5327
134. Corriu RJP, Masse JP (1972) J Chem Soc Chem Commun 144–145
135. Tamao K, Sumitani K, Kumada M (1972) J Am Chem Soc 94:4374–4376
136. Tamao K, Sumitani K, Kiso Y, Zembayashi M, Fujioka A, Kodama S, Nakajima I, Minato A, Kumada M (1976) Bull Chem Soc Jpn 49:1958–1969
137. Fürstner A, Leitner A, Mendez M, Krause H (2002) J Am Chem Soc 124:13856–13863
138. Fürstner A, Leitner A (2002) Angew Chem Int Ed 41:609–611
139. Fürstner A, Leitner A (2003) Angew Chem Int Ed 42:308–311
140. Ackerman L, Potukuchi HK, Kapdi AR, Schulzke C (2010) Chem Eur J 16:3300–3303
141. Liu N, Wang Z-X (2011) J Org Chem 76:10031–10038
142. Zhang X-Q, Wang Z-X (2013) Synlett 24:2081–2084
143. Nanashima Y, Shibata R, Miyakoshi R, Yokoyama A, Yokozawa T (2012) J Polym Sci Part A Polym Chem 50:3628–3640
144. Nanashima Y, Yokoyama A, Yokozawa T (2012) Macromolecules 45:2609–2613
145. Ghosh R, Sarkar A (2010) J Org Chem 75:8283–8286
146. Xi Z, Liu B, Chen W (2008) J Org Chem 73:3954–3957
147. Wolfe JP, Wagaw S, Marcoux J-F, Buchwald SL (1998) Acc Chem Res 31:80–818
148. Hartwig JF (1998) Angew Chem Int Ed 37:2046–2065
149. Muci AR, Buchwald SL (2002) Top Curr Chem 219:131–209
150. Hartwig JF (2006) Synlett 9:1283–1294
151. Shen Q, Ogata T, Hartwig JF (2008) J Am Chem Soc 130:6586–6596
152. Maiti P, Fros BP, Henderson JL, Nakamura Y, Buchwald SL (2011) Chem Sci 2:57–68
153. Tardiff BJ, Stradiotto M (2012) Eur J Org Chem 3972–3977
154. Tardiff BJ, Mcdonal R, Ferguson MJ, Stradiotto M (2012) J Org Chem 77:1056–1071
155. Moss TA, Addie MS, Nowak T, Waring MJ (2012) Synlett 23:285–289
156. Lipshutz BH, Nihan DM, Vinogradova E, Taft BR, Bošković ŽV (2008) Org Lett 10:4279–4282
157. Calhelha RC, Queiroz M-JRP (2010) Tetrahedron Lett 51:281–283
158. Doherty S, Knight JG, McGrady JP, Ferguson AM, Ward NAB, Harrington W, Clegg W (2010) Adv Synth Catal 352:201–211
159. Koley M, Wimmer L, Schnürch M, Mihovilovic MD (2011) Eur J Org Chem 2011:1972–1979
160. Jonckers THM, Maes BUW, Lemiere GLF, Dommisse R (2001) Tetrahedron 57:7027–7034
161. Wencewicz TA, Yang B, Rudloff JR, Oliver AG, Miller MJ (2011) J Med Chem 54:6843–6858

162. Doebelin C, Wagner P, Bertin I, Simonin F, Schmitt M, Bihel F, Bourguignon J-J (2013) RSC Adv 3:10296–10300
163. Abdul Khader KK, Sajith AM, Ali Padusha MS, Nagaswarupa HP, Muralidharan A (2014) Tetrahedron Lett 55:1778–1783
164. Liebeskind LS, Srogl J (2000) J Am Chem Soc 122:11260–11261
165. Myers BJ, Rigby JH (2001) Chemtracts 14:509–512
166. Lory P, Gilbertson SR (2005) Chemtracts 18:569–583
167. Prokopcova H, Kappe OC (2009) Angew Chem Int Ed 48:2276–2286
168. Liebeskind LS, Srogl J (2002) Org Lett 4:979–981
169. Prokopcova H, Kappe OC (2007) Adv Synth Catal 349:448–452
170. Prokopcova H, Kappe OC (2007) J Org Chem 72:4440–4448
171. Koley M, Wimmer L, Schnürch M, Mihovilovic MD (2013) J Heterocycl Chem 50:1368–1373
172. Hao L, Harrod JF, Lebuis A-M, Mu Y, Shu R, Samuel E, Woo H-G (1998) Angew Chem Int Ed 37:3126–3129
173. Harrod JF, Shu R, Woo H-G, Samuel E (2001) Can J Chem 79:1075–1085
174. Cook NC, Lyons JE (1965) J Am Chem Soc 87:3283–3284
175. Cook NC, Lyons JE (1966) J Am Chem Soc 88:3396–3403
176. Storer RI, Carrera DE, Ni Y, MacMillan DWC (2006) J Am Chem Soc 128:84–86
177. Lavilla R (2002) J Chem Soc Perkin Trans 1 :1141
178. Gutsulyak DV, Est AV, Nikonov GI (2011) Angew Chem Int Ed 50:1384–1387
179. Nielsen DU, Taaning RH, Lindhardt AT, Gøgsig TM, Skrydstrup T (2011) Org Lett 13:4454–4457
180. Metzger A, Melzig L, Despotopoulou C, Knochel P (2009) Org Lett 11:4228–4231
181. Metzger A, Melzig L, Knochel P (2010) Synthesis 2010:2853–2858
182. Melzig L, Metzger A, Knochel P (2010) J Org Chem 75:2131–2133
183. Melzig L, Metzger A, Knochel P (2011) Chem Eur J 17:2948–2956
184. Moon J, Jang M, Lee S (2009) J Org Chem 74:1403–1406
185. Qureshi Z, Deshmukh KM, Tambade PJ, Bhanage BM (2011) Synthesis 2011:243–250

Top Heterocycl Chem (2016) 45: 61–158
DOI: 10.1007/7081_2014_138

Published online: 1 April 2015

Metal Catalyzed Cross-Coupling Reactions in the Decoration of Pyrimidine, Pyridazine, and Pyrazine

Laurin Wimmer*, Lukas Rycek*, Moumita Koley, and Michael Schnürch

Abstract This chapter treats the cross-coupling chemistry of electron-deficient N-heterocycles containing two nitrogen atoms. The chapter is ordered according to heterocycle, coupling method, and the role of the N-heterocycle within the coupling process. The focus lies on new developments in the field, e.g., regarding new catalytic systems, and covers literature from 2008 until late 2013.

Keywords Cross-coupling • Metal catalysis • Palladium • Pyrazine • Pyridazine • Pyrimidine

Contents

*These authors have contributed equally to the chapter.

L. Wimmer, L. Rycek*, and M. Schnürch (✉)
Institute of Applied Synthetic Chemistry, Vienna University of Technology, Getreidemarkt 9/163, 1060 Vienna, Austria
e-mail: michael.schnuerch@tuwien.ac.at

M. Koley
Department of Organic Chemistry, Indian Institute of Science, Bangalore 560012, India

Abbreviations

9-BBN	3-Borabicyclo[3.3.1]nonane
acac	Acetylacetonate
BMIM	1-Butyl-3-methylimidazolium
BrettPhos	2-(Dicyclohexylphosphino)3,6-dimethoxy-2′,4′,6′-tri-*i*-propyl-1,1′-biphenyl
cBRIDP	Di-*tert*-butyl(2,2-diphenyl-1-methyl-1-cyclopropyl)phosphine
cod	Cyclooctadiene
CPhos	2-(2-dicyclohexylphosphanylphenyl)-*N1*,*N1*,*N3*,*N3*-tetramethylbenzene-1,3-diamine
CuMeSal	Copper(I) 3-methylsalicilate
CuTC	Copper(I) thiophene-2-carboxylate
Cy	Cyclohexyl
CyPF-*t*Bu	(*R*)-1-[(S_P)-2-(Dicyclohexylphosphino)ferrocenyl]ethyldi-*tert*-butylphosphine
DABCO	Diazabicyclo[2.2.2]octane
DavePhos	2-Dicyclohexylphosphino-2′-(*N*,*N*-dimethylamino)biphenyl
dba	Dibenzylideneacetone
DIPEA	*N*,*N*-Diisopropylethylamine
DMF	*N*,*N*-Dimethylformamide
DPE-Phos	Bis[(2-diphenylphosphino)phenyl] ether
dppf	1,1′-Bis(diphenylphosphino) ferrocene
dppp	1,3-Bis(diphenylphosphino)propane
DSC	Dye-sensitized solar cell
dtbpf	1,1′-Bis(di-*t*-butylphosphino)ferrocene
EDG	Electron donating group
EWG	Electron withdrawing group
HASPO	Heteroatom-substituted secondary phosphine oxide
IMes·HCl	1,3-Bis(2,4,6-trimethylphenyl)imidazolium chloride
IPrNi(allyl)Cl	Allyl[1,3-bis(2,6-diisopropylphenyl)imidazol-2-ylidene]chloropalladium(II)
JosiPhos	(*R*)-1-[(S_P)-2-(Dicyclohexylphosphino)ferrocenyl]ethyldi-*tert*-butylphosphine
MIDA	*N*-methyliminodiacetic acid

MTBE	Methyl *t*-butylether
NHC	N-heterocyclic carbene
NMP	*N*-methylpyrrolidone
OLED	Organic light-emitting diode
PEG	Polyethylene glycol
PEPPSI-IPent	Dichloro[1,3-bis(2,6-di-3-pentylphenyl)imidazol-2-ylidene](3-chloropyridyl)palladium(II)
PEPPSI-IPr	[1,3-Bis(2,6-diisopropylphenyl)imidazol-2-ylidene](3-chloropyridyl)palladium(II) dichloride
PMB	*p*-Methoxybenzyl
PTS	Polyoxyethanyl-α-tocopheryl sebacate
PyBroP	Bromotripyrrolidinophosphonium hexafluorophosphate
rt	Room temperature
RuPhos	2-Dicyclohexylphosphino-2′,6′-diisopropoxybiphenyl
Salen	2,2′-Ethylenebis(nitrilomethylidene)diphenol
SDPP	Silicadiphenyl phosphinite
SDS	Sodium dodecylsulfate
SES-NH_2	2-(trimethylsilyl)ethansulfonyl
S_NAr	Substitution nucleophilic aromatic
SPhos	2-Dicyclohexylphosphino-2′,6′-dimethoxybiphenyl
SPO	Secondary phosphine oxide
TBAB	Tetrabutylammonium bromide
TBAF	Tetrabutylammonium fluoride
TBS	*Tert*-butyldimethylsilyl
t-BuJosiPhos	(*R*)-1-[(S_P)-2-(Diphenylphosphino)ferrocenyl]ethyldi-*t*-butylphosphine
THF	Tetrahydrofuran
TMEDA	*N*,*N*,*N*′,*N*′-Tetramethylethylenediamine
tmp	Tetramethylpiperidide
TON	Turnover number
Xantphos	4,5-Bis(diphenylphosphino)-9,9-dimethylxanthene
XPhos	2-Dicyclohexylphosphino-2′,4′,6′-triisopropylbiphenyl

1 Pyrimidine in Cross-Coupling Reactions

1.1 Pyrimidine in Suzuki–Miyaura Reactions

1.1.1 Pyrimidine as (Pseudo)Halide

Catalyst Design

The coupling of 2-heteroaryl as well as polyfluoro-substituted (hetero)arylboronic acids is a long-standing problem due to the tendency of these compounds to

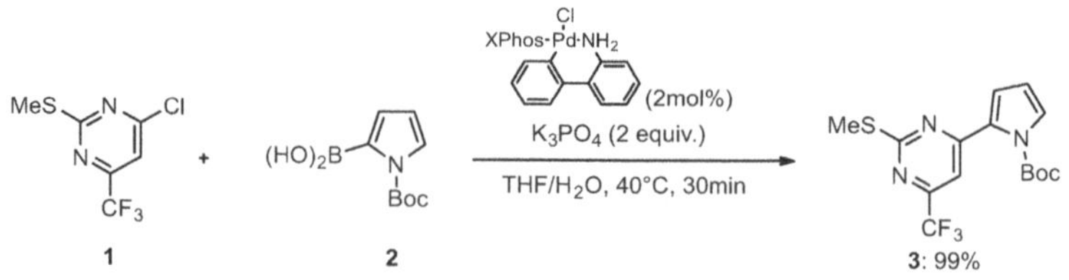

Scheme 1 Low temperature Suzuki coupling of pyrimidine chloride **1**

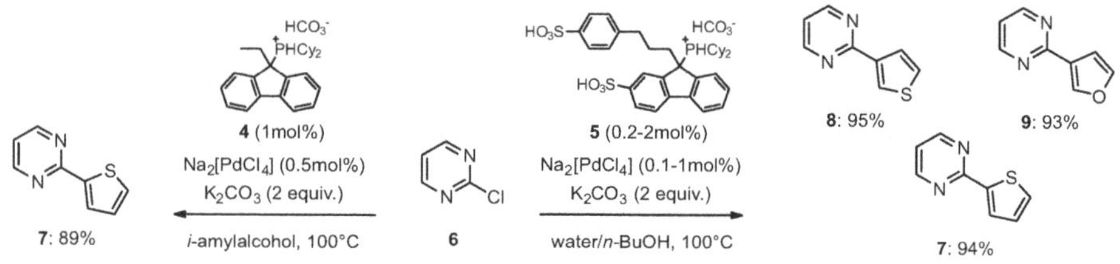

Scheme 2 Palladium-catalyzed cross-coupling of **6** towards biheteroaryls

undergo protodeborylation. In an effort to solve this problem, Buchwald and coworkers tried to find a catalytic system, where coupling occurs significantly faster than the concomitant decomposition of boronic acids [1]. They identified Pd(biphenyl-NH_2)Cl-XPhos as a suitable precatalyst, which forms the active XPhos-catalyst instantaneously in the presence of weak bases already at room temperature. Generally coupling products were obtained in excellent yields under these exceptionally mild conditions within 30 min up to 2 h. Substituted 4-chloropyrimidine **1** reacted smoothly with *N*-Boc-pyrrole boronic acid **2** and gave 99% yield of **3** (Scheme 1).

Fleckenstein and coworkers reported the coupling of furan- and thiopheneboronic acids under application of fluorenylphosphines **4** and **5** as highly active ligands [2]. Conditions employing the sulfonated and thus water-soluble ligand **5** proved highly effective, giving excellent yields for these challenging substrates (e.g., 2-chloropyrimidine with furan- and thiopheneboronic acid, 93–95% of products **7**–**9**, Scheme 2, right) (**5** is also active in the Sonogashira reaction; see Sect. 1.4.6). Under nonaqueous conditions, excellent yields were achieved for the coupling with many heteroaryl chlorides, but in the case of 2-chloropyrimidine, *i*-amyl alcohol had to be used instead of *n*-butanol to avoid the formation of the S_NAr product (Scheme 2, left).

Ackermann and Potukuchi reported heteroatom-substituted secondary phosphine oxides (HASPO, **10**) to be bench-stable and effective preligands for Suzuki–Miyaura coupling reactions of lithium 2-pyridylborates **11** [3]. During the formation of the active catalyst, the preligand undergoes a P–O hydride shift and forms a phosphorous-bound Pd–phosphite complex. A range of biaryls were obtained with typical yields in the range of 60–70%. Coupling of 5-bromopyrimidine with

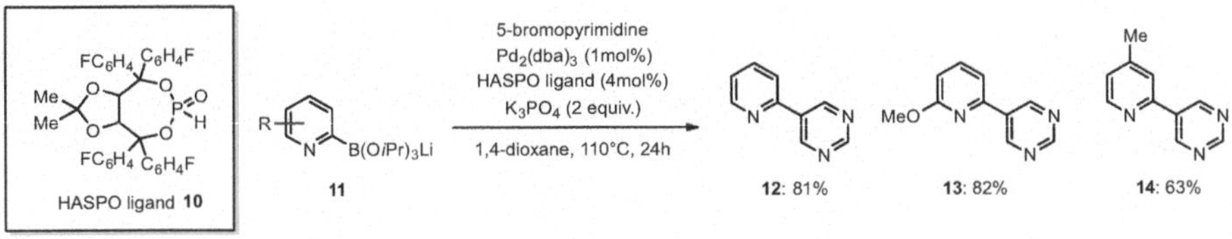

Scheme 3 HASPO-assisted Suzuki coupling of lithium 2-pyridylborates

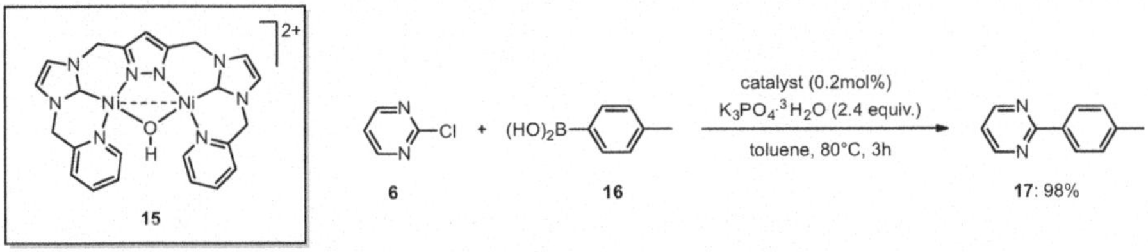

Scheme 4 An unusual nickel catalyst in Suzuki coupling of **6**

4-methylpyridinylborate and 6-methoxypyridinyborate gave 63% and 82% yield of **14** and **13**, respectively (Scheme 3).

An interesting example of a Ni-catalyzed Suzuki–Miyaura coupling was published by Zhou et al. [4]. The di-nickel complex **15** showed impressive catalytic activity: at catalyst loadings as low as 0.2–0.8 mol%, yields for the coupling of aryl chlorides and bromides were excellent, often quantitative (Scheme 4, **17**). This and similar complexes were also found to be a highly competent catalysts for the Kumada–Corriu reaction (see Sect. 1.6).

Liu et al. prepared air and moisture-stable NHC-preligand **18** bearing 3,3,3-trifluoropropyl groups for the tuning of electronic properties [5]. The authors rule out a steric effect and claim that the electron-withdrawing nature of the 3,3,3-trifluoropropyl substituents facilitates the formation of monocarbene-palladium complex as well as the reductive elimination by decreasing the electron density of the carbene. Upon treatment with a base in situ, the actual ligand is generated by elimination of HCl (Scheme 5a, **24**).

The synthesis of novel N-heterocyclic carbenes **19** with expanded rings (6- or 7-membered rings) was reported by Kolychev et al. [6]. These air and moisture-stable ligands showed good catalytic activity for the coupling of heteroaryl chlorides and tolylboronic acid. The diarylation product **23** of 4,6-dichloropyrimidine was obtained in only 61%. The arylation product **24** of 5-bromopyrimidine, however, was formed in 92% yield (Scheme 5b).

The benzimidazole-based NHC complex **20** was reported to efficiently catalyze the formation of 2-phenylpyrimidine **25** in 88% yield after 12 h at 40°C (Scheme 5c) [7].

The coupling of highly sterically demanding boronic acids was achieved by Tu et al. using acenaphthoimidazolylidene palladium complexes **21** with *t*-BuOK in dioxane at 80°C [8]. This catalyst tolerated a free NH_2 group and coupling of

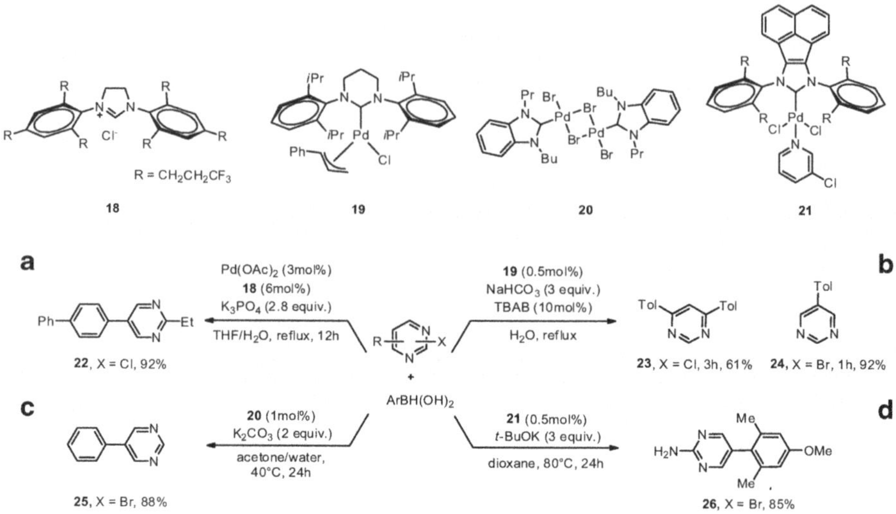

Scheme 5 Various ligands in palladium-catalyzed Suzuki couplings

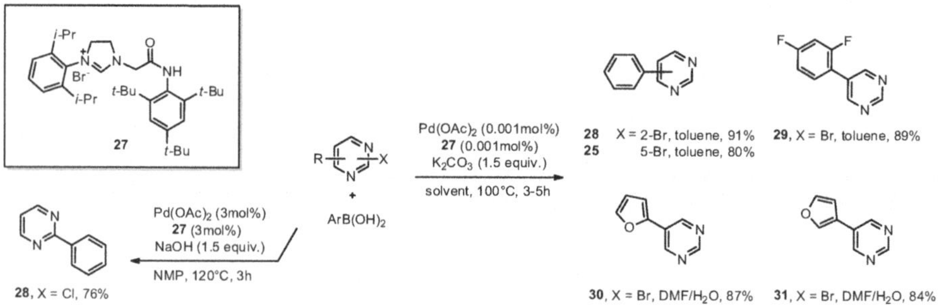

Scheme 6 Sterically demanding NHC ligands in the Suzuki coupling of pyrimidine halides

2-amino-5-bromopyrimidine with 2,6-dimethyl-4-methoxyphenylboronic acid gave 85% yield of **26** (Scheme 5d).

Kumar et al. reported the synthesis of palladium-amido-*N*-imidazolium salts and their application as NHC ligand precursors. Especially when these complexes contained ligands with great steric bulk (e.g., **27**), TON of up to 850,000 could be achieved in Suzuki–Miyaura coupling reactions (Scheme 6, catalyst loading 0.001%, 3 h, 100°C, products **25**, **28–31**) [9].

New Catalytic Systems

Yang et al. reported the combination of palladium acetate and XPhos together with cesium carbonate and aqueous butanol at room temperature to be a highly effective catalyst system for the coupling of chlorides and tosylates [10]. The authors

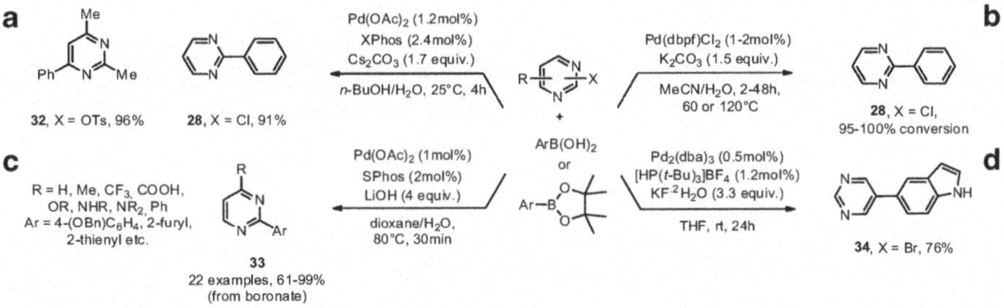

Scheme 7 Examples for palladium-catalyzed Suzuki coupling of pyridine halides and tosylates

demonstrated the applicability of their protocol for a wide range of heterocyclic substrates: 2-chloropyrimidine was converted into 2-phenylpyrimidine **28** in 91% yield within 4 h. In case of tosylates, sodium hydroxide proved to be slightly more effective. Under these modified conditions, 2,6-dimethylpyrimidinyl tosylate reacted smoothly with phenyl boronic acid to give 96% yield of **32** within 1 h (Scheme 7a).

Asano et al. developed a method for the efficient arylation of 2-chloropyrimidine [11]. He reported the use of a combination of palladium(II) acetate with the S-Phos ligand and lithium hydroxide in dioxane/water at 80°C to be highly effective for the coupling with a wide range of electron-rich boronic acid pinacolates. The authors showed that 2-heteroaryl boronates such as 2-furanyl and thienylboronic acid esters offer the advantage of being significantly more stable than the corresponding boronic acids (Scheme 7c, **33**).

Moseley and coworkers identified a robust protocol for the coupling of aryl bromides and chlorides utilizing the air-stable Pd(dbpf)Cl_2 catalyst in acetonitrile/water mixtures [12]. The protocol requires 1.5 equiv. of potassium carbonate and a slight excess of boronic acid. At 60°C the reaction reaches completion with a wide range of aromatic and heteroaromatic bromides and gave the coupling product **28** in excellent yield. In the case of aryl chlorides, long reaction times (up to 48 h at 60°C) were observed. This could be reduced to about 2 h by running the reaction in superheated solvent at 120°C (Scheme 7b).

Batches of boronic acids can contain significant amount of anhydrides, which goes in line with lower coupling efficiencies when no water is used as cosolvent [13]. Lou and Fu addressed this problem, by developing a protocol where simple $KF \cdot 2H_2O$ is employed, in order to introduce just enough water to hydrolyze potentially present anhydrides. At the same time, they suggest the use of a mixture of $Pd_2(dba)_3$ and *t*-butylphosphonium tetrafluoroborate as a convenient source of palladium and *t*-butylphosphine. Coupling products were obtained in good to excellent yields under mild conditions (THF, rt). Coupling of 5-bromopyrimidine with 5-indoleboronic acid afforded the product **34** in 76% yield (Scheme 7d).

An iron-catalyzed Suzuki cross-coupling of tetraarylborate salts with benzylbromides and 2-bromoheteroaryls was reported by Bedford et al. [14]. The catalytically active Fe(II) complex **35** requires diarylzinc as an additive, in which

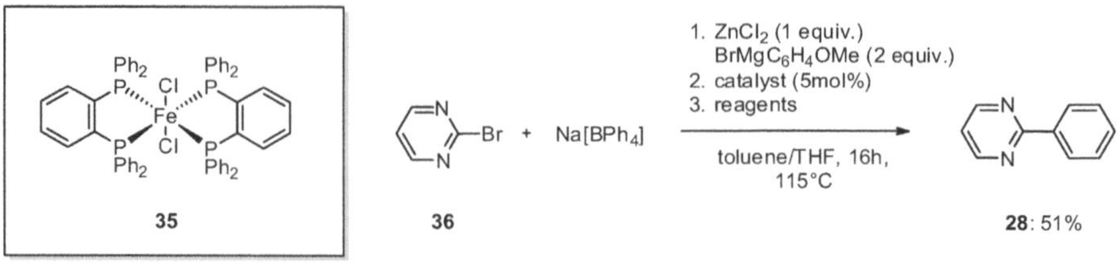

Scheme 8 Iron-catalyzed Suzuki coupling of **36**

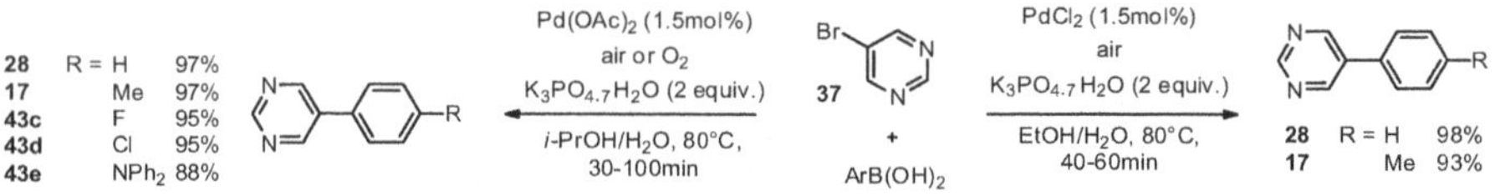

Scheme 9 Ligandless Suzuki coupling of pyrimidine bromide **37**

only minimal incorporation of these aryl residues into the product was found. Yields depend strongly on the combination of substrates and range from 38 to 88%. In case of the arylation of 2-bromopyrimidine **36** with sodium tetraphenylborate, a yield of 51% of **28** was achieved (Scheme 8).

Ligand-Free Coupling

An interesting example of an oxygen-promoted ligand-free coupling reaction was published by Liu and Wang [15]. N-heterocyclic bromides reacted very effectively with arylboronic acids in the presence of palladium acetate and air in absence of a ligand (Scheme 9, left). This protocol proved effective for the coupling of triarylamine-derived boronic acids as well [16]. Coupling also occurs if palladium(II) chloride and aqueous ethanol is employed [17], providing 5-phenylpyrimidine **28** in 98% yield (80°C, 40 min) (Scheme 9, right).

A protocol sparing the phosphine ligand was also published by Colombo and coworkers [18]. They used palladium acetate in combination with potassium fluoride under microwave conditions (120°C, MeOH, 20 min) to couple 2-chloro-5-bromopyrimidine with a variety of boronic acids with good yield. In the case of 3-nitrophenylboronic acid, only a low yield of 28% of **39a** could be achieved, and 4-dimethylaminophenylboronic acid gave only trace amounts of coupling product **39b** (Scheme 10).

Water as Reaction Medium

Lipshutz et al. reported the efficient coupling of heteroaryl bromides with boronic acids in water without cosolvent [19]. The coupling reaction takes place in micelles formed from the water-insoluble starting materials and tocopherol-modified PEG

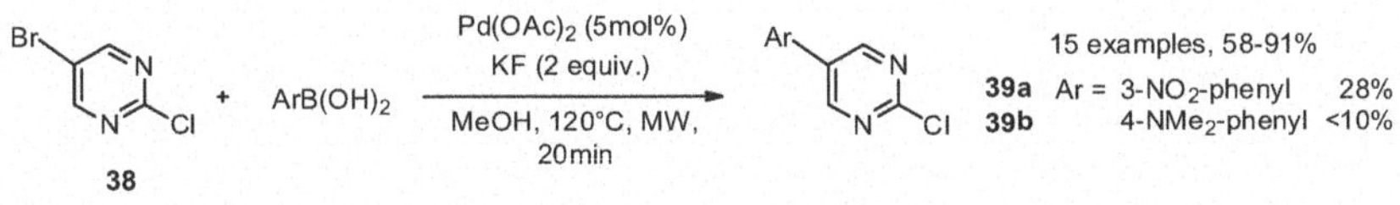

Scheme 10 Ligandless Suzuki coupling of pyrimidine chlorides

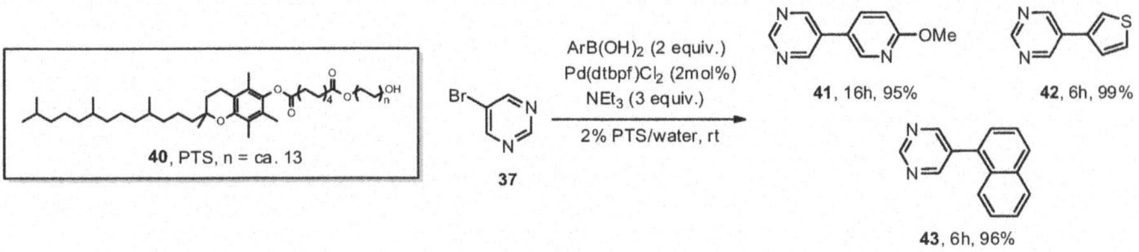

Scheme 11 Micellar catalysis in Suzuki couplings

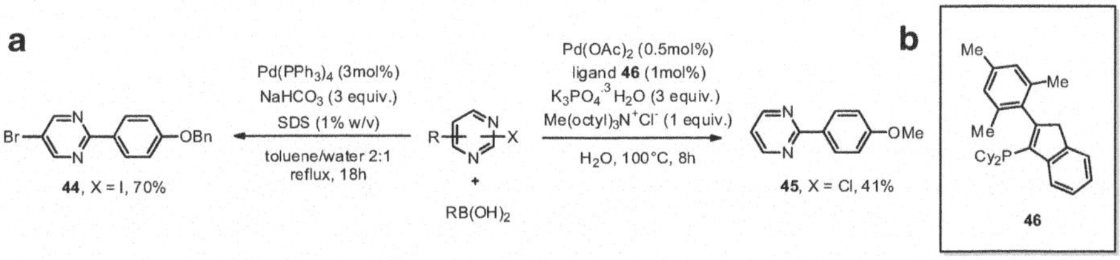

Scheme 12 Palladium-catalyzed Suzuki coupling in aqueous medium I

40 as tenside (2%) under mild conditions. As catalyst a bidentate ferrocenylphosphine was used. Among other examples, 5-bromopyrimidine **37** was coupled to naphthylboronic acid in 96% yield of **43** after 6 h at rt, showing that steric bulk is well tolerated (Scheme 11).

Vashchenko et al. demonstrated that the efficiency of coupling reactions in biphasic solvent systems such as toluene/water or DME/water can be improved, if surfactants like SDS and *n*-butanol are added [20]. These allow the formation of microemulsions which are believed to solubilize hydrophobic coupling partners. Under these conditions, coupling of benzyloxyphenylboronic acid with 4-bromo-2-iodopyrimidine gave the desired product **44** only in moderate 70% yield (Scheme 12a).

Mao and coworkers reported (mesitylindenyl)dicyclohexylphosphine **46** in the presence of $Me(octyl)_3N^+Cl^-$ as a remarkably active catalyst for the Suzuki–Miyaura reaction under aqueous conditions [21]. A catalyst loading as low as 0.0005 mol% gave the coupling product of chlorobenzene and 4-methoxyphenylboronic acid in 61%, which corresponds to a TON of 120,000. Good to excellent yields were obtained for a high number of examples with electron-rich or neutral boronic acids and a broad range of aryl- and heteroaryl chlorides. Coupling of 2-chloropyrimidine performed significantly worse than pyridine derivatives. In this case, the reaction required 0.5 mol% of catalyst and

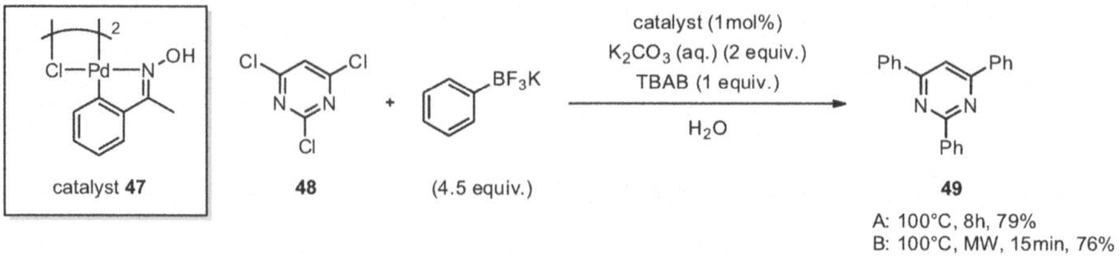

Scheme 13 Palladium-catalyzed Suzuki coupling in aqueous medium II

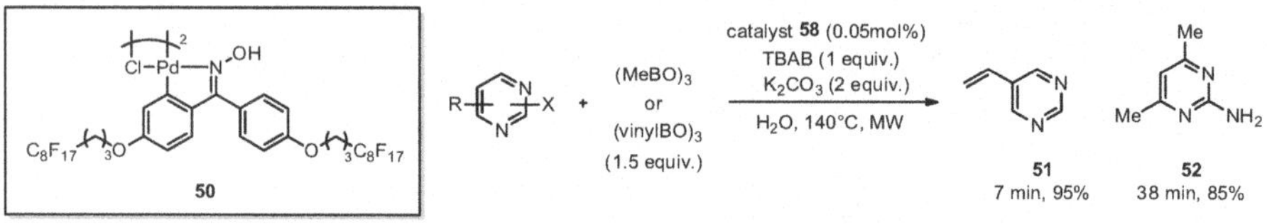

Scheme 14 Palladium-catalyzed Suzuki coupling in aqueous medium III

gave the product 2-(4-methoxyphenyl)pyrimidine **45** in only 41% yield (Scheme 12b).

Oxime-derived palladacycle **47** was reported to efficiently catalyze the coupling of aryltrifluoroborates under aqueous conditions [22]. In microwave heating with concomitant cooling of the reaction vessel in an airstream, shorten reaction times from several hours to minutes giving good yields of **49** were found (Scheme 13).

Lam and coworkers developed a fluorous, oxime-based palladacycle **50**, optimized for coupling chemistry in water under microwave irradiation [23, 24]. The catalyst was successfully employed for a range of Pd-catalyzed coupling reactions, e.g., Suzuki–Miyaura, Stille, Kumada, Heck, and Sonogashira coupling in aqueous or organic medium (see also Sect. 1.4.4). The catalyst can be recovered from the reaction mixture by fluorine solid-phase extraction and reused up to five times with no significant loss of activity. For the majority of coupling reactions, excellent recovery rates and very low levels of Pd-leeching were observed. Examples involving pyrimidine were the vinylation of 5-bromopyrimidine with trivinylboroxine (**51**, 95% yield) and the reaction of substituted 4-chloropyrimidine with trimethylboroxine to give 2-amino-4,6-dimethylpyrimidine **52** (140°C, 38 min, 85%) (Scheme 14).

Nanoparticles as Catalytically Active Species

Sing and coworkers synthesized an air and moisture-stable palladacycle precatalyst **54**, which was reported to form well-defined $Pd_{17}Se_{15}$ nanoparticles of ~8 nm size [25]. These efficiently promote the coupling of bromoaryls with boronic acids. While this protocol requires elevated temperatures, it allows, in the case of the reaction of 5-bromopyrimidine with phenyl boronic acid, the use of an

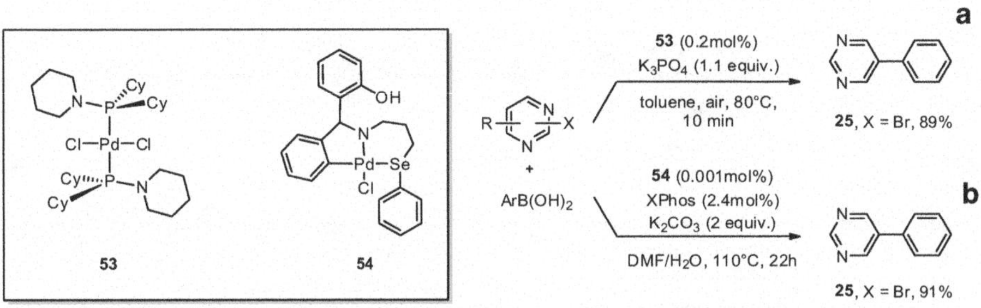

Scheme 15 Pd nanoparticles for Suzuki coupling reactions I

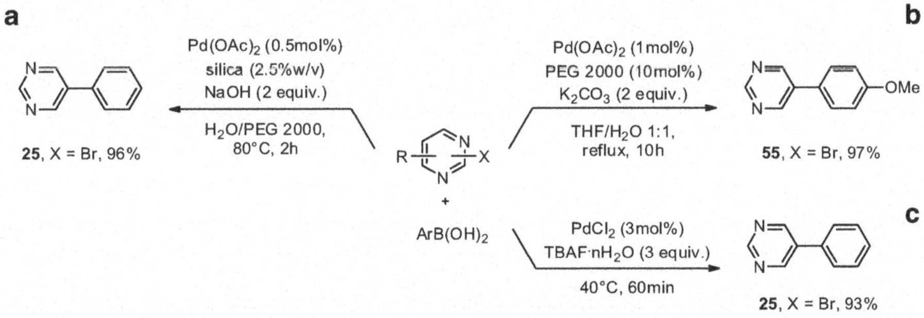

Scheme 16 In situ formation of Pd nanoparticles for Suzuki coupling reactions

exceptionally low catalyst loading of only 0.001 mol%. Product **25** was obtained with a yield of 91%, which corresponds to a TON of 91,000 (Scheme 15b).

Bolliger and Frech reported the use of dichlorobis[1-(dicyclohexylphosphanyl) piperidine]palladium **53** as a stable depot form of Pd nanoparticles with extremely high catalytic activity [26]. For the reaction of 5-bromopyrimidine with phenylboronic acid, 89% yield of **25** was achieved after 10 min at 80°C and with a catalyst loading of only 0.2 mol% (Scheme 15a).

Various methods for the in situ generation of catalytically active nanoparticles without the use of precatalysts have been published: $Pd(OAc)_2$ in water/PEG 2000 with addition of silica (Scheme 16a) [27], $PdCl_2$ with PEG 2000 in aqueous THF at reflux (Scheme 16b) [28], and $PdCl_2$ with TBAF at 40–100°C (Scheme 16c) [29]. The nanoparticles obtained by these methods efficiently catalyze Suzuki–Miyaura reactions.

Catalysts on Solid Support

Ul Islam et al. reported the synthesis of Pd nanoparticles **56** embedded in polyacetanilide and the use of this material in a ligand-free coupling reaction [30]. For heterocyclic substrates, this catalyst gave a yield of 72% of **25** in the reaction of 5-bromopyrimidine with phenylboronic acid (Scheme 17a).

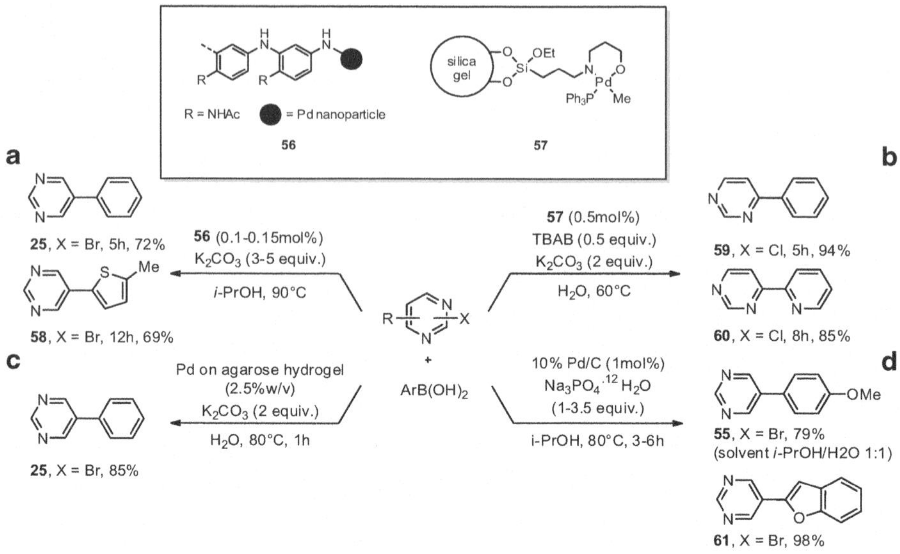

Scheme 17 Examples for solid supported catalysts in Suzuki coupling of pyrimidine derivatives

Lee et al. reported the immobilization of palladium on mesoporous zeolite [31]. Reaction of a silylated Pd complex with zeolite gave rise to catalyst **57**. The coupling reaction was conducted at 60°C in water and gave the products in excellent yield. The catalyst was recovered by simple filtration and could be reused at least ten times with no significant drop of catalytic activity (Scheme 17b). Later Lee et al. showed that this catalyst is also highly active in the Stille and the Sonogashira cross-coupling reaction [31].

Kitamura et al. showed that Pd/C can catalyze coupling reactions in combination with sodium phosphate in aqueous *i*-PrOH at 80°C [32]. Excellent yields were obtained for many heteroaryl bromides. Coupling of 5-bromopyrimidine with 4-methoxyphenylboronic acid proceeded only in 79% yield of **55** after 6 h. When heteroaryl boronic acids were employed, only low yields could be obtained under these conditions. This drawback was eliminated by using nonaqueous *i*-PrOH instead of the *i*-PrOH/H_2O mixture: 5-(benzofuran-2-yl)pyrimidine **61** was isolated in 98% with a catalyst loading of only 1 mol% Pd after 3 h. The authors showed that during the reaction, the palladium leaches into the reaction mixture until a maximum was reached after about 20 min. After that time, the dissolved palladium was again deposited on the active charcoal and a decrease of Pd-levels in solution could be observed. The catalyst could be reused for at least five runs with no loss of catalyst activity (Scheme 17d).

Firouzabadi et al. demonstrated that agarose hydrogel could be used as a stabilizing support and bioorganic ligand for Pd nanoparticles [33]. The hydrogel was prepared by heating $Pd(OAc)_2$ together with agarose in the presence of citric acid. It catalyzes the formation 5-phenylpyrimidine **25** from the corresponding

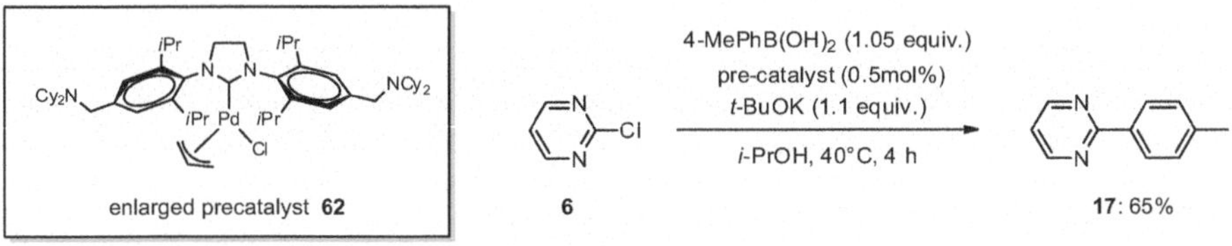

Scheme 18 Easily removable Pd catalyst for Suzuki coupling

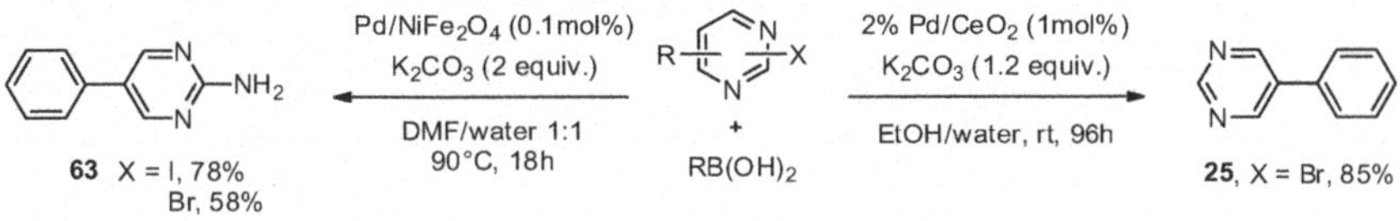

Scheme 19 Mixed metal species in Suzuki couplings

bromide in 85% yield (Scheme 17c). The same authors also published the use of an aminopropyl-functionalized clay as a solid support for nanoparticles [34]. The exfoliated layers of the clay are believed to wrap around the nanoparticles which are then easily dispersed in water. Compared to the immobilization in the agarose hydrogel, a similar high catalytic activity was achieved, and, additionally, an extension of the scope towards the Heck reaction was achieved. 5-Phenylpyrimidine was obtained in 88% yield after 10 h.

A different approach for recovering the catalyst was chosen by Schoeps et al. [35]. They designed imidazolium salts with a molecular weight of about 800 g/mol. The Pd-NHC complexes **62** prepared from these compounds were employed in Suzuki–Miyaura reaction as well as Buchwald–Hartwig reactions. After completion of the reaction, the catalyst could be successfully removed from the reaction mixture by solvent-resistant nanofiltration removing 97–99.9% of the Pd complex (Scheme 18, product **17** obtained in 65% yield).

Borhade et al. reported the application of Pd on ferrite (Fe_2O_3) as a heterogeneous catalyst in DMF/water at 90°C [36]. The reaction of boronic acids with aryliodides and bromides gave the coupling products in good to excellent yield. The coupling product **63** of 2-aminopyrimidin-5-boronic acid with bromobenzene was obtained in only 58% but with iodobenzene in 78% after 1 h. Due to the magnetic properties of ferrite, it is possible to recover the catalyst material from the reaction mixture by applying a magnetic field (Scheme 19, left).

Amoroso et al. developed conditions for the use of Pd/CeO_2 as cross-coupling catalyst [37]. Excellent yields were obtained for a selection of electron-rich and electron-poor substrates. 5-Bromopyrimidine, as the only heteroaromatic example, gave the arylation product **25** in 85% yield after a prolonged reaction time of 96 h (Scheme 19, right).

Lin and coworkers synthesized a heterogeneous catalyst for the Suzuki–Miyaura coupling reaction by depositing Pd on carbon nanotubes [38]. A catalyst loading of only 0.3% (based on Pd) was sufficient to achieve excellent yields for a variety of

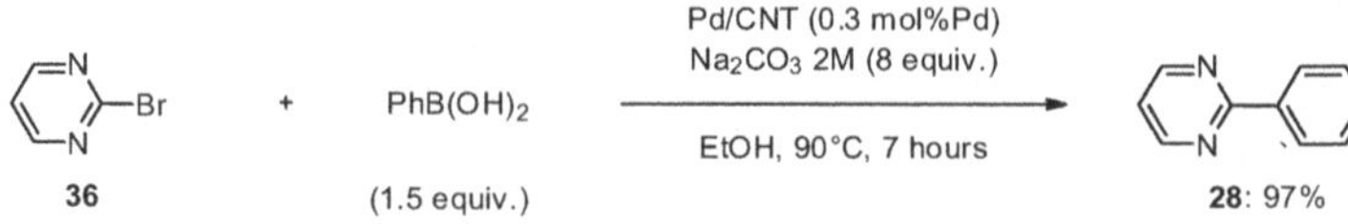

Scheme 20 Heterogeneous palladium in Suzuki couplings

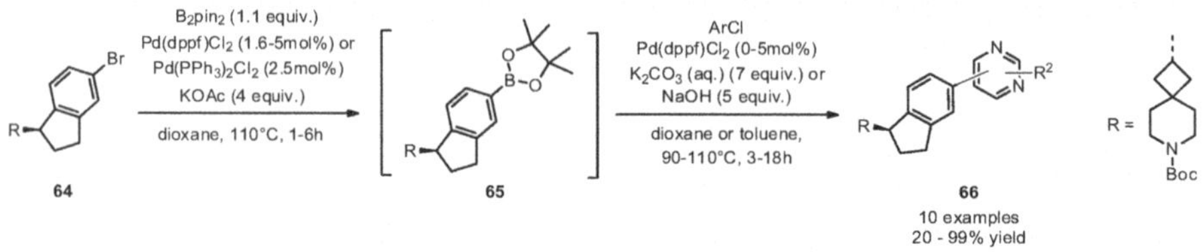

Scheme 21 Example for a one-pot borylation/coupling sequence

substrates. Reuse of the catalyst up to three times was demonstrated to proceed without significant loss of reaction efficiency. Coupling of 2-bromopyrimidine **36** with phenylboronic acid gave the coupling product **28** in 97% yield (Scheme 20).

Coupling Sequences

In the course of their effort to synthesize a library of indanes, workers at Pfizer developed a one-pot borylation/coupling sequence [39]. Starting from an amino-substituted 5-bromoindane **64**, first a borylation with bis(pinacolato)diboron gave the intermediate boronic acid ester **65**, which was subsequently coupled with heteroaromatic halides. Several substituted 2-chloro- and 4-chloropyrimidines reacted smoothly and gave the coupling products **66** in 81–99% yield with exception of the unsubstituted 4-chloropyrimidine (63% yield) and 6-chloro-4-cyanopyrimidine (20% yield) (Scheme 21).

Siddle et al. disclosed a sequential coupling procedure of 5-bromo-2-iodopyrimidine **67** and other aromatic dihalides [40]. In a first step, a Cu/phenanthroline-catalyzed C–N coupling in 2-position took place. The intermediate product **68** was isolated with yields ranging from 57% for the reaction with benzimidazol to 84% for pyrrole. In the second reaction step, the intermediate was coupled with boronic acids under standard Suzuki–Miyaura conditions to **69** in 43–79% yield. For two selected examples, the authors demonstrated that the reaction sequence can be performed in a one-pot manner with an overall yield of >50% (Scheme 22).

Anderson and Handy published a protocol for the sequential one-pot Suzuki–Miyaura coupling of 2,4-dichloropyrimidine **70**, utilizing the reactivity differences of the 2- and 4-position of pyrimidine [41]. The first step is carried out at 55°C in a mixture of ethanol, toluene, and water. Once full conversion was reached, the second boronic acid and additional base were added and the temperature was raised

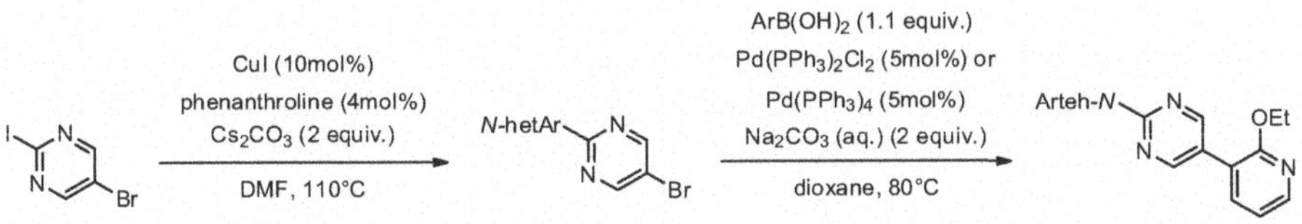

67 **68**, 4 examples 57 - 84% yield **69**, 4 examples 43-79% yield

Scheme 22 Sequential coupling of pyrimidine dihalide **67**

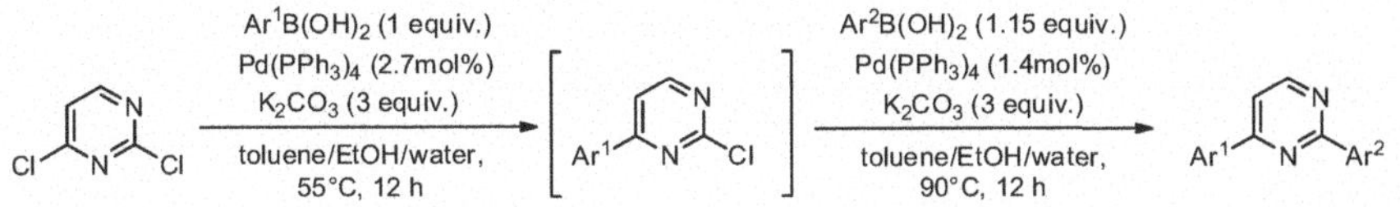

70 **71** **72**, 8 examples 15 - 95% yield

Scheme 23 Sequential coupling of pyrimidine dihalide **70**

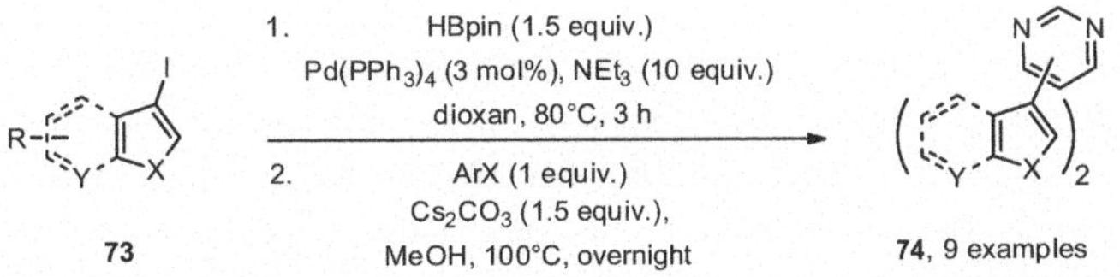

73 **74**, 9 examples 24 - 77% yield

Scheme 24 Borylation/cross-coupling sequence towards **74**

to 90°C. The reaction gave good to excellent yields over two steps (**72**, 62–95%) for different phenylboronic acids. The reaction proved sensitive to 2-substituted boronic acids; in these cases, the yields dropped to 15–21% (Scheme 23).

Tasch and coworkers developed a double Masuda borylation/Suzuki–Miyaura cross-coupling sequence for synthesizing symmetrically substituted bis(indolyl) diazines **74** [42]. The method was used to synthesize the marine alkaloid hyrtinadine A and a library of derivatives. Substituted 3-iodoindoles **73** were first reacted with pinacolborane to give the corresponding boronates. Then dry methanol to quench excess borane, dihalodiazines, and cesium carbonate was added for the coupling step. Depending on the substitution pattern of the indole coupling partner and the coupling position, yields of 24–77% of **74** were achieved when dihalopyrimidines were employed (Scheme 24).

The same group also applied very similar conditions to the synthesis of natural products meridianin A and meriolin 1 as well as a library of derivatives [43]. It is worth mentioning that coupling partners containing the 2-aminopyrimidine or 2-hydroxypyrimidine moiety are well tolerated without the need for a protecting group.

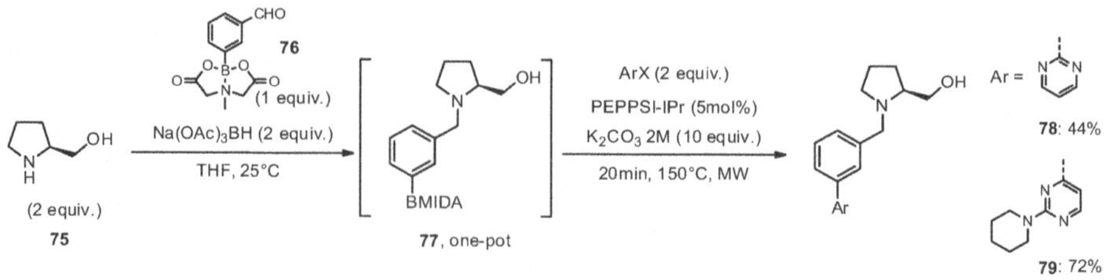

Scheme 25 Reductive amination/Suzuki–Miyaura sequence towards **78** and **79**

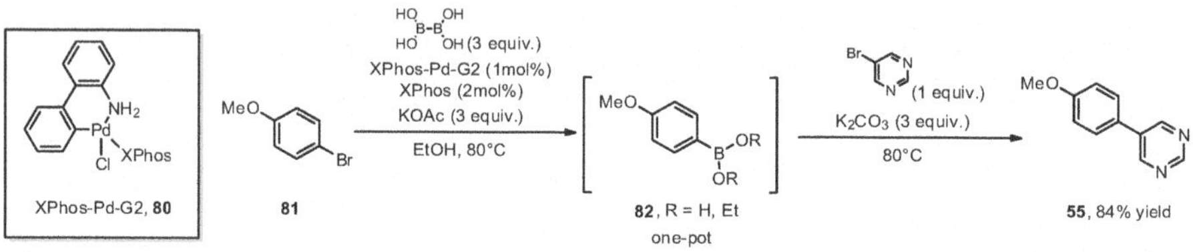

Scheme 26 One-pot borylation/Suzuki coupling towards **55**

Grob et al. used a one-pot reductive amination/Suzuki–Miyaura sequence for library synthesis [44]. For this purpose, they transformed formyl-substituted arylboronic acid into the corresponding MIDA boronates. These compounds were subjected to reductive amination conditions using triacetoxyborohydride. After completion of the reaction, the coupling partner, aq. potassium carbonate and PEPPSI-IPr, were added, and the reaction was heated to 150°C in the microwave. The reaction was assessed for a great variety of substrates with yields ranging from 14 to 63%. When 3-formylphenylboronic MIDA ester **76** was reacted with prolinol **75** and 2-chloropyrimidine, product **78** was obtained in 44% overall yield (Scheme 25).

An interesting one-pot borylation/Suzuki–Miyaura sequence was developed by Molander et al. using diboronic acid as borylation agent [45]. The method employs XPhos-Pd-G2 catalyst **80** and ethanol at 80°C. Efficient coupling was demonstrated for a large number of different substrates. Among the examples, the reaction of 4-bromopyrimidine and 4-bromoanisol **81** was reported to give the coupling product **55** in 84% yield (Scheme 26).

Kelson and coworkers published sequential borylation/Suzuki–Miyaura coupling for the synthesis of polyazatriaryl ligands [46]. Starting from bromoarenes, Pd(dppf)Cl_2-catalyzed borylation with bis(pinacolato)diboron as borylating agent gave the intermediate boronates **84**. After completion of the reaction, the coupling partner and sodium hydroxide were added. Among a number of other examples, the reaction of 1,3-dibromobenzene with 2-chloropyrimidine and 2,6-dibromopyridine with 5-chloropyrimidine gave the bisarylated product **85** in 56% and **86** in 59% yield, respectively (Scheme 27).

A coupling/cyclization sequence leading to (aza)indoles was developed by Hoelder and coworkers [47]. They coupled (ethoxyvinyl)borolane with aryl halides

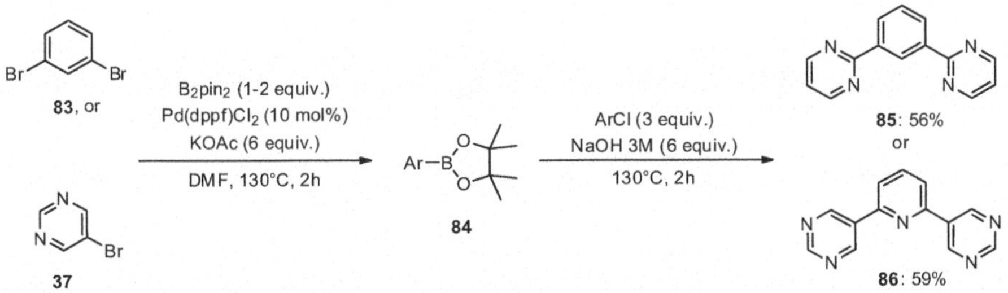

Scheme 27 Sequential borylation/Suzuki coupling towards **85** and **86**

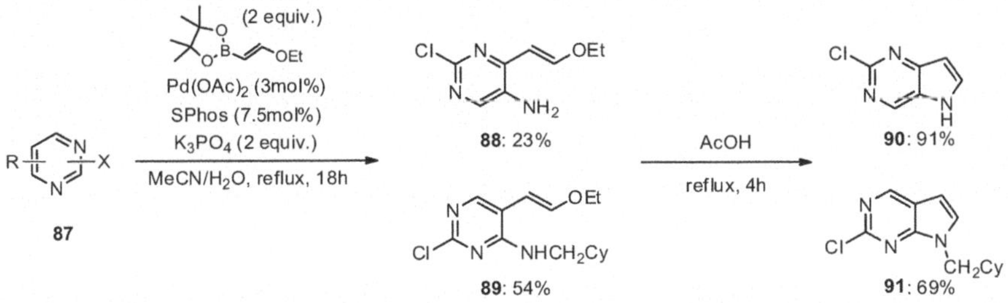

Scheme 28 A coupling/cyclization sequence leading to (aza)indoles

bearing an amino group in ortho-position to the halide. The coupling products **88** and **89** underwent then acid-catalyzed cyclization to (aza)indoles. This methodology was also applied to two isomeric dihalopyrimidine amines which gave the corresponding pyrrolopyrimidines in 91% (**90**) and 69% (**91**), respectively (Scheme 28).

Conditions for the coupling of an alkylborane generated in situ from benzylvinylcarbamate and 9-BBN were developed by Roy et al. [48]. The method was utilized for the synthesis of a small library of disubstituted N-heterocycles with moderate yields for the coupling with 2-chloropyrimidines but good to excellent yields for 2-bromopyridines (Scheme 29, products **93–95**).

In order to study their fluorescence properties, Ple and coworkers attempted to synthesize di- and trialkynylpyrimidines starting from di- or triiodopyrimidines **97** [49]. In case of 2,4-diiodo- and 4,6-diiodopyrimidine, coupling proceeded smoothly under standard Sonogashira coupling conditions. On the contrary, 2,4,6-triiodopyrimidine did not afford the expected trialkynylpyrimidine **96**, possibly due to the low stability of the starting material. The trialkynyl products **96** were then prepared successfully via Suzuki and Negishi alkynylation starting from 2,4,6-trichloropyrimidine (see also Sect. 1.4.6). Later this method was used to synthesize the star-shaped trialkynylpyrimidine **96e** with blue light-emitting properties (Scheme 30) [50].

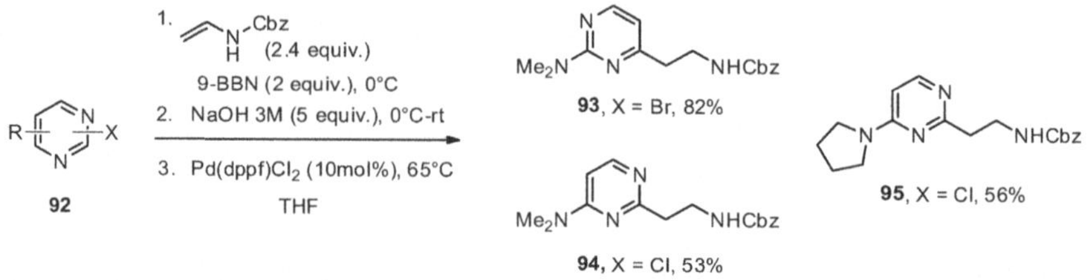

Scheme 29 In situ generation of alkylboranes and subsequent Suzuki coupling with pyrimidine halides

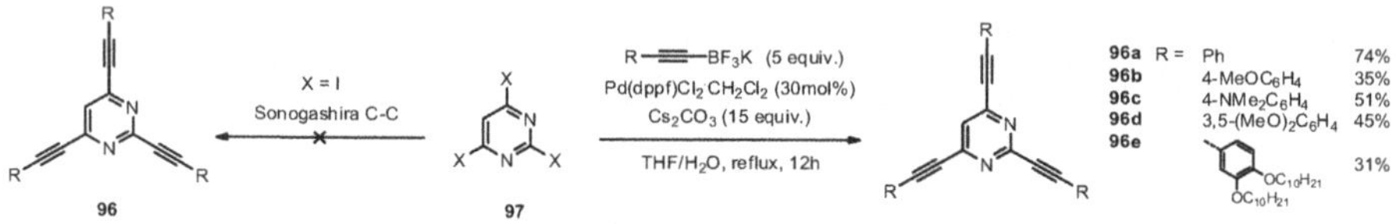

Scheme 30 Suzuki coupling as alternative to the Sonogashira protocol

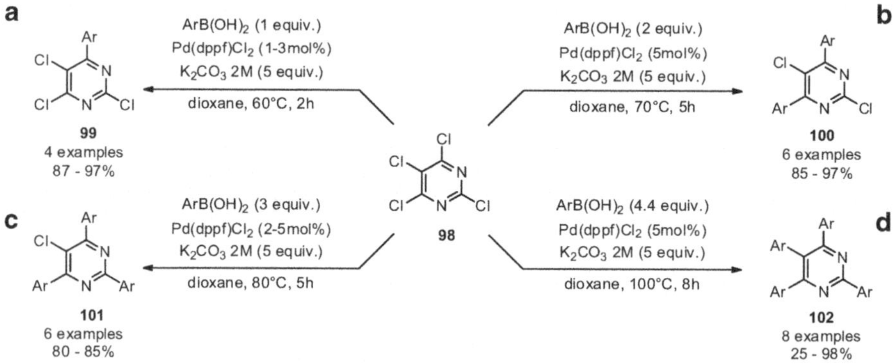

Scheme 31 Exploring site-selective Suzuki coupling of **98**

Selective Coupling

A site-selective coupling method starting from tetrachloropyrimidine **98** was published by Hussain et al. [51]. By careful tuning of the reagent loading, reaction time, and temperature, the authors synthesized the mono-, di-, tri- and tetraarylation products **99–102** in good to excellent yields (Scheme 31).

Molander et al. developed a "1,2-dianion equivalent" based on the addition of 9-BBN to vinyltrifluoroborate **103** [52]. The resulting diboraethane couples first on the borane side in the presence of KF and DavePhos to give **104**. After addition of potassium carbonate, RuPhos, and the second coupling partner, the remaining trifluoroborate was coupled. A broad scope of aromatic, heteroaromatic compounds, as well as bromoalkenes were coupled in one-pot fashion in 56–83% yield, including pyrimidine products **105** and **106** (Scheme 32).

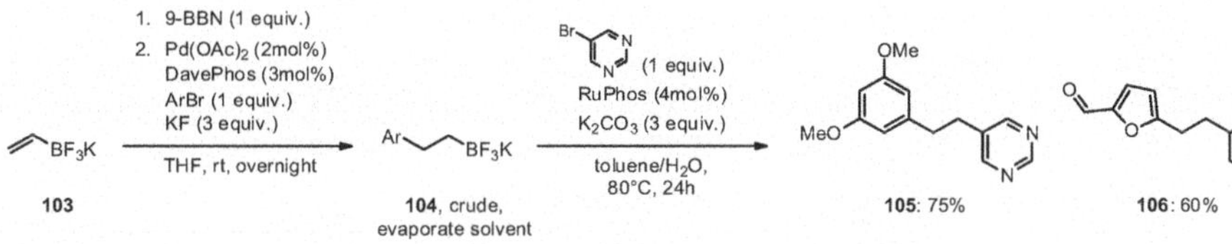

Scheme 32 Vinyltrifluoroborate **103** as 1,2-dianion equivalent

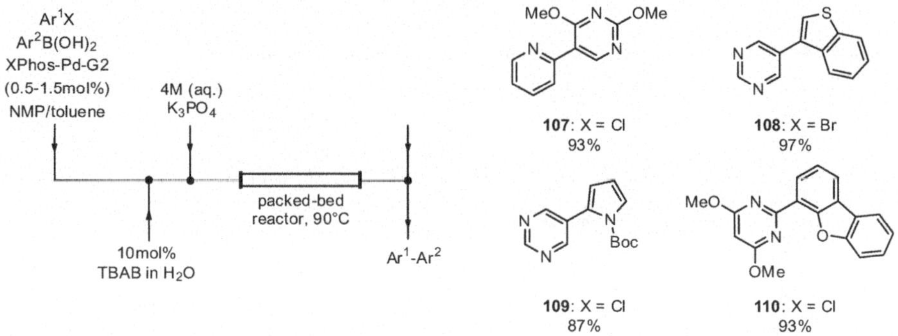

Scheme 33 Pyrimidine derivatives in continuous flow Suzuki couplings I

Continuous Flow

General conditions for the cross-coupling of heteroaryl chlorides under continuous flow conditions were developed by Noël and coworkers [53]. For this purpose, the homogenous XPhos precatalyst was used in a biphasic solvent mixture. A packed-bed reactor was used to increase the contact of the two immiscible phases. With a residence time of only 3 min, excellent yields of biaryls were achieved. For example, coupling of 2-chlorodimethoxypyrimidine with dibenzofuran gave the coupling product **110** in 94% yield (Scheme 33, **107–110**).

An impressive example of a lithiation/borylation/Suzuki–Miyaura cross-coupling reaction sequence in continuous flow was published by Shu et al. [54]. Starting from aryl bromides, in the first step, the aryllithium species was generated by metal-halogen exchange with residence times between 10 s up to 2 min. Alternatively, the aryllithium was generated by direct lithiation of thiophene or furan within 4–10 min. The organolithium species was trapped by combining the stream with triisopropylborate in THF for 1 min. After addition of base and a solution of XPhos precatalyst, the temperature was raised to 60°C for 10 min. Generally good to excellent yields were achieved. The reaction of 5-bromo-2-methoxyxylene and 4-bromopyrimidine yielded 97% of the coupling product **111** (Scheme 34). Later also a one-pot procedure with conditions optimized for running the reaction in batch was published by Oberli and Buchwald [55] (Scheme 35).

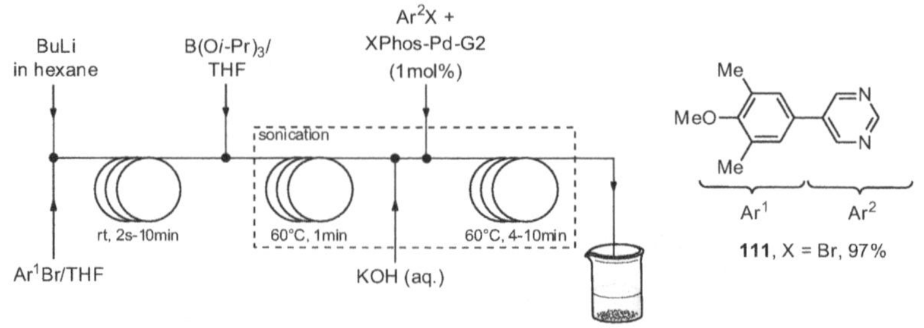

Scheme 34 Pyrimidine derivatives in continuous flow Suzuki couplings II

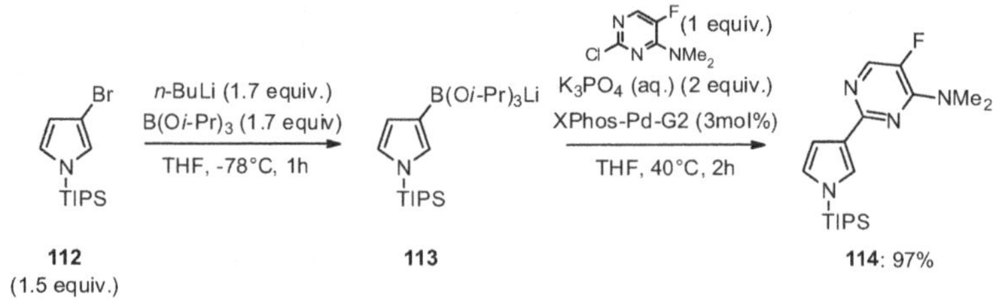

Scheme 35 Pyrimidine derivatives in continuous flow Suzuki couplings III

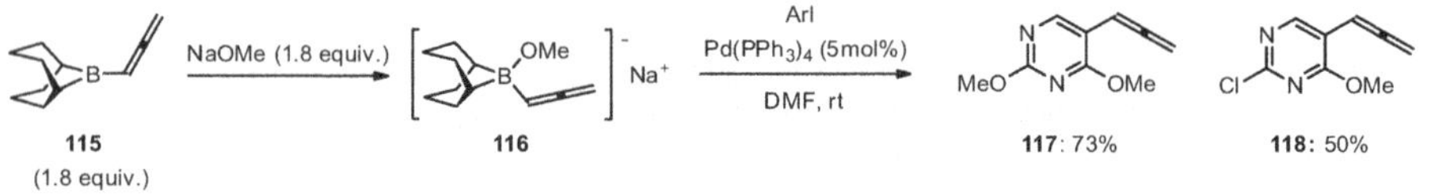

Scheme 36 Cross-coupling of pyrimidine iodides with **115**

Alternative Coupling Partners

Radkowski et al. discovered a new type of B-allenyl-9-BBN complex **115** which allows for the efficient allenylation of aryliodides [56]. The reaction requires sodium methoxide, $Pd(PPh_3)_4$, and proceeds smoothly at rt, giving the allenylation products in 50–95% yield. Two examples employing substituted 5-iodopyrimidines were reported giving the coupling products in 73% (**117**) and 50% yield (**118**) (Scheme 36).

Billingsley and Buchwald addressed a long-standing problem of Suzuki–Miyaura coupling of 2-pyridyl-boron compounds [57]. These substrates are particularly challenging due to their low reactivity in the transmetallation step as well as their propensity to decompose via a protodeborylation pathway. The authors found that lithium triisopropylborates in combination with diphenylphosphine oxide **119** as ligand react smoothly, giving the coupling products in moderate to excellent yields. When lithium 2-pyridylisopropylborate was coupled with

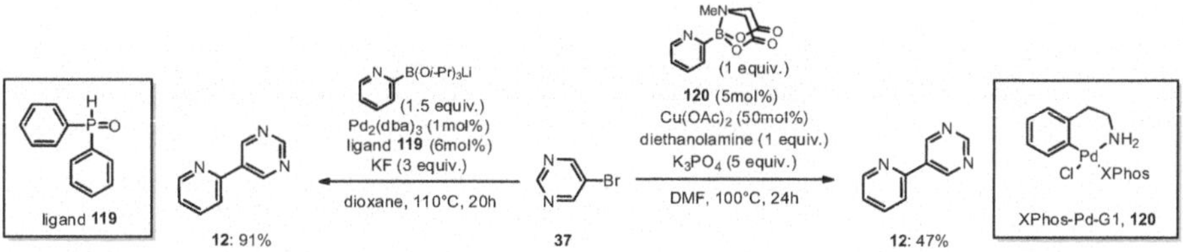

Scheme 37 Stable 2-pyridyl-boron compounds for Suzuki coupling reactions

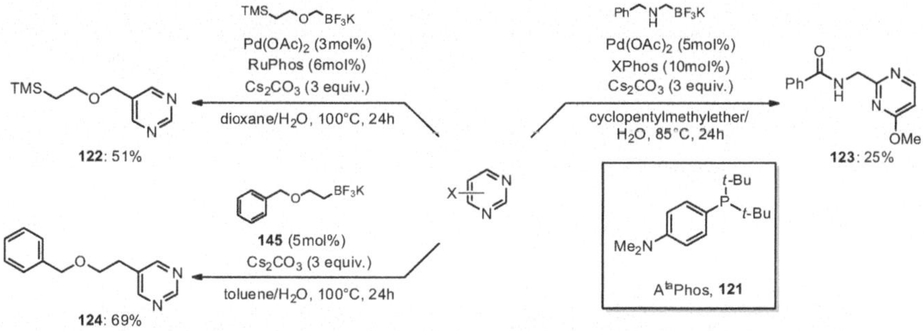

Scheme 38 Various ligands in Suzuki couplings of pyrimidine halides

5-bromopyrimidine **37**, the product **12** was obtained in 91% yield after 20 h (Scheme 37, left).

Very recently, Dick et al. reported another solution to that problem [58]. They converted the 2-pyridintriisopropylborate into its MIDA ester, an air-stable and storable compound. The ester is hydrolyzed under the reaction conditions, slowly releasing the boronic acid at a rate slower than the coupling reaction. The Cu-mediated XPhosPd-G1 (**120**)-catalyzed coupling reaction proceeds with aryl halides as well as triflates in generally good to excellent yields. For some examples, lower yields in the range of only 50% were achieved. For example, the coupling product **12** of 5-bromopyrimidine **37** was obtained in only 47% (Scheme 37, right).

Molander and coworkers reported the preparation and cross-coupling of alkoxymethyl [59], amidomethyl [60], and alkoxyethyl trifluoroborates [61]. Reaction conditions were optimized separately for these reagents: while a combination of $Pd(OAc)_2$/RuPhos was used for the alkoxymethyl reagent, $Pd(OAc)_2$/XPhos was employed for the amidomethyl trifluoroborates. In the case of the alkoxyethyl reagent, $PdCl_2A^{ta}$Phos **121** was identified as a suitable catalyst. Generally good yields could be achieved in the transformations, but halopyrimidines reacted only sluggishly and gave low or moderate yields (Scheme 38, **122–124**).

The preparation and coupling of substituted vinyl-BF_3K salts was investigated by Math et al. [62]. Internal vinyl-BF_3K salts **126** and a variety of electron-deficient (hetero)aryl bromides formed the coupling products in 67–88% yield. Three examples involving the coupling of 5-bromo-2-methoxypyrimidine **125** gave the

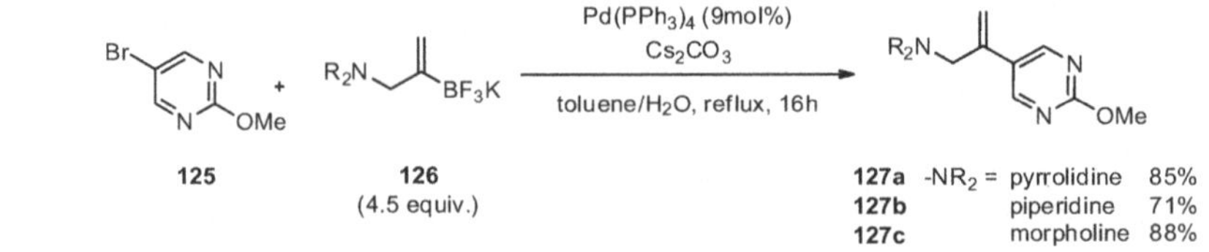

Scheme 39 Vinyl-BF_3K salts in the cross-coupling with **125**

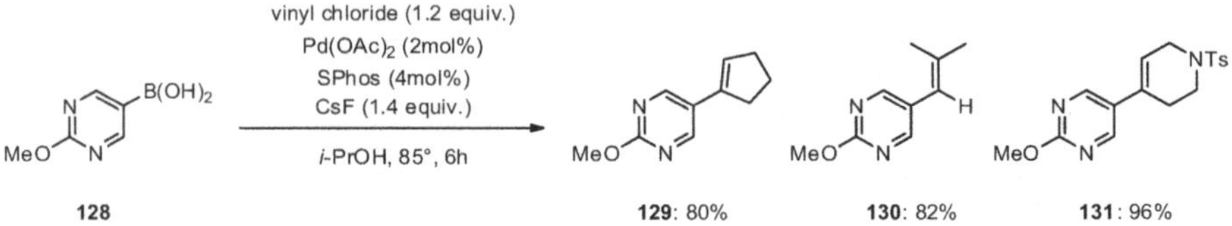

Scheme 40 Cross-coupling of **128** with various vinyl chlorides

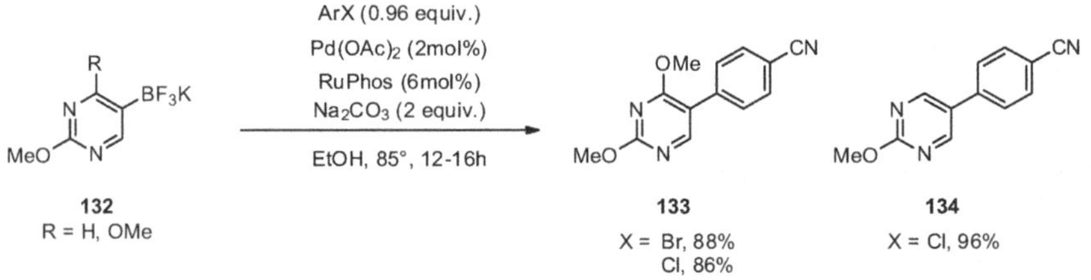

Scheme 41 Palladium-catalyzed Suzuki coupling of pyrimidine-BF_3K salts

products **127a–c** in 71–88% yield depending on the BF_3K-coupling partner (Scheme 39).

1.1.2 Pyrimidine Boron Compounds as Coupling Partners

Thakur et al. published the coupling of heteroaryl boronic acids with vinyl chlorides employing the SPhos ligand in conjunction with potassium fluoride in *i*-PrOH [63]. Under these conditions, minimal protodeboronation was observed. Electron-rich and electron-poor heterocycles as well as open-chain and cyclic vinyl chlorides are within the scope of the reaction. 4-Pyrimidineboronic acids such as **128** coupled with vinyl chlorides in 80–96% yield of **129–131** within 6 h (Scheme 40).

Molander et al. prepared a variety of heteroaromatic potassium trifluoroborate salts and explored their scope as coupling partners in Suzuki–Miyaura coupling [64]. As catalyst, a combination of RuPhos and $Pd(OAc)_2$ proved to be effective. Among other examples, they investigated coupling of substituted 5-pyrimidine trifluoroborates **132** with 4-halobenzonitrile. For these reactions, yields of 86–96% of **133** and **134** could be achieved (Scheme 41). Coupling of

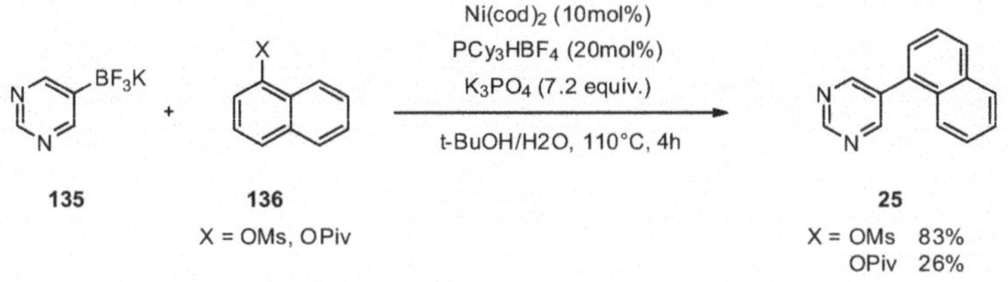

Scheme 42 Nickel-catalyzed Suzuki coupling of pyrimidine-BF_3K salts with aryl donors

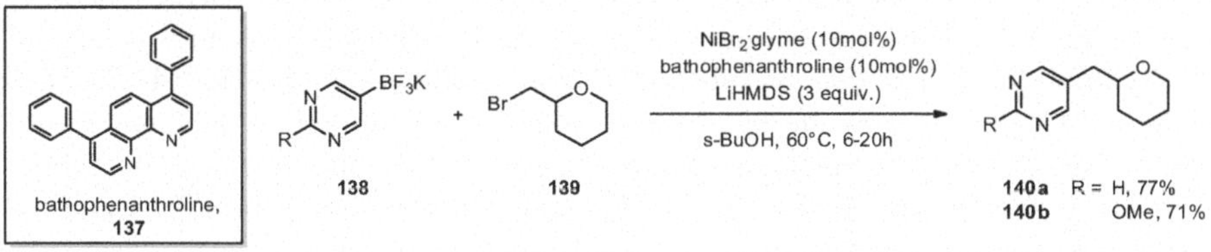

Scheme 43 Nickel-catalyzed Suzuki coupling of pyrimidine-BF_3K salts with alkyl halides

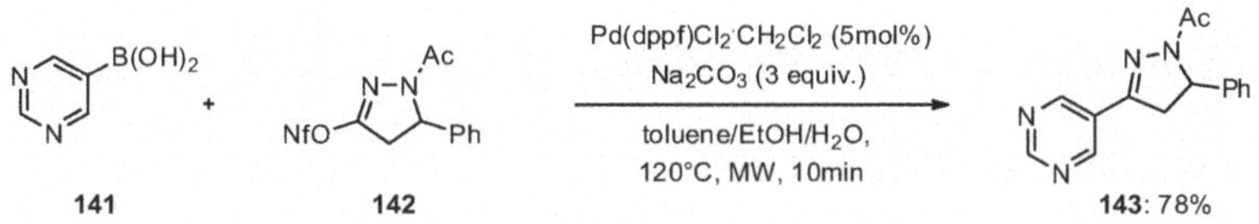

Scheme 44 Nonaflate as leaving group in Suzuki–Miyaura coupling of **141**

2-chloropyrimidine with thiophene-3-trifluoroborate and furan-2-trifluoroborate gave the coupling products in 52% and 92%, respectively.

The group of Molander also reported conditions for the coupling of phenolmethanesulfonates with aryl tetrafluoroborates [65]. Coupling was effected in the presence of $Ni(cod)_2/PCy_3HBF_4$ in generally good to excellent yields. Arylation of 5-pyrimidine potassium tetrafluoroborate **135** with 1-naphtol methanesulfonate **136** gave the product **25** in 83% after 4 h at 110°C (Scheme 42).

The same group found that alkyl halides are efficiently coupled with (hetero)aryl tetrafluoroborates in the presence of $Ni(II)Br_2$ and bathophenanthroline **137** [66]. The use of the more stable potassium tetrafluoroborates **138** allows the coupling of challenging substrates such as 2-thiophenetetrafluoroborates. Reported yields are typically in the range of 50–80% (Scheme 43, **140**a, b).

An example of a nonaflate as the coupling partner was reported by Grimm et al. [67]. The substituted pyrazoline nonaflate **142** reacted smoothly under microwave conditions and gave 78% of the coupling product **143** within 10 min (Scheme 44).

Menard and Lautens reported a rhodium-catalyzed desymmetrization of 2,3-bicyclic hydrazines such as **145** with arylboronic acids [68, 69]. The reaction could be tuned to give either cyclopentenes after ring-opening, or, especially in case

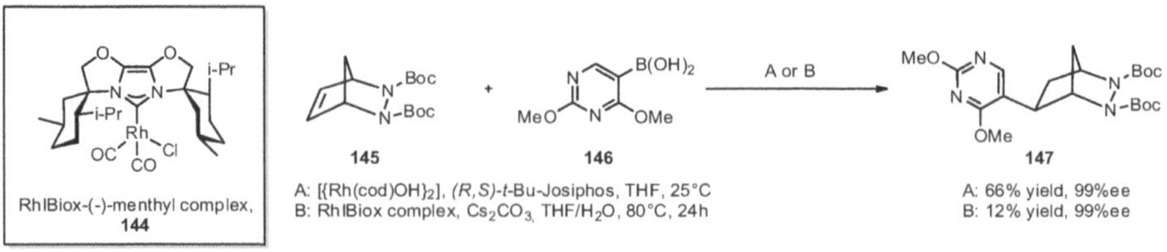

Scheme 45 Rhodium-catalyzed hydroarylation of **146**

of heteroaromatic coupling partners, the hydroarylation product with excellent stereoselectivity. The combination of $[Rh(cod)OH]_2$ with *t*-Bu-Josiphos catalyzed the hydroarylation reaction of substituted 5-pyrimidineboronic acid **146** in 66% yield of **147** with 99% *ee*. In the initial protocol, 2-substituted boronic acids were required in order to obtain high stereoselectivity. This limitation was overcome with the use of a Rh-IBiox-catalyst **144**, now allowing for a broader range of compounds to be coupled with 94–99% *ee* [70] (Scheme 45).

1.2 Pyrimidine in Stille Reactions

1.2.1 Pyrimidine as (Pseudo)Halide

Intrastrand cross linking-forming oligonucleotides are an interesting class of molecules, since they are capable of blocking DNA replication; thus, gene expression can be suppressed. Therefore, such compounds have found application as chemotherapeutic agents. Hattori et al. have designed and synthesized oligodeoxinucleoside, which was in a later stage incorporated into the DNA chain and cross-linked to thymine and served purpose of gene expression blocker [71]. The synthesis of oligodeoxinucleoside started with Stille coupling between pyrimidine derivative **148** and tributyl(vinyl)stannane which was without purification treated with octanethiol, giving rise to compound **150** in 77% yield over two steps (Scheme 46). Octanethiol served the purpose of a protecting group of the vinylic double bond, which was restored in a later stage, since it was anticipated not to tolerate the synthesis conditions for the nucleotide (Scheme 46).

Wicke and Engels have investigated Stille coupling as potential tool for post-synthetic modification of synthetic RNA oligonucleotides [72]. Initially, optimization of the reaction conditions was performed on the monomeric uridine **151**, and under the optimal conditions (Pd_2dba_3, $P(fur)_3$, DMF, 60°C), they were able to couple tributyl(vinyl)stannane with 5-iodouridine in 90% yield of **152a**. Furthermore, 2-furyl, 2-thienyl and 2-benzothienyl moieties were introduced in the same way, providing derivatized mononucleotides in yields up to 99% (Scheme 47, **152b–d**). Moreover, they were able to carry out the developed modification on the oligonucleotide as well. Penta- and dodecamers were decorated with the functionalities mentioned above at both terminal and nonterminal positions.

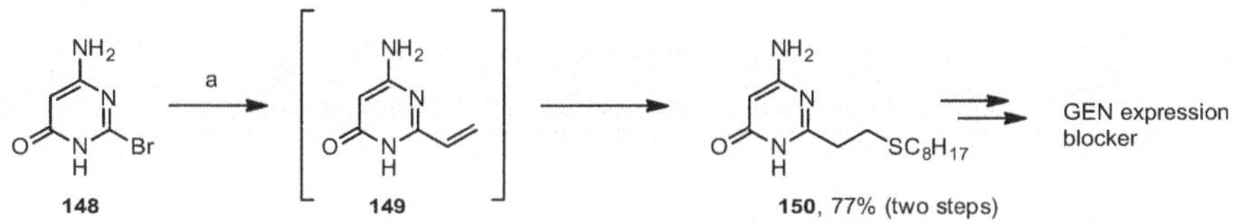

Scheme 46 (*a*) (i) $Bu_3SnCHCH_2$, $Pd(PPh_3)_2Cl_2$, DMF; (ii) $C_8H_{17}SH$

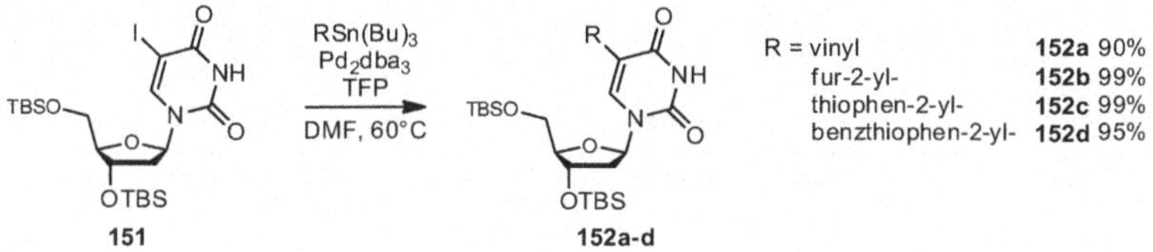

Scheme 47 Post-synthetic notification of RNA nucleotide

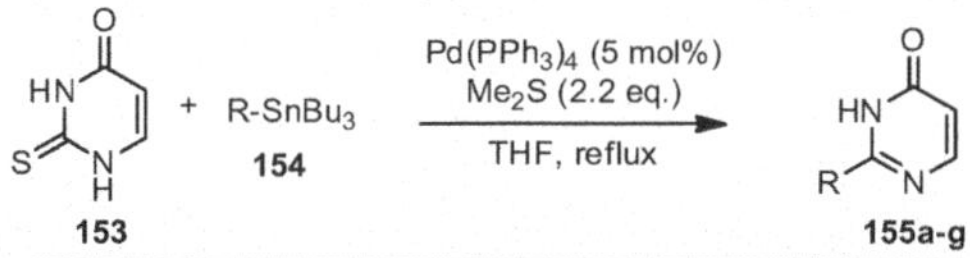

Entry	R	time (h)	Product	Yield (%)
1	Phenyl	24	**155a**	68
2	3-nitrophenyl	24	**155b**	63
3	4-methoxyphenyl	24	**155c**	52
4	2-furyl	14	**155d**	64
5	2-thienyl	14	**155e**	62
6	3-pyridinyl	24	**155f**	53
7	PhCH=CH-	48	**155g**	36

Scheme 48 Desulfative Stille cross-coupling

Desulfitative cross-coupling of 2-thiouracil **153** and its derivatives was described by Sun et al. [73]. The method does not require protection of free nitrogen (unlike in the case, when boronic acid is used for coupling). Under optimized conditions, tributylstannylbenzene **154** couples with 2-thiouracil in the presence of $Pd(PPh_3)_4$ and $CuBr{\cdot}Me_2S$ yielding 68% of product **155a**. Electron donating and electron-withdrawing substitution of the organotin reagent is tolerated, as well as utilization of diverse heterocycles and styrene as tin electrophile, providing derivatized uracil in yields from 36 to 64% (**155b–g**). Substitution of uracil is tolerated as well (Scheme 48).

Kazmier and Bukovec have investigated the synthetic utility of stannylated ally carbonates [74]. They have demonstrated that a carbonate moiety can serve as leaving group for palladium-catalyzed amination reactions and resulting stannylated allyl amines can undergo further coupling under Stille conditions. Moreover, the amination/Stille coupling sequence can be performed in a one-pot fashion. 5-bromopyrimidine was utilized for the second step of the sequence as well. Initially, ethyl (2-(tributylstannyl)allyl) carbonate **156** is coupled with piperidine **157**, providing 2-(tributylstannyl)allylpiperidine, which subsequently undergoes Stille

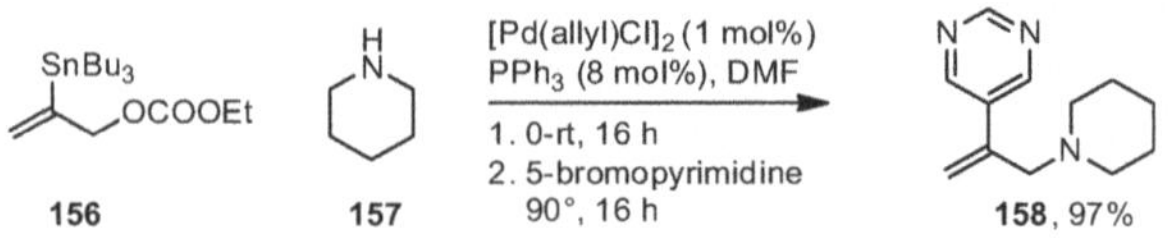

Scheme 49 Stille coupling of pyrimidine with stannylated allyl amines

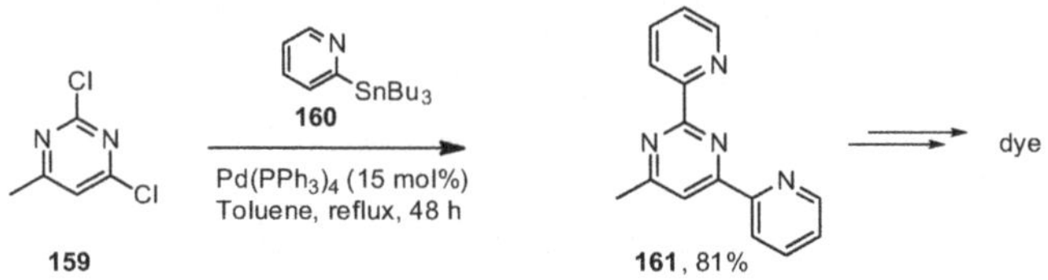

Scheme 50 Derivatization of pyrimidine via Stille coupling in the synthesis of dyes

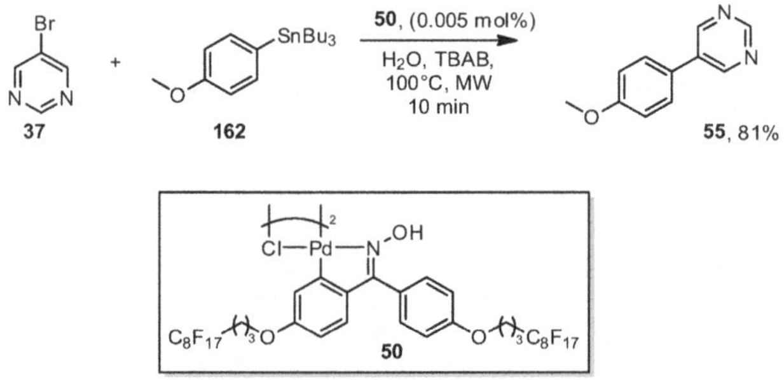

Scheme 51 Microwave-assisted Stille coupling of **37**

coupling upon addition of 5-bromopyrimidine. The final product **158** was obtained in 97% of overall yield (Scheme 49).

Synthesis and evaluation of optical properties of 4-arylvinyl-2,6-di(pyramid-2-yl)pyrimidines were performed by Hadad et al. [75]. The synthetic strategy comprises of double Stille coupling between 2,4-dichloro-6-methylpyrimidine **159** with 2-tributylstannylpyridine **160** and subsequent aldol condensation of formed 4-methyl-2,6-di(pyramid-2-yl)pyrimidine. Under the reaction conditions (15 mol%, $Pd(PPh_3)_4$, toluene), double Stille coupling provided 81% of **161** (Scheme 50).

Few new catalytic systems were developed and evaluated for Stille coupling. Susanto et al. described oxime-based palladacycle **50** [23]. Extremely low catalyst loading (0.005 mol%) proved to be sufficient for the reaction to proceed successfully. 5-Bromopyrimidine **37** provided 91% yield of **55** in the reaction with tributyl-4-methoxyphenylstannane **162** (Scheme 51).

A silica-anchored catalyst **57** was reported by Lee et al. [31]. Ketoiminatophosphan-Pd complex was able to catalyze Stille coupling between

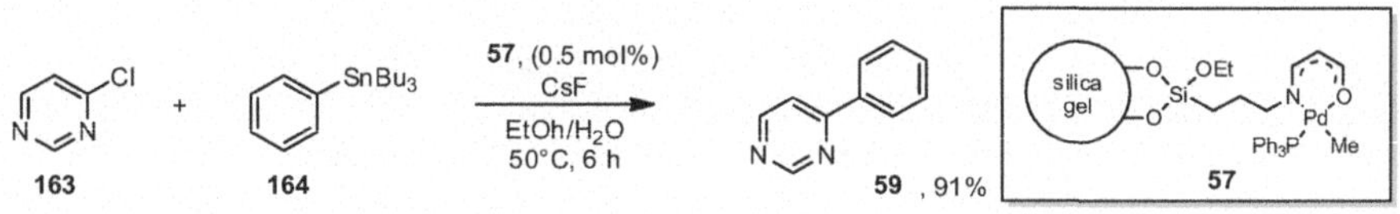

Scheme 52 Stille coupling via immobilized catalyst **166**

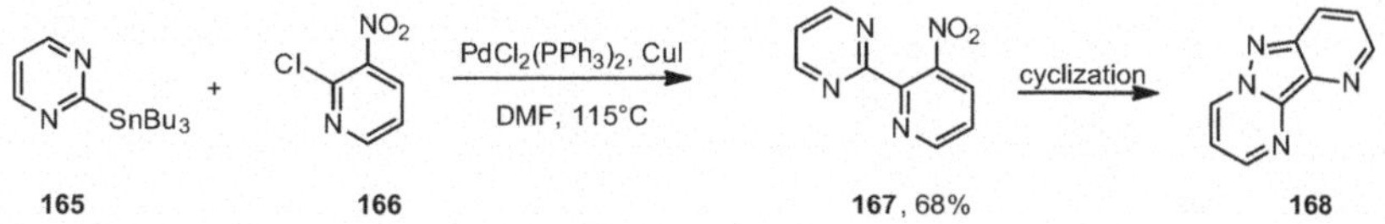

Scheme 53 Two-step synthetic approach towards polyaza tricycles via Heck reaction and subsequent nitrene-mediated cyclization

4-chloropyrimidine **163** and tributylphenyl stannane **164** towards **59** in 91% yield (Scheme 52).

1.2.2 Pyrimidine as Organometal Species

Many papers describe the synthesis of various structural motives, involving Stille coupling of pyrimidine, as a main or one of the main steps. Herein, pyrimidine could play a role of both, electrophile and nucleophile.

Nyffenegger et al. have developed a two-step synthesis towards polynitrogen-fused tricycles [76]. The sequence consists of Stille coupling of diverse electrophiles with various *o*-nitro(hetero)aryl chlorides and subsequent nitrene-mediated cyclization in the presence of triethylphosphite. Employing 2-tributylstannylpyrimidine **165** in the reaction with 2-chloro-3-nitropyridine **166** provided 68% of **167**, which upon treatment with triethylphosphite provided final fused tricyclic structure **168** in 86% yield (Scheme 53).

With the aim of developing a method for the production of ^{11}C-labeled radio pharmaceuticals for positron emission tomography (PET), Doi and coworkers have investigated the methylation of common heterocycles with methyl iodide via Stille coupling.[77] Due to the short half-life of ^{11}C ($t_{1/2} = 20.4$ min), fast synthesis of labeled species is a prerequisite. The strategy, reported by Doi, relies on the coupling of iodomethane with excess of the stannylated heteroaryl. However, previously developed reaction conditions (CH_3I/stannane/Pd_2dba_3/P(*o*-tolyl)$_3$/CuCl/K_2CO_3 in a ratio of 1:40:0.5:2:2:2) turned out to be inefficient when applied to the selected heterocycles. After reoptimization of conditions (CH_3I/stannane/Pd_2dba_3/P(*o*-tolyl)$_3$/CuCl/K_2CO_3 in ratio 1:40:0.5:16:2:5 at 100°C in NMP for 5 min), trapping of iodomethane **170** with 5-tributylstannylpyrimidine **169** provided 87% of 5-methylpyrimidine **171** (Scheme 54).

Hirano et al. have investigated the properties of hypervalent pentacoordinated boron compounds [78]. For this purpose, compound **175** was synthesized. Within the synthetic sequence, Stille coupling on intermediate **172** was reported with

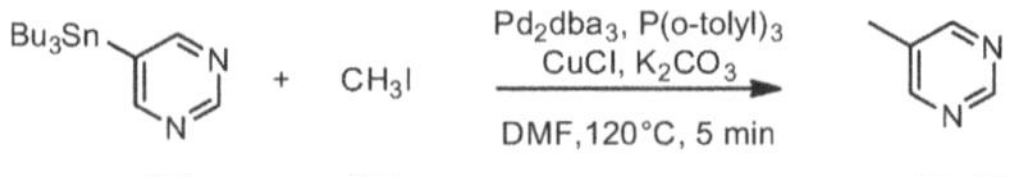

Scheme 54 Methylation of pyrimidine with MeI via Stille coupling

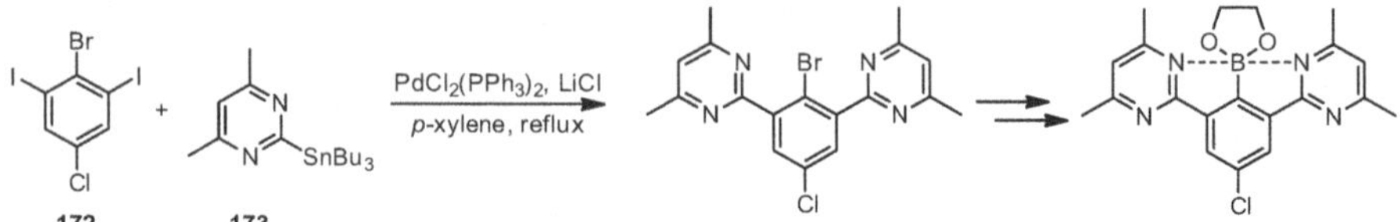

Scheme 55 Synthesis of the hypervalent-boron species via Stille coupling

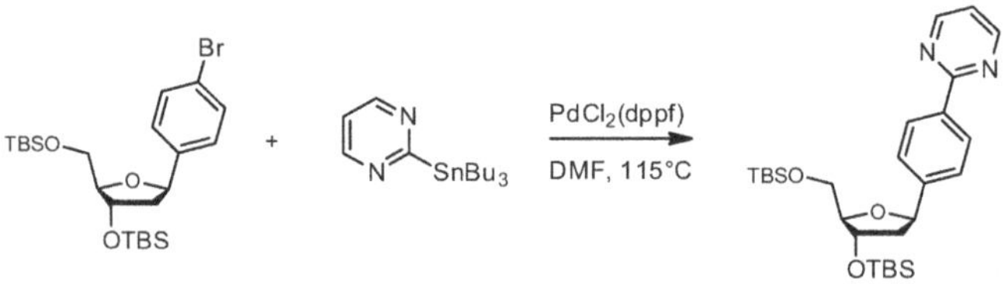

Scheme 56 Synthesis of 4-aryl substituted phenylene *C*-2′-deoxyribonucleosides

2-tributylstannyl-4,6-dimethylpyrimidine **173** as coupling partner in the presence of $PdCl_2(PPh_3)_2$ and LiCl. Compound **174** was obtained in rather poor yield of 13%. Low efficiency of the last step was ascribed to insufficient reactivity of the starting material and instability of the product at the high reaction temperature. **174** was then further elaborated to the desired hypervalent-boron compound **175** (Scheme 55).

Several reports have been published, describing either modification of pyrimidine containing nucleic bases or even a synthesis of artificial nucleosides, taking advantage of Stille coupling either as one step in the linear synthesis or as tool to introduce diversity at a specific position.

Hocek and coworkers have developed a synthesis of 4-arylsubstituted benzene *C*-2′-deoxyribonucleosides [79]. As a starting point for the derivatization at position 4 of the benzene ring, compound **176** was synthesized (Scheme 56) first and subsequently submitted to cross-coupling. Several aryl groups were introduced via coupling with **165**, including 2-pyrimidinyl. This reaction yielded 54% of **177**, which placed pyrimidine among the challenging substrates.

In 2008 Bargadí and Rossi have developed a novel approach towards 6-substituted uracils [80]. Synthesis was carried out as a one-pot procedure, consisting of stannylation of 4-chloro-2,6-dimethoxypyrimidine **178**, subsequent Stille coupling with aryliodide or benzoyl chloride, and hydrolysis. Stille coupling was carried out in the presence of $PdCl_2(PPh_3)_2$. The overall yields of the 6-substituted uracils **179a–b** were in the range between 30 and 57% (after three steps, Scheme 57).

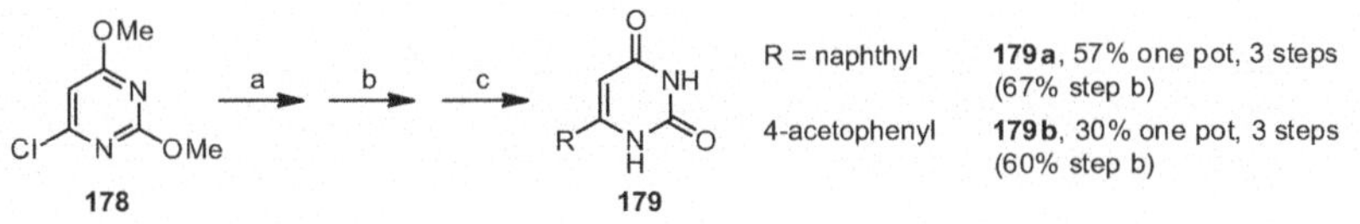

Scheme 57 One-pot synthesis of substituted uracils; (*a*) stannylation (*b*) $PdCl_2(PPh_3)_2$, toluene, reflux (*c*) hydrolysis

1.3 Pyrimidine in the Negishi Reaction

1.3.1 Pyrimidine as (Pseudo)Halide

Catalyst Development

Several new catalytic systems were investigated for Negishi coupling. Organ and coworkers have shown that Pd-PEPPSI-IPent (**183**), a NHC-based palladium complex, is a potent catalyst capable of coupling of a variety of (hetero)aryls, including sterically hindered 2,6-substituted substrates [81]. The corresponding Grignard reagents were first treated with $ZnCl_2$ and subsequently mixed with the (hetero) aryl halide coupling partner. Reaction between 2,6-dimethylphenylzincbromide **180** with 2,4-dimethoxy-5-bromopyrimidine **182** led to 89% of **184a**. Reaction of the same pyrimidine derivate with 4-bromoisoquinoline **181** provided 70% of product **184b** (Scheme 58a).

Another palladium-based catalyst was designed by Boliger and Frech [82]. Treatment of $Pd(cod)_2Cl_2$ with 1-(dicyclohexylphosphanyl)piperidine led to the formation of dichloro{bis[1-(dicyclohexylphosphanyl)piperidine]} complex **53**, which was subsequently evaluated in Negishi coupling. Various 2- or 5-bromopyrimidines were coupled with different diarylzinc reagents, giving yields in the range of 86–96% (**186a–e**, **187a**, **b**). Attractive features of this method are very low catalyst loadings (0.01 mol%) and short reaction times (in the range of minutes for all pyrimidines) (Scheme 58b).

Besides the investigation on palladium as a catalyst for Negishi coupling, also nickel has attracted some attention. Binuclear cationic nickel catalyst **15** ligated by NHC was described by Xi et al. as the first Ni/NHC catalyst for Negishi coupling [83]. At the relatively mild reaction conditions (80°C, 5 h), various (hetero)aryls were coupled with diverse arylzinc species. 2-chloropyrimidine **6** provided 99% of **189a**, when coupled with *p*-tolyl zinc chloride. Reaction with *o*-tolyl zinc chloride provided 97% of the biaryl **189b** (Scheme 59a).

Gerber and Frech have synthesized an aminophosphine-based nickel catalyst and applied it in Negishi coupling [84]. $NiCl_2$ was treated with 2 equiv. of tri (piperidin-1-yl)phosphine, and the catalyst formed was found to be highly active in the coupling reaction already at low catalyst loading of 0.1 mol%. The reaction was carried out at a temperature of 60°C for up to 2 h. Several 5-bromopyrimidines were tested with various diarylzincs giving yields between 54 and 97% (Scheme 59b).

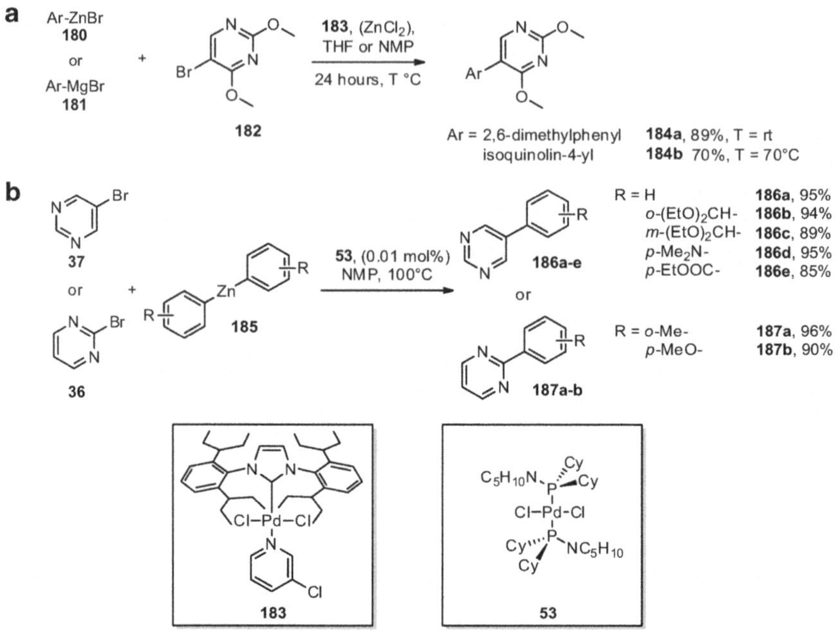

Scheme 58 New catalytic systems based on palladiums (**183**) and (**53**) and their application

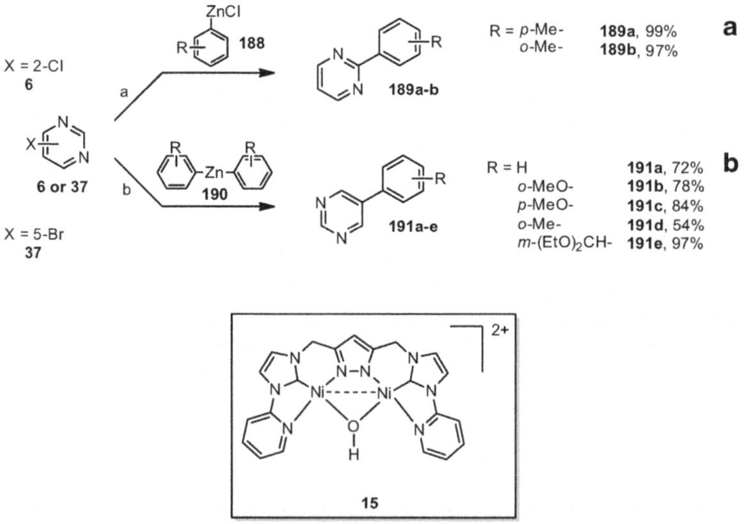

Scheme 59 Negishi coupling catalyzed by nickel complex 15; (*a*) 15 (0.1 mol%), THF/NMP (1:1), 80°C, 2 or 3 h; (*b*) $NiCl_2$ (0.1 mol%), $P(NC_5H_{10})_3$ (0.2 mol%), THF/NMP, (1:2), 60°C, 0.5–2 h

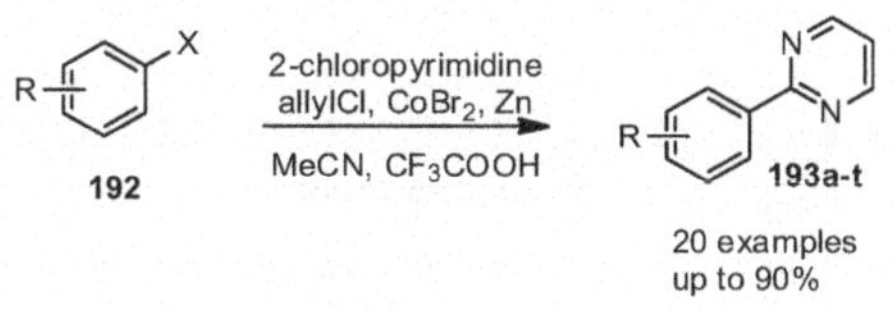

Scheme 60 Cobalt-catalyzed Negishi coupling

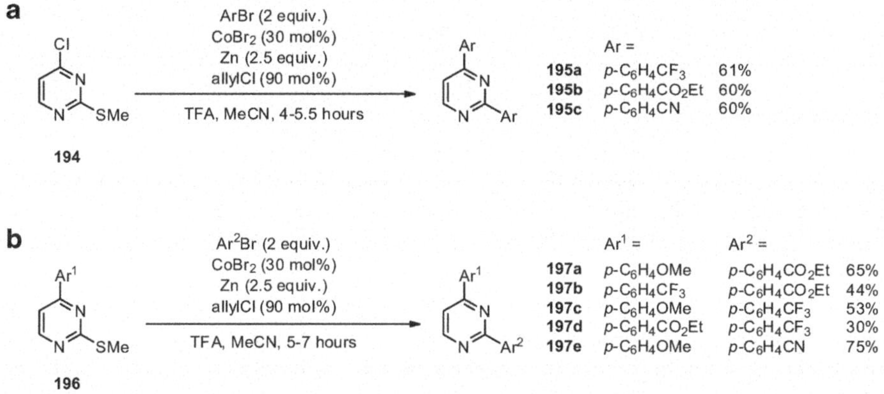

Scheme 61 $CoBr_2$-catalyzed desulfitative Negishi coupling

Cobalt as a catalyst for Negishi coupling was reported by Bégouin and Gosmini [85]. Various electron-rich and few electron-poor aryl halides were in situ transformed into arylzinc reagents and coupled with 2-chloropyrimidine in the presence of the catalytic system, comprising of $CoBr_2$, allyl chloride (of which the role within the reaction was not clarified yet), and traces of trifluoroacetic acid. At 50°C in acetonitrile, the reaction provided satisfactory yields up to 90% when electron-poor or electron-rich substrates were reacted. *O*-substitution of aryl halides caused a slight drop in reaction yields. Aryl chlorides were investigated in the coupling with 2-chloropyrimidine as well. The final products were obtained in yields from 50 to 60%, with no obvious effect of electronic properties of aryls (Scheme 60).

Begouin et al. also disclosed the $CoBr_2$-catalyzed desulfitative coupling of heteroaromatic methylthioethers and arylzinc reagents [86]. The authors found trace amounts of bis-coupling product when they treated 4-chloro-2-methylthio-pyrimidine **194** with various arylzinc reagents in the presence of 10 mol% $CoBr_2$. When the catalyst loading was increased to 30 mol%, the bis-coupling product became the major product, even if only one equiv. of coupling partner was employed. With 2 equiv. of coupling partner, yields of about 60% were achieved (Scheme 61a). Also the synthesis of unsymmetrically substituted pyrimidines was successful: when the crude reaction mixture of the first arylation was subjected to bis-arylation conditions (30 mol% $CoBr_2$), products **197a–e** were isolated in 30–75% yield (Scheme 61b).

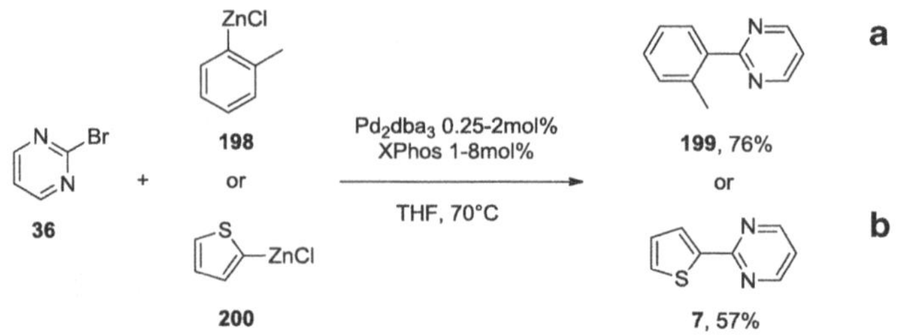

Scheme 62 Pd-XPhos-catalyzed arylation of 2-bromopyrimidine

Methodology

In 2004, the group of Buchwald published a Negishi coupling between arylzinc chlorides and aryl bromides (Scheme 62a) [87]. Scope study included reaction between 2-tolylzinc chloride **198** with 2-bromopyridine **36**. Employing a proper catalytic system, consisting of Pd_2dba_3 and XPhos as ligand, allowed using conveniently low catalyst loadings as low as 0.25% of palladium source with 4 equiv. of ligand. The coupling product was obtained in 76% yield. Similarly, only using 2 mol% catalyst loading, 2-bromopyridine **36** was coupled with thiophen-2-yl zinc chloride **200**, yielding 57% of **7** as shown by Luzung et al. [88] (Scheme 62b).

It was also demonstrated that cross-coupling between benzylic zincates and aryl halides can be mediated by nickel. Both pyrimidine chlorides and pyrimidine bromides were employed, providing the desired products in yields up to 98% using Ni(acac)$_2$ and PPh_3 as ligand. Moreover, employing thioether as leaving group is possible as well when PPh_3 is replaced with DPE-Phos. 4,6-dimethoxy benzylzinc chloride reacted with 2-methylthiopyrimidine **206b** in 85% yield (Scheme 63).

Dunst and Knochel developed LiCl-mediated insertion of Mg into an aryl–Cl bond and its subsequent transmetalation with $ZnCl_2$ and Negishi coupling [89]. In this reaction, 1-chloro-3-fluorobenzene was reacted with 4-iodo-2,6-dimethoxypyrimidine **209**, providing the desired product **210** in 74% yield (Scheme 64).

In the same group, direct metalation of protected uracils and thiouracils was developed [90]. Similarly as in the previous case, a Grignard reagent was transmetalated with $ZnCl_2$ and subjected to Negishi coupling. 3-Trifluoromethyl iodide **212a** and 4-(ethoxycarbonyl)phenyl iodide **212b** underwent coupling with 2,6-dimethoxypyrimidin-4-yl magnesium chloride **211** with 80% and 71% yield respectively (Scheme 65).

Knochel and coworkers also reported the first example of a direct insertion of aluminum into C–X bonds [91]. This process is rendered possible by the presence of certain metal salts (e.g., $InCl_3$/LiCl or $BiCl_3$/LiCl). The organoaluminum reagents **215** were transmetalated using Zn(OAc)$_2$. The resulting organozinc reagents smoothly underwent Pd-catalyzed cross-couplings with aryliodides. Among many other examples, it has been demonstrated that a substituted 6-iodopyrimidine bearing an ester (**216**) coupled with 4-iodotrifluorotoluene **214**

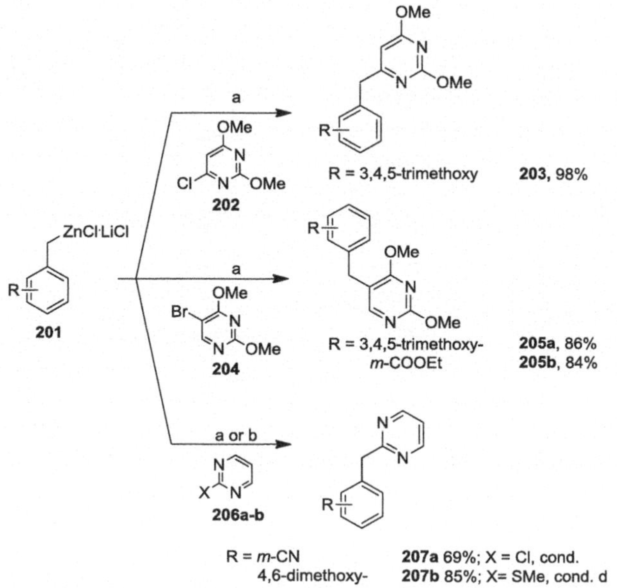

Scheme 63 Coupling of (pseudo)halopyrimidine with benzylic zinc reagents; (*a*) $Ni(acac)_2$ 0.5 mol %), PPh_3 (2 mol%), THF, 60°C; (*b*) $Ni(acac)_2$ (2.5 mol%), DPE-Phos (5 mol%), THF, 25°C

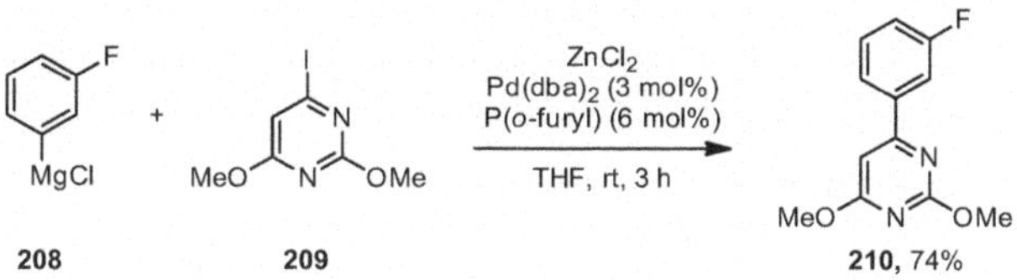

Scheme 64 Coupling of organozinc reagents, generated in situ via magnesium insertion and transmetalation

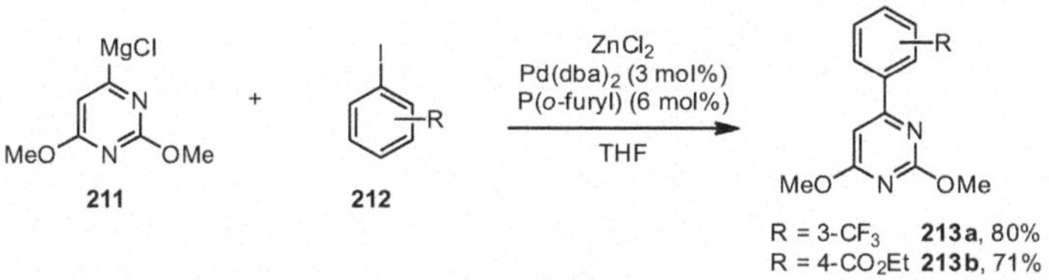

Scheme 65 Synthesis of uracil derivatives via Negishi coupling

in 73% yield after 4.5 h at 30°C (Scheme 66a). Later, this methodology was extended to substrates containing benzylic chlorides (Scheme 66b) [92].

Synthesis of hydroxymmethyl-2-(2′-pyrimidyl)ferrocene was described by Omedes et al. [93]. After acetal protection of the carbonyl group, the compound was treated with *s*-BuLi and $ZnBr_2$. Addition of $PdCl_2(PPh_3)_2$ and 2-iodopyrimidine led to the formation of racemic product **224a** in 75%. Hydrolysis

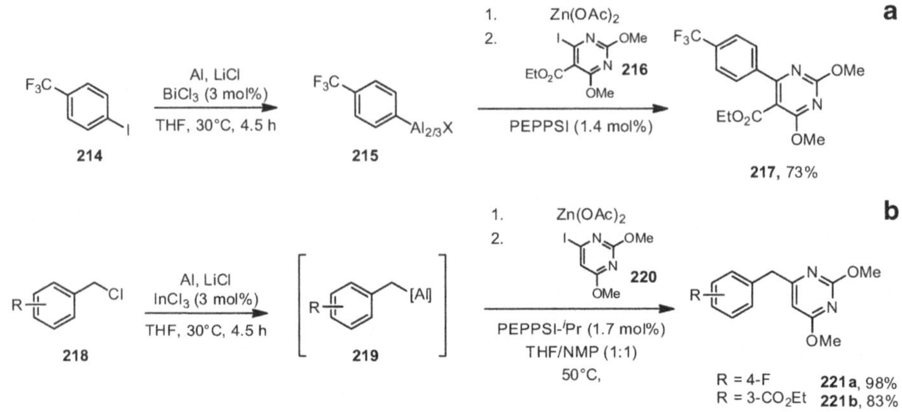

Scheme 66 Coupling of organozinc reagents, prepared in situ via direct Al-insertion and transmetalation

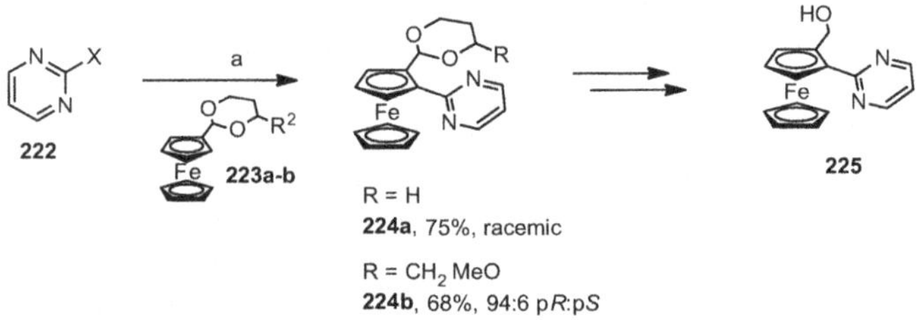

Scheme 67 Negishi coupling of ferrocenylzinc bromides; (*a*) (1) ferrocene, *s*-BuLi, THF, −78°C; (2) $ZnBr_2$, THF, −30°C to rt, 1 h; (3) 2-iodopyrimidine, $PdCl_2(PPh_3)_2$, THF, reflux, overnight

and reduction provided the corresponding alcohol **225**. When enantiopure starting material **223b** was used instead, the reaction gave enantiomerically enriched **224b** in 68% yield with an enantiomeric ratio 94:6 (*pR*:*pS*) (Scheme 67).

Hasnik et al. have reported hydroxymethylation of (hetero)aryls via Negishi coupling of the corresponding (hetero)aryl halide with (benzoxymethyl)zinc iodide **227** [94]. Coupling with 5-bromo-2-iodopyrimidine in the presence of $Pd(PPh_3)_4$ proceeds selectively into the iodine position and **228** is isolated in 96%. Further deprotection yields 75% of 5-bromo-2-hydroxymethyl pyrimidine **229** (Scheme 68a).

Suhartono et al. have developed a synthetic method towards nonnatural amino acids containing an aryl moiety [95]. Key intermediate **230** was obtained from glutamic acid in four steps. Treatment of **230** with Zn and I_2 and subsequent $NiCl_2(PPh)_3$ promoted Negishi coupling with with 2-bromopyrimidine towards arylated intermediate **231** in 86% yield. **231** was further converted to the desired amino acids **232** (Scheme 68b).

Preparation of 3-aryl-2,2-dimethylpropanoates via Negeshi coupling was described by Kwak et al. [96]. Firstly, methyl-3-iodo-2,2-dimethylpropanoate **233**

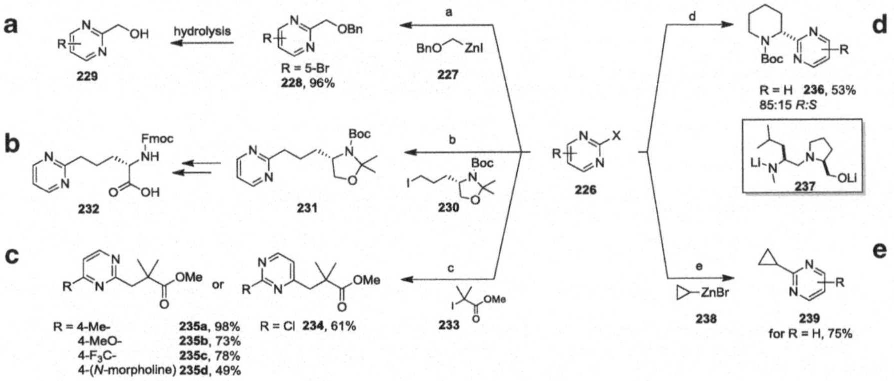

Scheme 68 Negishi coupling of alkylzinc reagents with halopyrimidines; (*a*) $Pd(PPh_3)_4$, THF, rt (*b*) (1) 268, Zn, I_2; (2) 2-bromopyrimidine, $NiCl_2(PPh_3)_2$ (*c*) X=Cl; Zn-Cu, toluene, DMA, Pd $(PPh_3)_4$ 110°C (*d*) X=Br; (1) N-Boc-piperidine, s-BuLi, TMDEA (4 eq.), Et_2O, −78°C; (2) (*S,S*)-237, −45°C, 5 h; (3) −78°C, $ZnCl_2$ warm to rt; (4) 2-bromopyrimidine, $Pd(OAc)_2$, *t*-$Bu_3P·HBF_4$ (*e*) X=Br; $Pd(OAc)_2$ (1 mol%), CPhos (2 mol%), THF, rt, 30 min

is treated by Zn-Cu couple, and the resulting zincated species is subjected to Negishi conditions ($Pd(PPh_3)_4$, 110°C) and addition of (hetero)aryl chloride). Series of diversely substituted 2-chloropyrimidines was tested, and corresponding products **235a–d** were isolated in 49–98% yields. The reaction tolerates both electron-withdrawing and electron-donating substituents (Scheme 68c).

Beng and Gawley reported highly enantioselective (hetero)arylation of Boc-protected piperidine [97]. Treatment of the starting material with *s*-BuLi and chiral additive **237**, followed by addition of $ZnCl_2$ and finally submission of the formed zinc species to the coupling condition ($Pd(OAc)_2$/*t*-$Bu_3P·HBF_4$) in the presence of 2-bromopyrimidine, leads to the formation of **236** in 53% yield. Enantiomeric ratio (*R*:*S*) was determined to be 85:15 (Scheme 68d).

Negishi coupling of a secondary sp^3-C-zinc center with (hetero)aromatic halides was investigated by Han and Buchwald [98]. Out of the tested ligands, CPhos performed the best. Reaction of cyclopropylzinc chloride **238** with 2-bromopyrimidine, catalyzed by $Pd(OAc)_2$ and CPhos in a THF/toluene mixture, provided 75% of **239**. Although carried out under very mild conditions (room temperature), full conversion was reached after only 30 min (Scheme 68e).

Thaler et al. have investigated diastereoselective $C(sp^3)$-$C(sp^2)$ Negishi coupling [99]. Related to this chapter, iodomenthol was firstly converted to the corresponding zincated species by treatment with zinc and lithium chloride. 240 was then reacted with 5-bromopyrimidine in the presence of $Pd(PPh_3)_4$. Product **242** was obtained in 76% with 98:2 diastereomeric ratio (Scheme 69).

Negishi coupling of pyrimidine was applied within the synthesis of DSC sensitizers [100]. First, 2-(*N,N*-diaryl)thiophene **243** was metalated using *n*-BuLi and subsequent addition of $ZnCl_2$. Reaction with 5-bromo-2-iodopyrimidine **345** in the presence of $Pd(PPh_3)_4$ provides formation of desired products **246a–c** in yields between 56 and 81%. Subsequent modifications led to the desired derivatives, which were investigated for their optical properties (Scheme 70).

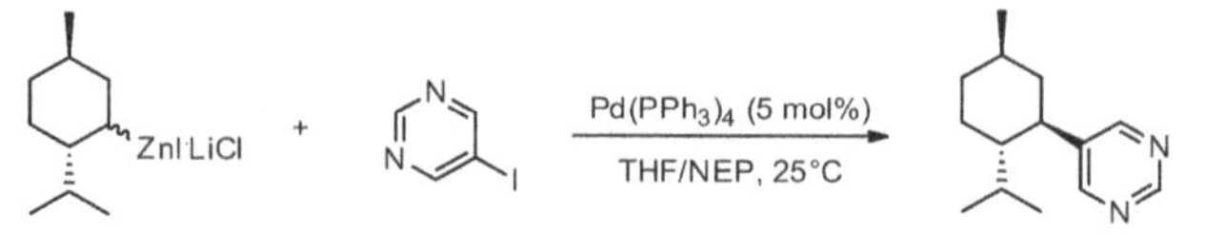

240 **241** **242**, 76%, d.r. 98:2

Scheme 69 Diastereoselective introduction of pyrimidine to menthol

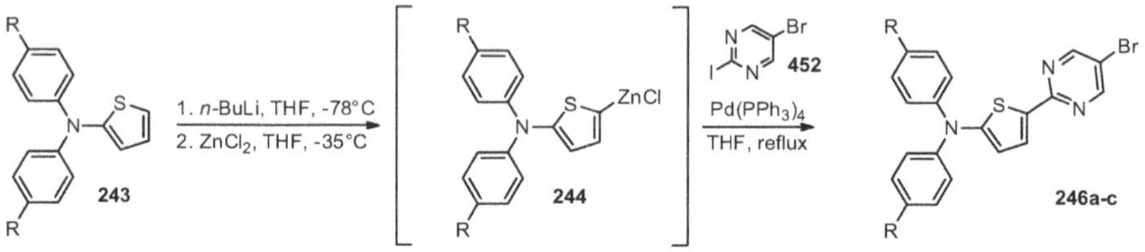

246a R = H		71%
246b	OMe	56%
246c	OHex	81%

Scheme 70 Synthesis of the precursor of dye for DSSC

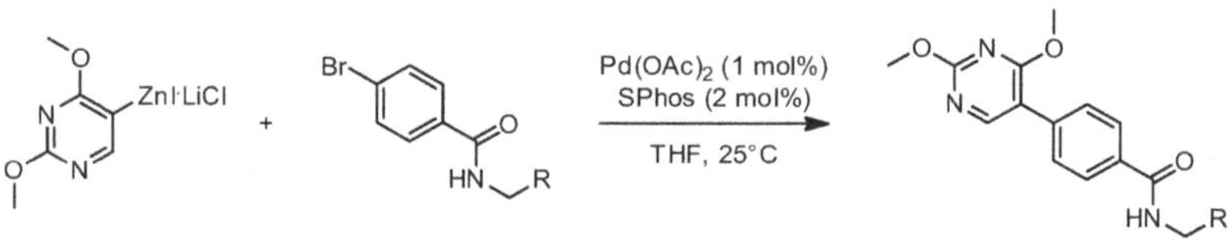

247 **248a-b**

R = cyclopropyl- **249a**, 87%
CH$_2$-*t*-Bu **249b**, 97%

Scheme 71 Negishi cross-coupling in the presence of free amide

Monolikakes et al. have investigated the possibility of Negishi coupling between various zinc nucleophiles and aryl halides bearing unprotected amides [101]. Using Pd(OAc)$_2$/S-Phos and mild reaction conditions (25°C), reaction between 2,4-dimethoxypyrimid-5-yl zinc iodide **247** and 4-bromo-*N*-(cyclopropylmethyl) benzamide **248a** provided 87% of the desired biaryl **249a**. In the reaction with 4-bromo-*N*-(3,3-dimethylbutyl)benzamide **248b**, the same nucleophile provided 97% of **249b** (Scheme 71).

Several reports were published dealing with direct metalation of aromatic species and subsequent Negishi coupling. Utilizing the proper metalating agent allows regioselective metalation even in the presence of chlorides or bromides. In most cases, pyrimidine is used as electrophile; nevertheless, direct metalation of pyrimidine is described as well.

Staben and coworkers have developed a method of regioselective metalation of azine *N*-oxides with TMPZnCl·LiCl [102]. Subsequent treatment of the metalated species with heteroaryl bromides in the presence of Pd catalyst leads to formation of biaryls. A mixture of 2-bromopyrimidine with 3-fluoropyridine *N*-oxide **250a**,

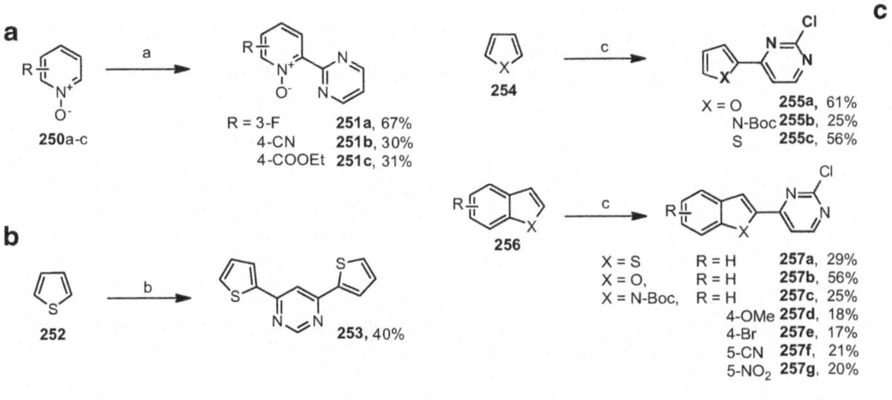

Scheme 72 Direct metalation and Negishi coupling of heterocyclic substrates (*a*) TMPZnCl·LiCl, 2-bromopyrimidine, $PdCl_2$(dppf)·CH_2Cl_2, THF, rt to 60°C (*b*) (1) $ZnCl_2$·TMEDA (0.5 eq.), Li (tmp), THF, rt, 2 h; (2) heteroaryl chloride, $PdCl_2$ (2 mol%), dppf (2 mol%), reflux 24 h (*c*) (1) LiTMP. THF, −78° or −90°C, 1 h; (2) $ZnCl_2$·TMEDA, rt. 1 h; (3) $PdCl_2$ (2 mol%), dppf (2 mol %), heteroaryl chloride; (4) Hydrolysis

4-(methoxycarbonyl)pyridine *N*-oxide **250b** or 4-cyanopyridine *N*-oxide **250c** was initially treated with metalation agent and subsequently $PdCl_2$(dppf)-CH_2Cl_2 was added. The resulting biaryls **251a–c** were isolated in 67%, 31%, and 30%, respectively. Compared to other investigated heterocycles, pyrimidines turned out to be difficult substrates to couple (Scheme 72a).

The same conclusion can be drawn from the work of Snégaroff et al. [103], who have developed a deprotonative metalation using amino/alkyl mixed lithium-zinc combinations. When thiophene was subjected to the metalation conditions with $ZnCl_2$·TMEDA and Li(tmp) and subsequently reacted with 4,6-dichloropyrimidine, the desired 4,6-(dithiophen-2-yl)pyrimidine **253** was isolated in 40% yield (Scheme 72b).

Seggio et al. investigated deprotonative lithiation of several heterocycles followed by transmetalation with $ZnCl_2$·TMDEA and cross-coupling with (hetero)aryl chlorides [104]. Furan, thiophene, and their benzo-fused analogs were metalated and reacted with 2,4-dichloropyrimidine. All couplings proceeded into position 4 of pyrimidine. In general, fused heterocycles provided comparable yields. 2-Chloro-4-(2-furyl)pyrimidine **255a** was obtained in 61%, while the yield of 2-chloro-4-(2-benzofuryl)pyrimidine **257b** was slightly lower with 56%. 2-Chloro-4-(2-thienyl)pyrimidine **255c** was isolated in 56% yield as well and analogous 2-chloro-4-(2-benzothiophenyl)pyrimidine **257a** in rather poor 29% yield (Scheme 72c).

The scope of the transformation was extended in a follow-up work [105]. Boc-protected imidazole, benzimidazole, and variously substituted benzimidazole derivatives were coupled as well. However, obtained yields did not exceed 25%, independently on the electronic properties of the substrate (Scheme 72c).

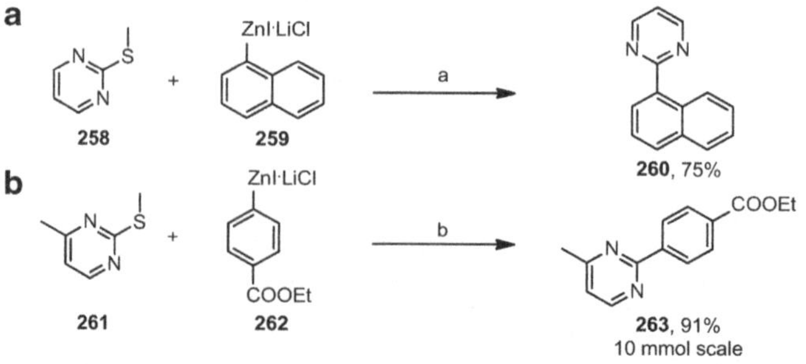

Scheme 73 Desulfitative Negishi coupling of methylthiopyrimidines; (*a*) Pd_2dba_3 (0.25–2 mol%), XPhos (1–8 mol%), THF, 70°C; (*b*) $Pd(OAc)_2$ (2.5 mol%), SPhos (5 mol%), THF, 25°C

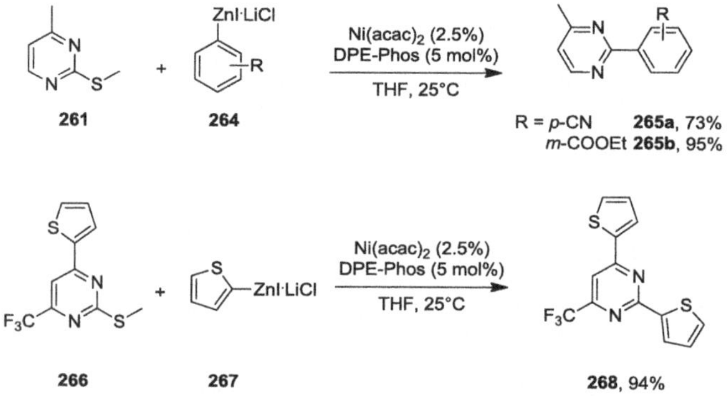

Scheme 74 Nickel-catalyzed Negishi cross-coupling, involving pyrimidine sulfides

Pyrimidine Sulfides in Negishi Reactions

More recently, coupling between arylzincates and aromatic sulfides was developed in the group of Knochel [106, 107]. The reaction is catalyzed by Pd(OAc)/SPhos as catalytic system, employing 2.5 mol% of catalyst and 5 mol% of ligand. Reaction between 2-thiomethylpyrimidine **258** and naphthylzinc iodide **259** yields 75% of **260** and reaction between 4-(ethoxycarbonyl)phenylzinc iodide **262** with 4-methyl-2-thimethylpyrimidine **261** provides 91% of the biaryl **263** (Scheme 73a, b).

Nickel-catalyzed coupling of arylzincates and aromatic thiols was reported as well, again by Knochel and coworkers [106, 107]. Employing $Ni(acac)_2$ and DPE-Phos leads to coupling of various 2-thiomethylpyrimidines with arylzinc chlorides in yields up to 95% (Scheme 74).

Besides $C(sp^2)$-nucleophiles, benzylic $C(sp^3)$-zincates were investigated as well [108–110]. Under palladium catalysis, diverse 2-methylthiopyrimidines (**258**, **261**, **272**) couple with substituted benzylzinc chlorides (**269**) in yields up to 94% under mild conditions (25–50°C, Scheme 75a, b, c). It has been shown that the reaction conditions are compatible with substrates bearing an ester functionality (Scheme 75c).

In addition, a method for selective decoration of the pyrimidine ring was described. Employing different starting materials **274** and **276** led to the formation

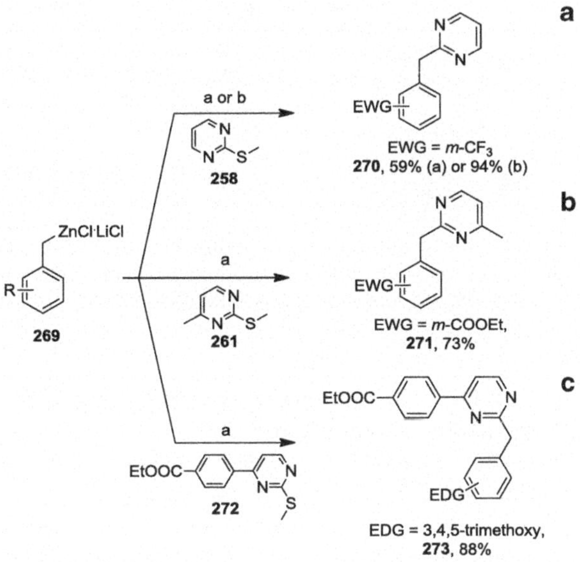

Scheme 75 Desulfitative Negishi coupling of benzylic zinc reagents; (*a*) $Pd(OAc)_2$ (2.5 mol%), SPhos (5 mol%), THF, 25°C; (*b*) $Pd(OAc)_2$ (2.5 mol%), SPhos (5 mol%), THF, 50°C

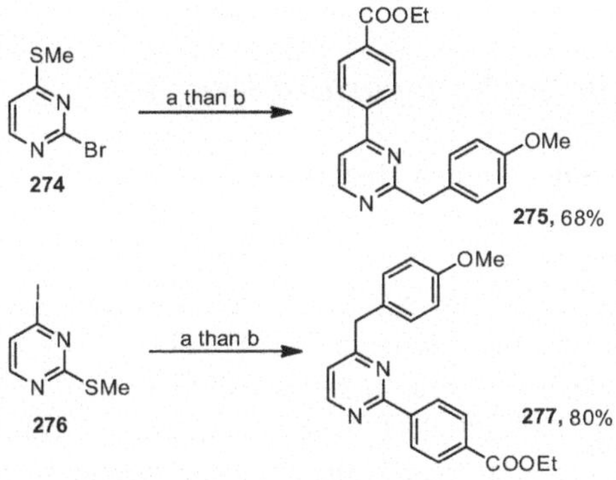

Scheme 76 Selective, sequential Negishi couplings on pyrimidine; (*a*) $Pd(dba)_2$ (2 mol%), TFP (5 mol%), 25°C, *p*-(methoxy)benzylzinc chloride; (*b*) $Pd(OAc)_2$ (2.5 mol%), S-Phos (5 mol%), 25°C, *p*-(ethoxycarbonyl)phenylzinc chloride

of regioisomers **275** and **277** via consecutive introduction of benzylic and aromatic moieties in yields of 68% and 80% over two steps (Scheme 76).

1.3.2 Pyrimidine as Organometal Species

In the same report of Snégaroff et al. discussed above (see section "Methodology") [103], pyrimidine **278** was treated with $ZnCl_2 \cdot TMEDA$ and Li(tmp) and subsequently coupled with 2-chloropyridine or 4-chloropyrimidine in the presence of

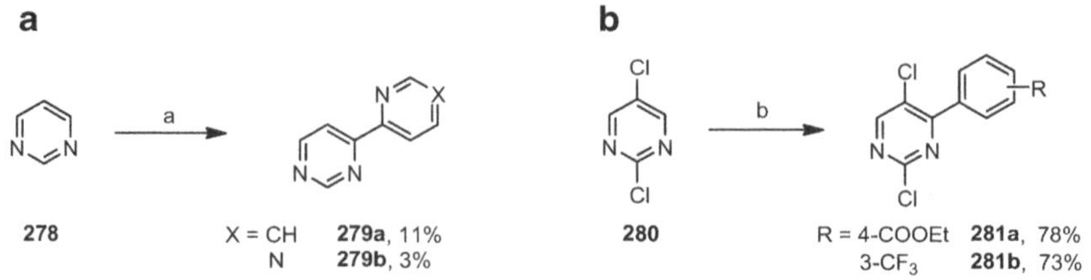

Scheme 77 Direct metalation and cross-coupling of pyrimidine; (*a*) (1) $ZnCl_2$·TMEDA (0.5 eq.), Li(tmp), THF, rt, 2 h, (2) heteroaryl chloride, $PdCl_2$ (2 mol%), dppf (2 mol%), reflux 24 h; (*b*) (1) TMP_2Zn·$2MgCl_2$·2LiCl, THF, 25°C, 45 min, (2) ArI, $Pd(dba)_2$ (3 mol%), $P(o\text{-furyl})_3$ (6 mol%), THF, 65°C, 45 min

$PdCl_2$ and dppf under reflux. The resulting products **279a** and **279b** were obtained in 11% and 3%, respectively (Scheme 77a).

A more successful method of direct regioselective metalation of 2,5-dichloropyrimidine **280** (into position 4) and subsequent Pd-catalyzed cross-coupling was reported by Morsin and Knochel [111]. Using TMP_2ZnCl_2·2LiCl at 25°C, followed by treatment with either ethyl 4-iodobenzoate or 3-3-iodotrifluoromethylbezene in the presence of $Pd(dba)_2$ and $P(o\text{-furyl})_3$, led to the coupled products **281a** and **281b** in 78% and 73% yield, respectively (Scheme 77b).

1.4 Pyrimidine in the Sonogashira Reaction

1.4.1 Copper-Free Sonogashira Reactions

Several Cu-free variants have been published throughout the last years. If performed in the ionic liquid BMIM tetrafluoroborate **284**, high yields in absence of Cu-salts have been obtained (Scheme 78a) [112].

Highly efficient und Cu-free coupling was also facilitated by tetradentate *N,N,N′,N′*-tetra(diphenylphosphinomethyl)-1,2-ethylenediamine **286**/$[Pd(C_3H_5)Cl]_2$. In the presence of this catalyst, 5-bromopyrimidine reacted with phenylacetylene **283** to give 98% yield after 20 h (Scheme 78b) [113].

Under conditions developed by Ngassa et al., a variety of aryl bromides and phenylacetylene reacted in the presence of $Pd(PPh_3)_4$, ligand 2-(di-*t*-butylphosphino)-1,1′-biphenyl **288**, and cesium carbonate. 2- and 5-bromopyrimidine gave the coupling products **289** in 82% and 65%, respectively (Scheme 78c) [114].

Recently, Pu et al. reported $[p\text{-}Me_2NC_6H_4(t\text{-}Bu)_2P]_2Pd(0)$ **292** and $[p\text{-}Me_2NC_6H_4(t\text{-}Bu)_2P]_2PdCl_2$ to be excellent precatalysts for Cu-free alkynylations [115]. Aryl bromides reacted with phenylacetylene in the presence of cesium carbonate within 4 h in acetonitrile at 80°C (e.g., 2-bromopyrimidine, 93%; 5-bromopyrimidine, 93%). Some aryl chlorides required DMF as a solvent, reaction temperatures 90–110°C and up to 8 h (e.g., methyl 6-chloronicotinate and

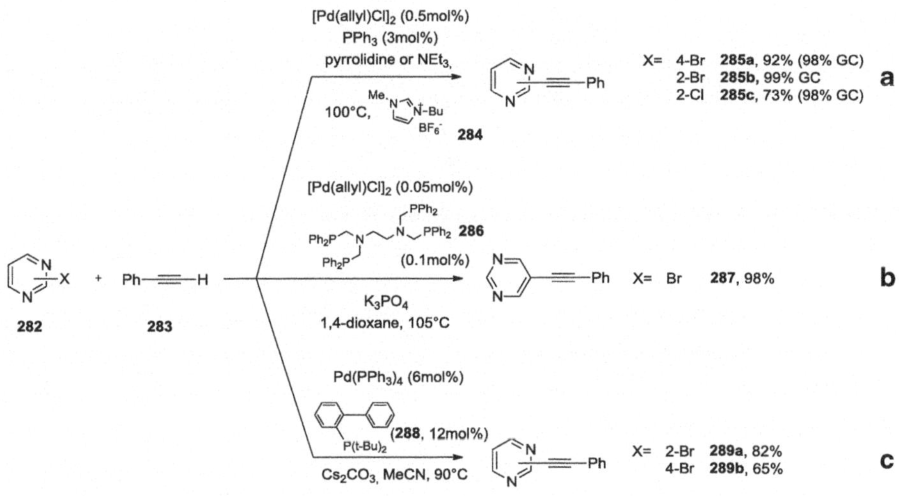

Scheme 78 Copper-free Sonogashira reactions of halopyrimidines

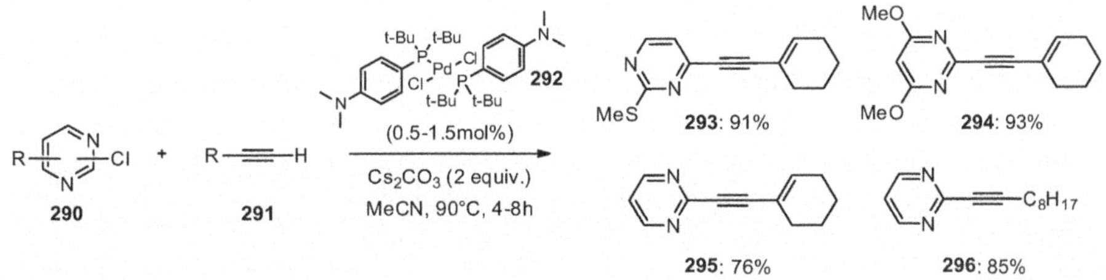

Scheme 79 Copper-free Sonogashira coupling of chloropyrimidines

1-decyne, 95%). However, chloropyrimidines **290** reacted smoothly in acetonitrile, yielding the coupling products **293–296** in 76–93% yield (Scheme 79).

1.4.2 Aqueous Sonogashira Reactions

Alkynylation of a number of pyrimidinylamidines **297** in the presence of $PdCl_2(PPh_3)_2$/DABCO was achieved in acetonitrile as solvent [116]. For substrates containing other heterocycles, coupling could also be carried out in aqueous medium (Scheme 80).

Fleckenstein and Plenio employed the doubly sulfonated and thus water-soluble fluorenyldicyclohexylphosphine ligand **5** for the coupling reaction of (hetero)aryl chlorides and bromides with a variety of terminal alkynes [117]. Using environmentally benign potassium carbonate as base and a water/isopropanol mixture as solvent, the reaction proceeded smoothly with excellent yields (e.g., 2-chloropyrimidine and 1-octyne, 94% yield of **300**) (Scheme 81a).

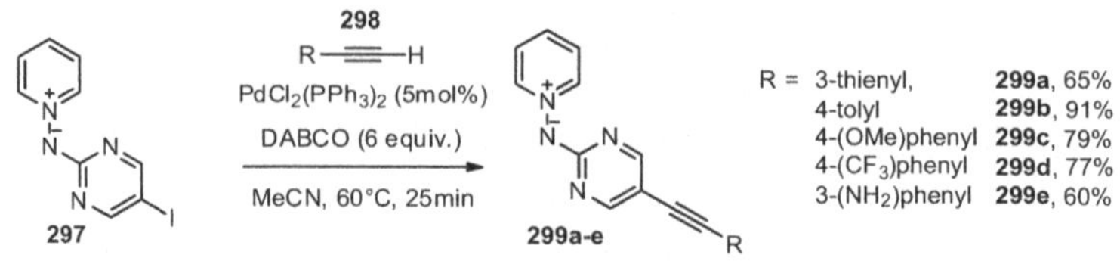

Scheme 80 Sonogashira coupling reactions of pyrimidinylamidines

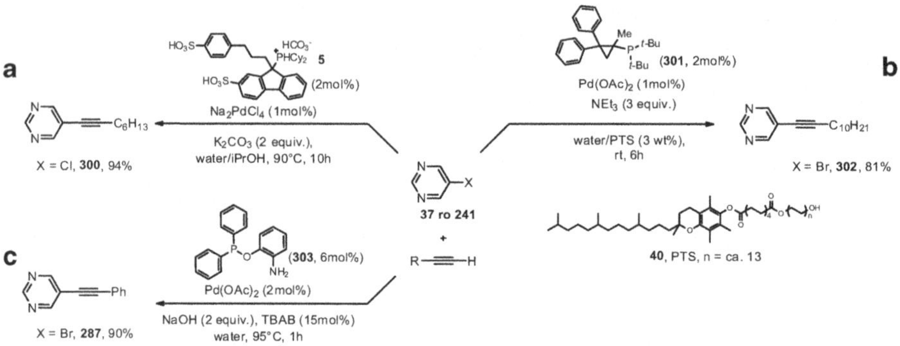

Scheme 81 Novel catalysts for the Sonogashira reaction in aqueous solutions

In order to overcome the problem of substrate solubility in aqueous reaction media, Lipshutz et al. used the nonionic amphiphile PTS as an agent for the formation of nano-micelles [118]. Among other catalysts Pd(dtbpf)Cl_2 and Pd $(OAc)_2$/cBRIDP **301**, a cyclopropylphosphine ligand catalyzed the coupling reaction efficiently at room temperature in the absence of a Cu source and gave excellent yields for most examples (5-bromopyrimidine and 1-decyne, 6 h, 81% of **302**) (Scheme 81b).

Successful coupling in neat water was achieved with the ligand 2-aminodiphenyl phosphinite **303** in a heterogeneous reaction mixture [119]. The catalyst could be recovered by centrifugation and was reused up to six times without significant loss of activity (Scheme 81c).

1.4.3 Ligand-Free Sonogashira Reactions

Several catalytic systems have been reported to couple halopyrimidines in absence of ligands (Scheme 82): The combination of PdEnCat™/KOH/EtOH represents an environmentally benign catalytic system [120]. Isolated yields vary widely (22–94%), with 52% for the reaction between 5-bromopyrimidine and phenylacetylene. The relatively harsh conditions limit the applicability of this method to robust coupling partners (Table 1, Entry 1).

An investigation of gelatin as supporting material for Pd nanoparticles showed that these particles constitute an effective catalyst in the presence of KOAc/TBAB

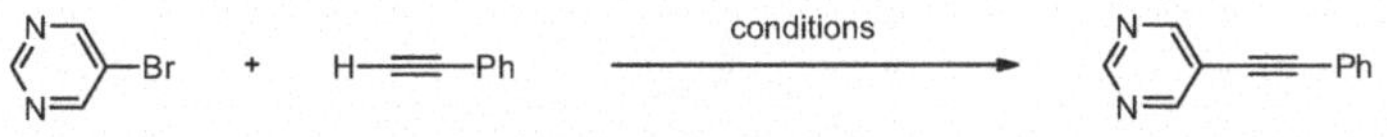

37 283 287

Scheme 82 Sonogashira reaction under ligand-free conditions

Table 1 Examples for the Sonogashira reaction under ligand-free conditions

Entry	Conditions	Time (h)	Yield (%)	References
1	PdEnCat™ 30 (2 mol%), KOH (2 equiv.), EtOH, reflux	24	52	[120]
2	Pd(0) nanoparticles on gelatin, KOAc (1.5 equiv.), TBAB, 100°C	2	77	[121]
3	Fe_2O_3 nanoparticles (5 mol%), K_2CO_3 (2 equiv.), ethylene glycol, 125°C	72	76	[122]
4	CuI (10 mol%), 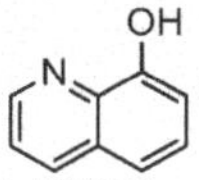 (20 mol%), Cs_2CO_3 (2 equiv.), DMF, 130°C	24	96	[123]

[121]. Among 17 examples, reactivity decreased in the order aryliodides (80–94% yield), bromides (74–85%), and chlorides (53–60%). 5-Bromopyrimidine was coupled with phenylacetylene in 77% (Table 1, Entry 2).

Under conditions developed by Hoseini and coworkers, coupling occurs in the presence of magnetite nanoparticles as the sole catalyst in ethylene glycol without ligand or Cu source present [122]. Due to their magnetic properties, catalyst particles can easily be separated from the reaction mixture and reused. Good to excellent yields were obtained after extended reaction times (e.g., 5-bromopyrimidine and phenylacetylene, 72 h, 76%) (Scheme 82) (Table 1, Entry 3).

8-Hydroxyquinoline was used as a bifunctional N,O-ligand in conjunction with CuI, allowing for coupling reactions to take place in the absence of Pd in moderate to very good yields (5-bromopyrimidine and phenylacetylene, 130°C, 24 h, 96%) [123] (Table 1, Entry 4).

1.4.4 New Catalysts

Beller and coworkers identified the indolylphosphine ligand CatacXium®PIntB **305** in conjunction with Na_2PdCl_4, CuI, and TMEDA as a powerful catalytic system, especially suited for the Sonogashira reaction of electron-rich heterocycles [124]. Nevertheless, coupling of 5-bromopyrimidine with trimethylsilylacetylene gave the coupling product **306** in only 77% yield (Scheme 83).

The oxime-based Pd-cycle **50**, which is also a competent catalyst in Suzuki–Miyaura, Heck, Stille, and Kumada–Corriu–Tamao coupling reactions (see also section "Water as Reaction Medium"), was successfully employed for the coupling

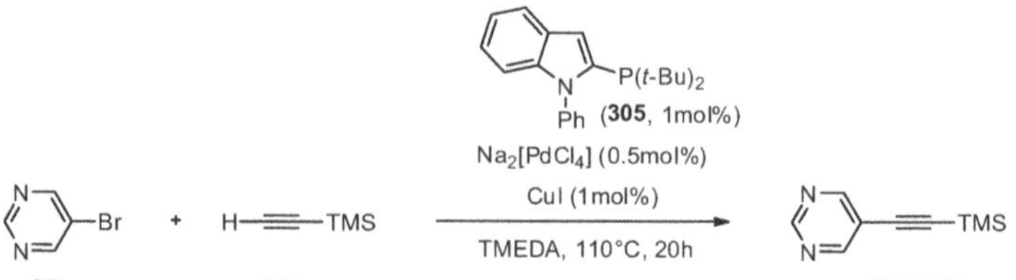

Scheme 83 Application of CatacXium®PlntB ligand

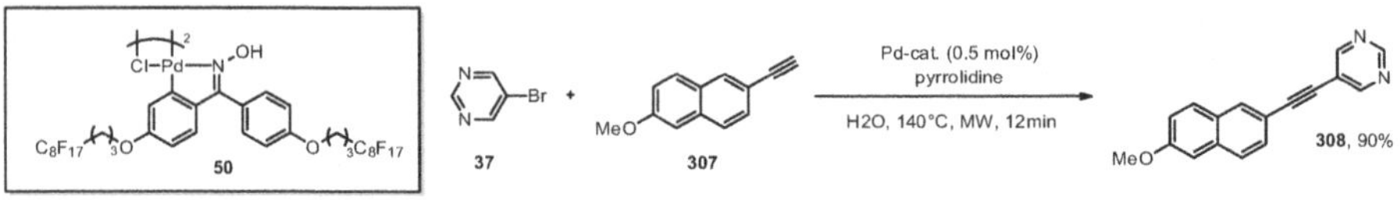

Scheme 84 Coupling catalyzed by an oxime-based palladacycle

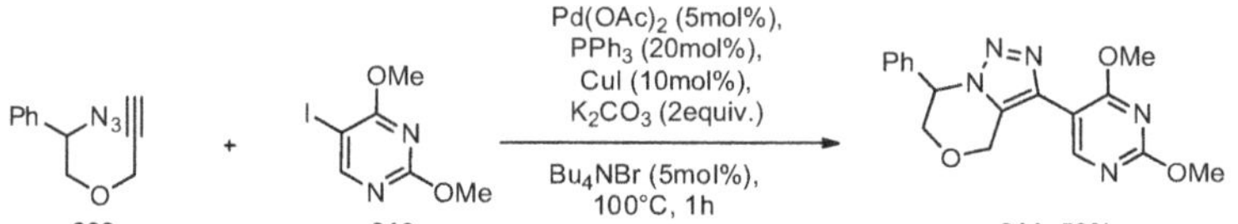

Scheme 85 Sonogashira-click cascade

of 5-bromopyrimidine and 2-ethynyl-6-methoxynaphthalene **307** (90% yield) [23, 24] (Scheme 84).

1.4.5 New Reactions

Achari and coworkers combined a Sonogashira coupling reaction with an azide-click reaction in a one-pot protocol to synthesize a library of triazole-fused morpholines in 53–90% yield [125]. The combination of $Pd(OAc)_2/PPh_3/CuI$ proved effective, catalyzing the coupling reaction as well as the heteroannulation reaction. The authors showed that the corresponding acyclic intermediate coupling products can be isolated when the reaction is run in the absence of CuI. They therefore proposed a mechanism, where coupling occurs first, followed by the copper(I)-catalyzed heteroannulation. When 2,4-dimethoxy-5-iodopyrimidine **310** was employed as coupling partner, the fused ring system **311** was obtained in 59% yield after 1 h at 100°C (Scheme 85).

Gendron et al. described a novel protocol for the synthesis of diarylidene acetone **313** via a Sonogashira/isomerization reaction of propargylic alcohols **312** [126]. Coupling was effected with $PdCl_2(PPh_3)_2/CuI/NEt_3$ in THF, heated to 120°C under microwave irradiation (Scheme 86).

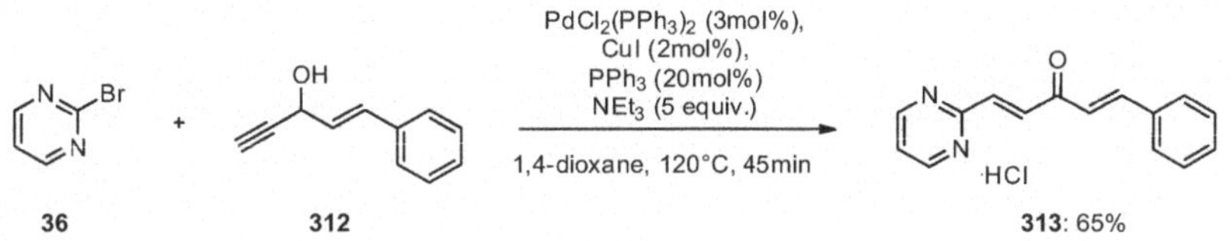

Scheme 86 Sonogashira/isomerization reaction

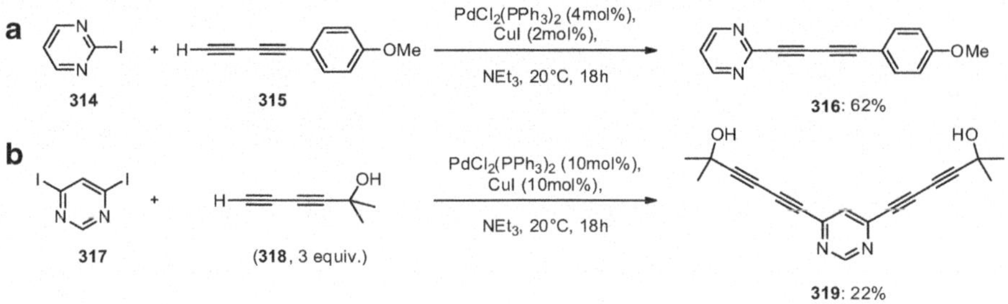

Scheme 87 Sonogashira coupling of diynes

1.4.6 Various

Bryce and coworkers reported the synthesis of terminal aryldiynes [127]. Heteroatoms on the aryl group lead to a significant increase in stability of these compounds due to H-bonding with the terminal alkyne proton. Unlike other compounds of the same class, the reported examples could be isolated and purified. As a prototypical example of how these compounds can be employed, the coupling of 2-iodopyrimidine **314** with (4-methoxy)phenylbutadiyne **315** under classical Sonogashira conditions was carried out in 62% yield. In an effort to synthesize oligo-butadiynes as molecular wires, 2-hydroxypropyl-protected butadiyne **318** was reacted with 4,6-diiodopyrimidine **317** under similar conditions and afforded the bis-coupled product **319** in 22% yield [128] (Scheme 87).

Plé and coworkers synthesized a number of dialkynylpyrimidines as precursors for organic electronic materials starting from the diiodopyrimidines **320** [49]. However, coupling of 2,4,6-triiodopyrimidine failed under Sonogashira conditions, presumably due to the limited stability of the starting material. This problem was circumvented by employing 2,4,6-trichloropyrimidine instead, which was successfully alkynylated under Suzuki–Miyaura or Negishi conditions (see section "Coupling Sequences") (Scheme 88).

Sequential coupling of 2,4,5,6-tetrachloropyrimidine with arylacetylene gave rise to di-, tri-, and tetraynepyrimidines with a high degree of regioselectivity [129]. The obtained products showed promising fluorescence properties (Scheme 89).

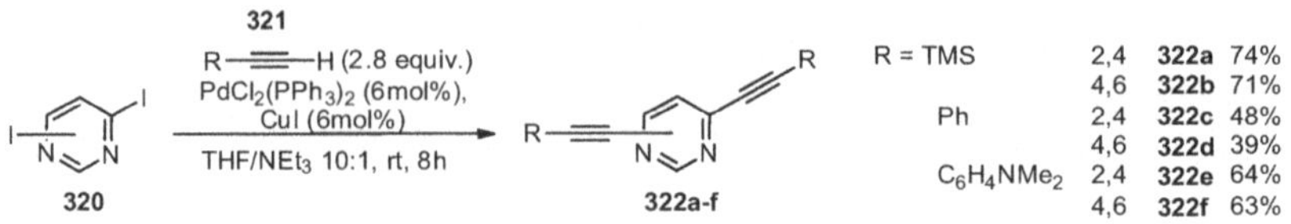

Scheme 88 Double alkynylation of diiodopyrimidines

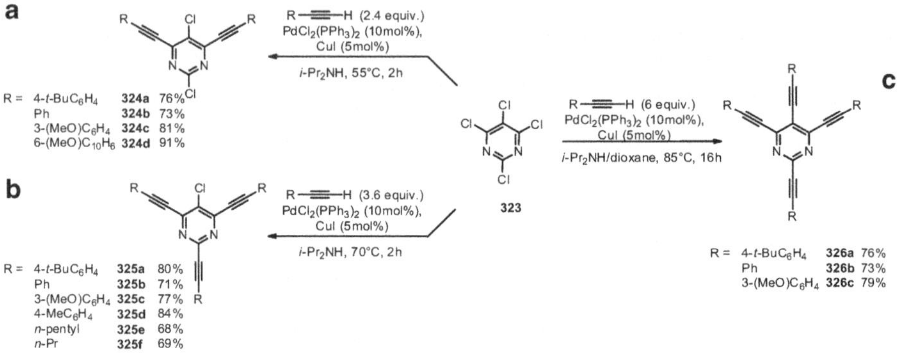

Scheme 89 Regioselective di-, tri-, and tetraalkynylation of tetrachloropyrimidine

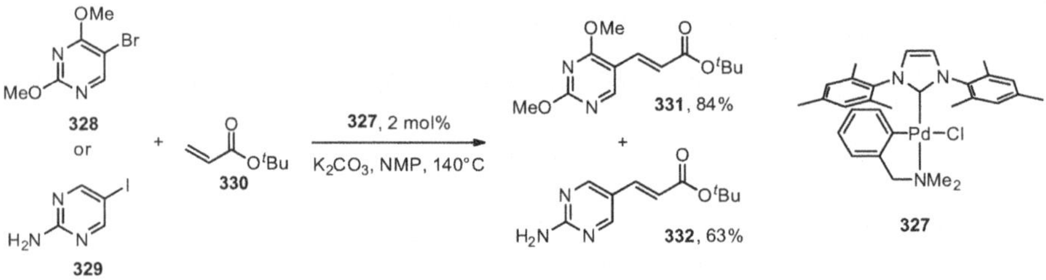

Scheme 90 NHC-based palladium catalyst for the Heck reaction and its application in coupling involving various pyrimidines

1.5 Pyrimidine in the Heck Reaction

1.5.1 Catalyst Development

Several new catalytic systems were developed for the Heck reaction. Ying and coworkers have developed an NHC-ligated palladacycle **327** which can be prepared in a simple one-pot three-component reaction starting from *N,N*-dimethybenzylamine, $PdCl_2$, and IMes·HCl [130, 131]. This precatalyst can be successfully utilized in the Heck reaction between *t*-butylacrylate **330** and 5-bromo-2,4-dimethoxypyrimidine **328** or 5-iodo-2-aminopyrimidine **329**. At 140°C in NMP and in the presence of K_2CO_3, the reaction provided 84% and 63% yield, respectively (Scheme 90).

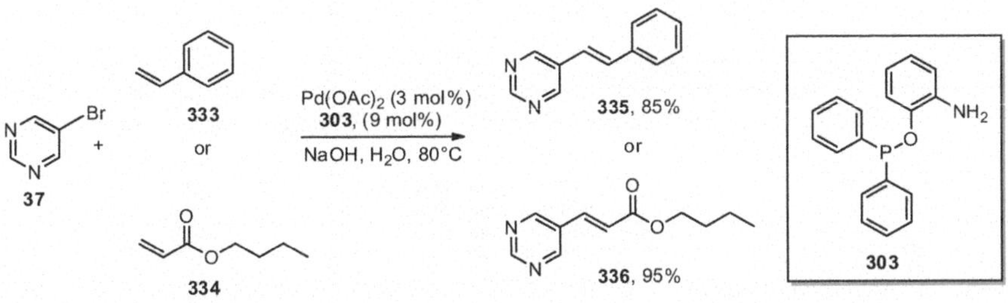

Scheme 91 Application of phosphinite ligand 303 in alkenylation reactions of 5-bromopyrimidine

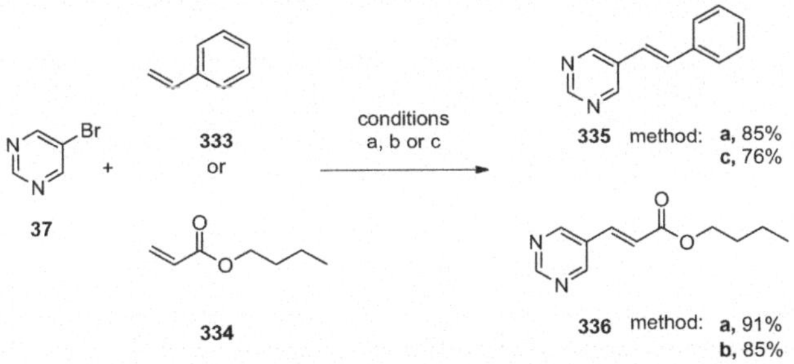

Scheme 92 Gels as support for Pd nanoparticles; (*a*) $PdCl_2$ on gelatin, $n$$Pr_3N$, neat, 140°C, 100–200 min; (*b*) Pd(0) on SDPP, $n$$Pr_3N$, neat, 130°C, 100 min; (*c*) $Pd(OAc)_2$ on agarose, Et_3N, 120°C, 12 h

The novel ligand 2-aminophenyldiphenylphosphinite **303** was synthesized from 2-aminophenol and chlorodiphenyl phosphine and evaluated in Heck coupling by Gholinejad et al. [132]. Its utility has been demonstrated in reactions between (hetero)aryl halides and either styrene **333** or *n*-butylacrylate **334**. Using 3 mol% of $Pd(OAc)_2$ and 6 mol% of ligand in water and with NaOH or Cs_2CO_3 as base, **37** provided 85% yield of product **335** and 75% yield in case of 5-*n*-butylacrylate-pyrimidine **336** (Scheme 91).

Several nanoparticle-based catalysts were developed by the Iranpoor group (Scheme 92). The effect of different solid supports for palladium nanoparticles was investigated. In the earliest work, nanoparticles were deposited on gelatin, a nontoxic, degradable, and easily accessible natural product support [133]. Heck reaction could then be conducted under solvent-free conditions at 140°C using tri (*n*-propyl)amine as base. 5-Bromopyrimidine **37** was coupled with styrene **333** or *n*-butylacrylate **334**, providing 91% and 85% yield, respectively.

Silicadiphenylphosphinite (SDPP) is another solid support for palladium nanoparticles which was investigated [134]. Its utilization in the Heck coupling reaction between **37** and **334** under the defined conditions (solvent-free, 130°C, *n*-Pr_3N) provided 85% of the desired product.

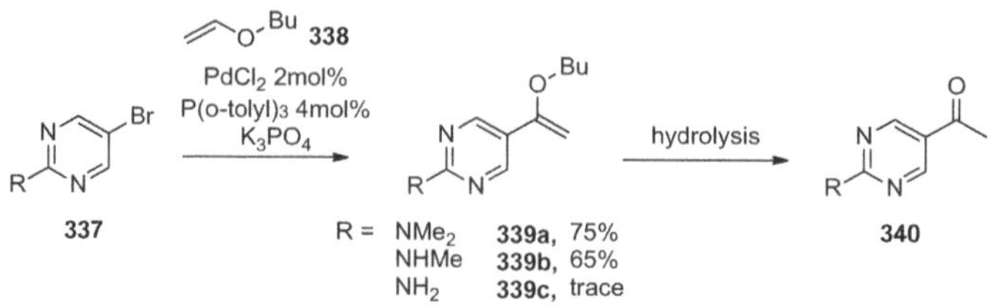

Scheme 93 Acylation of arenes: Pd-catalyzed vinylation followed by hydrolysis

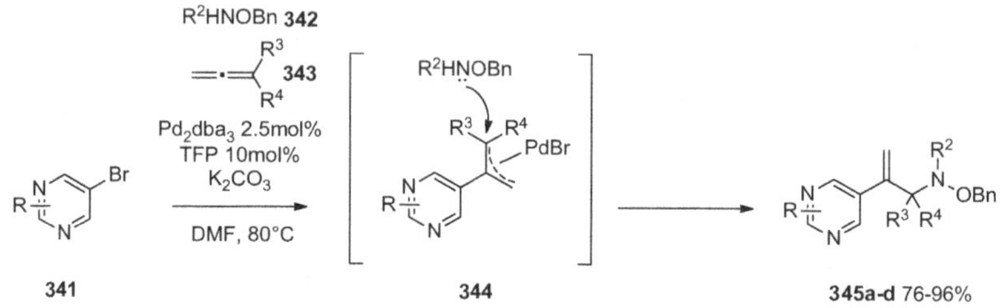

Scheme 94 Trapping of allene-derived Pd-allyl complexes with hydroxylamines

The last investigated support material was agarose [135]. Even though it was possible to lower the temperature to 120°C, a slight drop of the reaction yield in the reaction between **37** with **333** (solvent-free, Et_3N) to 76% was observed. Moreover, significantly longer time was required for the completion of the reaction than in case of gelatin (12 h in case of agarose vs. 200 min in case of gelatin).

1.5.2 Synthetic Methods

He et al. have developed a method for preparation of acetylated heteroaryls via Heck reaction of heteroaryl bromides with vinyl(*n*-butyl)ether and subsequent acid hydrolysis of the Heck products [136]. Substituted 2-amino-4-bromopyrimidines (**337**) were evaluated as substrates for this transformation. Under the reaction conditions, 75% of desired 5-acetyl-2-(*N*,*N*-dimethylamino)pyrimidine **339a** was isolated. The drop of the yield to 65% (**339b**) was observed when monomethylated 5-bromo-2-(*N*-methylamino)pyrimidine was used. Only traces of the product were detected when 4-bromo-2-aminopyrimidine was coupled (Scheme 93).

Elboray et al. have developed a palladium-catalyzed three-component reaction between (hetero)aryl halides, allenes, and protected hydroxylamines, formally consisting of Heck and Tsuji–Trost allylation [137]. In the first step, (hetero)aryl halide **341** couples with allene **343**, providing η^3-coordinated allylic species **344**, which undergoes a nucleophilic attack of hydroxylamine giving rise to 1,1-disubstituted olefines **345**. Several pyrimidine-based substrates were also submitted to this reaction, providing reaction yields in the range from 76% to 96% (Scheme 94).

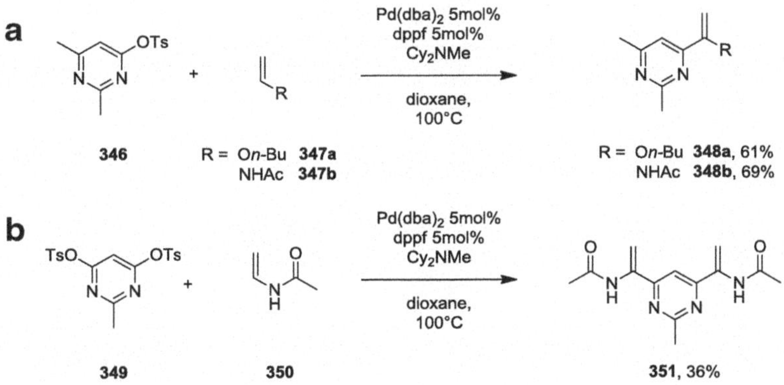

Scheme 95 Vinylation of pyrimidine tosylates

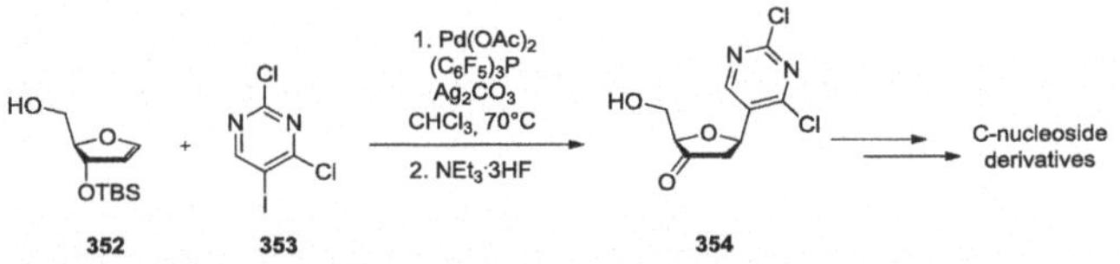

Scheme 96 Arylation of glycals towards artificial C-nucleosides

Skrydstrup and coworkers have investigated regioselective Heck coupling of heteroaryl tosylates with electron-rich olefins [138]. Several pyrimidine derivatives were coupled with electron-rich vinyl(*n*-butyl)ether or *N*-vinylacetamide, using Pd_2dba_3, Cy_2NMe as base, and dppf as ligand in dioxane at 100°C. All the couplings took place on the more substituted olefinic carbon. 2,4-Dimethyl-6-tosylpyrimidine **346** provided 61% of desired product when coupled with vinyl(*n*-butyl)ether **347a** and 69% when *N*-vinylacetamide **347b** was used. 2-Methyl-4,6-ditosylpyrimidine **349** underwent bis-olefination under the reaction conditions, providing product **351** in 36% yield (Scheme 95).

Heck reaction was utilized to synthesize artificial nucleosides by Kubelka et al. [139, 140] (Scheme **96**). Synthesis started with the reaction between 2,4-dichloro-5-iodopyrimidine **353** and 3′-TBS-protected glycal **352**. Under the reaction conditions ($Pd(OAc)_2$, $(C_6F_5)_3P$, Ag_2CO_3 in $CHCl_3$ at 70°C), partial desilylation was observed. Therefore, the Heck product was directly deprotected by the treatment with triethylamine trihydrofluoride. Reaction yielded acceptable 42% (after two steps) and could be successfully upscaled to 5 g. Halopyrimidine **354** was further elaborated towards artificial nucleosides (see also Sect. 1.8.4).

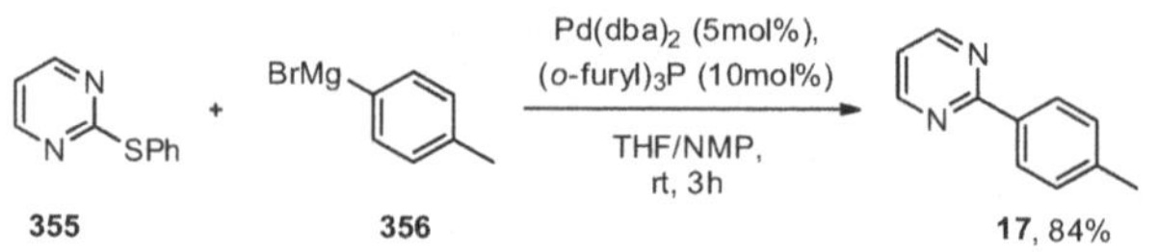

Scheme 97 Desulfidative variant of Kumada coupling

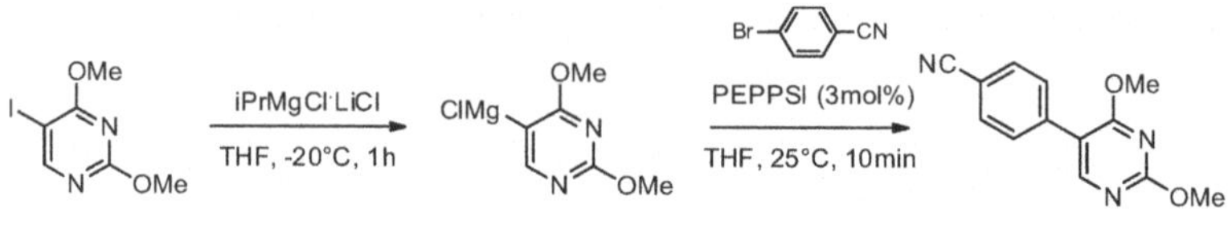

Scheme 98 Kumada coupling of iodopyrimidines, employing *i*-PrMgCl

1.6 Pyrimidine in Kumada–Corriu–Tamao Reactions

Van der Eycken and coworkers reported a development of a desulfitative variant of the Kumada coupling reaction [141]. Pd(dba)$_2$/(*o*-furyl)$_3$P efficiently catalyzed the addition of aryl Grignard reagents to thioesters and 2-heteroarylsulfides. 2-Phenylthiopyrimidine **355** reacted smoothly with 4-tolylmagnesium chloride to give the coupling product **17** in 84% yield after 3 h. In contrast to many other desulfitative couplings (i.e., Liebeskind–Srogl coupling [142, 143]), the reaction does not require a Cu source as an additive (Scheme 97).

Manolikakes and Knochel found that the rate of the Kumada coupling reaction was greatly enhanced, when reagents derived from *i*-PrMgCl via Mg/I-exchange were used [144]. Following this finding, the role of isopropyl iodide which is formed in the reaction was investigated. The authors reasoned that a change in the mechanism towards a radical mechanism was responsible for the increase in reaction rate. Among other examples, 5-iodo-2,4-dimethoxypyrimidine **357** was transformed into the corresponding Grignard reagent. Coupling with 4-bromobenzonitrile catalyzed by the PEPPSI catalyst (3 mol%) afforded the product **359** in 83% yield after only 10 min at room temperature(Scheme 98).

The complex (IPr)Ni(allyl)Cl **360** was employed as a catalyst for the Kumada coupling of aryl chlorides [145]. The reaction gave excellent yields for a range of products, but 2-chloropyrimidine performed worse than most substrates under the reaction conditions: when reacted with phenyl or tolyl Grignard reagent, it gave 72% or 74% yield, respectively. More importantly, the catalyst also proved to be competent in activating arylethers, but no examples employing pyrimidine were shown.

As a continuation of their work in the Suzuki–Miyaura cross-coupling reaction [146], Xi et al. reported nickel(II) complex **361** of a pyridine-functionalized NHC ligand to be highly active catalysts for the coupling of aryl- and vinyl chlorides in Kumada coupling [147]. Products were obtained in very good to excellent yields after 12 h at room temperature. When 2-chloropyrimidine was reacted with

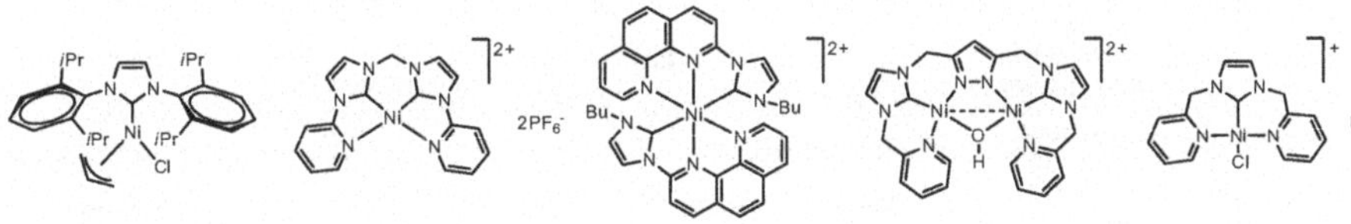

360 361 362 15 363

Scheme 99 New nickel-based complexes, used to catalyze Kumada coupling of pyrimidines

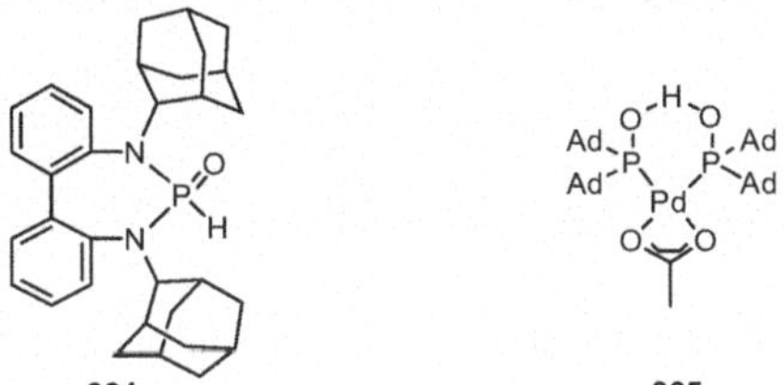

364 365

Scheme 100 New phosphine oxide-based ligands

o-tolylmagnesium bromide, 89% of the coupling product could be isolated. Also unsymmetrical pincer complex **362** showed a comparable activity [148]. Binuclear nickel complex **15** proved to be highly active for the coupling of 2-chloropyrimidine with tolylmagnesium chloride [4]. Even the sterically demanding *o*-tolylpyrimidine coupling product could be isolated in quantitative yield after 12 h at rt. Complex **15** also catalyzes the Suzuki–Miyaura coupling reaction (see section “Catalyst Design”).

Ni(II)-complex **363**, which was derived from a trinuclear Cu-complex, lead to a further improvement in reaction efficiency, especially for chloro-N-hetercyclic coupling partners: 2-chloropyrimidine gave a quantitative yield of the coupling product when reacted with *p*-tolylmagnesium bromide [149] (Scheme 99).

Schulzke and coworkers developed a coupling protocol employing bisadamantylphosphine oxide **364** as ligand which allowed for the coupling of 2-pyridinylmagnesium bromide [150]. Using $Pd(dba)_2$/ligand (1 mol%), good to excellent yields were achieved within 20 h at 60°C in THF. 5-(Pyridin-2-yl) pyrimidine was isolated in 81% yield.

Secondary phosphine oxides (SPOs) **365** (Scheme 100) were also used as air-stable preligands for nickel(II) by Fang and coworkers [151]. During the formation of the active catalyst, the preligand undergoes a P–O hydride shift and forms phosphorous-bound Ni-phosphite complex. Very good to excellent yields were achieved in the coupling of aryl chlorides, fluorides, and tosylates. When 4,6-dichloropyrimidine was reacted with 2 equiv. of phenylmagnesium bromide, the bisarylated product was obtained in 88% yield after 20 h at room temperature.

The group of Skrydstrup reported – among other heterocyclic substrates – the employment of pyrimidine tosylates and phosphonates (**372**) in iron-catalyzed cross-coupling reactions with alkyl Grignard reagents [152]. Coupling occurred at −10 to −15°C within minutes and very good to excellent yields were achieved. As catalyst, either $Fe(acac)_3$, $FeCl_3$, or a Fe-salen complex was employed. Pyrimidine

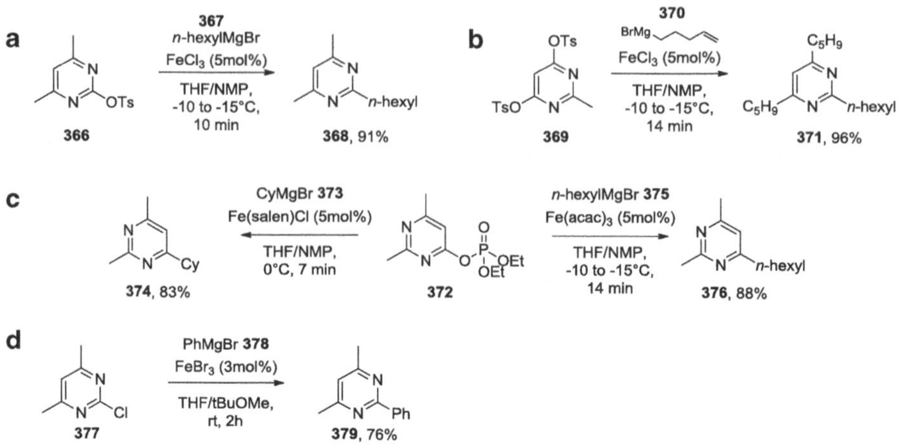

Scheme 101 Utilization of iron catalysts in Kumada reaction on pyrimidines

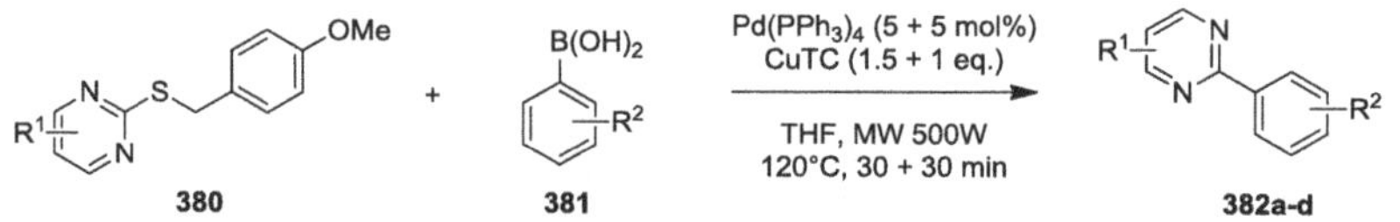

382a	R^1 = H	R^2 = 4-OMe	73%
382b	H	4-Me	85%
382c	4,6-dimethoxy	4-OMe	88%
382d	4,6-dimethoxy	4-Me	73%

Scheme 102 MW-accelerated LSC utilizing *p*-methoxybenzyl sulfide as leaving group

containing substrates performed particularly well in the reaction. 4,6-Dimethylpyrimidyltosylate **366** reacted with *n*-hexylmagnesium bromide in 91% yield and 2-methylpyrimidin-4,6-ditosylate **369** bis-coupled with 4-pentenylmagnesium bromide in 96% yield (Scheme 101a–c).

Later, Knochel and coworkers found that in a mixture of THF and MTBE, iron (III)bromide acts as a highly efficient catalyst for the arylation of N-heterocyclic halides [153]. 4,6-Dimethyl-2-phenylpyrimidine **379** was obtained in 76% after 2 h at room temperature, from corresponding 4,6-dimethyl-2-chloropyrimidine and **377** and phenylmagnesium bromide **378**.

1.7 Pyrimidine in Liebeskind–Srogl Coupling Reactions

Modha et al. investigated microwave-accelerated LSC utilizing *p*-methoxybenzyl sulfide as leaving group [154]. Several pyrimidine derivatives were subjected to the reaction conditions ($Pd(PPh_3)_4$, CuTC) and coupled with boronic acids. Reaction provided desired coupling products in yields between 73 and 88% (Scheme 102).

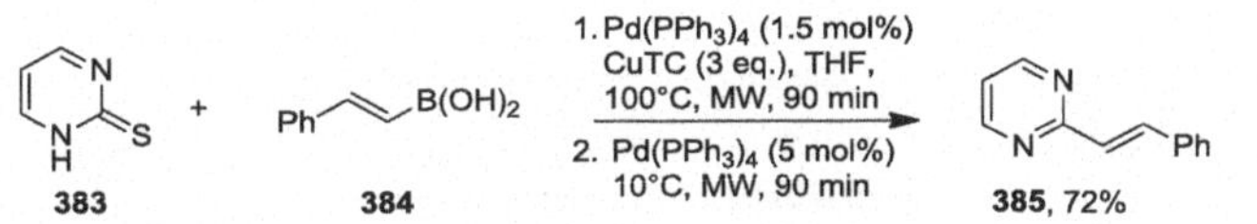

Scheme 103 MW-accelerated desulfitative LSC of pyrimidinethiones and vinylboronic acids

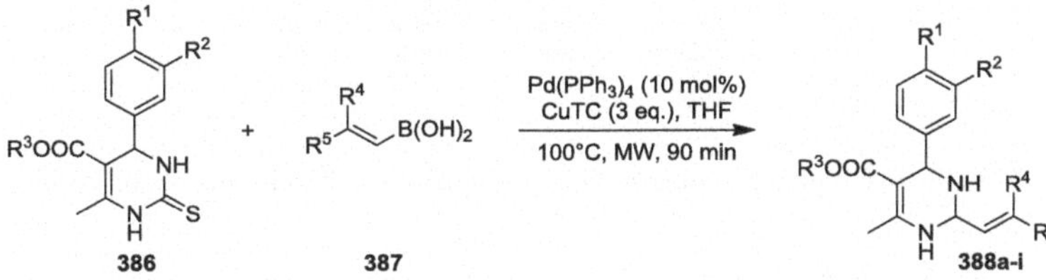

Entry	Dihydropyrimidine	Boronic acid	Product	Yield single mode [%]	Yield multimode [%]
1	($R^1 = R^2 = H, R^3 = Et$)	($R^4 = H, R^5 = Ph$)	**388a**	65	65
2	($R^1 = R^2 = H, R^3 = Et$)	($R^4 = H, R^5 = Ph$)	**388b**	64	66
3	($R^1 = R^2 = H, R^3 = Et$)	($R^4 = R^5 = Me$)	**388c**	55	60
4	($R^1 = R^2 = Me, R^3 = H$)	($R^4 = H, R^5 = Ph$)	**388d**	68	65
5	($R^1 = R^2 = Me, R^3 = H$)	($R^4 = H, R^5 = {}^tBu$)	**388e**	71	69
6	($R^1 = R^2 = Me, R^3 = H$)	($R^4 = R^5 = Me$)	**388f**	61	62
7	($R^1 = H, R^2 = OH, R^3 = Et$)	($R^4 = H, R^5 = Ph$)	**388g**	65	63
8	($R^1 = H, R^2 = OH, R^3 = Et$)	($R^4 = H, R^5 = {}^tBu$)	**388h**	60	67
9	($R^1 = H, R^2 = OH, R^3 = Et$)	($R^4 = R^5 = Me$)	**388i**	63	65

Scheme 104 MW-accelerated desulfitativ LSC of 3,4-dihydropyrimidine-2(1H)-thiones and vinylboronic acids

Another protocol, employing microwave irradiation upon the LSC of pyrimidine, was developed by Arshad et al. [155]. Various vinyl boronic acids could undergo desulfitative coupling with pyrimidinethiones in the presence of palladium catalyst and copper(I) cofactor. Under the optimized conditions, utilizing $Pd(PPh_3)_4$ and CuTC, styrylboronic acid **384** could be coupled with pyrimidine-2(*1H*)-thione **383** in 72% yield (Scheme 103).

Moreover, series of highly substituted 3,4-dihydropyrimidine-2(*1H*)-thiones **386** was examined in the LSC as well. In the reaction with substituted vinyl boronic acids **387**, desired products were obtained in 55–71% when single-mode microwave irradiation was applied and in 60–69% when reaction was carried out in multimode, demonstrating that the application of both modes is feasible and not significant difference is observed (Scheme 104).

Additionally, pyrimidine-2(*1H*)-thione **383** can undergo Liebeskind–Srogl type coupling with tributylstannanes **389a–d** and trimethoxyphenylsilane **390**. Desired products are obtained in yields between 60 and 80% (Scheme 105).

A reaction with an unusual regioselectivity was reported by Farahat and Boykin [156]. Submitting 2,4-dimethylthiopyrimidine **392** and diverse boronic acids or arylstannanes to the Liebeskind–Srogl conditions ($Pd(PPh_3)_4$, CuTC or CuMeSal) can lead potentially to three different products, 2-arylated **393**, 4-arylated **394**, or

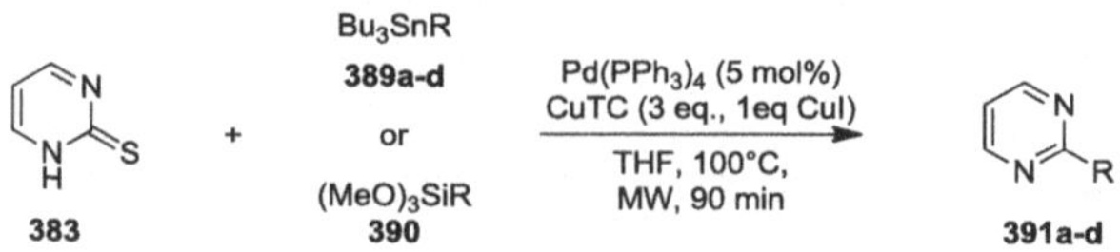

Entry	Reagent	R	Product	Yield [%]
1	**389a**	Ph	**391a**	80
2	**389b**	2-pyridzl	**391b**	72
3	**389c**	2-furyl	**391c**	69
4	**389d**	vinyl	**391d**	63
5	**390**	Ph	**391a**	60

Scheme 105 MW-accelerated desulfitative LSC of pyrimidinethiones and stannanes or silanes

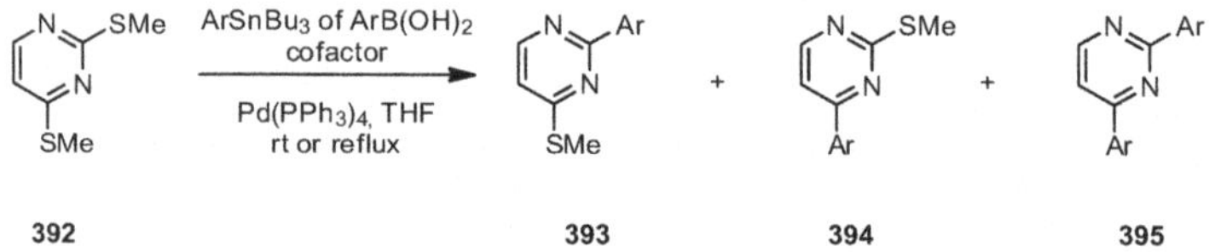

Entry	Reactant	Temperature	Cofactor	393 [%]	394 [%]	395 [%]
1	PhB(OH)$_2$	rt	CuMeSal	55	3	2
2			CuTC	1	0	0
3	4-MeOPhB(OH)$_2$	rt	CuMeSal	57	4	5
4			CuTC	2	0	0
5	4-CNPhB(OH)$_2$	rt	CuMeSal	59	4	5
6			CuTC	2	0	0
7	tributyl(2-furyl)stannan	rt	CuMeSal	63	0	6
8			CuTC	15	0	0
9	tributyl(2-thienyl)stannan	rt	CuMeSal	61	0	5
10			CuTC	20	0	0
11	PhB(OH)$_2$	reflux	CuMeSal	4	0	41
12			CuTC	3	3	49
13	4-MeOPhB(OH)$_2$	reflux	CuMeSal	2	0	70
14			CuTC	5	3	48
15	4-CNPhB(OH)$_2$	reflux	CuMeSal	4	0	69
16			CuTC	4	2	51
17	tributyl(2-furyl)stannan	reflux	CuMeSal	5	0	79
18			CuTC	2	0	55
19	tributyl(2-thienyl)stannan	reflux	CuMeSal	4	0	74
20			CuTC	3	0	52

Scheme 106 Unusual regioselective reaction of bismethylsulfanylated pyrimidines

bisarylated **395**. In spite of the usually observed enhanced reactivity of position four over position two in pyrimidine, the reaction provided mostly 2-arylated products at rt. When the temperature was increased to reflux (in THF), 2,4-bisarylated products were formed (Scheme 106).

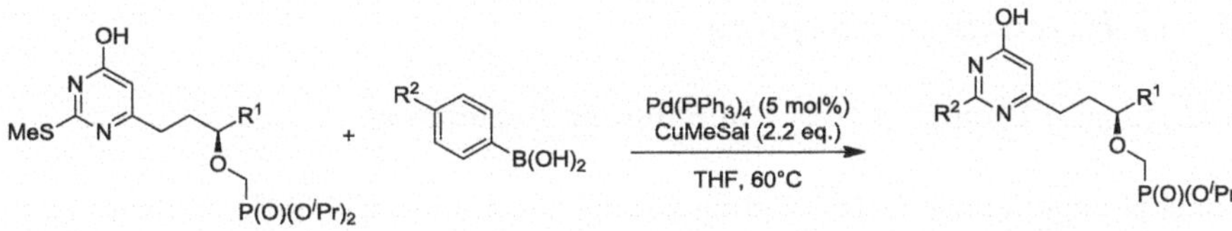

R = H, **396a**
R = *(R)*-CH_2OH **396b**

397a-l

398a-l

Entry	R^1	R^2	Product	Yield [%]
1	H	H	**398a**	89
2	H	F	**398b**	64
3	H	OMe	**398c**	57
4	H	NO_2	**398d**	56
5	H	CF_3	**398e**	50
6	H	OH	**398f**	41
7	*(R)*-CH_2OH	H	**398g**	75
8	*(R)*-CH2OH	F	**398h**	54
9	*(R)*-CH2OH	OMe	**398i**	60
10	*(R)*-CH2OH	NO_2	**398j**	41
11	*(R)*-CH2OH	CF_3	**398k**	65
12	*(R)*-CH2OH	OH	**398l**	63

Scheme 107 Synthesis of acyclic nucleoside phosphonates via LSC

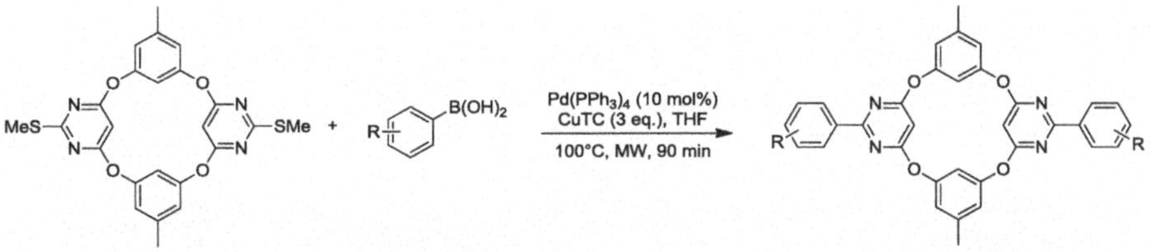

399

400a R = 4-^{t}Bu
400b 4-OMe
400c 4-CN
400d 2-Me

401a R = 4-^{t}Bu 68%
401b 4-OMe 78%
401c 4-CN 75%
401d 2-Me 76%

Scheme 108 Decoration of oxacalix[2]arene[2]pyrimidines via LSC

An efficient synthesis of acyclic nucleoside phosphonates via LSC was reported by Brehova et al. [157]. Optimization of the reaction conditions revealed CuMeSal to be superior to CuTC. Under optimized conditions, intermediates **396a**, **b** were coupled with several phenylboronic acid derivatives providing desired phosphonate esters in yields up to 89% (Scheme 107). Subsequent hydrolysis led to the desired phosphonates.

Post-macrocyclization functionalization of oxacalix[2]arene[2]pyrimidines via LSC was investigated by Van Rosson et al. [158]. Under the standard conditions employing $Pd(PPh_3)_4$ and CuTC, pyrimidine-based macrocycle **399** can be decorated with various phenylboronic acids with yields between 68 and 78% (Scheme 108).

1.8 C–Heteroatom Coupling

1.8.1 Ligand Promoted C–N Couplings on Pyrimidine

Shen et al. reported the C–N coupling of aliphatic amines with different heteroaryl halides in their comprehensive study of CyPF-*t*Bu ligand (**402**, also called JosiPhos) [159]. In general, JosiPhos, in combination with $Pd(OAc)_2$, mediated a coupling of various substrates with strikingly low catalyst loadings, as low as 0.0005 mol%. However, upon the couplings of 2- or 5-bromopyrimidines (**36 or 37**) with different electrophiles, catalyst loading had to be increased to 1 mol%. Coupled products **404** and **406** were obtained in yields up to 96% (Scheme 109).

Several phosphine ligands were developed in the group of Buchwald and tested in a variety of C–N couplings. Two new ligands were reported BrettPhos (Schemes 110 and 111) for couplings with primary amines and RuPhos for coupling with secondary amines (Scheme 112). Their application included coupling of pyrimidines with both aliphatic and aromatic amines [160]. A general advantage of these ligands over JosiPhos, reported by Hartwig, is a increase of reaction rate. Nevertheless, direct comparison of the three ligands in the C–N coupling on pyrimidines (and also in general) is not possible due to the fact that different

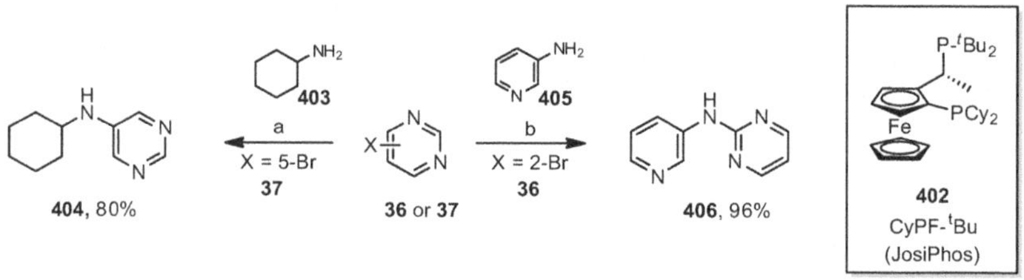

Scheme 109 Structure and application of Hartwig's JosiPhos (*a*) $Pd(OAc)_2$ (1 mmol%), CyPF-*t*Bu (1 mol%), 100°C or 110°C, 48 h or 20 h

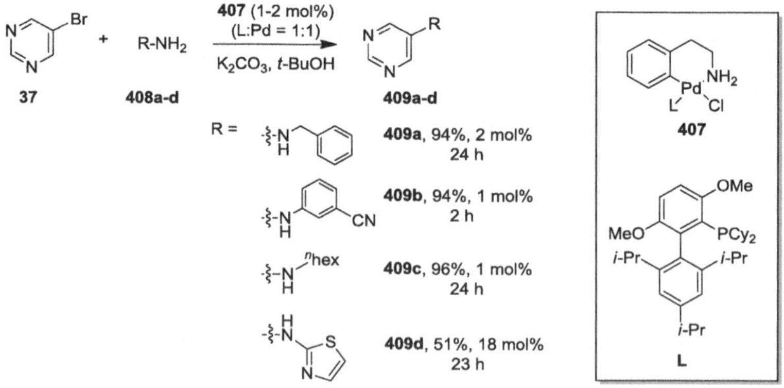

Scheme 110 Structure and application of Buchwald's BrettPhos

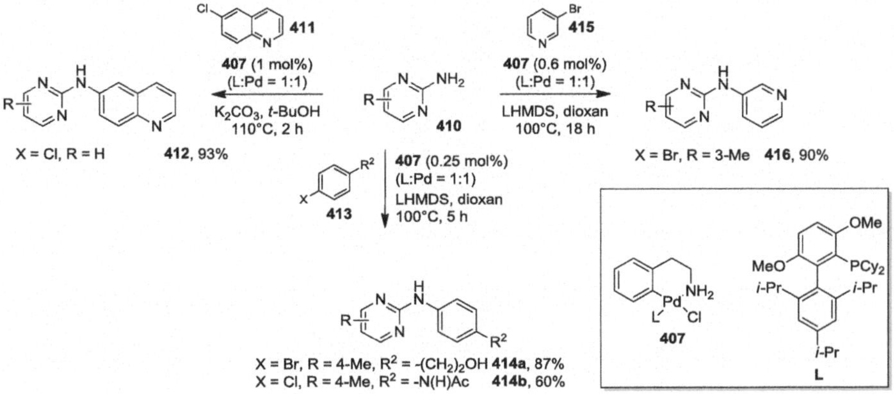

Scheme 111 Application of Buchwald's BrettPhos

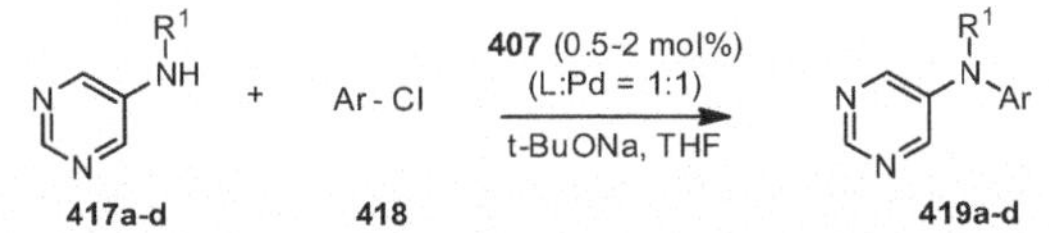

Entry	R^1	Ar	Product	mol %	Time [h]	Yield [%]
1	*n*-hexyl-	2-Py	**419a**	2	18	95
2	*p*-(ethoxycarbonyl)phenyl-	pyrazinyl-	**419b**	0.5	18	90[a]
3	benzyl-	6-quinolinyl-	**419c**	2	18	88
4	*p*-methoxybenzyl-	2-quinolinyl-	**419d**	2	12	75

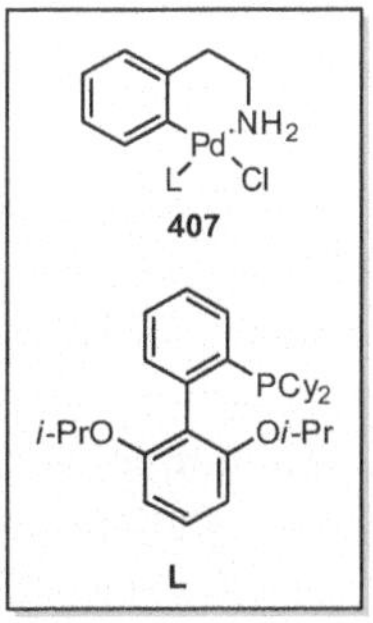

Scheme 112 Structure and application of Buchwald's RuPhos, [a]Cs_2CO_3/^{t}BuOH

substrates and conditions were applied. Moreover, when aliphatic benzylamine **408a** and *n*-hexylamine **408c** were utilized in the coupling with 5-bromopyrimidine **37** (Scheme 110), shortening of the reaction time to 24 h could be possible only at the cost of increasing the catalyst loading to 2 mol%. The reaction yielded 94% and 96% of desired products, respectively. When aromatic 3-cyanoaniline **408b** was used, 94% of product was isolated after 2 h, utilizing 1 mol% of catalyst. Coupling with 2-aminothiazole **408d** was also possible, but very difficult. 18 mol% of catalyst was necessary and the reaction provided only 51% of desired 5-*N*-(2-aminothizolo)pyrimidine **409d** (Scheme 110).

Additionally, 2-aminopyrimidine and 4-methyl-2-aminopyrimidine were investigated as nucleophiles. 0.25–1 mol% of catalyst loading was required for the coupling with different aryl chlorides. In the range of 2–18 h, reaction afforded 60–93% of desired products (Scheme 111).

A series of secondary amines (**417a-d**) was investigated as well [160]. They were coupled with different heteroaryl chlorides, providing yields from 75 to 95% with catalyst loadings in the range from 0.5 to 2 mol%. Reactions were completed within 12–18 h (Scheme 112).

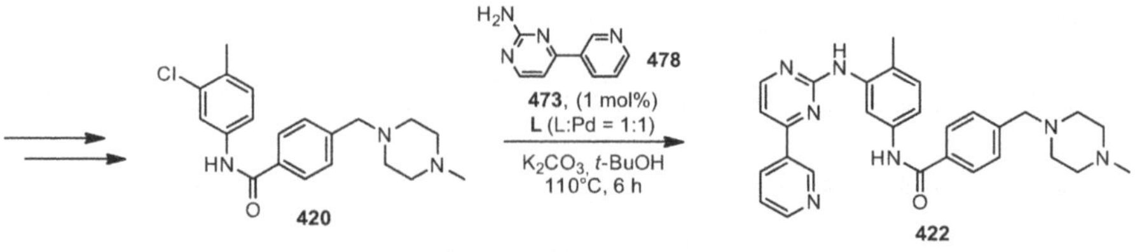

Scheme 113 Synthesis of Imatinib (Gleevec®) using Buchwald's BrettPhos

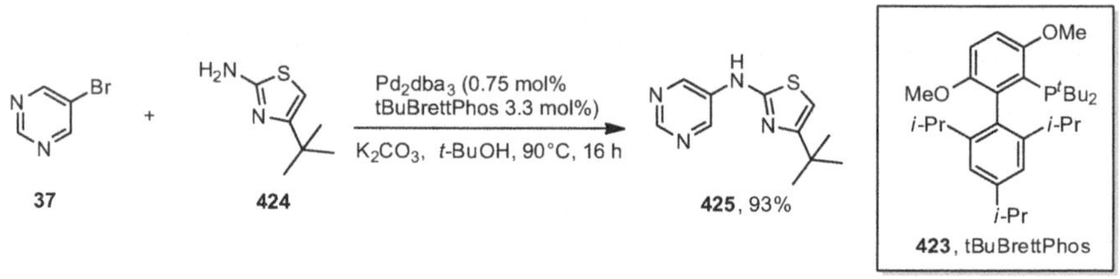

Scheme 114 *t*-BuBrettPhos as a ligand for aminothiazolization of pyrimidine

Finally, the methodology was applied to the synthesis of Imatinib (Gleevec®) **422**, a compound used for the treatment of chronic myelogenous leukemia (Scheme 113).

Another report from Buchwald's laboratory describes *N*-arylation of 2-aminothiazoles utilizing *t*-BuBrettPhos **423**, [161] a ligand initially developed for C–N coupling of amides [162]. The pK_A values of 2-aminothiazoles are closer to amides than to anilines or other (hetero)aromatic amines; therefore, application of **423** proved to be beneficial. Coupling of 4-*t*-butyl-2-aminothiazole **424** with 5-bromopyrimidine **37** catalyzed by $Pd(OAc)_2$ and **423** in water afforded 93% of product (Scheme 114).

Fors et al. showed that BrettPhos can be applied successfully for C–N bond-forming processes, starting from various amines and aryliodides [163]. It has to be mentioned that in Pd-catalyzed C–C cross-coupling reactions, (hetero)aryliodides are the most reactive aryl halides, whereas the same substrates are the least reactive in Pd-catalyzed C–N couplings. Employing BrettPhos turned out to be beneficial, and its performance is superior to the performance of some other common Buchwald-type ligands (DavePhos, XPhos, and RuPhos). Under the optimized conditions, 2-aminopyrimidine **427** coupled with 3,5-dimethyliodobenzene **426** in 90% yield. Employing 5-iodopyrimidine **241** in the coupling with aminopyrazine **430** led to the formation of the desired product in 86% yield (Scheme 115).

Udea et al. developed a method for regioselective (hetero)arylation of triazoles [164]. In the presence of Pd_2dba_3, Me$_4$*t*BuXPhos **440**, and K_3PO_4, coupling between 5-bromopyrimidine **37** and 4-(2,4-difluorophenyl)-2*H*-1,2,3-triazole **431**

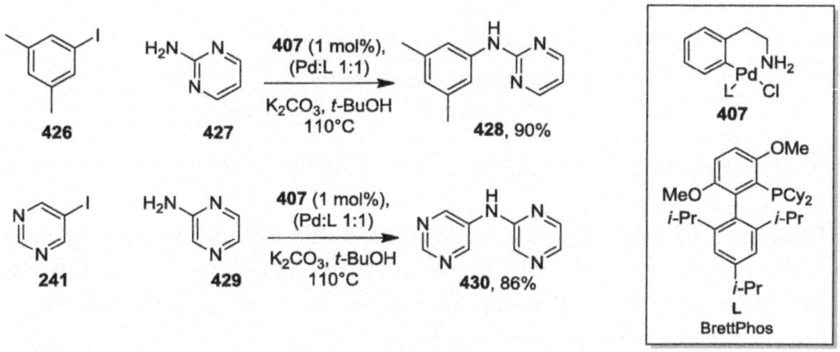

Scheme 115 Utilization of BrettPhos for C–N coupling of aryliodides

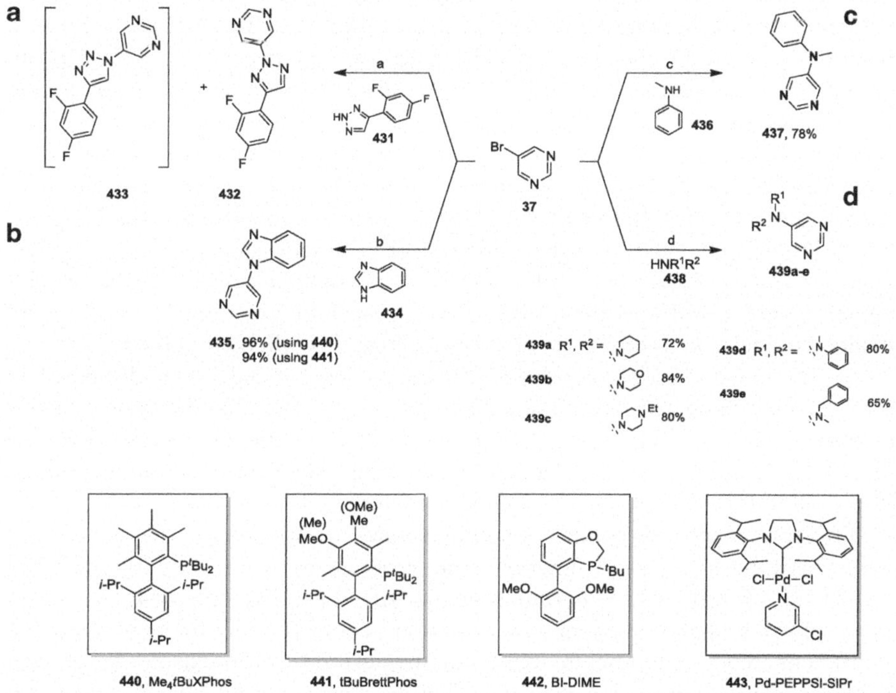

Scheme 116 (*a*) Pd_2dba_3 (0.5 mol%), Me$_4$tBuXPhos (**440**, 1.0 mol%), K_3PO_4, toluene, 120°C, 5 h; (*b*) Pd_2dba_3 (0.5 mol% Pd:L 1:1), Me$_4$tBuXPhos (**440**) or Me$_3$(OMe)tBuXPhos (**441**), K_3PO_4, toluene/dioxane (5:1), 120°C, 5 h; (*c*) Pd_2dba_3 (2.5 mol%), BI-DIME (**442**, 10 mol%), NaOt-Bu, toluene, 110°C, 20 h; (*d*) Pd-PEPPSI-SIPr (**443**, 4 mol%) , Cs_2CO_3, DME, 80°C, 24 h

was realized in 90% yield with 98:2 ratio of N^2 to N^1 substituted triazole (**432**:**433**, Scheme 116a).

In the report of Ueda et al., the synthesis and application of Me$_3$(OMe)*t*BuXPhos (**441**) is discussed [165]. **441** is described as a surrogate for Me$_4$*t*BuXPhos (**440**) previously reported as efficient ligand for various C–N bond formations.

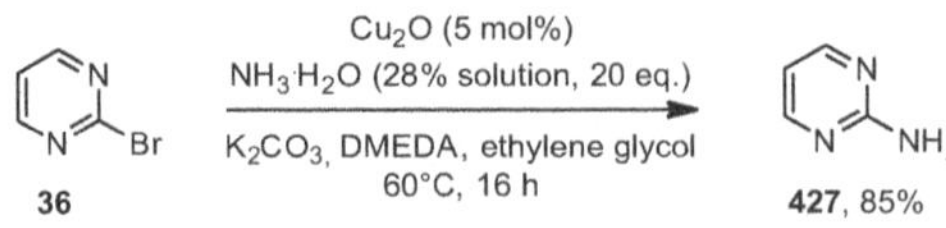

Scheme 117 Copper-catalyzed amination of 2-bromopyridine using aqueous ammonium

The performance of both ligands was comparable for all examined examples, involving the coupling of **37** with benzimidazole **434** (96% using **440** and 94% using **441**, Scheme 116b). The advantage of the new ligand **441** is a better availability of the precursors for its synthesis.

Rodriguez et al. described the synthesis and application of several oxaphosphole-based monophosphorus ligands [166]. Among them, ligand **442** shown excellent efficiency in C–N coupling of sterically hindered starting materials with low catalyst loadings (0.05–2.5 mol% Pd_2dba_3 and 0.3–10 mol% ligand). Reaction between **37** and *N*-methylaniline **436** was carried out as well. Herein, 2.5 mol% of catalyst together with 10 mol% of ligand were necessary, giving the final product **437** in 78% yield (Scheme 116c).

N-heterocyclic carbens proved to be very successful alternatives to phosphine ligands for various kinds of cross-couplings such as Suzuki–Miyaura or Buchwald–Hartwig amination [167]. Organ et al. [168] synthesized and characterized a series of new Pd-NHC precatalysts, where **443** (Pd-PEPPSI-*i*-Pr) showed the best performance in Buchwald–Hartwig amination. **37** was coupled with several secondary amines. Using 4 mol% of precatalyst, the corresponding coupling products were obtained in yields between 65 and 84% in 24 h at 80°C (Scheme 116d).

Several ligand classes were developed for a copper metal center. Elmkaddem et al. reported copper-catalyzed amination of 2-bromopyridine **36** using aqueous ammonium [169]. Employing Cu_2O and bidentate dimethylethylenediamine (DMEDA) as ligand, they were able to convert **36** into 2-aminopyrimidine **427** in 85% yield under relatively mild conditions (Scheme 117).

Different ligands with similar structural features, containing a pyridine moiety in close proximity to carbonyl function, were reported by Xi et al. [170] and Chen et al. [171]. Xi et al. utilized pyridine-functionalized 1,3-diketone (**444**) as a ligand for CuI. Such a catalytic system turned out to be a powerful tool for an introduction of different arenes and/or heteroarenes into the position 1 of imidazole. Employing 2-chloropyrimidine provided 96% of 2-*N*-imidazolopyrimidine **448a** (Scheme 118, entry 1). Among the tested substrates, 2-chloropyrimidine belonged to the best-performing (hetero)arens, even comparable to (hetero)aryl bromides. Coupling products served as precursors for the synthesis of NHC as ligands for a metal center.

Chen et al. investigated **445** as a ligand for copper (I) bromide. Application of such a catalytic system on the reaction between imidazole **447** and 5-chloro-2-bromopyrimidine provided 76% of 5-chloro-2-*N*-imidazolopyrimidine **448b** (Scheme 118, entry 2).

Cao et al. [172]. reported a catalytic system, based on CuI and hexamethyltetramine. This allowed the formation of 2-*N*-imidazolopyrimidine **448a** from

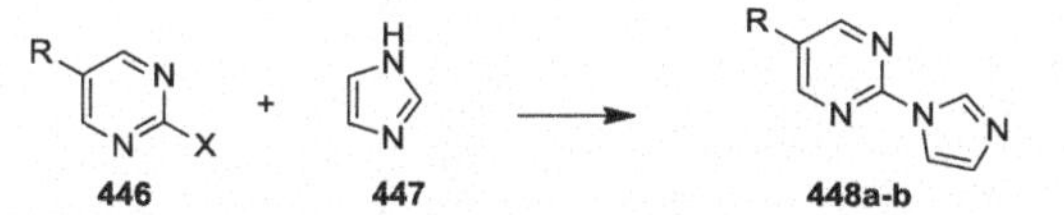

Entry	R	X	Ligand	Conditions	Product	Yield [%]
1	H	Cl	**444**	a	**448a**	96
2	Cl	Br	**445**	b	**448b**	76

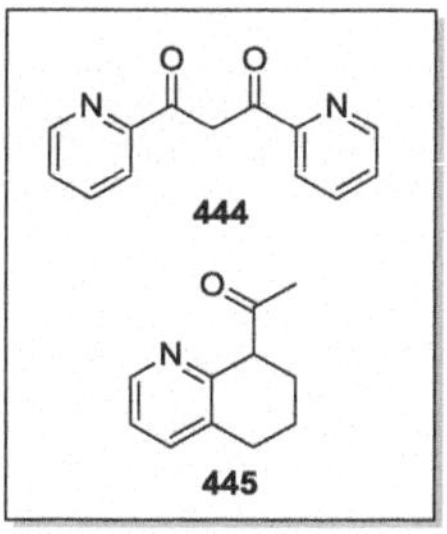

Scheme 118 Ligands for copper metal center for an effective Ullmann-type amination. (*a*) CuI (10 mol%), **446** (10 mol%), K_2CO_3, DMF, 110°C, 24 h, under N_2; (*b*) CuBr (5 mol%), 446 (10 mol %), Cs_2CO_3, DMSO, 60°C, 5 h, under N_2

Entry	X	Product	Ligand	Conditions	Yield [%]
1	Cl	**448a**	HMTA	a	96
2	Cl	**448a**	D-Glucseamine	b	95
3	Br	**448a**	per-6-ABCD	c	98

Scheme 119 Amination of pyrimidine, utilizing nanoparticles on solid support, (*a*) CuI (5 mol%), HMTA (5 mol%), K_2CO_3, DMF, 140°C, 20 h; (*b*) CuI (20 mol%), D-Glucosamine (40 mol%), Cs_2CO_3, 110°C, 24 h; (*c*) CuI (20 mol%), per-6-ABCD (10 mol%), K_2CO_3, DMSO, 110°C, 24 h

2-chloropyrimidene and imidazole in very good 96% yield. 5 mol% of CuCl and the additive were necessary, and compared to other tested substrates, 2-chloropyrimidine gave very good results (Scheme 119, entry 1).

Natural sugars or sugar derivatives were examined as environmentally friendly ligands for *N*-arylation of imidazoles by Cheng et al. [173]. Among the investigated catalytic systems CuI/D-Glucosamine promoted *N*-arylation most efficiently. Coupling between 2-chloropyrimidine with imidazole, catalyzed with this system, provided 95% of desired 2-imidazolopyrimidine **448a** (Scheme 119, entry 2).

Per-6-amino-B-cyclodextrine (per-6-ABCD) was reported by Surech and Pitchumani as another sugar-based ligand capable of enhancing the performance of copper catalysts [174]. Utilizing CuI/per-6-ABCD, 5-bromopyrimidine was coupled to position 1 of imidazole in 98% yield. Besides its role as a ligand for the copper metal center, per-6-ABCD acts as a cavity for transitory binding of the substrate as well as via combined electrostatic and hydrophobic interactions (Scheme 119, entry 3).

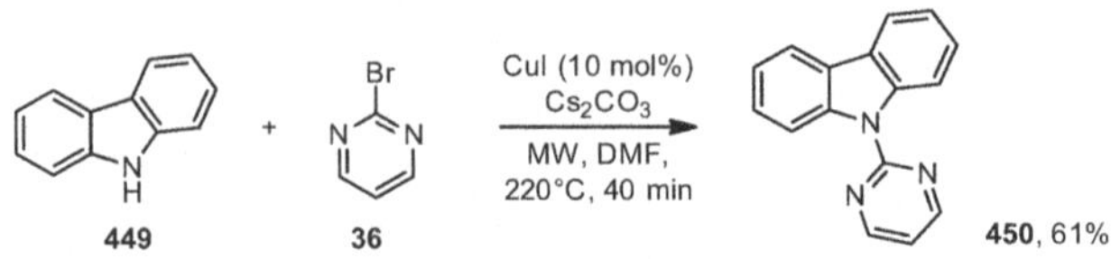

Scheme 120 Ligand-free *N*-(hetero)arylation of carbazole

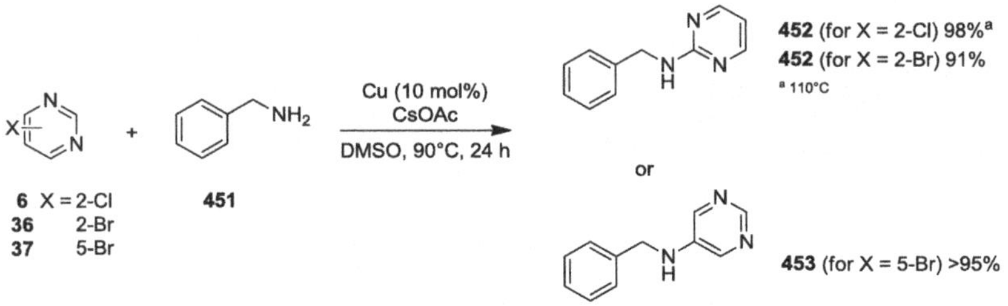

Scheme 121 Cu-catalyzed ligand-free amination of pyrimidine

1.8.2 Ligand-Free Copper-Catalyzed C–N Coupling of Pyrimidines

Several examples of ligand-free copper-catalyzed C–N bond-forming processes were described as well. Kwon et al. reported CuI-catalyzed, ligand-free microwave-accelerated *N*-(hetero)arylation of carbazoles with various (hetero) aryl halides [175]. A general drawback of the reported method is a very high temperature (220°C) and still long reaction time. Coupling of carbazole **449** with 2-bromopyrimidine **36** provided 61% of **450**. Compared to other investigated examples within the study, 2-bromopyrimidine turned out to be rather a difficult substrate (Scheme 120).

Another example of copper-catalyzed, ligand-free C–N coupling was reported by Bolm and coworkers [176]. They investigated coupling of various primary and secondary aliphatic or aromatic amines with heteroaryl halides. In this case, reactions were heated conventionally, resulting in prolongation of reaction times (24 h). However, transformations could be carried out at significantly lower temperatures (90°C in most cases). Relevant examples involve is coupling of benzylamine **451** with 2-chloropyrimidine 6 (required 110°C), 2-bromopyrimidine **36**, and 5-bromopyrimidine **37**. The corresponding products could be isolated in 98%, 91%, and >95% yield, respectively (Scheme 121).

1.8.3 Other Catalytic Systems

Seechurn et al. have examined a series of π-allyl-based palladium complexes (most of them showing good air stability) in C–C/N couplings [177]. They found complex **454** as the best-performing catalyst and carried out a wide substrate scope investigation. Coupling of 2-bromopyrimidine **36** with *N*-methylaniline **436** provided 83%

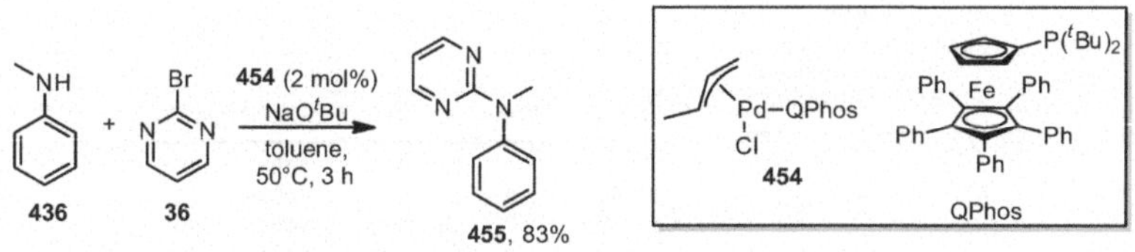

Scheme 122 New palladium complex for Buchwald–Hartwig amination

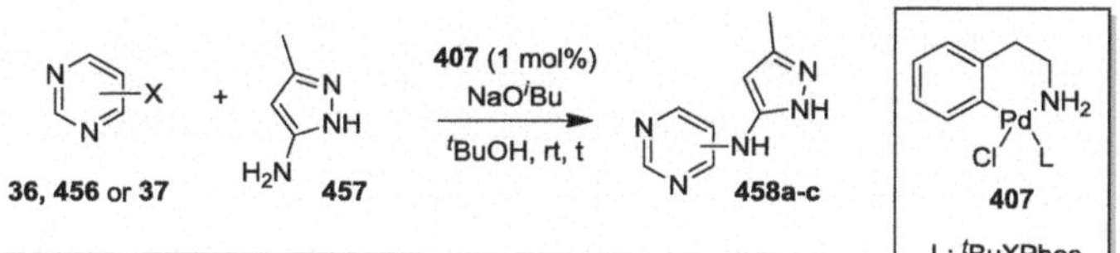

Entry	Starting material	X	Amine position	Time [h]	Product	Yield [%]
1	**36**	2-Br	2	6	**458a**	61
2	**465**	4-Br	4	6	**458b**	0
3	**37**	5-Br	5	4	**458c**	70

Scheme 123 Buchwald–Hartwig amination of pyrimidine, utilizing palladacycle 407 with *t*BuXPhos as the ligand

of the coupling product **455** using 2 mol% of the complex. The reaction could be carried out at relatively mild conditions, requiring only 50°C and 3 h (Scheme 122).

3-Methyl-5-aminopyrazole **457** can be coupled with 2-, 4-, and 5-bromopyrimidine, employing precatalyst **407** in combination with *t*BuXPhos, NaO*t*-Bu in *t*-butanol, as reported by Waring et al. [178]. Reaction takes place at room temperature. While coupling of 2- and 5-bromoryrimidine proceeded in good yields (61 and 70%, resp.), 4-bromopyrimidine did not undergo the desired transformation at all (Scheme 123).

Cubic copper(I) oxide nanoparticles turned out to be a useful catalytic system for C–N bond-forming transformations in combination with 1,10-phenantroline [179]. Series of imidazole and indole derivatives were coupled with either 2-bromo- or 2-chloropyrimidine (**36** or **6**). 2-bromopyrimidine generally performed better. Reaction between imidazole and 2-bromopyrimidine **36** yielded 86% of biaryl, whereas 2-chloropyrimidine afforded only 72% of coupling product (Scheme 124a, entries 1 and 2). A methyl group in position 2 of imidazole had a positive influence on the reaction with **6**, and the coupling product **460b** was obtained in 98% (Scheme 124a, entry 3). In the case of 2-phenylimidazole, only traces of product were detected (Scheme 124a, entry 4). A fused benzene ring as in benzimidazole did not influence the reaction significantly (77% in the coupling with **6**, Scheme 124b, entry 1). Switching to *N*-methyl-benzimidazole gave a notably lower yield of 66% (Scheme 124b, entry 2). Employing indole provided good results as well as giving 85% and 68% yield in reactions with 2-bromo- and 2-chloropyrimidine, respectively (Scheme 124b, entries 3 and 4).

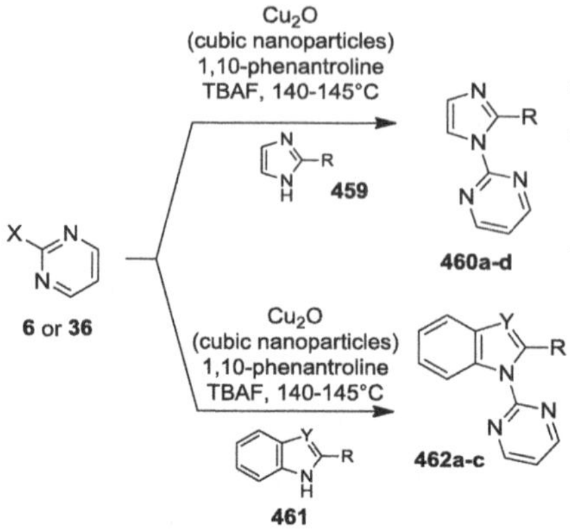

a

Entry	X	R	Product	Yield [%]
1	Br	H	**460a**	86
2	Cl	H	**460a**	72
3	Cl	Me	**460b**	98
4	Cl	Ph	**460c**	trace

b

Entry	X	Y	R	Product	Yield [%]
1	Cl	N	H	**462a**	77
2	Cl	N	Me	**462b**	66
3	Br	CH	H	**462c**	85
4	Cl	CH	H	**462c**	68

Scheme 124 Utilization of copper oxide nanoparticles on C–N coupling involving pyrimidine

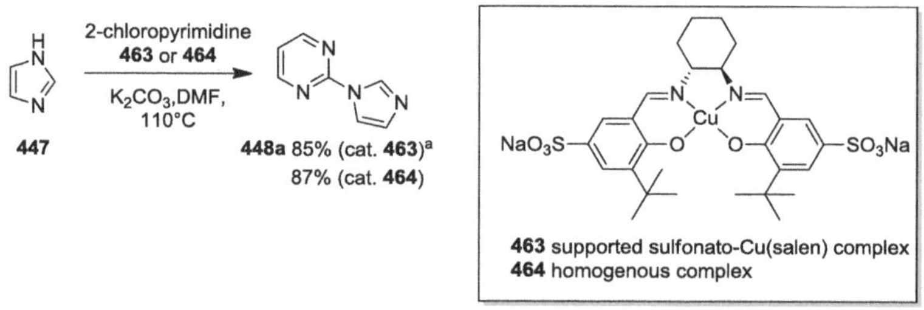

Scheme 125 New copper complex and its utilization in Ullmann-type reaction. [a]Stated yield after the third catalytic cycle, after the recovery by filtration

Kantam et al. reported a Merrifield resin-supported sulfonato-Cu(salen) complex (**463** and **464**) for the C–N coupling reactions of (hetero)aryl chlorides [180]. Applying this catalytic system to the reaction between 2-chloropyrimidine and imidazole led to the formation of the desired product **448a** in 85% yield (third cycle after recovery by filtration). At 110°C, with 1 mol% of the catalyst, the reaction was finished in 3 h. For comparison, the corresponding homogeneous reaction with unsupported catalyst gave 87% yield (Scheme 125).

1.8.4 Miscellaneous C–N Couplings on Pyrimidines

Vimolratana et al. investigated palladium-catalyzed introduction of various amides into position 2 of 2-chloropyrimidine [181]. Employing bidentate ligands turned out to be crucial, and particularly XantPhos provided an excellent yield of 94% in the coupling reaction between benzamide and 2-chloropyrimidine (Scheme 126, entry

Scheme 126 Coupling of amides and 2-chloropyrimidine

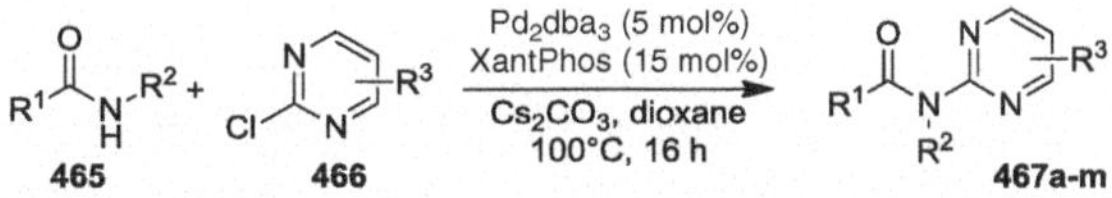

Entry	R^1	R^2	R^3	Product	Yield [%]
1	Ph	H	H	**467a**	94 (71)[a]
2	*p*-F_3C-phenyl	H	H	**467b**	71
3	*p*-MeO-phenyl	H	H	**467c**	87
4	*o*-tolyl	H	H	**467d**	80
5	Me	H	H	**467e**	91
6	^{t}Bu	H	H	**467f**	96
7	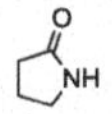		H	**467g**	93
8	Me	Me	H	**467h**	38
9	Ph	Me	H	**467i**	32
10	Ph	H	5-Et	**467j**	65
11	Ph	H	4-CF_3	**467k**	96
13	Ph	H	4-OMe	**467l**	82
13	Ph	H	4-Me	**467m**	27

1). A series of primary (entries 1–6, 10–13) and secondary (entries 7–9), aromatic (entries 1–4, 9–13) and aliphatic (entries 5–8) amides were examined. Notably, both electron-rich and electron-deficient aromatic amides coupled efficiently (entries 2–4). Also primary aliphatic amides provided good results (entries 5–6). Cyclic pyrrolidin-2-one underwent coupling smoothly as well, providing **467g** in 93% (entry 7). However, the coupling of acyclic secondary amides suffered from lower yields (entries 8–9). Variation of the halopyrimidine component was permitted as well, but 2-bromopyrimidine provided a lower yield when coupled to benzamide (71% vs. 94% of **467a**, entry 1). Both electron-rich and electron-deficient pyrimidines were well accepted (entries 9–13).

Sulfonamides are particularly important functionalities in pharmaceutical industry. Several reports were published, dealing with metal-catalyzed C–N cross-coupling reactions between (hetero)aryl halides and sulfonamides as an alternative to the commonly applied coupling of sulfonylchlorides with aromatic amines.

In 2010 Baffoe et al. published a copper-catalyzed Ullman-type reaction between various (hetero)aryl bromides and different sulfonamides (Scheme 127, entries 1 and 3) [182]. Among the tested substrates, 2-bromopyrimidine and 4-bromopyrimidine coupled with rather poor yields, providing 35% of **469a** and 60% of **470** in respective manner. Wang et al. reported coupling between 2-bromopyrimidine and benzensulfonamide catalyzed by CuI/DMEDA catalytic system [183]. In this case, the coupling product **469b** was obtained in 86% yield (Scheme 127, entry 2).

Anjanappa et al. investigated palladium-catalyzed 2-(trimethylsilyl) ethanesulfonyl (SES-NH_2) amidation of (hetero)arenes [184]. SES-NH_2 can be used as an ammonia surrogate, since it can be easily cleaved to liberate free amine by reaction with fluoride ions. Applying $Pd(OAc)_2$ and XantPhos as ligand,

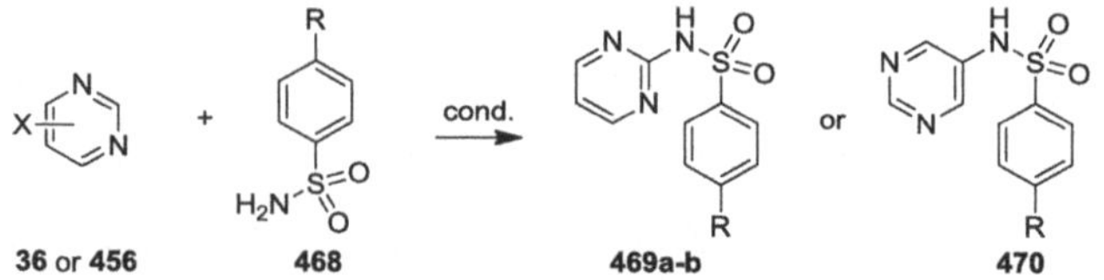

Entry	X	R	Conditions	Product	Yield [%]
1	2-Br	H	a	**469a**	35[a]
2	2-Br	Me	b	**469b**	86[b]
3	4-Br	H	a	**470**	60[a]

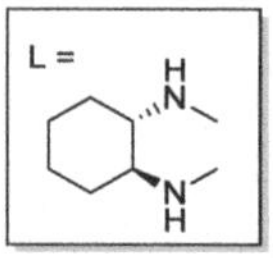

Scheme 127 Decoration of sulfonamides via Ullmann-type reaction. (*a*) CuI (5 mol%), DMEDA (0.5 eq.), K_2CO_3, rt; (*b*) CuI (15 mol%), L (30 mol%), K_2CO_3, DMF, 100°C

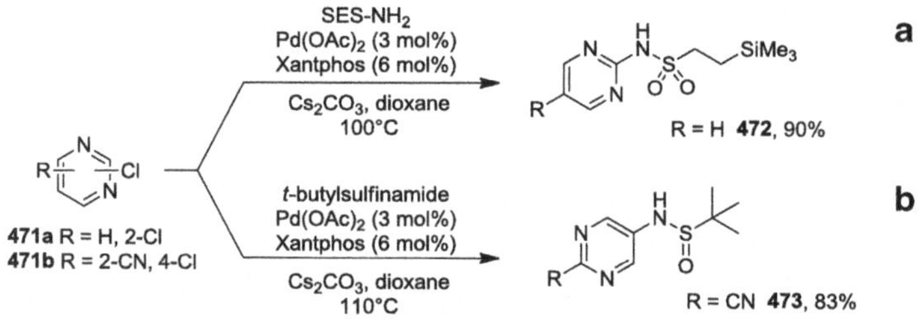

Scheme 128 Pd-catalyzed amination using ammonia surrogates

introduction of SES-NH_2 into position 2 of 2-chlropyrimidine provided 90% of product **472** (Scheme 128a).

Another ammonium surrogate is *t*-butylsulfinamide which was developed in the same research group [185]. In this moiety, the central sulfur atom occurs in oxidation state two lower than in sulfonamides. Anjanappa et al. have developed palladium-catalyzed C–N coupling between various (hetero)aryl halides and *t*-butylsulfinamide, allowing an easy introduction of divers (hetero)aryls. Reaction with 2-cyano-5-chloropyrimidine provided 83% of desired product **473** (Scheme 128b).

Sulfonimidamides are aza analogs of sulfonamides, in which one oxygen atom is replaced by nitrogen. This enables further derivatization on the additional nitrogen atom. Arvidsson and coworkers have described a synthesis of sulfonimidamides and developed a method to arylate the nitrogen atom via palladium-catalyzed C–N coupling [186]. Employing precatalyst **407** (L=RuPhos), coupling between sulfonimidamide **474** and 5-bromopyrimidine proceeded with 86% yield (Scheme 129).

A method for preparation of unsymmetrically *N,N*-diarylated ureas was described by Breitler et al. [187]. The process comprises of a one-pot Pd-catalyzed arylation/deprotection of mono-protected urea, followed by a second arylation. Under the optimized reaction conditions, 2-chloro-4,5-dimethoxypyrimidine **476** and *N*-PMB-urea **477** provided in the coupling/deprotection sequence 84% of the

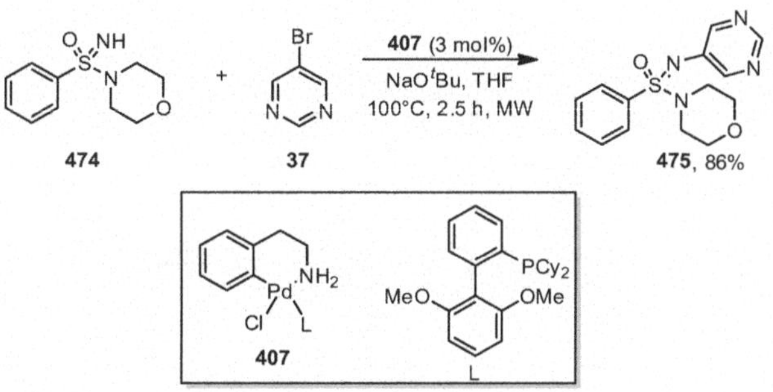

Scheme 129 Decoration of sulfonimidamides

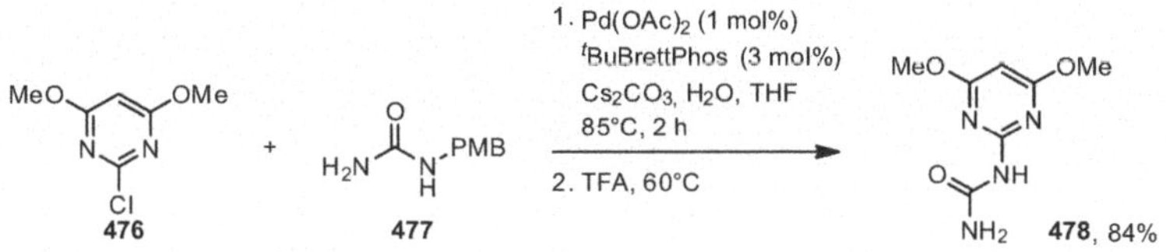

Scheme 130 Monoarylation o urea via Buchwald–Hartwig amination

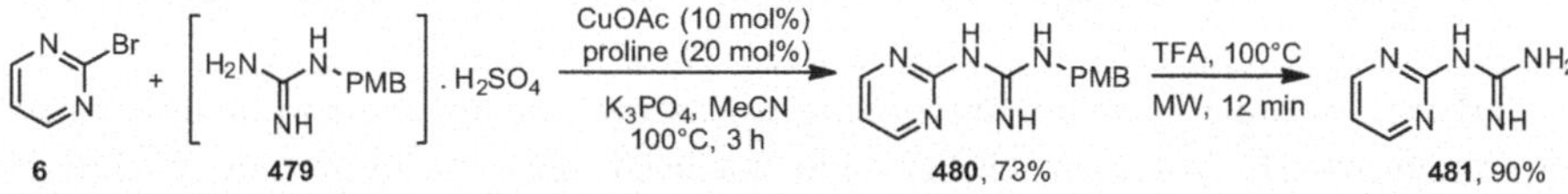

Scheme 131 Monoarylation of guanidine

desired mono-arylated product **478** (Scheme 130). Unfortunately, this substrate was not submitted to the further *N*-derivatization.

Monoarylation of guanidine was described by Hammoud et al. [188]. They found a copper catalyst to be superior to palladium catalysts. The combination of CuOAc with proline as additive turned out to be crucial. Employing mono-PMB-guanidine **479** in guanidination of 2-bromopyrimidine provided 73% of the desired product **480** which was successfully deprotected to give **481** (Scheme 131).

Another moiety able to undergo palladium-catalyzed C–N coupling with pyrimidine, is benzaldehyde-derived hydrazone **482**. Such a reaction yields *N*,*N*′-diarylated hydrazones, which under oxidative conditions are able to cyclize to provide triazolopyrimidines [189]. Four pyrimidine derivatives were examined within the substrate scope investigation in the reaction with benzophenone hydrazone **482**. Among them, 2-chloropyrimidine provided 63% of the *N*,*N*′-diarylated hydrazone **483**. 4-chloro-2-methylthio- and 4-chloro-2,6-dimethoxypyrimidine provided 73% and 90% of product respectively (**485a**, **b**, Scheme 132, entries 1 and 2). Coupling of 2,4-dichloropyrimidine proceeded into position 4 giving 58% yield (**485c**, Scheme 132, entry 3). A clear trend for the preference of electron-rich substrates

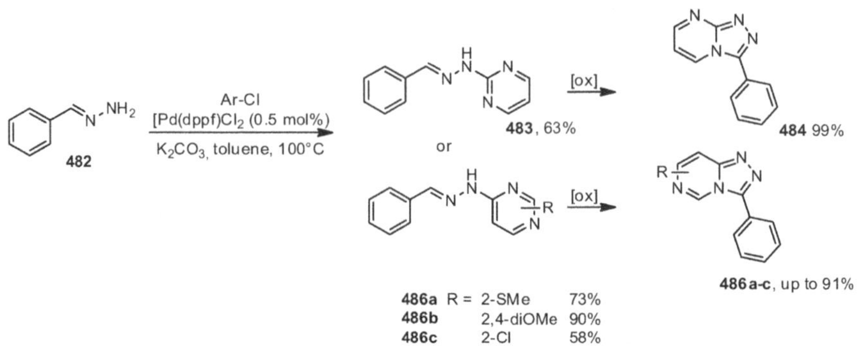

Scheme 132 Synthesis of triazolopyrimidines via coupling of hydrazones and subsequent oxidation

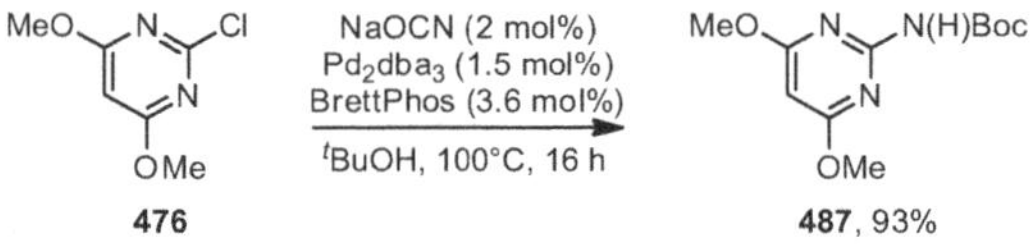

Scheme 133 Unconventional preparation of Boc-protected 2-aminopyrimidine

was observed. All intermediates underwent smooth oxidations towards the corresponding heterocycles in yields between 77 and 99% (Scheme 132).

Carbamates are important functionalities, as they occur in pharmacologically active compounds, polymers and last but not least, they are important protective groups in organic synthesis. Vinogradova et al. have reported one pot multicomponent synthesis of various carbamates via palladium catalyzed reaction of (hetero)arylhalides or triflates with sodium isocyanide and an alcohol as a trapping nucleophile[190]. Employing 2-chloro-4,6-dimethoxypyrimidine **476**, sodium isocyanide and *t*-butanol led to the formation of Boc-protected 2-amino-4,6-dimethoxypyrimidine **487** in 93% yield (Scheme 133).

Several reports were published describing the synthesis of organic compounds, utilizing C–N cross-coupling reactions as a key step. Das et al. synthesized a series of 3-(heteroaryl)aminocoumarines via Buchwald–Hartwig amination of 3-aminocoumarin **488** and halopyrimidines [191]. Utilizing $Pd(OAc)_2$ and BINAP as catalytic system and Cs_2CO_3 as base five examples was reported giving the corresponding products in 77–90% yield (Scheme 134).

A practical four step route towards 2-dialkylamino-4-arylamino-6-aminopyrimidines was developed by Li and Rosenau [15]. At a late stage, Buchwald–Hartwig amination into position 6 was carried out before Boc-deprotection gave the final products. Cross-coupling was conducted in the presence of Pd_2dba_3, XPhos, and Cs_2CO_3, giving yields between 71 and 92%. In addition, obtained structures can be utilized in a second Buchwald–Hartwig amination as nucleophiles in reaction with aryl bromides. Applying the same reaction conditions (at slightly

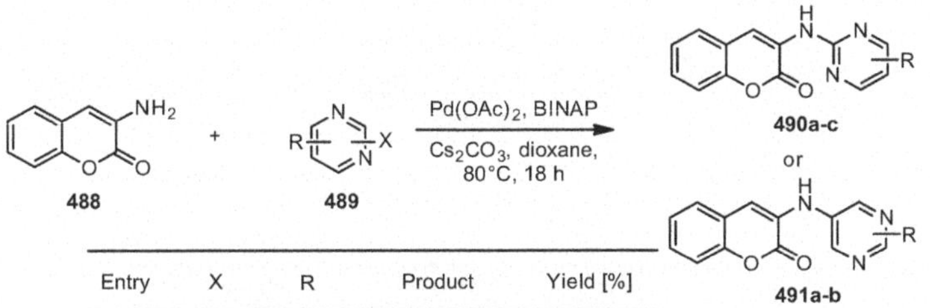

Entry	X	R	Product	Yield [%]
1	2-Br	H	**490a**	77
2	2-Cl	4-OMe	**490b**	88
3	2-Br	4-Me	**490c**	80
4	5-Br	H	**491a**	86
5	5-Cl	5-N(Me)$_2$	**491b**	90

Scheme 134 Decoration of aminocoumarines with pyrimidines

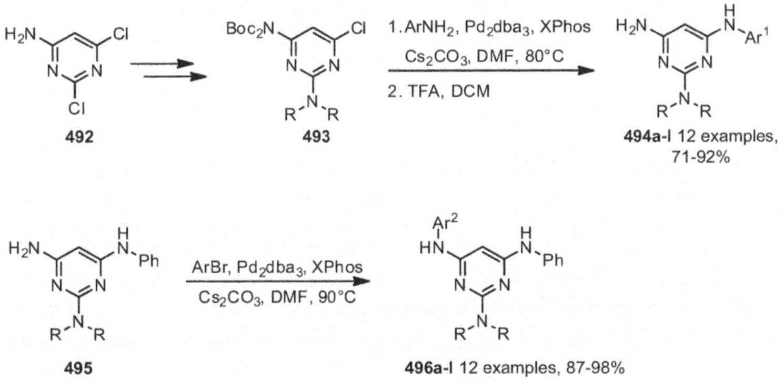

Scheme 135 Synthesis of unsymetric triaminopyrimidines

elevated temperature) led to interesting nonsymmetrical diarylaminopyrimidines **496a–l** in yields up to 98% (Scheme 135).

Artificial *C*-nucleosides are an important compound class. Kubelka et al. investigated the synthesis of 2,4-disubstituted pyrimidin-5-yl *C*-2-deoxyriboses via selective functionalization of 2,4-dichloropyrimidin-5-yldeoxyribose or its TBS-protected analog **497** [139]. The substrate was subjected either to conditions suitable for nucleophilic substitution or metal-catalyzed cross-coupling. The selectivity of the transformation is caused by the difference in reactivity of positions 2 and 4 of the pyrimidine ring. A dimethylamino group was introduced via Buchwald–Hartwig amination, utilizing Pd(OAc)$_2$, JohnPhos, and sodium *t*-butanoate (TBS-protected nucleoside was used). Notably, at room temperature, coupling proceeded preferentially into position 4. However, a minor amount of the regioisomer was isolated as well (71% vs. 16% respectively, Scheme 136, entry 1). At higher temperature, the amination took place in both positions giving diaminated product in 84% yield (Scheme 136, entry 2).

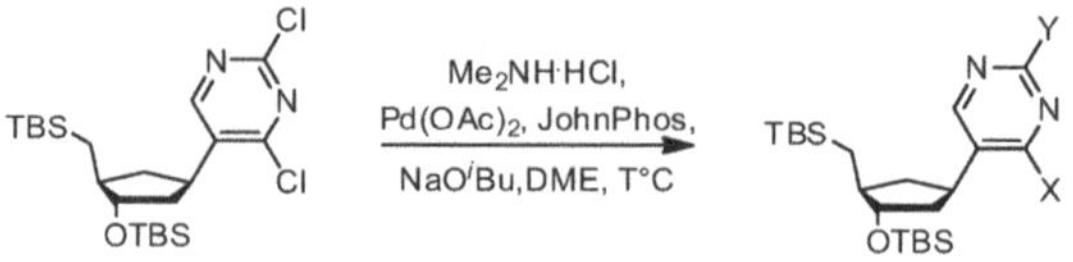

497 **498a-c**

Entry	Temperature [°C]	X	Y	Product	Yield [%]
1	rt	N(Me)$_2$	Cl	**498a**	71
		Cl	N(Me)$_2$	**498b**	16
2	80°C	N(Me)$_2$	N(Me)$_2$	**498c**	84

Scheme 136 Derivatization of 2,4-disubstituted pyrimidin-5-yl *C*-2-deoxyriboses

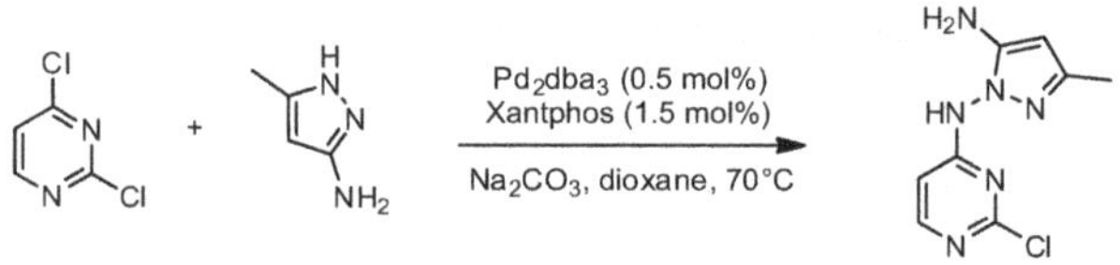

70 **499** **500**, 71%
only regioisomer

Scheme 137 Regioselective coupling of 2,4-dichloropyrimidine and 3-amino-1*H*-pyrazoles

Shen et al. investigated the regioselectivity of cross-coupling between 2,4-dichloropyrimidine **70** and 3-amino-1*H*-pyrazoles [192]. They discovered that submission of the substrates to cross-coupling conditions (Pd$_2$dba$_3$, Xantphos, Na$_2$CO$_3$) led to coupling with the endocyclic secondary amino group. It was demonstrated that coupling of 2,4-dichloropyrimidine proceeded exclusively into position 4 of pyrimidine and position 2 of the pyrazole yielding 71% of **500**, (Scheme 137). Alternatively, treatment of reactants with HCl in dioxane led to S$_N$Ar of chloride preferably with the exocyclic primary amino group of 3-amino-1*H*-pyrazole into the position 4 of pyrimidine.

The method for preparation of pyridinium heteroaryl-stabilized amidines based on palladium-catalyzed C–N cross-coupling utilizing Pd$_2$dba$_3$, BINAP, and NaO*t*-Bu was developed by Cordoba et al. [193]. *N*-aminopyridinium iodide **501** and 4-methyl-2-metylthiopyrimidine **503** or 5-bromopyrimidine **37** provided 50% of the desired products **504** and **505** in both cases (Scheme 138).

A method to couple electron-deficient heteroaryl amines with heteroaryl halides was developed in the group of Yin [194]. Optimization of reaction conditions revealed that the combination of Pd$_2$dba$_3$ with bidentate dppf ligand and Cs$_2$CO$_3$ in toluene was the best condition. In the study, pyrimidines were employed as electrophiles and nucleophiles as well. Aminopyrimidines **506** and **508** were coupled with different heteroaryls, including pyrimidine halides, providing coupling products in 62–87% yield (Scheme 139a, b).

Lach et al. developed a general route to unsubstituted *N*-(heteroaryl)aryl-aminobenzensulfonamides, relying on aminobenzensulfonamide and (hetero)aryl halides coupling [195]. The synthetic route consists of synthesis of bis-PMB protected 3- or 4-aminobenzensulfonamide **514** (PMB group protects the

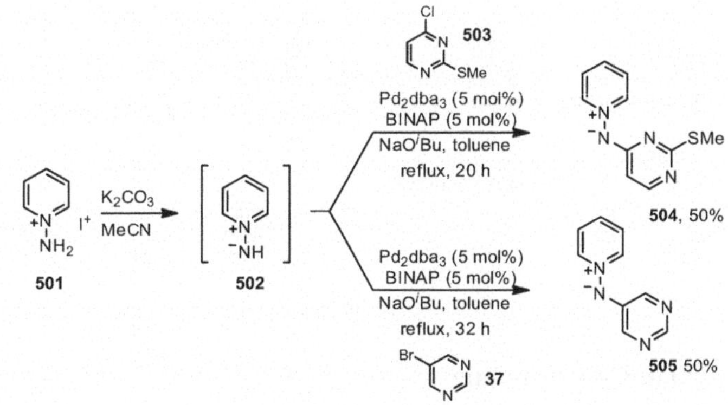

Scheme 138 Preparation of pyridinium heteroaryl-stabilized amidines via Pd-catalyzed C–N cross-coupling

a

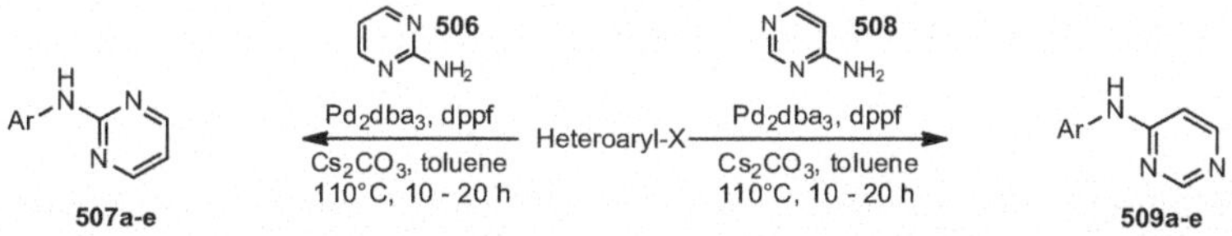

Entry	Heteroaryl	X	Product	Yield [%]
1	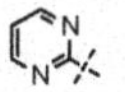	Br	**507a**	86
2		Cl	**507a**	82
3	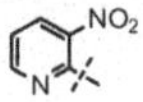	Cl	507b	76
4	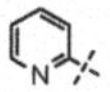	Br	507c	63
5	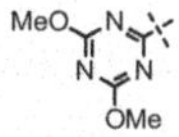	Cl	**507d**	83
6	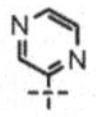	Br	507e	78

Entry	Heteroaryl	X	Product	Yield [%]
1	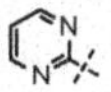	Br	**509a**	87
2		Cl	**509a**	82
3	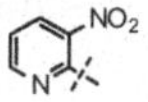	Cl	509b	75
4	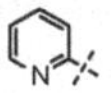	Br	509c	80
5	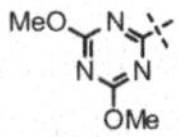	Cl	**509d**	76
6	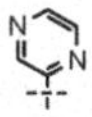	Br	509e	67

b

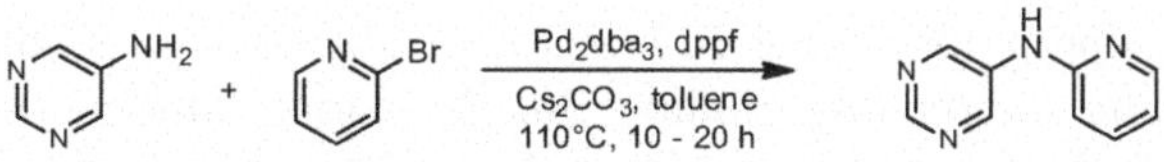

510 **511** **512**, 62%

Scheme 139 Coupling of electron-deficient heteroaryl amines with heteroaryl halides

sulfonamide moiety) and its subsequent Buchwald–Hartwig amination/deprotection sequence. Employing 5-bromopyrimidine and 2-methoxycarbonyl-5-bromopyrimidine afforded the desired products **515 a–c** in up to 60% yield (Scheme 140).

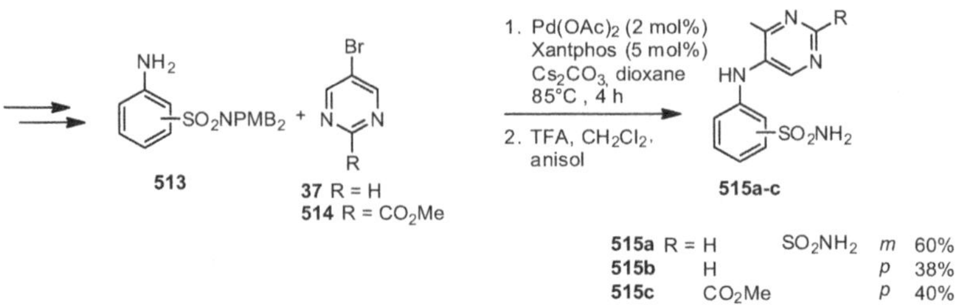

Scheme 140 A general route to unsubstituted *N*-(heteroaryl)aryl-aminobenzensulfonamides

1.8.5 Other than C–N Coupling

Besides the reports, dealing with C–N coupling, several papers were published, discussing reactions between carbon-based electrophiles and nucleophiles other than nitrogen. Herein, carbon–phosphorus, carbon–sulfur, and carbon–oxygen cross-coupling reactions are summarized.

C–P Coupling

Since Hirao's pioneering work in the area of C–P coupling in 1982 [196], many follow-up papers have been published modifying the original procedure, extending the scope of reagents, etc., and research in the field continues till today. In some of the reports, pyrimidine halides served as substrate for coupling with different phosphorus-based moieties.

Balabassi et al. carried out some reinvestigation of the original Hirao conditions [197]. They focused on the elimination of the main problems of the transformation, namely, rather high catalyst loadings and dealkylation of the desired products occurring as a side reaction. The latter problem was solved by replacing the original base, triethylamine, with more sterically hindered secondary amine (*N*,*N*-diisopropylethyl amine), since the dealkylation proceeds via a S_N2 mechanism. Applying DMF or MeCN as solvent and $Pd(OAc)_2$ in combination with the dppf ligand as a catalytic system allowed to reduce the catalyst loading from 5 to 1 mol%. Moreover, the first example of a (hetero)aryl chloride able to undergo C-P bond formation was reported. Diisopropylphosphite **517** was then reacted with many (hetero)aryl halides, including 2-bromopyrimidine **36**, 2-chloropyrimidine **6**, 5-bromopyrimidine **37**, and 5-chloropyrimidine **516**, yielding desired products in up to 83% (Scheme 141).

Another report from the Montchamp laboratory dealt with the coupling of (hetero)aryl chlorides and phosphorus nucleophiles, in this particular case *H*-phosphinate esters [198]. Under the optimized conditions (2 mol% of $Pd(OAc)_2$ and Xantphos, *i*-Pr_2Net), reaction between 2-chloropyrimidine **6** and ethyl octylphosphinate **520** provided 52% of **521**. 2-Chloropyrimidine turned out to be a rather difficult substrate in comparison to others (Scheme 142).

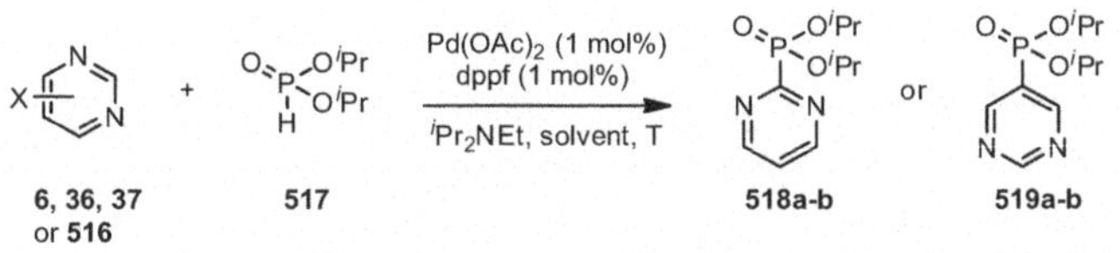

Entry	X	Solvent	Temperature	Product	Yield [%]
1	2-Br	CH_3CN	reflux	**518a**	67
2	2-Cl	DMF	85°C	**518b**	28
3	5-Br	CH_3CN	reflux	**519a**	83
4	5-Cl	DMF	105°C	**519b**	62

Scheme 141 Reinvestigation of the original Hirao conditions

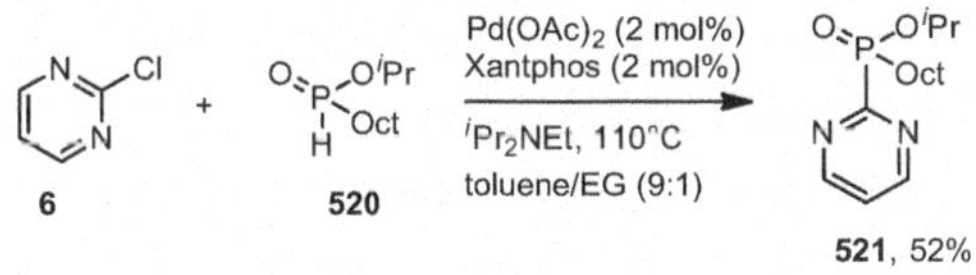

Scheme 142 Coupling of *H*-phosphinate esters with 2-chloropyrimidine

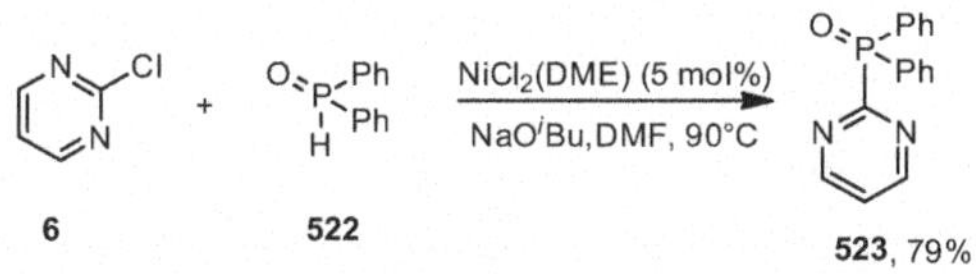

Scheme 143 Cross-coupling of diphenylphosphine oxide with 2-chloropyrimidine

Zhang et al. developed nickel-catalyzed cross-coupling of diphenylphosphine oxide with various (hetero)aryl chlorides [199]. In the presence of $NiCl_2$(DME) and *t*-BuONa at 90°C, 2-chloropyrimidine **6** was able to undergo reaction with diphenylphosphine oxide **522** to yield 79% of corresponding diphenylpyrimidin-2-ylphosphine oxide **523** (Scheme 143).

C–S and C–O Coupling

In general, utilizing sulfur nucleophiles in cross-coupling reactions is rather challenging due to the ability of sulfur to strongly coordinate to the metal center, which can possibly lead to catalyst poisoning. First reports of C–S couplings appeared in the 1980s when Migita et al. published palladium-catalyzed cross-coupling of aryl bromides or iodides and stannyl sulfides leading to arylsulfides [200]. The transformation has been subject to many investigations, and nowadays C–S coupling is still an active area of research.

In 2009, Liu and coworkers developed Fe/Cu co-catalyzed cross-coupling of aryl halides and thiols [201]. A variety of substrates, including aromatic and aliphatic thiols and different (hetero)aryl halides, were coupled under the optimized conditions (10 mol% Fe_2O_3, $Cu(OAc)_2$, 20 mol% TMEDA). Among them,

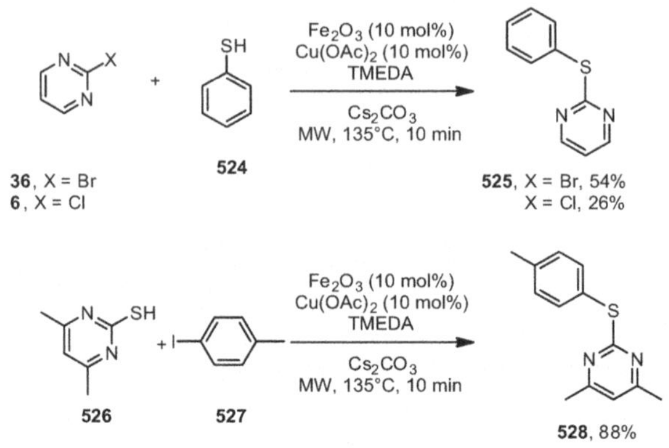

Scheme 144 Cu/Fe catalyzed C–S bond formation reaction on pyrimidines

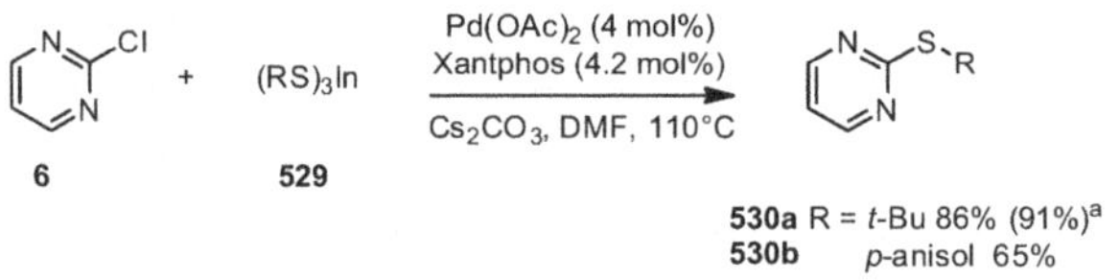

Scheme 145 Indium catalyzed cross-coupling on pyrimidine, [a]DIPEA used

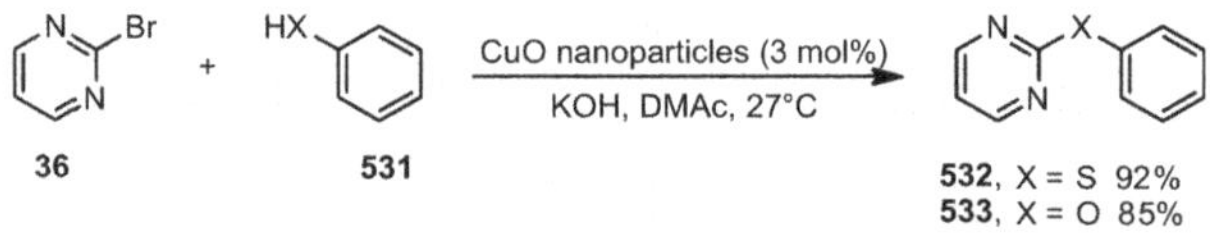

Scheme 146 Copper(II)oxide nanoparticles for C–S/C–O cross-coupling on pyrimidine

2-bromopyrimidine and 2-chloropyrimidine were coupled with thiophenol **524**, yielding 54% and 26% of **525**, respectively. In addition, 4,6-dimethylpyrimidine-2-thiol **526** was coupled with *p*-iodotoluene **527**, providing 88% of cross-coupling product **528** (Scheme 144).

Mo et al. developed palladium-catalyzed cross-coupling, utilizing indium tris (organothiolates) as nucleophiles and aryl chlorides [202]. Under optimized conditions, 2-chloropyrimidine underwent coupling with indium tris(*t*-butylthiolate) **529** yielding 91% of sulfide **530a**. Reaction of 2-chloropyrimidine with aromatic indium tris(*p*-anisylthiolate) provided 65% of sulfide **530b** (Scheme 145).

Recently, Kervembu and Babu have demonstrated the utilization of CuO nanoparticles as an active catalyst for cross-coupling reactions between (hetero) aryl bromides and aryl chlorides with (thio)phenols leading to C–S or C–O bond formation [203]. The reactions were carried out in absence of ligand at rt. This makes the procedure attractive since additionally a relatively low catalyst loading (3 mol%) could be applied. Thiophenol gave in the reaction with 2-bromopyrimidine 92% of **532**, whereas phenol delivered 85% of the desired ether **533** (Scheme 146).

In 2012, Dash et al. developed palladium-catalyzed d_3-methoxylation of diverse (hetero)aryl bromides [204]. An optimization study revealed that using $Pd(OAc)_2$,

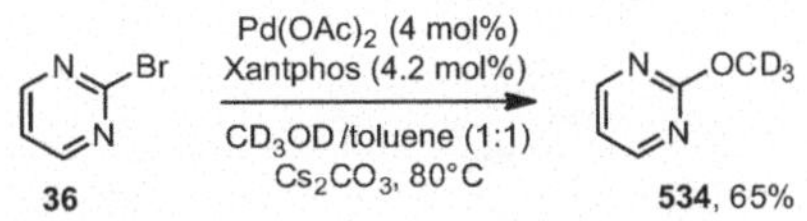

Scheme 147 Pd-catalyzed d_3-methoxylation of pyromodine

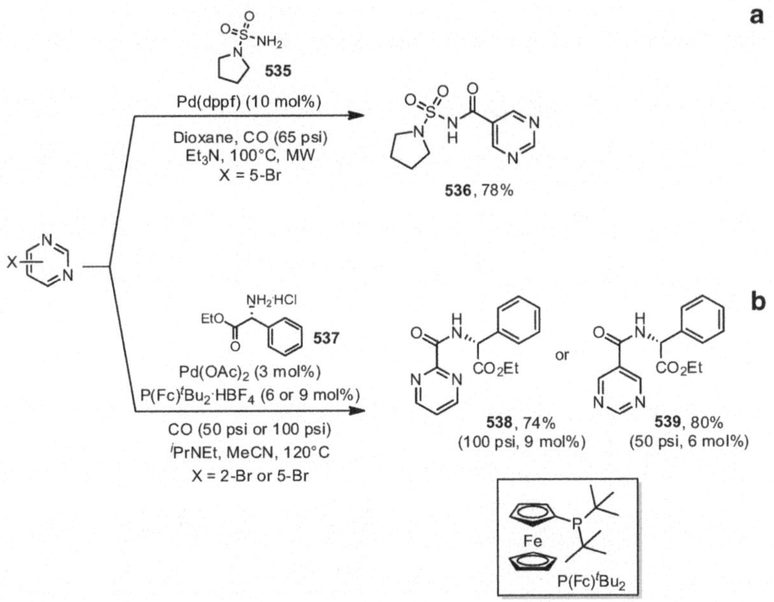

Scheme 148 Carbonylation of pyrimidine using gaseous CO

*t*BuXPhos, and Cs_2CO_3 leads to the best results. Applying such conditions to the reaction of 2-bromopyrimidine with an excess of MeOD led to formation of the desired product **534** in 65% (Scheme 147).

1.9 Carbonylation

Carbonylations proved to be synthetically very useful reactions since they allow simple access towards a variety of chemical functionalities such as amides, esters, and others. Since the pioneering work published by Heck in 1974 [205], many variations of the reaction were carried out and research in the field continues. In recent times, several papers were published, dealing with carbonylation reactions on the pyrimidine core.

Roberts et al. have reported microwave-accelerated carbonylation of (hetero) aryl halides in the presence of sulfamide **535** and 65 psi gaseous CO [206]. The reaction was catalyzed by Pd/dppf and coupling of 5-bromopyrimidine provided the corresponding compound **536** in 78% yield (Scheme 148a).

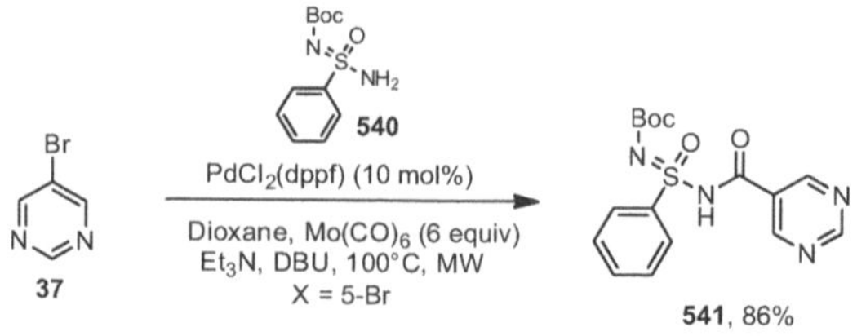

Scheme 149 Carbonylation of pyrimidine using gaseous $Mo(CO)_6$ as CO source

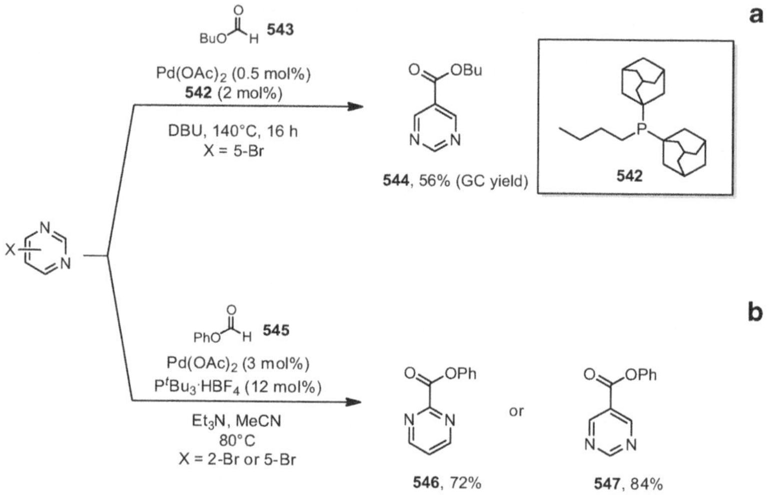

Scheme 150 Carbonylation of pyrimidine using formates

Aminocarbonylation of pyrimidine was described by Qu et al. [207] applying $Pd(OAc)_2$ together with monodentate di-*t*-butyl ferrocene phosphine. In the presence of (*R*)-ethyl 2-amino-2-phenylacetate hydrochloride **537** and under 50 or 100 psi pressure of CO, 2-bromopyrimidine or 5-bromopyrimidine provided the corresponding products **538** and **539** in 74% and 80% yield, respectively (Scheme 148b).

Developing a surrogate for gaseous CO is of interest mainly because of the toxicity of carbon monoxide. Borhade et al. have utilized $Mo(CO)_6$ as CO source for palladium-catalyzed carbonylation of various (hetero)aryl halides in the presence of nucleophilic sulfonamides [208]. 5-Bromopyrimidine was treated with mono Boc-protected benzenesulfonamide **540** in the presence of Pd catalyst and the molybdenum complex. Desired aminocarbonylated product **541** was obtained in 86% (Scheme 149).

Besides using gaseous CO or surrogate, carbonylation reactions utilizing formates were reported as well. Beller and coworkers investigated carbonylation of (hetero)aryl halides using butyl formate **543** [209]. The reaction was catalyzed by $Pd(OAc)_2$/(diadamantyl)butylphosphine (**542**). 5-Bromopyrimidine provided 56% of **544** (GC yield, Scheme 150a).

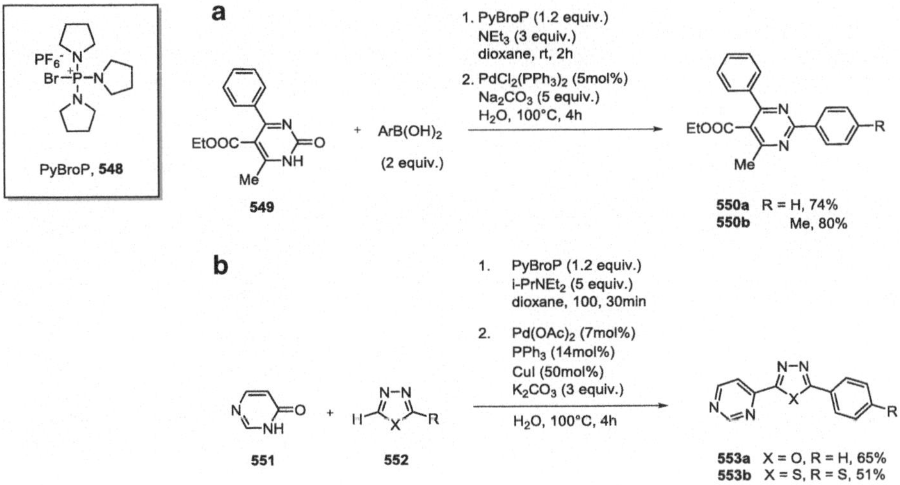

Scheme 151 Direct arylation of tautomerizable hydroxyl pyrimidines

5-Bromopyrimidine and 2-bromopyrimidine were carbonylated with phenyl formate by Udea et al. as well [210]. Under the catalysis of $Pd(OAc)_2/P^tBu_3$, products **546** and **547** were obtained in 84% and 72%, respectively (Scheme 150b).

1.10 Miscellaneous Reactions

An interesting direct arylation reaction of tautomerizable hydroxy-N-heterocycles with arylboronic acids was reported by Kang et al. [211]. Biaryl products were obtained in good to excellent yields. The authors assume that the OH group is activated via formation of a phosphonium salt with PyBroP **548** followed by insertion of the Pd catalyst into the C–O bond. Two examples of the reaction of highly substituted 2-pyrimidone **549** were reported, giving the coupling products **550** in 74% and 80% yield respectively (Scheme 151a).

Sharma et al. disclosed a variant of the reaction described above. Here, an organocuprate, which is formed in situ by metalation of an acidic methylene group with CuI, serves as the organometallic coupling partner. As substrates benzoxazoles, oxadiazoles, and thiadiazoles were reported (Scheme 151b) [212].

Faul and coworkers have developed a one-pot protocol for the synthesis of 2-allyl-2-arylcyanoacetates. The process consists of a sequential $Pd(OAc)_2$/dppf-catalyzed arylation of the enolate, followed by in situ trapping of the tertiary carbanion with an alkyl halide [213]. 4-Bromopyrimidine **37** reacted smoothly, giving the final products **555** and **556** in 81–85% yield (Scheme 152).

Mosquera et al. investigated the Pd-catalyzed coupling of triarylindium reagents with 5-bromo-2-chloropyrimidine **38** [214]. Double arylation towards **558** as well

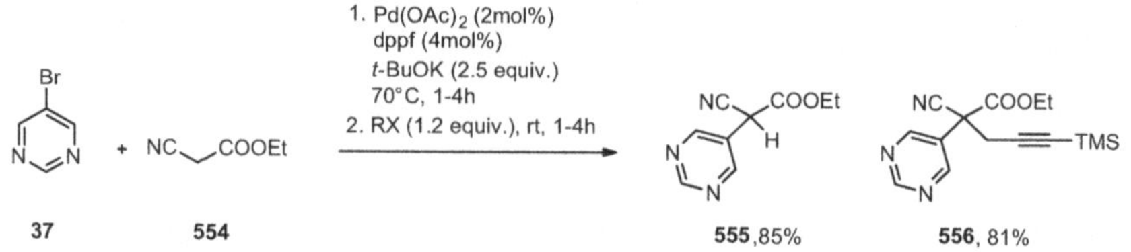

Scheme 152 Synthesis of 2-allyl-2-arylcyanoacetates in one pot

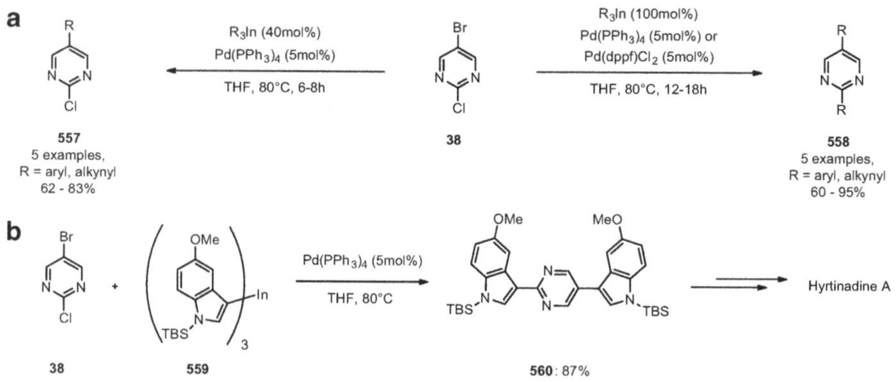

Scheme 153 Pd-catalyzed cross-coupling of pyrimidines with triaryl indium

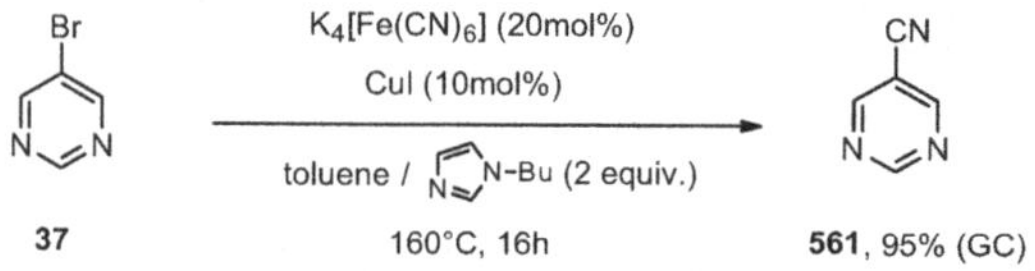

Scheme 154 Copper-catalyzed cyanation of 4-bromopyrimidine

as selective arylation in 4-position towards **557** was achieved in 60–95% yield (Scheme 153a). The method was applied to the synthesis of **560**, a key intermediate in the synthesis of hyrtinadine A (Scheme 153b).

Schareina et al. reported a Cu(I)-catalyzed coupling reaction of aryl bromides with potassium hexacyanoferrate as a nontoxic cyanide source [215]. The authors identified alkylimidazoles as highly effective ligands for the Cu catalyst in this transformation. Although harsh reaction conditions are required (140–180°C, 16 h) for most substrates, the reaction proceeded cleanly and gave the corresponding benzonitriles in 60–99% yield, e.g., 5-cyanopyrimidine **561** was obtained in 95% yield (Scheme 154).

Extending his previous work on the preparation of organoindium compounds [216] by direct insertion of In(0) into C–X bonds, Knochel and coworkers reported a general method for the preparation of benzylic In(III) reagents [217]. The insertion was found to be much faster for benzylbromides (0°C, up to

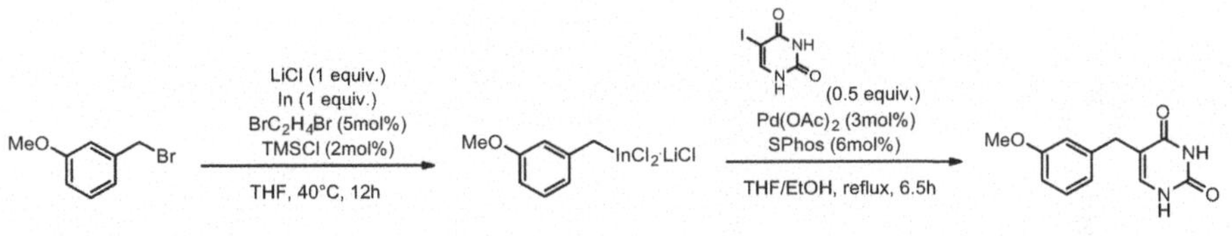

Scheme 155 Preparation and application of benzylic In(III) reagents

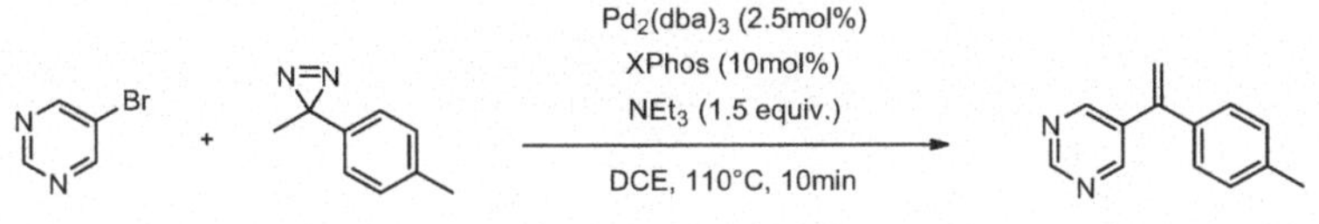

Scheme 156 Pd-catalyzed coupling of diaziridine and 5-bromopyridine

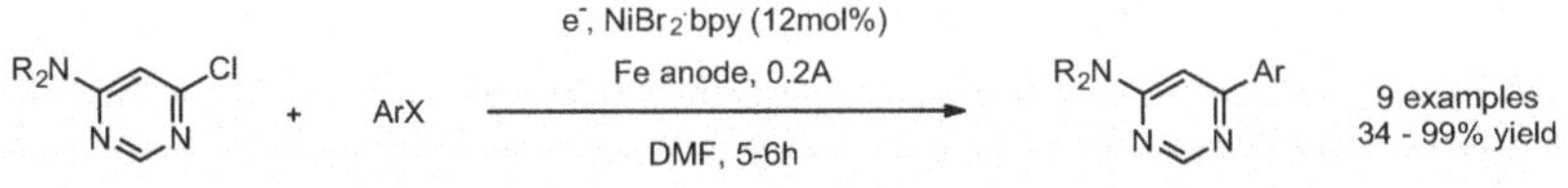

Scheme 157 Electrochemical reductive coupling of pyrimidine

30 min) compared to benzyl chlorides (up to 16 h, 40°C). The reactivity of the resulting In(I) reagents towards cross-coupling reactions could be significantly increased by a transmetallation step with *i*PrMgCl · LiCl, giving rise to a mixed alkylbenzyl In(III) species. These reagents (e.g., **563**) react smoothly with aryliodides and bromides in the presence of $Pd(OAc)_2$/SPhos at 25–40°C. 5-Iodouracil was coupled with the substituted benzylindium compound **563** giving 72% yield of **564** (Scheme 155).

The use of diaziridines such as **565** as coupling partners in a Pd-catalyzed cross-coupling reaction was investigated by Zhao et al. [218]. The products of this reaction are 1,1-diarylalkenes. The combination of $Pd_2(dba)_3$/XPhos in conjunction with microwave heating was found to be highly effective in this transformation towards 1-arylstyrenes. Products were generally obtained in good to excellent yields after only 10 min. 5-Bromopyrimidine, as the only heteraromatic example, gave a moderate 56% yield of **566** (Scheme 156).

An example of electrochemical reductive coupling was reported by Sengmany et al. [219]. 4-Amino-6-chloropyrimidines **567** were coupled with aryl halides in the presence of $NiBr_2$bpy and a sacrificial iron anode. Yields vary strongly depending on the nature of the amine substituent and the coupling partner giving compounds of type 568 in 34–99% yield (Scheme 157).

A three-component reaction of amino(thio)phenols **570** with isocyanide **569** and an aryl halide was published by Lang and coworkers [220]. In the presence of a Pd

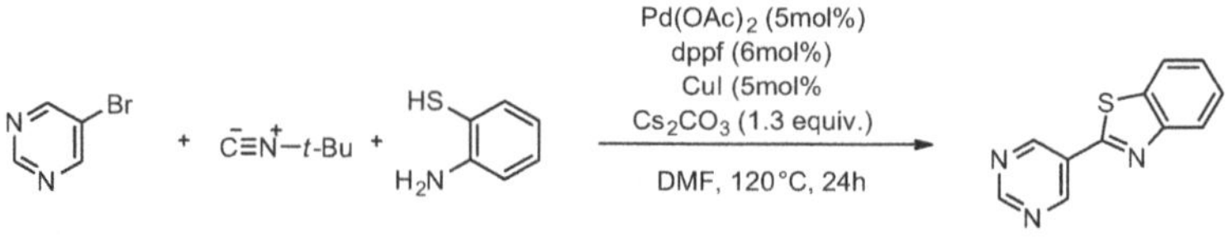

37 569 (1.8 equiv.) 570 (1.5 equiv.) **571**, 29%

Scheme 158 Pd- and Cu-catalyzed 3-component coupling

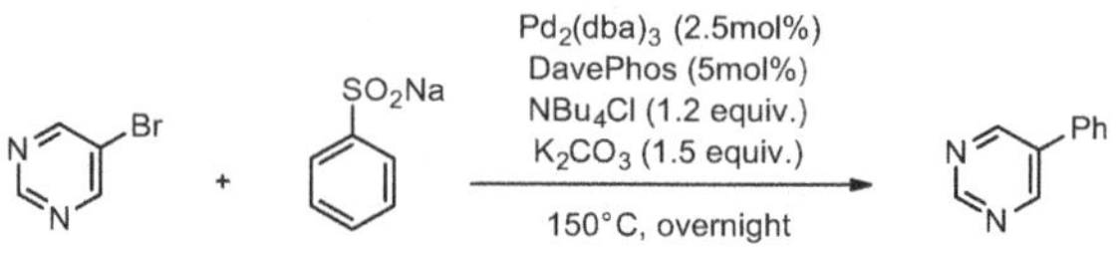

37 572 (1.5 equiv.) **28**, 64%

Scheme 159 Pd-catalyzed desulfitative cross-coupling of phenylsulfinate and bromo pyrimidine

catalyst, 2-aryl benzoxazoles and benzothiazoles were obtained. The authors noted an interesting difference in the reaction mechanism: aminophenols are believed to react via isocyanide insertion into the aryl-Pd bond followed by coupling with the amine nucleophile. The reaction conditions used for aminophenols proved ineffective for aminothiophenols: Here the starting material is believed to undergo a noncatalyzed reaction with the isocyanide, giving rise to 2-unsubsituted benzothiazoles. When CuI was added, formation of organocuprates and subsequent coupling with an aryl halide gave 2-arylbenzothiazoles. One example involving pyrimidine is presented in Scheme 158.

The Pd-catalyzed desulfitative cross-coupling of phenylsulfinate **572** with aryl halides was reported by Colomb and Billard [221], thus extending the scope of a previously reported reaction with arlytriflates [222] towards heteroaromatic substrates under similar conditions (Scheme 159).

2 Pyrazine in Cross-Coupling Reactions

Pyrazines are widely used intermediates in medicinal chemistry [223–230]. Hence, their decoration to increase complexity at a late stage of a synthetic sequence is of significant interest to synthetic chemists in general and medicinal chemists in particular. Naturally, cross-coupling reactions have been used in pyrazine chemistry, however not to the same extent as in pyrimidine chemistry. In the following chapters, the recent progress in this field is reviewed (2008 to mid-2013).

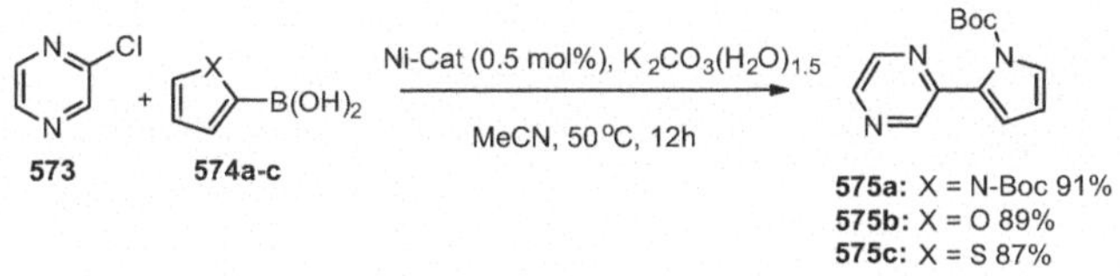

Scheme 160 Ni-catalyzed Suzuki coupling on pyrazine

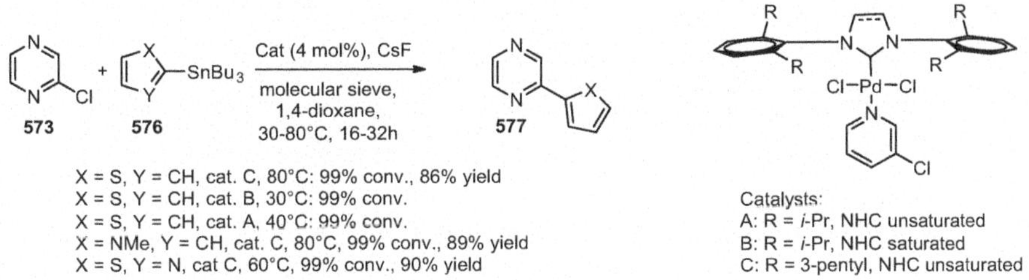

Scheme 161 NHC-Pd complex-catalyzed Stille coupling of pyrazine

2.1 Pyrazine in Suzuki–Miyaura Reactions

The replacement of the precious metal palladium with the first-row, abundant metal nickel for Suzuki–Miyaura couplings could significantly reduce the cost of the catalyst. Recently, Ge and Hartwig reported a single-component nickel catalyst, the dppf-ligated cinnamylnickel(II) chloride [(dppf)Ni(cinnamyl)Cl], which promoted coupling of heteroaryl halides with heteroaryl boronic acids towards biheteroaryls (Scheme 160). Among the more than three-dozen examples, pyrazine was used as well: pyrazine chloride **573** was coupled with *N*-Boc-2-pyrroleboronic acid, 2-benzofuranboronic acid, and 2-benzothiopheneboronic acid to give **575a** in 91%, **575b** in 89%, and **575c** in 87% yield. These high yields were obtained with 0.05% nickel catalyst only and no additional ligand was required [231].

2.2 Pyrazine in Stille Reactions

Stille cross-coupling reactions with NHC-palladium complexes as catalysts have been reported on **573** with a series of heterocyclic stannanes **576** (Scheme 161). Catalyst C, containing 3-pentyl residues, proved to be most general which was attributed to the flexible steric bulk of the 3-pentyl group. However, Pd-PEPPSI precatalysts were equally efficient in coupling of **573** with a thiophene stannane. Generally, all examples reported for pyrazine gave full conversion and high yield of **577** at relatively low temperatures as low as 30°C in one example. Low temperatures are beneficial since decomposition of organostannanes or heteroaryl halides is minimized [232].

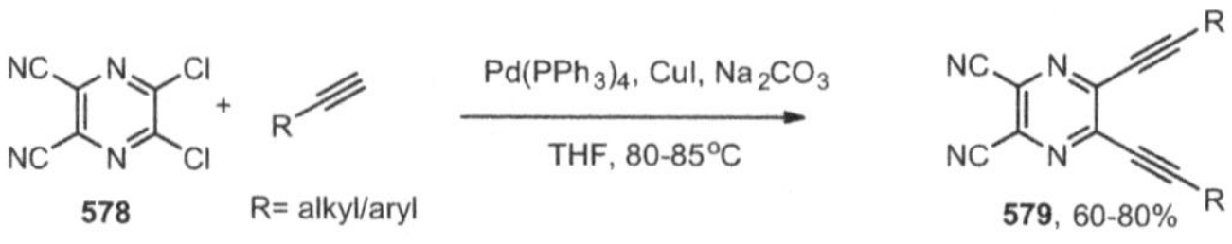

Scheme 162 Sonogashira coupling of substituted pyrazine

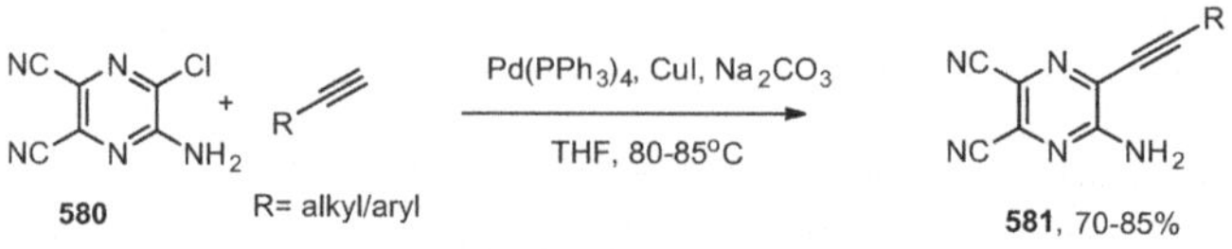

Scheme 163 Sonogashira coupling of pyrazine and alkyl/aryl alkyne

2.3 Pyrazine in Sonogashira Reactions

The palladium-catalyzed coupling of terminal alkynes with aryl or vinyl halides and triflates, usually in the presence of a copper co-catalyst, is commonly known as the Sonogashira reaction [233–236].

Van Lier published a paper dedicated to the investigation of Sonogashira cross-coupling on pyridazine derivatives [237]. Readily available 2,3-dicyano-5,6-bis-chloropyrazine **578** was coupled with a series of alkynes in positions 2 and 3 (Scheme 162). In a typical reaction, **578** was coupled with 1-hexadecyne using $Pd(PPh_3)_4$, copper iodide, and sodium carbonate as a base in THF at 80–85°C for 4 h giving **579** in 60% yield. A variety of different alkynes was applied including aliphatic (1-hexyne, hex-5-yn-1-ol, undec-10-yn-1-ol) and aromatic ones (phenylacetylene, 4-*n*-Bu-phenylacetylene, 4-CF_3-phenylacetylene, 4-NMe_2-phenylacetylene, 2-naphthylacetylene, and ferrocenylacteylene). Yields were in the range from 60 to 80%.

As second substrate 5-amino-6-chloro-2,3-dicyanopyrazine **580** was submitted to Sonogashira couplings with a similar range of alkynes (Scheme 163). In this case, coupling took place in position 6 and the free amino group was well tolerated. Yields of **581** up to 85% were reported in many cases.

2.4 Pyrazine in Heck Reactions

Langer and coworkers reported Heck cross-coupling reactions of 2,3-dichloropyrazine **582**. Three different reaction conditions were disclosed giving rise to three different product classes, namely, 2,3-dialkenyl-, 2-alkenyl-3-alkyl-, or 2,3-dialkylpyrazines **583**, **584**, and **585**, respectively (Scheme 164) [238]. The reaction of **582** with an alkene (ethyl acrylate and various styrenes) and a catalytic system comprising of $Pd(OAc)_2$ (5 mol%) and XPhos or SPhos (10 mol%) afforded

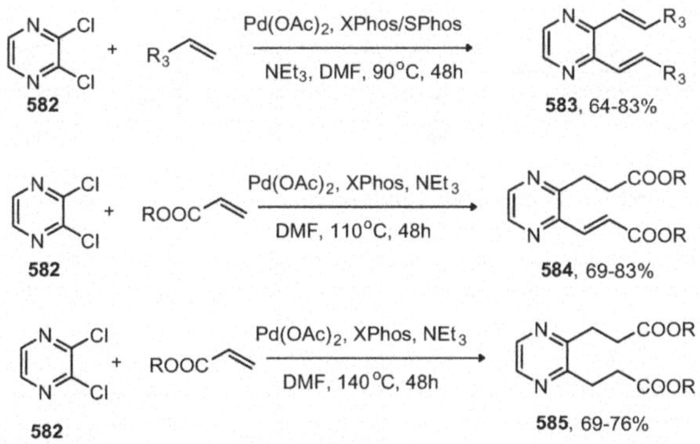

Scheme 164 Heck reaction on pyrazine

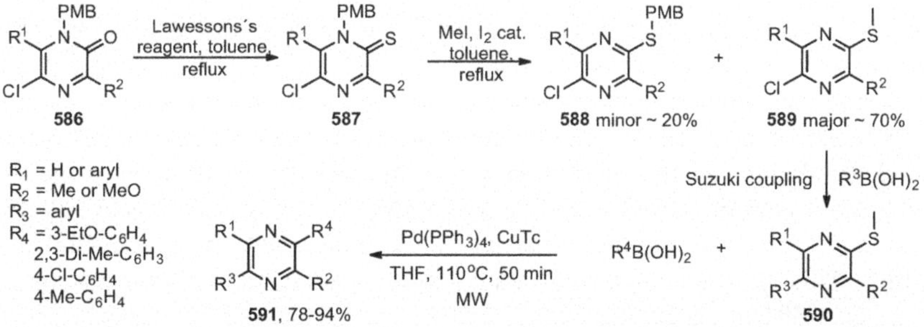

Scheme 165 Example of orthogonality: Suzuki and Liebeskind–Srogl reaction on pyrazine

the 2,3-dialkenylpyrazine **583** in good yield. Already in these reactions, partial dehydrogenation was observed, especially at higher temperatures. Using various acrylates (2.5 equiv.) as coupling partners and 110°C rather than 90°C, 2-alkenyl-3-alkylpyrazines **584** were obtained in yields ranging from 69 to 83%. As explanation for the reduction, protodemetalation was proposed. Raising the temperature further to 140°C gave complete reduction to **585**, again tentatively via protodemetalation.

2.5 *Pyrazine in Liebeskind–Srogl Reactions*

The Liebeskind–Srogl reaction is the coupling of a thioester or thioether with an organometal compound, in most cases a boronic acid. A palladium catalyst and a copper salt (in stoichiometric amount, most often copper(I) thiophene-2-carboxylate) are required. Due to the absence of base in this transformation, it is orthogonal to Suzuki–Miyaura cross-couplings since in the absence of base, the thiospecies acts as a leaving group while C-X bonds remain unaffected.

This orthogonality of methods was exploited in the synthesis of asymmetrically 2,3,5,6-tetrasubstituted pyrazines **591**, starting from N1-PMB-protected 3,5-dichloro-2(*1H*) pyrazinone **586** (Scheme 165). Initially, the thioether moiety

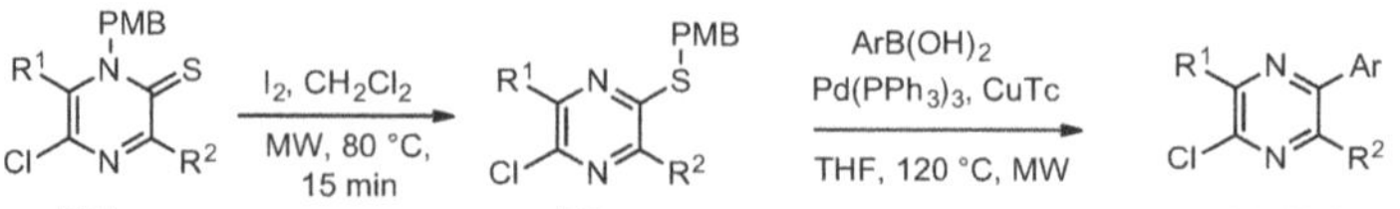

Scheme 166 Liebeskind–Srogl reaction of pyrazine

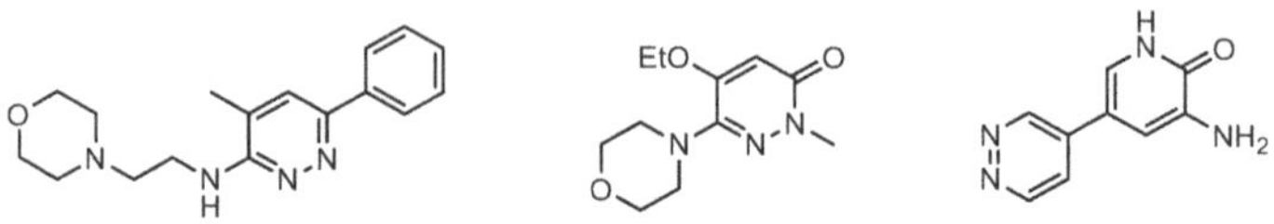

Scheme 167 Pyridazine and pyridazinone-based drugs

had to be introduced which was achieved via formation of the thio-pyrazinone intermediate using Lawesson's reagent to give **587**, followed by iodine-catalyzed methylation and concomitant aromatization to generate the thiomethyl-substituted pyrazines **589** which can be used as substrate for the Liebeskind–Srogl reaction [239]. Methylation reaction was always accompanied by the transfer of the PMB group to form para methoxylbenzyl thioether **588**. Coupling reactions of **589** under Suzuki conditions (or Sonogashira) gave substrates **590** for Liebeskind–Srogl coupling. This last step of the sequence works very well with various boronic acids to give **591** in excellent yields.

The protocol described above could be streamlined by eliminating the methylation step and directly using PMB-thioether **588** as substrate instead (Scheme 166). Liebeskind–Srogl coupling between **588** and arylboronic acids towards **592** proceeded in high yields also with this starting material. Initially, the reactions were sluggish due to reagent decomposition. This difficulty was overcome by adding the reagents in two portions aryl boronic acids of different electronic properties could be applied [240].

3 Pyridazine in Cross-Coupling Reactions

Pyridazine derivatives display diverse biological activities such as anticancer [241] or analgesic effects [242] and can be applied for the treatment of urinary incontinence [243], inflammatory pain [244], obesity [245, 246], and neurodegenerative diseases potentially [247–249]. Examples of some pyridazine and pyridazinone-based drugs are shown in Scheme 167.

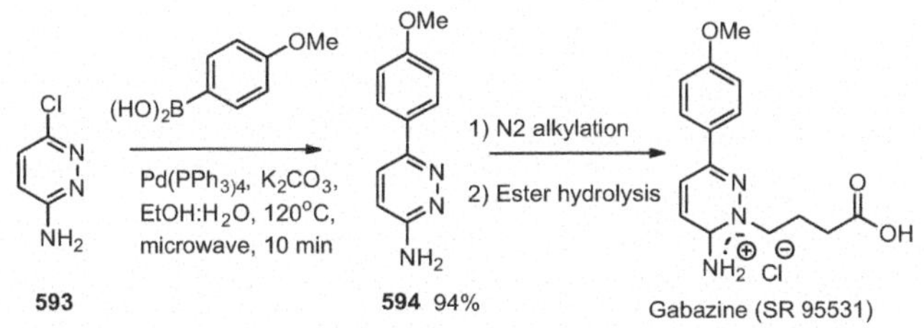

Scheme 168 Suzuki coupling of pyridazine

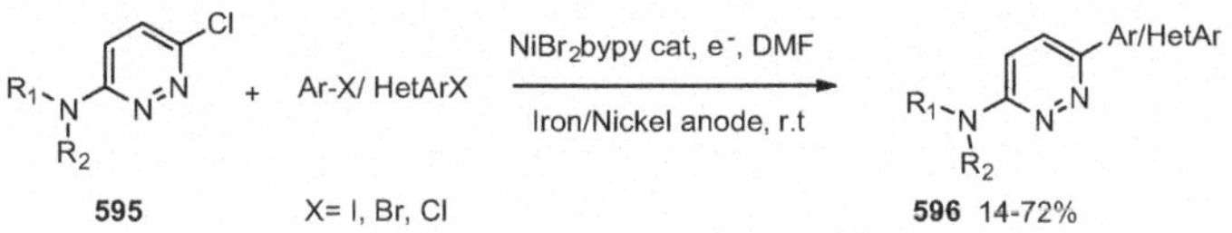

Scheme 169 Ni-catalyzed electrochemical arylation of pyridazine

3.1 Suzuki Coupling on Pyridazine

Broad substrate scope and high efficiency for the Suzuki coupling of unprotected 3-amino-6-chloropyridazine **593** can be achieved using a microwave-assisted protocol (Scheme 168) [250]. This method offers excellent reactivity for a variety of electron-deficient, electron-rich, and sterically hindered arylboronic acids, as well as heteroaromatic boronic acids with yields being uniformly good to excellent (17 examples, 75–95%). Chloro- and amino-substituents on the boronic acid and the amino substituent on the pyridazine ring are well tolerated under the reaction conditions, which can extend the scope for further functionalization of the pyridazine scaffold. In Scheme 168, synthesis of **594** is displayed which is a precursor for gabazine (SR-95531), which shows high specificity and potency towards both $GABA_A$ and $GABA_C$ receptors [251, 252].

Recently, a nickel-catalyzed electrochemical arylation of 3-amino-6-chloropyridazines **595** at r.t. is described, using an iron/nickel electrode as the sacrificial anode (Scheme 169). In total 27 examples of **596** were disclosed, and also the scope of the amine residue was investigated (e.g., morpholino, pyrrolo, imidazole, and series of dialkylamino). Substituents on the halide coupling partner such as methoxy, methyl, CF_3, chloro, cyano, and COOEt were tolerated. Pyridine and quinoline halides gave relatively low yield. 3-Bromothiophene gave a good yield of 71%, but additionally 1 equiv. of halide was added after 2.5 h. Comparable transformations involving classical reactions, such as Suzuki or Stille cross-couplings, revealed that the electrochemical method constitutes an alternative tool due to a broad substrate scope, use of ambient reaction condition, and a cheaper catalyst as compared to palladium [253].

References

1. Kinzel T, Zhang Y, Buchwald SL (2010) A new palladium precatalyst allows for the fast Suzuki–Miyaura coupling reactions of unstable polyfluorophenyl and 2-heteroaryl boronic acids. J Am Chem Soc 132(40):14073–14075
2. Fleckenstein CA, Plenio H (2008) Efficient Suzuki–Miyaura coupling of (Hetero)aryl chlorides with thiophene- and furanboronic acids in aqueous n-butanol. J Org Chem 73(8):3236–3244
3. Ackermann L, Potukuchi HK (2009) Palladium-catalyzed cross-coupling reactions of 2-pyridylborates with air-stable HASPO preligands. Synlett 17:2852–2856
4. Zhou Y et al (2008) Dinickel(II) complexes of bis(N-heterocyclic carbene) ligands containing [Ni2(μ-OH)] cores as highly efficient catalysts for the coupling of aryl chlorides. Organometallics 27(22):5911–5920
5. Liu T et al (2012) General and highly efficient fluorinated-N-heterocyclic carbene-based catalysts for the palladium-catalyzed Suzuki–Miyaura reaction. Tetrahedron 68(32):6535–6547
6. Kolychev EL et al (2013) Expanded ring diaminocarbene palladium complexes: synthesis, structure, and Suzuki–Miyaura cross-coupling of heteroaryl chlorides in water. Dalton Trans 42(19):6859–6866
7. Gupta S, Basu B, Das S (2013) Benzimidazole-based palladium-N-heterocyclic carbene: a useful catalyst for C–C cross-coupling reaction at ambient condition. Tetrahedron 69(1):122–128
8. Tu T et al (2012) Robust acenaphthoimidazolylidene palladium complexes: highly efficient catalysts for Suzuki–Miyaura couplings with sterically hindered substrates. Org Lett 14 (16):4250–4253
9. Kumar MR, Park K, Lee S (2010) Synthesis of amido-N-imidazolium salts and their applications as ligands in Suzuki–Miyaura reactions: coupling of hetero- aromatic halides and the synthesis of milrinone and irbesartan. Adv Synth Catal 352(18):3255–3266
10. Yang J et al (2012) Room-temperature Suzuki–Miyaura coupling of heteroaryl chlorides and tosylates. Eur J Org Chem 2012(31):6248–6259
11. Asano S, Kamioka S, Isobe Y (2012) Suzuki–Miyaura cross-coupling reaction of aryl and heteroaryl pinacol boronates for the synthesis of 2-substituted pyrimidines. Tetrahedron 68 (1):272–279
12. Moseley JD et al (2012) A mild robust generic protocol for the Suzuki reaction using an air stable catalyst. Tetrahedron 68(30):6010–6017
13. Lou S, Fu GC (2010) Palladium/tris(tert-butyl)phosphine-catalyzed Suzuki cross-couplings in the presence of water. Adv Synth Catal 352:2081–2084
14. Bedford RB et al (2009) Simple mixed Fe-Zn catalysts for the Suzuki couplings of tetraarylborates with benzyl halides and 2-halopyridines. Chem Commun 2009(42):6430–6432
15. Liu C, Yang W (2009) A fast and oxygen-promoted protocol for the ligand-free Suzuki reaction of 2-halogenated pyridines in aqueous media. Chem Commun 2009(41):6267–6269
16. Liu C, Ni Q, Qiu J (2011) Very fast, ligand-free and aerobic protocol for the synthesis of 4-aryl-substituted triphenylamine derivatives. Eur J Org Chem 16:3009–3015
17. Liu C et al (2011) Oxygen-promoted PdCl2-catalyzed ligand-free Suzuki reaction in aqueous media. Org Biomol Chem 9(4):1054–1060
18. Colombo M, Giglio M, Peretto I (2008) Simple microwave-assisted ligand-free Suzuki Cross-coupling: functionalization of halo-pyrimidine moieties. J Heterocycl Chem 45 (4):1077–1081
19. Lipshutz BH, Abela AR (2008) Micellar catalysis of Suzuki–Miyaura cross-couplings with heteroaromatics in water. Org Lett 10(23):5329–5332
20. Vashchenko V et al (2008) Palladium-catalyzed Suzuki cross-coupling reactions in a microemulsion. Tetrahedron Lett 49(9):1445–1449

21. Mao S-L et al (2012) A highly active catalytic system for Suzuki–Miyaura cross-coupling reactions of aryl and heteroaryl chlorides in water. Org Biomol Chem 10(47):9410–9417
22. Alacid E, Najera C (2008) First cross-coupling reaction of potassium aryltrifluoroborates with organic chlorides in aqueous media catalyzed by an oxime-derived palladacycle. Org Lett 10(21):5011–5014
23. Susanto W et al (2012) Development of a fluorous, oxime-based palladacycle for microwave-promoted carbon–carbon coupling reactions in aqueous media. Green Chem 14(1):77–80
24. Susanto W et al (2012) Fluorous oxime palladacycle: a precatalyst for carbon–carbon coupling reactions in aqueous and organic medium. J Org Chem 77(6):2729–2742
25. Rao GK et al (2010) Palladacycle containing nitrogen and selenium: highly active pre-catalyst for the Suzuki–Miyaura coupling reaction and unprecedented conversion into nano-sized Pd17Se15. Chem Commun 46(32):5954–5956
26. Bolliger JL, Frech CM (2010) Dichloro-bis(aminophosphine) complexes of palladium: highly convenient, reliable and extremely active Suzuki–Miyaura catalysts with excellent functional group tolerance. Chem Eur J 16(13):4075–4081
27. Shi S, Zhang Y (2008) Silica-assisted Suzuki–Miyaura reactions of heteroaryl bromides in aqueous media. Green Chem 10(8):868–872
28. Razler TM et al (2009) A preparatively convenient ligand-free catalytic PEG 2000 Suzuki–Miyaura coupling. J Org Chem 74(3):1381–1384
29. Tao L et al (2009) Generation of Pd nanoparticles in situ from $PdCl_2$ in TBAF: an efficient and reusable catalytic system for the Suzuki–Miyaura reaction under ligand- and solvent-free conditions. Chin J Chem 27(7):1365–1373
30. Islam RU et al (2011) Conjugated polymer stabilized palladium nanoparticles as a versatile catalyst for Suzuki cross-coupling reactions for both aryl and heteroaryl bromide systems. Catal Sci Technol 1(2):308–315
31. Lee D-H, Jung J-Y, Jin M-J (2010) Highly active and recyclable silica gel-supported palladium catalyst for mild cross-coupling reactions of unactivated heteroaryl chlorides. Green Chem 12(11):2024–2029
32. Kitamura Y et al (2010) Ligand-free and heterogeneous palladium on carbon-catalyzed hetero-Suzuki–Miyaura cross-coupling. Adv Synth Catal 352(4):718–730
33. Firouzabadi H et al (2011) Agarose hydrogel as an effective bioorganic ligand and support for the stabilization of palladium nanoparticles. Application as a recyclable catalyst for Suzuki–Miyaura reaction in aqueous media. RSC Adv 1(6):1013–1019
34. Firouzabadi H et al (2011) Palladium nanoparticles supported on aminopropyl-functionalized clay as efficient catalysts for phosphine-free C–C bond formation via Mizoroki–Heck and Suzuki–Miyaura reactions. Bull Chem Soc Jpn 84(1):100–109
35. Schoeps D et al (2009) Solvent-resistant nanofiltration of enlarged (NHC)Pd(allyl)Cl complexes for cross-coupling reactions. Organometallics 28(13):3922–3927
36. Borhade SR, Waghmode SB (2011) Studies on Pd/$NiFe_2O_4$ catalyzed ligand-free Suzuki reaction in aqueous phase: synthesis of biaryls, terphenyls and polyaryls. Beilstein J Org Chem 7(41):310–319
37. Amoroso F et al (2009) An efficient and reusable catalyst based on Pd/CeO_2 for the room temperature aerobic Suzuki–Miyaura reaction in water/ethanol. J Mol Catal A Chem 315 (2):197–204
38. Zhang P-P et al (2009) Pd-CNT-catalyzed ligand-less and additive-free heterogeneous Suzuki–Miyaura cross-coupling of aryl bromides. Tetrahedron Lett 50(31):4455–4458
39. Fernando DP et al (2012) Spiroazetidine-piperidine bromoindane as a key modular template to access a variety of compounds via C–C and C–N bond-forming reactions. Tetrahedron Lett 53(47):6351–6354
40. Siddle JS, Batsanov AS, Bryce MR (2008) Sequential metal-catalyzed N-heteroarylation and C–C cross-coupling reactions: an expedient route to tris(hetero)aryl systems. Eur J Org Chem 16:2746–2750

41. Anderson SC, Handy ST (2010) One-pot double Suzuki couplings of dichloropyrimidines. Synthesis 16:2721–2724
42. Tasch BOA, Merkul E, Mueller TJJ (2011) One-pot synthesis of diazine-bridged bisindoles and concise synthesis of the marine alkaloid hyrtinadine A. Eur J Org Chem 2011(24):4532–4535
43. Merkul E, Schaefer E, Mueller TJJ (2011) Rapid synthesis of bis(hetero)aryls by one-pot Masuda borylation-Suzuki coupling sequence and its application to concise total syntheses of meridianins A and G. Org Biomol Chem 9(9):3139–3141
44. Grob JE et al (2011) One-pot reductive amination and Suzuki–Miyaura cross-coupling of formyl aryl and heteroaryl MIDA boronates in array format. J Org Chem 76(12):4930–4940
45. Molander GA, Trice SLJ, Kennedy SM (2012) Scope of the two-step, one-pot palladium-catalyzed borylation/Suzuki cross-coupling reaction utilizing bis-boronic acid. J Org Chem 77(19):8678–8688
46. Avitia B et al (2011) Single-flask preparation of polyazatriaryl ligands by sequential borylation/Suzuki–Miyaura coupling. Tetrahedron Lett 52(14):1631–1634
47. Whelligan DK et al (2010) Two-step synthesis of aza- and diazaindoles from chloroamino-N-heterocycles using ethoxyvinylborolane. J Org Chem 75(1):11–15
48. Roy S et al (2010) Direct synthesis of Cbz-protected (2-amino)-6-(2-aminoethyl)pyridines. Tetrahedron 66(11):1973–1979
49. Achelle S et al (2008) Bis- and tris(arylethynyl)pyrimidine oligomers: synthesis and light-emitting properties. Tetrahedron 64(12):2783–2791
50. Achelle S et al (2009) Star-shaped ethynylpyrimidine with long alkoxyl side chains: synthesis, fluorescence and 2D self-assembling. Tetrahedron Lett 50(50):7055–7058
51. Hussain M et al (2010) Synthesis of aryl-substituted pyrimidines by site-selective Suzuki–Miyaura cross-coupling reactions of 2,4,5,6-tetrachloropyrimidine. Adv Synth Catal 352 (9):1429–1433
52. Molander GA, Sandrock DL (2009) Utilization of potassium vinyltrifluoroborate in the development of a 1,2-dianion equivalent. Org Lett 11(11):2369–2372
53. Noel T, Musacchio AJ (2011) Suzuki–Miyaura cross-coupling of heteroaryl halides and arylboronic acids in continuous flow. Org Lett 13(19):5180–5183
54. Shu W et al (2011) Continuous-flow synthesis of biaryls enabled by multistep solid-handling in a lithiation/borylation/Suzuki–Miyaura cross-coupling sequence. Angew Chem Int Ed 50 (45):10665–10669
55. Oberli MA, Buchwald SL (2012) A general method for Suzuki-Miyaura coupling reactions using lithium triisopropyl borates. Org Lett 14(17):4606–4609
56. Radkowski K, Seidel G, Fuerstner A (2011) Suzuki–Miyaura cross coupling reactions of B-allenyl-9-BBN. Chem Lett 40(9):950–952
57. Billingsley KL, Buchwald SL (2008) A general and efficient method for the Suzuki–Miyaura coupling of 2-pyridyl nucleophiles. Angew Chem Int Ed 47(25):4695–4698
58. Dick GR, Woerly EM, Burke MD (2012) A general solution for the 2-pyridyl problem. Angew Chem Int Ed 51(11):2667–2672, S2667/1-S2667/80
59. Molander GA, Canturk B (2008) Preparation of potassium alkoxymethyltrifluoroborates and their cross-coupling with aryl chlorides. Org Lett 10(11):2135–2138
60. Molander GA, Hiebel M-A (2010) Synthesis of amidomethyltrifluoroborates and their use in cross-coupling reactions. Org Lett 12(21):4876–4879
61. Fleury-Bregeot N et al (2012) Suzuki-Miyaura cross-coupling of potassium alkoxyethyltrifluoroborates: access to aryl/heteroarylethyloxy motifs. J Org Chem 77(22):10399–10408
62. Math SK et al (2012) Substituted potassium internal vinyltrifluoroborates: preparation and use in Suzuki–Miyaura cross-coupling reactions. Tetrahedron Lett 53(23):2847–2849
63. Thakur A, Zhang K, Louie J (2012) Suzuki–Miyaura coupling of heteroaryl boronic acids and vinyl chlorides. Chem Commun 48(2):203–205
64. Molander GA, Canturk B, Kennedy LE (2009) Scope of the Suzuki–Miyaura cross-coupling reactions of potassium heteroaryltrifluoroborates. J Org Chem 74(3):973–980

65. Molander GA, Beaumard F (2010) Nickel-catalyzed C–O activation of phenol derivatives with potassium heteroaryltrifluoroborates. Org Lett 12(18):4022–4025
66. Molander GA et al (2010) Nickel-catalyzed cross-coupling of potassium aryl- and heteroaryltrifluoroborates with unactivated alkyl halides. Org Lett 12(24):5783–5785
67. Grimm JB, Wilson KJ, Witter DJ (2009) A divergent approach to the synthesis of 3-substituted-2-pyrazolines: Suzuki cross-coupling of 3-sulfonyloxy-2-pyrazolines. J Org Chem 74(16):6390–6393
68. Menard F, Lautens M (2008) Chemodivergence in enantioselective desymmetrization of diazabicycles: ring-opening versus reductive arylation. Angew Chem Int Ed 47(11):2085–2088
69. Panteleev J, Menard F, Lautens M (2008) Ligand control in enantioselective desymmetrization of bicyclic hydrazines: Rhodium(I)-catalyzed ring-opening versus hydroarylation. Adv Synth Catal 350(18):2893–2902
70. Bexrud J, Lautens M (2010) A rhodium IBiox[(-)-menthyl] complex as a highly selective catalyst for the asymmetric hydroarylation of azabicyles: an alternative route to epibatidine. Org Lett 12(14):3160–3163
71. Hattori K et al (2009) Formation of highly selective and efficient interstrand cross-linking to thymine without photo-irradiation. Chem Commun 2009(42):6463–6465
72. Wicke L, Engels JW (2012) Postsynthetic on column RNA Labeling via stille coupling. Bioconjugate Chem 23(3):627–642
73. Sun Q, Suzenet F, Guillaumet G (2010) Desulfitative cross-coupling of protecting group-free 2-thiouracil derivatives with organostannanes. J Org Chem 75(10):3473–3476
74. Bukovec C, Kazmaier U (2011) Stannylated allyl carbonates as versatile building blocks for the diversity oriented synthesis of allylic amines and amides. Org Biomol Chem 9(8):2743–2750
75. Hadad C et al (2011) 4-Arylvinyl-2,6-di(pyridin-2-yl)pyrimidines: synthesis and optical properties. J Org Chem 76(10):3837–3845
76. Nyffenegger C et al (2008) An efficient route to polynitrogen-fused tricycles via a nitrene-mediated N–N bond formation under microwave irradiation. Tetrahedron 64(40):9567–9573
77. Suzuki M et al (2009) Pd0-Mediated rapid coupling between methyl iodide and heteroarylstannanes: an efficient and general method for the incorporation of a positron-emitting 11C radionuclide into heteroaromatic frameworks. Chem Eur J 15(45):12489–12495
78. Hirano Y, Kojima S, Yamamoto Y (2011) A hypervalent pentacoordinate boron compound with an N-B-N three-center four-electron bond. J Org Chem 76(7):2123–2131
79. Joubert N et al (2008) Modular synthesis of 4-aryl- and 4-amino-substituted benzene C-2-′-deoxyribonucleosides. Synthesis 12:1918–1932
80. Bardagi JI, Rossi RA (2008) A novel approach to the synthesis of 6-substituted uracils in three-step, one-pot reactions. J Org Chem 73(12):4491–4495
81. Calimsiz S et al (2010) Pd-PEPPSI-IPent: low-temperature negishi cross-coupling for the preparation of highly functionalized, tetra-ortho-substituted biaryls. Angew Chem Int Ed 49(11):2014–2017
82. Bolliger JL, Frech CM (2010) Pd(Cl)2{P(NC5H10)(C6H11)2}2 – a highly effective and extremely versatile palladium-based negishi catalyst that efficiently and reliably operates at low catalyst loadings. Chem Eur J 16(36):11072–11081
83. Xi Z, Zhou Y, Chen W (2008) Efficient negishi coupling reactions of aryl chlorides catalyzed by binuclear and mononuclear nickel-N-heterocyclic carbene complexes. J Org Chem 73(21):8497–8501
84. Gerber R, Frech CM (2011) Negishi cross-coupling reactions catalyzed by an aminophosphine-based Nickel system: a reliable and general applicable reaction protocol for the high-yielding synthesis of biaryls. Chem Eur J 17(42):11893–11904
85. Begouin J-M, Gosmini C (2009) Cobalt-catalyzed cross-coupling between in situ prepared arylzinc halides and 2-chloropyrimidine or 2-chloropyrazine. J Org Chem 74(8):3221–3224

86. Begouin J-M, Rivard M, Gosmini C (2010) Cobalt-catalyzed C-SMe bond activation of heteroaromatic thioethers. Chem Commun (Cambridge, UK) 46(32):5972–5974
87. Milne JE, Buchwald SL (2004) An extremely active catalyst for the negishi cross-coupling reaction. J Am Chem Soc 126(40):13028–13032
88. Luzung MR, Patel JS, Yin J (2010) A mild Negishi cross-coupling of 2-heterocyclic organozinc reagents and aryl chlorides. J Org Chem 75(23):8330–8332
89. Dunst C, Knochel P (2011) Selective Mg insertion into substituted mono- and dichloro arenes in the presence of LiCl: a new preparation of boscalid. Synlett (14):2064–2068
90. Mosrin M, Boudet N, Knochel P (2008) Regio- and chemoselective magnesiation of protected uracils and thiouracils using TMPMgClÂ · LiCl and TMP2MgÂ · 2LiCl. Org Biomol Chem 6(18):3237–3239
91. Bluemke T et al (2010) Preparation of functionalized organoaluminiums by direct insertion of aluminum to unsaturated halides. Nat Chem 2(4):313–318
92. Bluemke TD et al (2011) New preparation of benzylic aluminum and zinc organometallics by direct insertion of aluminum powder. Org Lett 13(24):6440–6443
93. Omedes M et al (2008) Diastereoselective addition of organozinc and organomagnesium reagents to 2-(2′-pyrimidyl)ferrocenecarbaldehyde. Tetrahedron 64(18):3953–3959
94. Hasnik Z, Silhar P, Hocek M (2008) Hydroxymethylations of aryl halides by Pd-catalyzed cross-couplings with (benzoyloxy)methylzinc iodide. Scope and limitations of the reaction. Synlett 4:543–546
95. Suhartono M et al (2010) Synthetic aromatic amino acids from a Negishi cross-coupling reaction. Synthesis 2:293–303
96. Kwak Y-S et al (2009) Efficient and convenient preparation of 3-aryl-2,2-dimethylpropanoates via Negishi coupling. Chem Commun (Cambridge, UK) 2009(16):2145–2147
97. Beng TK, Gawley RE (2011) Application of catalytic dynamic resolution of N-Boc-2-lithiopiperidine to the asymmetric synthesis of 2-aryl and 2-vinyl piperidines. Org Lett 13 (3):394–397
98. Han C, Buchwald SL (2009) Negishi coupling of secondary alkylzinc halides with aryl bromides and chlorides. J Am Chem Soc 131(22):7532–7533
99. Thaler T et al (2010) Highly diastereoselective Csp3-Csp2 Negishi cross-coupling with 1,2-, 1,3- and 1,4-substituted cycloalkylzinc compounds. Nat Chem 2(2):125–130
100. Lin L-Y et al (2011) Efficient organic DSSC sensitizers bearing an electron-deficient pyrimidine as an effective π-spacer. J Mater Chem 21(16):5950–5958
101. Manolikakes G et al (2009) Negishi cross-couplings compatible with unprotected amide functions. Chem Eur J 15(6):1324–1328
102. Gosselin F et al (2012) Heteroarylation of azine N-oxides. Org Lett 14(3):862–865
103. Snegaroff K et al (2010) Deprotonative metalation of substituted benzenes and heteroaromatics using amino/alkyl mixed lithium-zinc combinations. Chem Eur J 16 (27):8191–8201
104. Seggio A et al (2008) Synthesis of unsymmetrical heterobiaryls using palladium-catalyzed cross-coupling reactions of lithium organozincates. Synlett 19:2955–2960
105. Seggio A et al (2009) Palladium-catalyzed cross-couplings of lithium arylzincates with aromatic halides: synthesis of analogues of isomeridianin G and evaluation as GSK-3β, inhibitors. Synthesis 21:3617–3632
106. Melzig L, Metzger A, Knochel P (2011) Pd- and Ni-catalyzed cross-coupling reactions of functionalized organozinc reagents with unsaturated thioethers. Chem Eur J 17(10):2948–2956
107. Metzger A, Melzig L, Knochel P (2010) Scaled-up transition-metal-catalyzed cross-coupling reactions of thioether-substituted N-heterocycles with organozinc reagents. Synthesis 16:2853–2858
108. Metzger A et al (2009) Pd-catalyzed cross-coupling of functionalized organozinc reagents with thiomethyl-substituted heterocycles. Org Lett 11(18):4228–4231

109. Metzger A et al (2008) A general preparation of polyfunctional benzylic zinc organometallic compounds. Chem Asian J 3(8–9):1678–1691
110. Schade MA et al (2008) Nickel-catalyzed cross-coupling reactions of benzylic zinc reagents with aromatic bromides, chlorides and tosylates. Chem Commun (Cambridge, UK) 2008 (26):3046–3048
111. Mosrin M, Knochel P (2009) Regio- and chemoselective metalation of chloropyrimidine derivatives with TMPMgCl · LiCl and TMP2Zn · 2MgCl2 · 2LiCl. Chem Eur J 15(6):1468–1477
112. Saleh S et al (2009) A straightforward copper-free palladium methodology for the selective alkynylation of a wide variety of S-, O-, and N-based mono- and diheterocyclic bromides and chlorides. Tetrahedron 65(34):7146–7150
113. Yi T et al (2012) Highly efficient Pd/tetraphosphine catalytic system for copper-free Sonogashira reactions of aryl bromides with terminal alkynes. Catal Lett 142(5):594–600
114. Ngassa FN, Lindsey EA, Haines BE (2009) The first Cu- and amine-free Sonogashira-type cross-coupling in the C-6-alkynylation of protected 2′-deoxyadenosine. Tetrahedron 65 (21):4085–4091
115. Pu X, Li H, Colacot TJ (2013) Heck alkynylation (copper-free Sonogashira coupling) of aryl and heteroaryl chlorides, using Pd complexes of t-Bu2(p-NMe2C6H4)P: understanding the structure-activity relationships and copper effects. J Org Chem 78(2):568–581
116. Cordoba M et al (2013) Sonogashira reaction on pyridinium N-haloheteroarylaminides: regioselective synthesis of N-alkyl-3-alkynyl-5-arylpyridin-2-yl amines. Tetrahedron 69 (11):2484–2493
117. Fleckenstein CA, Plenio H (2008) Aqueous/organic cross coupling: Sustainable protocol for Sonogashira reactions of heterocycles. Green Chem 10(5):563–570
118. Lipshutz BH, Chung DW, Rich B (2008) Sonogashira couplings of aryl bromides: room temperature, water only, no copper. Org Lett 10(17):3793–3796
119. Firouzabadi H, Iranpoor N, Gholinejad M (2010) Recyclable palladium-catalyzed Sonogashira–Hagihara coupling of aryl halides using 2-aminophenyl diphenylphosphinite ligand in neat water under copper-free condition. J Mol Catal A Chem 321(1–2):110–116
120. Barros JC et al (2011) Sonogashira coupling using PdEnCat: a copper-, phosphine-, amine- and microwave-free alternative to the preparation of arylalkynes. Appl Organomet Chem 25 (11):820–823
121. Firouzabadi H, Iranpoor N, Ghaderi A (2011) Gelatin as a bioorganic reductant, ligand and support for palladium nanoparticles. Application as a catalyst for ligand- and amine-free Sonogashira–Hagihara reaction. Org Biomol Chem 9(3):865–871
122. Firouzabadi H et al (2011) Magnetite (Fe3O4) nanoparticles-catalyzed Sonogashira-Hagihara reactions in ethylene glycol under ligand-free conditions. Adv Synth Catal 353 (1):125–132
123. Wu M et al (2008) The use of a bifunctional copper catalyst in the cross-coupling reactions of aryl and heteroaryl halides with terminal alkynes. Eur J Org Chem 23:4050–4054
124. Torborg C, Zapf A, Beller M (2008) Palladium catalysts for highly selective Sonogashira reactions of aryl and heteroaryl bromides. ChemSusChem 1(1–2):91–96
125. Chowdhury C et al (2009) Expedient and rapid synthesis of 1,2,3-triazolo[5,1-c]morpholines through Palladium-copper catalysis. J Org Chem 74(9):3612–3615
126. Gendron T, Davioud-Charvet E, Muller TJJ (2012) Versatile synthesis of dissymmetric diarylideneacetones via a palladium-catalyzed coupling-isomerization reaction. Synthesis 44(24):3829–3835
127. West K et al (2008) Synthesis, structures and reactions of isolable terminal aryl/biaryl-butadiynes (Ar-C≡C-C≡CH). Eur J Org Chem 30:5093–5098
128. West K et al (2008) Carbon-rich molecules: synthesis and isolation of aryl/heteroaryl terminal bis(butadiynes) (HC≡C-C≡C-Ar-C≡C-C≡CH) and their applications in the synthesis of oligo(arylenebutadiynylene) molecular wires. Org Biomol Chem 6(11):1934–1937

129. Malik I et al (2011) Synthesis and photophysical properties of alkynylated pyrimidines by site-selective Sonogashira reactions of 2,4,5,6-tetrachloropyrimidine; first synthesis of tetraalkynyl-pyrimidines. Eur J Org Chem 11:2088–2093
130. Kantchev EAB et al (2008) Practical Heck–Mizoroki coupling protocol for challenging substrates mediated by an N-heterocyclic carbene-ligated palladacycle. Org Lett 10 (18):3949–3952
131. Peh G-R et al (2009) N-Heterocycle carbene (NHC)-ligated cyclopalladated N, N-dimethylbenzylamine: a highly active, practical and versatile catalyst for the Heck–Mizoroki reaction. Org Biomol Chem 7(10):2110–2119
132. Firouzabadi H, Iranpoor N, Gholinejad M (2009) 2-Aminophenyl diphenylphosphinite as a new ligand for heterogeneous palladium-catalyzed Heck–Mizoroki reactions in water in the absence of any organic co-solvent. Tetrahedron 65(34):7079–7084
133. Firouzabadi H, Iranpoor N, Ghaderi A (2011) Solvent-free Mizoroki–Heck reaction catalyzed by palladium nanoparticles deposited on gelatin as the reductant, ligand and the non-toxic and degradable natural product support. J Mol Catal A Chem 347(1–2):38–45
134. Iranpoor N et al (2012) Palladium nanoparticles supported on silicadiphenyl phosphinite (SDPP) as efficient catalyst for Mizoroki–Heck and Suzuki–Miyaura coupling reactions. J Organomet Chem 708–709:118–124
135. Firouzabadi H et al (2012) Palladium nano-particles supported on agarose as efficient catalyst and bioorganic ligand for C–C bond formation via solventless Mizoroki–Heck reaction and Sonogashira–Hagihara reaction in polyethylene glycol (PEG 400). J Mol Catal A Chem 357:154–161
136. He T et al (2008) Acetylation of N-heteroaryl bromides via PdCl2/(o-tolyl)3P catalyzed Heck reactions. Synthesis 6:887–890
137. Elboray EE, Gao C, Grigg R (2012) Skeletal diversity via Pd(0) catalysed three-component cascades of allene and halides or triflates with protected hydroxylamines and formamide. Tetrahedron 68(14):3103–3111
138. Goegsig TM et al (2009) Heteroaromatic tosylates as electrophiles in regioselective Mizoroki–Heck-coupling reactions with electron-rich olefins. Chem Eur J 15(24):5950–5955, S5950/1-S5950/89
139. Kubelka T, Slavetinska L, Hocek M (2012) A general regioselective approach to 2,4-disubstituted pyrimidin-5-yl C-2-deoxyribonucleosides. Synthesis 44(6):953–965
140. Kubelka T et al (2010) Synthesis of 2,4-disubstituted pyrimidin-5-yl C-2-′-deoxyribonucleosides by sequential regioselective reactions of 2,4-dichloropyrimidine nucleosides. Eur J Org Chem 2100(14):2666–2669
141. Mehta VP, Modha SG, Van der Eycken E (2009) Mild room-temperature palladium-catalyzed C3-arylation of 2(1H)-pyrazinones via a desulfitative Kumada-type cross-coupling reaction. J Org Chem 74(17):6870–6873
142. Prokopcova H, Kappe CO (2009) The Liebeskind–Srogl C–C cross-coupling reaction. Angew Chem Int Ed 48(13):2276–2286
143. Liebeskind LS, Srogl J (2002) Heteroaromatic thioether-boronic acid cross-coupling under neutral reaction conditions. Org Lett 4(6):979–981
144. Manolikakes G, Knochel P (2009) Radical catalysis of Kumada cross-coupling reactions using functionalized Grignard reagents. Angew Chem Int Ed 48(1):205–209
145. Iglesias MJ, Prieto A, Nicasio MC (2012) Kumada–Tamao–Corriu coupling of heteroaromatic chlorides and aryl ethers catalyzed by (IPr)Ni(allyl)Cl. Org Lett 14 (17):4318–4321
146. Xi Z et al (2007) Synthesis and structural characterization of nickel(II) complexes supported by pyridine-functionalized N-heterocyclic carbene ligands and their catalytic activities for Suzuki coupling. Organometallics 26(26):6636–6642
147. Xi Z, Liu B, Chen W (2008) Room-temperature Kumada cross-coupling of unactivated aryl chlorides catalyzed by N-heterocylic carbene-based nickel(II) complexes. J Org Chem 73 (10):3954–3957

148. Gu S, Chen W (2009) Pincer complexes of palladium- and nickel-containing 3-butyl-1-(1,10-phenanthrolin-2-yl)imidazolylidene as efficient aqueous Sonogashira and Kumada coupling reactions. Organometallics 28(3):909–914
149. Chen C, Qiu H, Chen W (2012) Trinuclear copper(I) complex of 1,3-bis(2-pyridinylmethyl) imidazolylidene as a carbene-transfer reagent for the preparation of catalytically active nickel (II) and palladium(II) complexes. J Organomet Chem 696(26):4166–4172
150. Ackermann L et al (2010) Kumada–Corriu cross-couplings with 2-pyridyl grignard reagents. Chem Eur J 16(11):3300–3303
151. Jin Z et al (2012) Biphenyl-based diaminophosphine oxides as air-stable preligands for the nickel-catalyzed Kumada–Tamao–Corriu coupling of deactivated aryl chlorides, fluorides, and tosylates. Chem Eur J 18(2):446–450, S446/1-S446/93
152. Goegsig TM, Lindhardt AT, Skrydstrup T (2009) Heteroaromatic sulfonates and phosphates as electrophiles in iron-catalyzed cross-couplings. Org Lett 11(21):4886–4888
153. Kuzmina OM et al (2012) Iron-catalyzed cross-coupling of N-heterocyclic chlorides and bromides with arylmagnesium reagents. Org Lett 14(18):4818–4821
154. Modha SG et al (2011) An expeditious route toward pyrazine-containing nucleoside analogues. J Org Chem 76(3):846–856
155. Arshad N, Hashim J, Kappe CO (2009) Palladium(0)-catalyzed, copper(I)-mediated coupling of cyclic thioamides with alkenylboronic acids, organostannanes, and siloxanes. J Org Chem 74(14):5118–5121
156. Farahat AA, Boykin DW (2011) Unusual regioselective reactions of 2,4-bis(methylsulfanyl) pyrimidine under modified Suzuki and Stille cross-coupling conditions. Synthesis 44(1):120–124
157. Brehova P et al (2011) The efficient synthesis of 2-arylpyrimidine acyclic nucleoside phosphonates using Liebeskind–Srogl cross-coupling reaction. Tetrahedron 67(38):7379–7385
158. Van Rossom W et al (2008) Efficient post-macrocyclization functionalizations of oxacalix[2] arene[2]pyrimidines. Org Lett 10(4):585–588
159. Shen Q, Ogata T, Hartwig JF (2008) Highly reactive, general and long-lived catalysts for palladium-catalyzed amination of heteroaryl and aryl chlorides, bromides, and iodides: scope and structure–activity relationships. J Am Chem Soc 130(20):6586–6596
160. Maiti D et al (2011) Palladium-catalyzed coupling of functionalized primary and secondary amines with aryl and heteroaryl halides: two ligands suffice in most cases. Chem Sci 2(1):57–68
161. McGowan MA, Henderson JL, Buchwald SL (2012) Palladium-catalyzed N-arylation of 2-aminothiazoles. Org Lett 14(6):1432–1435
162. Fors BP et al (2009) An efficient system for the Pd-catalyzed cross-coupling of amides and aryl chlorides. Tetrahedron 65(33):6576–6583
163. Fors BP, Davis NR, Buchwald SL (2009) An efficient process for Pd-catalyzed C–N cross-coupling reactions of aryl iodides: insight into controlling factors. J Am Chem Soc 131 (16):5766–5768
164. Ueda S, Su M-J, Buchwald SL (2011) Highly N2-selective palladium-catalyzed arylation of 1,2,3-triazoles. Angew Chem Int Ed 50(38):8944–8947
165. Ueda S et al (2012) Me3(OMe)tBuXPhos: a surrogate ligand for Me4tBuXPhos in palladium-catalyzed C–N and C–O bond-forming reactions. J Org Chem 77(5):2543–2547
166. Rodriguez S et al (2011) Oxaphosphole-based monophosphorus ligands for palladium-catalyzed amination reactions. Adv Synth Catal 353(4):533–537
167. Marion N et al (2006) Modified (NHC)Pd(allyl)Cl (NHC=N-heterocyclic carbene) complexes for room-temperature Suzuki–Miyaura and Buchwald–Hartwig reactions. J Am Chem Soc 128(12):4101–4111
168. Organ MG et al (2008) Pd-catalyzed aryl amination mediated by well defined, N-heterocyclic carbene (NHC)-Pd precatalysts, PEPPSI. Chem Eur J 14(8):2443–2452

169. Elmkaddem MK et al (2010) Efficient synthesis of aminopyridine derivatives by copper catalyzed amination reactions. Chem Commun (Cambridge, UK) 46(6):925–927
170. Xi Z et al (2008) CuI/L (L=pyridine-functionalized 1,3-diketones) catalyzed C–N coupling reactions of aryl halides with NH-containing heterocycles. Tetrahedron 64(19):4254–4259
171. Chen H et al (2010) Mild conditions for copper-catalyzed N-arylation of imidazoles. Synthesis 9:1505–1511
172. Cao C et al (2012) Cheap Cu(I)/hexamethylenetetramine (HMTA) catalytic system for C–N coupling reactions. Synth Commun 42(2):279–284
173. Cheng D et al (2008) D-Glucosamine – a natural ligand for the N-arylation of imidazoles with aryl and heteroaryl bromides catalyzed by CuI. Green Chem 10(2):171–173
174. Suresh P, Pitchumani K (2008) Per-6-amino-β-cyclodextrin as an efficient supramolecular ligand and host for Cu(I)-catalyzed N-arylation of imidazole with aryl bromides. J Org Chem 73(22):9121–9124
175. Kwon JK et al (2011) N-Arylation of carbazole by microwave-assisted ligand-free catalytic CuI reaction. Tetrahedron 67(26):4820–4825
176. Liu Z-J et al (2010) Ligand-free copper-catalyzed amination of heteroaryl halides with alkyl- and arylamines. Adv Synth Catal 352(18):3158–3162
177. Johansson Seechurn CCC, Parisel SL, Colacot TJ (2011) Air-stable Pd(R-allyl)LCl (L=Q-Phos, P(t-Bu)3, etc.) systems for C-C/N couplings: insight into the structure–activity relationship and catalyst activation pathway. J Org Chem 76(19):7918–7932
178. Moss TA et al (2012) Room-temperature palladium-catalyzed coupling of heteroaryl amines with aryl or heteroaryl bromides. Synlett 23(2):285–289
179. Tang B-X et al (2008) N-arylations of nitrogen-containing heterocycles with aryl and heteroaryl halides using a copper(I) oxide nanoparticle/1,10-phenanthroline catalytic system. Synthesis 11:1707–1716
180. Kantam ML, Ramani T, Chakrapani L (2008) N-Arylation of heterocycles with chloro- and fluoroarenes using resin-supported sulfonato-Cu(salen) complex. Synth Commun 38(4):626–636
181. Vimolratana M, Simard JL, Brown SP (2011) Palladium-catalyzed amidation of 2-chloropyrimidines. Tetrahedron Lett 52(9):1020–1022
182. Baffoe J, Hoe MY, Toure BB (2010) Copper-mediated N-heteroarylation of primary sulfonamides: synthesis of mono-N-heteroaryl sulfonamides. Org Lett 12(7):1532–1535
183. Wang X et al (2012) Copper-catalyzed N-arylation of sulfonamides with aryl bromides under mild conditions. Tetrahedron Lett 53(1):7–10
184. Anjanappa P et al (2008) 2-(Trimethylsilyl)ethanesulfonyl amide as a new ammonia equivalent for palladium-catalyzed amination of aryl halides. Tetrahedron Lett 49(31):4585–4587
185. Prakash A et al (2011) Efficient indoles and anilines syntheses employing tert-butyl sulfinamide as ammonia surrogate. Tetrahedron Lett 52(43):5625–5628
186. Funes Maldonado M et al (2012) Synthesis and arylation of unprotected sulfonimidamides. Tetrahedron 68(36):7456–7462
187. Breitler S et al (2011) Synthesis of unsymmetrical diarylureas via Pd-catalyzed C-N cross-coupling reactions. Org Lett 13(12):3262–3265
188. Hammoud H et al (2012) Direct guanidinylation of aryl and heteroaryl halides via copper-catalyzed cross-coupling reaction. J Org Chem 77(1):417–423
189. Thiel OR et al (2010) Palladium-catalyzed coupling of aldehyde-derived hydrazones: practical synthesis of triazolopyridines and related heterocycles. Angew Chem Int Ed 49 (45):8395–8398
190. Vinogradova EV et al (2013) Palladium-catalyzed synthesis of N-Aryl carbamates. Org Lett 15(6):1394–1397
191. Das AR, Medda A, Singha R (2010) Synthesis of biologically potent new 3-(heteroaryl) aminocoumarin derivatives via Buchwald–Hartwig C–N coupling. Tetrahedron Lett 51 (7):1099–1102

192. Shen Z et al (2010) Switching the chemoselectivity in the amination of 4-chloroquinazolines with aminopyrazoles. Org Lett 12(3):552–555
193. Cordoba M, Izquierdo ML, Alvarez-Builla J (2008) New approaches to the synthesis of pyridinium N-heteroarylaminides. Tetrahedron 64(34):7914–7919
194. Zhang G et al (2013) Palladium-catalyzed cross-coupling of electron-deficient heteroaromatic amines with heteroaryl halides. Synth Commun 43(3):456–463
195. Lach F, Pasquet M-J, Chabanne M (2011) A general route to unsubstituted N-aryl and heteroarylaminobenzenesulfonamides. Tetrahedron Lett 52(16):1882–1887
196. Hirao T et al (1982) Palladium-catalyzed new carbon-phosphorus bond formation. Bull Chem Soc Jpn 55(3):909–913
197. Belabassi Y, Alzghari S, Montchamp J-L (2008) Revisiting the Hirao cross-coupling: improved synthesis of aryl and heteroaryl phosphonates. J Organomet Chem 693 (19):3171–3178
198. Deal EL, Petit C, Montchamp J-L (2011) Palladium-catalyzed cross-coupling of H-phosphinate esters with chloroarenes. Org Lett 13(12):3270–3273
199. Zhang H-Y et al Nickel-catalyzed C–P cross-coupling of diphenylphosphine oxide with aryl chlorides. Org Biomol Chem 10(48):9627–9633
200. Kosugi M et al (1985) Palladium-catalyzed reaction of stannyl sulfide with aryl bromide. Preparation of aryl sulfide. Bull Chem Soc Jpn 58(12):3657–3658
201. Xin K et al (2009) Efficient iron/copper cocatalyzed S-arylations of thiols with aryl halides. J Comb Chem 11(3):338–340
202. Mo J et al (2011) Palladium-catalyzed carbon-sulfur cross-coupling reactions of aryl chlorides with indium tris(organothiolates). Chem Lett 40(9):980–982
203. Babu SG, Karvembu R. Room temperature Ullmann type C–O and C–S cross coupling of aryl halides with phenol/thiophenol catalyzed by CuO nanoparticles. Tetrahedron Lett 54 (13):1677–1680
204. Dash P, Janni M, Peruncheralathan S (2012) Trideuteriomethoxylation of aryl and heteroaryl halides. Eur J Org Chem 26:4914–4917
205. Schoenberg A, Heck RF (1974) Palladium-catalyzed amidation of aryl, heterocyclic, and vinylic halides. J Org Chem 39(23):3327–3331
206. Roberts B, Liptrot D, Alcaraz L (2010) Novel aryl and heteroaryl acyl sulfamide synthesis via microwave-assisted palladium-catalyzed carbonylation. Org Lett 12(6):1264–1267
207. Qu B et al (2009) Palladium-catalyzed aminocarbonylation of heteroaryl halides using di-tert-butylphosphinoferrocene. Tetrahedron Lett 50(45):6126–6129
208. Borhade SR, Sandstroem A, Arvidsson PI (2013) Synthesis of novel aryl and heteroaryl acyl sulfonimidamides via Pd-catalyzed carbonylation using a nongaseous precursor. Org Lett 15 (5):1056–1059
209. Schareina T et al (2010) An improved protocol for palladium-catalyzed alkoxycarbonylations of aryl chlorides with alkyl formates. Adv Synth Catal 352(7):1205–1209
210. Ueda T, Konishi H, Manabe K (2012) Palladium-catalyzed carbonylation of aryl, alkenyl, and allyl halides with phenyl formate. Org Lett 14(12):3100–3103
211. Kang F-A, Sui Z, Murray WV (2008) Pd-catalyzed direct arylation of tautomerizable heterocycles with aryl boronic acids via C-OH bond activation using phosphonium salts. J Am Chem Soc 130(34):11300–11302
212. Sharma A, Vachhani D, Van der Eycken E (2012) Direct heteroarylation of tautomerizable heterocycles into unsymmetrical and symmetrical biheterocycles via Pd/Cu-catalyzed phosphonium coupling. Org Lett 14(7):1854–1857
213. Wang X et al (2008) Palladium-catalyzed one-pot synthesis of 2-alkyl-2-arylcyanoacetates. J Org Chem 73(4):1643–1645
214. Mosquera A et al (2008) Cross-coupling reactions of indium organometallics with 2,5-dihalopyrimidines: synthesis of hyrtinadine A. Org Lett 10(17):3745–3748
215. Schareina T et al (2008) A bio-inspired copper catalyst system for practical catalytic cyanation of aryl bromides. Synthesis 20:3351–3355

216. Chen Y-H, Knochel P (2008) Preparation of Aryl and heteroaryl indium(III) reagents by the direct insertion of indium in the presence of LiCl. Angew Chem Int Ed 47(40):7648–7651
217. Chen Y-H, Sun M, Knochel P (2009) LiCl-mediated preparation of functionalized benzylic indium(III) halides and highly chemoselective palladium-catalyzed cross-coupling in a protic cosolvent. Angew Chem Int Ed 48(12):2236–2239
218. Zhao X et al (2010) Microwave-assisted, Pd(0)-catalyzed cross-coupling of diazirines with aryl halides. Org Lett 12(23):5580–5583
219. Sengmany S, Le Gall E, Leonel E (2011) An electrochemical synthesis of functionalized arylpyrimidines from 4-amino-6-chloropyrimidines and aryl halides. Molecules 16:5550–5560
220. Bochatay VN et al (2013) Mechanistic exploration of the palladium-catalyzed process for the synthesis of benzoxazoles and benzothiazoles. J Org Chem 78(4):1471–1477
221. Colomb J, Billard T (2013) Palladium-catalyzed desulfitative arylation of 3-haloquinolines with arylsulfinates. Tetrahedron Lett 54(11):1471–1474
222. Zhou C et al (2012) Palladium-catalyzed desulfitative arylation by C–O bond cleavage of aryl triflates with sodium arylsulfinates. J Org Chem 77(22):10468–10472
223. Asaki T et al (2007) Structure-activity studies on diphenylpyrazine derivatives: a novel class of prostacyclin receptor agonists. Bioorg Med Chem 15(21):6692–6704
224. Buron F et al (2005) Synthesis of pyrazine alkaloids from *Botryllus leachi*. Diazines 43. J Org Chem 70(7):2616–2621
225. Buron F et al (2007) Towards a biomimetic synthesis of barrenazine A. Tetrahedron Lett 48 (25):4327–4330
226. Corbett JW et al (2007) Heteroatom-linked indanylpyrazines are corticotropin releasing factor type-1 receptor antagonists. Bioorg Med Chem Lett 17(22):6250–6256
227. Dembitsky VM, Gloriozova TA, Poroikov VV (2007) Natural peroxy anticancer agents. Mini-Rev Med Chem 7(6):571–589
228. Geiger C et al (2007) Synthesis of bicyclic σ receptor ligands with cytotoxic activity. J Med Chem 50(24):6144–6153
229. Martinez MM, Sarandeses LA, Sestelo JP (2007) Enantioselective synthesis of (-)-barrenazines A and B. Tetrahedron Lett 48(48):8536–8539
230. Kim KB, Crews CM (2008) Chemical genetics: exploring the role of the proteasome in cell biology using natural products and other small molecule proteasome inhibitors. J Med Chem 51(9):2600–2605
231. Ge S, Hartwig JF (2012) Highly reactive, single-component nickel catalyst precursor for Suzuki–Miyuara cross-coupling of heteroaryl boronic acids with heteroaryl halides. Angew Chem Int Ed 51(51):12837–12841
232. Dowlut M, Mallik D, Organ MG (2010) An efficient low-temperature Stille-Migita cross-coupling reaction for heteroaromatic compounds by Pd-PEPPSI-IPent. Chem Eur J 16 (14):4279–4283, S4279/1-S4279/53
233. Brandsma L, Vasilevsky SF, Verkruijsse HD (1997) Application of transition metal catalysts in organic synthesis. Springer-Verlag/Heidelberg, Berlin/New York
234. Diederich F, Stang PJ (eds) (1998) Metal-catalyzed cross-coupling reactions. Wiley-VCH, Weinheim, 517 pp
235. Pattenden G (ed) (1992) Comprehensive organic synthesis: selectivity, strategy and efficiency in modern organic chemistry, volume 3: carbon–carbon σ-bond formation. Pergamon, New York
236. Rossi R, Carpita A, Bellina F (1995) Palladium- and/or copper-mediated cross-coupling reactions between 1-alkynes and vinyl, aryl, 1-alkynyl, 1,2-propadienyl, propargyl and allylic halides or related compounds. A review Org Prep Proced Int 27(2):127–160
237. Ali H, van Lier JE (2012) An easy route for the synthesis of pyrazine-2,3-dicarbonitrile 5,6-bis-substituted derivatives using a palladium catalyst. Tetrahedron Lett 53(36):4824–4827

238. Malik I et al (2010) Synthesis of 2,3-disubstituted pyrazines and quinoxalines by Heck cross-coupling reactions of 2,3-dichloropyrazine and 2,3-dichloroquinoxaline. Influence of the temperature on the product distribution. Tetrahedron 66(9):1637–1642
239. Mehta VP et al (2008) A novel and versatile entry to asymmetrically substituted pyrazines. J Org Chem 73(6):2382–2388
240. Modha SG et al (2012) Efficient preparation of tetrasubstituted pyrazines starting from pyrazin-2(1H)-ones. Synthesis 44(11):1614–1624
241. Won Y-H, Park M-S (2010) Synthesis and anticancer activities of new 3-allylthio-6-(mono or disubstituted)aminopyridazines. Arch Pharmacal Res 33(2):189–196
242. Rohet F et al (1997) Synthesis and analgesic effects of 3-substituted 4,6-diarylpyridazine derivatives of the arylpiperazine class. Bioorg Med Chem 5(4):655–659
243. Allerton CMN et al (2009) Design and synthesis of pyridazinone-based 5-HT2C agonists. Bioorg Med Chem Lett 19(19):5791–5795
244. Gleave RJ et al (2010) Synthesis and evaluation of 3-amino-6-aryl-pyridazines as selective CB2 agonists for the treatment of inflammatory pain. Bioorg Med Chem Lett 20(2):465–468
245. Isabel E et al (2011) Biological activity and preclinical efficacy of azetidinyl pyridazines as potent systemically-distributed stearoyl-CoA desaturase inhibitors. Bioorg Med Chem Lett 21(1):479–483
246. Liu G et al (2007) Discovery of potent, selective, orally bioavailable stearoyl-CoA desaturase 1 inhibitors. J Med Chem 50(13):3086–3100
247. Wan Z et al (2011) Pyridazine-derived γ-secretase modulators. Bioorg Med Chem Lett 21 (13):4016–4019
248. Contreras J-M et al (1999) Aminopyridazines as acetylcholinesterase inhibitors. J Med Chem 42(4):730–741
249. Contreras J-M et al (2001) Design, synthesis, and structure–activity relationships of a series of 3-[2-(1-benzylpiperidin-4-yl)ethylamino]pyridazine derivatives as acetylcholinesterase inhibitors. J Med Chem 44(17):2707–2718
250. Gavande N et al (2010) Microwave-enhanced synthesis of 2,3,6-trisubstituted pyridazines: application to four-step synthesis of gabazine (SR-95531). Org Biomol Chem 8(18):4131–4136
251. Woodward RM, Polenzani L, Miledi R (1993) Characterization of bicuculline/baclofen-insensitive (ρ-like) γ-aminobutyric acid receptors expressed in *Xenopus* oocytes. II Pharmacology of γ-aminobutyric acidA and γ-aminobutyric acidB receptor agonists and antagonists. Mol Pharmacol 43(4):609–625
252. Zhang J, Xue F, Chang Y (2008) Structural determinants for antagonist pharmacology that distinguish the ρ1 GABAC receptor from GABAA receptors. Mol Pharmacol 74(4):941–951
253. Sengmany S et al (2013) An electrochemical nickel-catalyzed arylation of 3-amino-6-chloropyridazines. J Org Chem 78(2):370–379

Top Heterocycl Chem (2016) 45: 159–230
DOI: 10.1007/7081_2015_140

Published online: 24 March 2015

Metal-Catalysed Cross-Coupling Reactions in the Synthesis and Transformations of Quinolones and Acridones

Raquel S.G.R. Seixas*, Vera L.M. Silva*, and Artur M.S. Silva

Abstract Metal-catalysed cross-coupling reactions have had a large impact on synthetic organic chemistry and have found many applications in target-oriented synthesis. Their widespread use in organic synthesis is due to the mild conditions associated with the reactions together with their tolerance of a wide range of functional groups. The cross-coupling reactions have been applied to the synthesis of a large number of natural products and bioactive compounds of complex molecular structures. This chapter presents an overview of the applications of cross-coupling reactions in the synthesis and transformations of quinolones and acridones. These compounds are widely recognized by their diverse bioactivity being useful structural moieties for drug candidates. Furthermore they hold significant interest due to their host–guest chemistry; applications in chemical, biochemical and environmental analyses and utility in synthetic methods' development.

Keywords Acridones · Carbonylation · Copper · Cross-coupling · Cyclization · Metal-catalysed · Palladium · Quinolones · Reaction mechanisms

Contents

*These authors have contributed equally to the chapter.

R.S.G.R. Seixas, V.L.M. Silva, and A.M.S. Silva (✉)
Department of Chemistry and QOPNA, University of Aveiro, 3810-193 Aveiro, Portugal
e-mail: artur.silva@ua.pt

Abbreviations

Ac	Acetyl
acac	Acetylacetonate
Alk	Alkyl
BINAP	2,2′-Bis(diphenylphosphino)-1,1′-binaphthyl
Bn	Benzyl
Boc	*tert*-Butoxycarbonyl
bppm	*tert*-Butoxycarbonyl-4-diphenylphosphino-2-diphenylphosphinomethylpyrrolidine
Bu	Butyl
CH	Classical heating
cod	Cyclooctadiene
CuTC	Copper(I)-thiophene-2-carboxylate
Cy	Cyclohexyl
DABCO	1,4-Diazabicyclo[2.2.2]octane
dba	Dibenzylideneacetone
DBU	1,8-Diazabicyclo[5.4.0]undec-7-ene
DEAD	Diethyl azodicarboxylate
DIOP	2,3-*O*-Isopropylidene-2,3-dihydroxy-1,4-bis(diphenylphosphino) butane
DIPEA	Diisopropylethylamine
DMA	Dimethylacetamide
DMEDA	Dimethylethylenediamine
DMF	Dimethylformamide
DMPU	Dimethylpropylene urea
DMSO	Dimethyl sulfoxide
DPEphos	Bis[(2-diphenylphosphino)phenyl]ether
dppe	Bis(diphenylphosphino)ethane
dppf	1,1′-Bis(diphenylphosphino)ferrocene

dppp	1,3-Bis(diphenylphosphino)propane
DTBMP	2,6-Di-*tert*-butyl-4-methylpyridine
ee	Enantiomeric excess
er	Enantiomeric ratio
Et	Ethyl
HMPA	Hexamethylphosphoramide
Hx	Hexyl
LDA	Lithium diisopropylamide
Me	Methyl
MOM	Methoxymethyl
Ms	Mesyl
MW	Microwave
NBS	*N*-Bromosuccinimide
Pent	Pentyl
Ph	Phenyl
Piv	Pivaloyl
PMHS	Polymethyl-hydrosiloxane
Pr	Propyl
PTSA	*p*-Toluenesulfonic acid
Py	Pyridine
R,*R*-BDPP	(2*S*,3*S*)-(−)-Bis(diphenylphosphino)butane
rac	Racemic
R-JOSIPHOS	(*R*)-1-[(*S*)-2-(diphenylphosphino)ferrocenyl] ethyldicyclohexylphosphine
rt	Room temperature
SPHOS	2-Dicyclohexylphosphino-2′,6′-dimethoxy-1,1′-biphenyl
TBAB	Tetrabutylammonium bromide
THF	Tetrahydrofuran
THP	Tetrahydropyran
TMEDA	*N*,*N*,*N*′,*N*′-tetramethyl-1,2-ethylenediamine
TMS	Trimethylsilyl
TMSA	Trimethylsilylacetylene
tol	4-Methylphenyl(tolyl)
Xantphos	4,5-Bis(diphenylphosphino)-9,9-dimethylxanthene
XPhos	2-Dicyclohexylphosphino-2′,4′,6′-triisopropylbiphenyl

1 Introduction

Carbon–carbon bond formation through transition metal-catalysed cross-coupling reactions remains one of the most powerful methods in organic synthesis. In this chapter a review of metal-catalysed cross-coupling reactions applied to the synthesis and transformation of quinolin-2(1*H*)- and quinolin-4(1*H*)-ones up to 2012 is presented, highlighting the reaction conditions, including the catalysts, ligands, bases and selectivity, evidencing the representative C–C bond-forming reactions

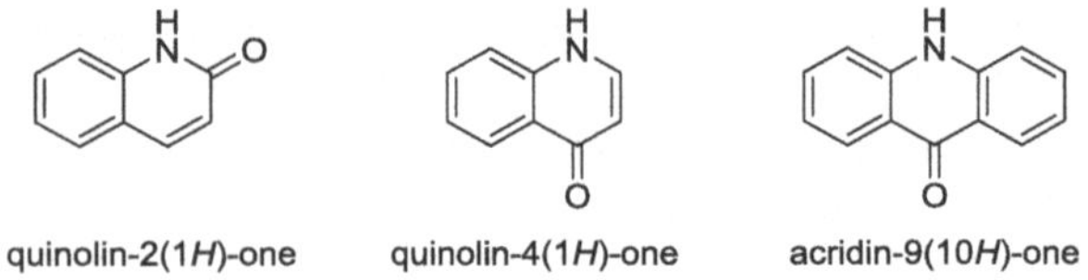

Fig. 1 Compounds discussed in this chapter

(e.g. Sonogashira, Suzuki, Buchwald–Hartwig, Stille, Heck reaction, among others). In the same period, publications concerning these types of reactions in the synthesis of acridin-9(10*H*)-ones are very scarce, with a combined Buchwald–Hartwig aryl amination-directed remote metalation (DreM) protocol, an aryl amination followed by an intramolecular Friedel–Crafts acylation using polymer-supported Pd and Sc catalysts and more recently the electrocyclization of (*E*,*E*)-2,3-distyrylquinolin-4(1*H*)-ones being only known. To the best of our knowledge, there is no report about their chemical transformations.

Quinolin-2(1*H*)- and quinolin-4(1*H*)-ones are benzo-α- and benzo-γ-pyridones, respectively, since they are constituted by a α- or γ-pyridone *ortho*-fused with a benzene ring (Fig. 1). The quinolone motif is common in nature, particularly in alkaloids from Rutaceae family but can also be produced by different animal and bacterial species (e.g. [1, 2]). These compounds represent privileged moieties in medicinal chemistry and are ubiquitous substructures associated with biologically active natural products. Quinolin-4(1*H*)-ones are well-known drugs used as antibiotics (e.g. fluoroquinolones) due to their excellent antimicrobial activity (e.g. [3]), but they also possess other interesting biological properties such as antimalarial (e.g. [4]) and antitumoural (e.g. [5, 6]) activities, among others. Considerable interest in studying the isomeric quinolin-2(1*H*)-ones is also related to the different kinds of important biological activities demonstrated. These compounds were reported as antiulcer (e.g. rebamipide), antihistaminic (e.g. repirinast) (e.g. [7]) and anticancer (e.g. tipifarnib) (e.g. [8, 9]) agents and have also been evaluated, for example, as inhibitors of HIV-1 reverse transcriptase (e.g. [10, 11]) and antivirals (e.g. [12]). Up to now several methods for the synthesis of quinolones have been reported in the literature, but the most commonly used involve the condensation of anilines with β-ketoesters followed by cyclization to give quinolin-4(1*H*)-ones (Conrad–Limpach synthesis) or quinolin-2(1*H*)-ones (Conrad–Limpach–Knorr synthesis). Other classical methods leading to the formation of the C3–C4 bond include the Friedländer synthesis, Camps modification and Niementowski reaction [13–15].

Acridin-9(10*H*)-ones, also called dibenzo-γ-pyridones, are tricyclic compounds presenting a γ-pyridone with two *ortho*-fused benzene rings (Fig. 1). Several naturally occurring and synthetic acridin-9(10*H*)-one derivatives are known due to their biomedical potential including antiviral [16], antimalarial (e.g. [17]), antitumoural and anticancer activities (e.g. [18, 19]). Besides medicinal applications, acridin-9(10*H*)-ones are also used in host–guest interactions and in chemical, biochemical, and environmental analysis as fluorescence probes and as analytical tools in biomimetic chemistry [20–26]. Acridin-9(10*H*)-ones are commonly prepared by the acid-induced ring closure of *N*-arylanthranilic acids, themselves usually obtained from the Ullmann condensation of anilines with *ortho*-halogen-substituted benzoic acids [27–31]. Other methods involve the intermolecular nucleophilic coupling of arynes with *ortho*-aminobenzoates and subsequent intramolecular nucleophilic cyclization [32, 33] and the anionic *N*-Fries rearrangement of *N*-carbamoyldiarylamines to anthranylamides followed by cyclization with triflic anhydride [34].

Most of these methods often require harsh conditions, tedious workup and purification procedures and are regioselectivity compromised, and the variety of substrates is limited. The drawbacks of the conventional methods can be overcome by metal-catalysed reactions in the synthesis and transformation of these types of compounds. This is a recent topic but several methods have already been established meaning that it is an important and upward research field.

2 Synthesis of Quinolin-4(1*H*)-ones

2.1 Sonogashira Reaction

2-Iodoanilides **1** reacted with terminal propargyl alcohols **2** under Pd-catalysed conditions to yield the corresponding 2-substituted anilides **3** [35, 36]. Among several catalysts, $PdCl_2(PPh_3)_2$ was found to be the best, and the addition of CuI was not found to give any additional advantage. Propargyl alcohols **2** in the presence of Et_3N appear to be capable to reduce Pd(II) to Pd(0) before the oxidative addition of Pd(0) to the aryl halide could occur. Only in one case (R^1=CF_3; R^2=R^3=H) the corresponding indole, obtained by spontaneous cyclization of 2-substituted anilides **3**, was isolated.

2-Aryl-2,3-dihydroquinolin-4(1*H*)-ones **5** were synthesized from the 2-substituted anilides **3** by acid-catalysed Meyer–Schuster rearrangement leading to *N*-substituted 2′-aminochalcones **4** followed by alkaline hydrolysis, deprotection and cyclization under acidic conditions (Scheme 1).

Recently, Spivey and co-workers reported a similar general and straightforward method to prepare 2,3-dihydroquinolin-4(1*H*)-ones **9** starting from 2-(pseudo) halogenated anilines **6** by a two-step procedure involving a Sonogashira coupling with propargyl alcohols **7** followed by a Brønsted acid-catalysed cyclization (Scheme 2) [37]. Anilines **8** were converted into quinolin-4(1*H*)-ones **9** by acid-catalysed tandem Rupe rearrangement–Donnelly–Farrell ring-closure reaction upon treatment with concentrated HCl, followed by basic workup. The Rupe rearrangement involves a regioselective hydration–dehydration rearrangement of the alkyne moiety, probably via aldol **10**, to give the corresponding α,β-unsaturated ketone, which by acid-catalysed 6-*endo*-trig Michael-type ring closure is converted into the dihydroquinolin-4(1*H*)-ones **9** (Scheme 2). For the cyclization of the *N*-acetyl derivatives, the authors proposed the Rupe rearrangement followed by the in situ acetamide hydrolysis immediately prior to cyclization [37].

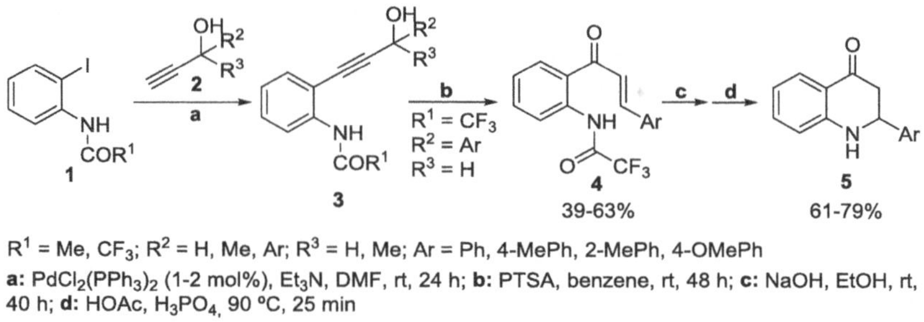

Scheme 1 Synthesis of 2,3-dihydroquinolin-4(1*H*)-ones **5** from 2-substituted anilides **3** obtained by Pd-catalysed reaction of 2-iodoanilides **1** with propargyl alcohols **2** [35, 36]

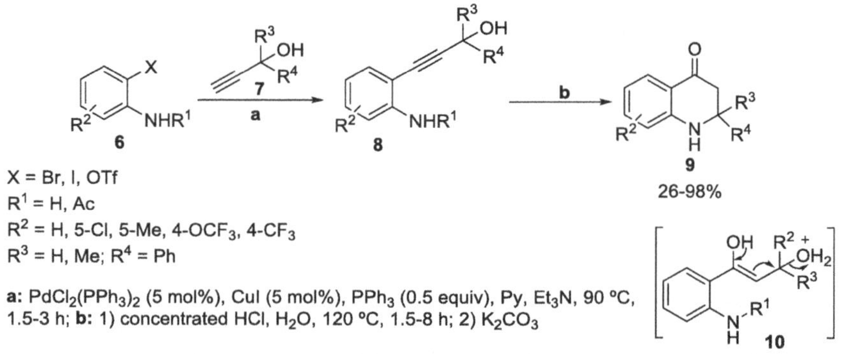

Scheme 2 Synthesis of 2,3-dihydroquinolin-4(1*H*)-ones **9** via Sonogashira coupling of 2-(pseudo)haloanilines **6** with propargyl alcohols **7** followed by acid-catalysed cyclization [37]

This reaction allows the introduction of different groups at C-2 of the quinolin-4(1*H*)-one skeleton (e.g. R^3=H, Me; R^4=Ph), thus providing access to a wider variety of 2-substituted 2,3-dihydroquinolin-4(1*H*)-ones **9**.

1,2-Disubstituted quinolin-4(1*H*)-ones **15** have been prepared by Cu-catalysed heterocyclization of 1-(2-halophenyl)-2-en-3-amin-1-ones **14**, readily obtained by conjugate addition of primary amines to α,β-ynones **13** prepared by Sonogashira cross-coupling of terminal alkynes **12** with commercially available 2-bromo- and 2-chlorobenzoyl chlorides **11** (Scheme 3) [38]. Best results were obtained when the cyclization of 1-(2-bromophenyl)-2-en-3-amin-1-ones **14** was carried out in the presence of CuI, DMEDA and K_2CO_3 in DMSO at 80°C. Under the same conditions, 1-(2-chlorophenyl)-2-en-3-amin-1-ones **14** can also be used for this synthesis although reaction rates are lower. The reaction tolerates a variety of useful functionalities (ester, keto-, cyano- and chloro-substituents), although alkyl substituents are tolerated with some limitations and enaminones derived from primary alkylamines require a stronger base (NaO*t*-Bu) and those containing 3-alkyl substituents did not afford quinolin-4(1*H*)-ones. Quinolin-4(1*H*)-ones **15** can also be

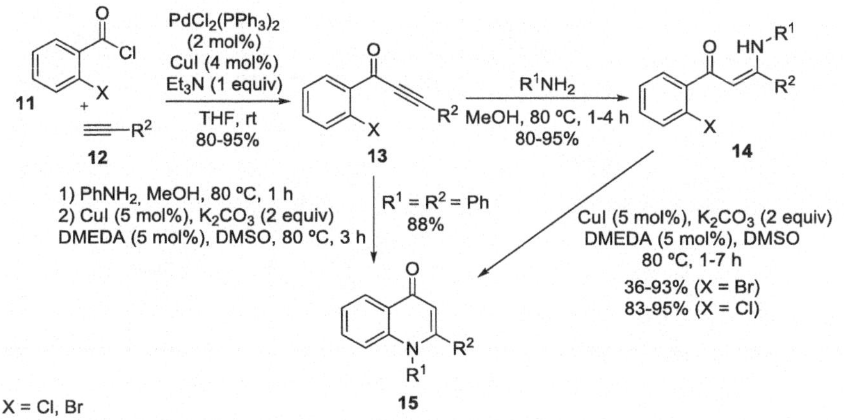

X = Cl, Br
R^1 = Ph, 3-CF_3Ph, 4-OMePh, 4-AcPh, 4-Cl-naphthalen-1-yl, 4-CNPh, 2,4-F_2Ph, 3,4,5-$(OMe)_3$Ph, 4-ClPh, 3-OMePh, 4-FPh, 4-CO_2MePh, Bu, Bn, Cy, 4-MePh.
R^2 = Ph, 4-ClPh, 3-OMePh, 4-AcPh

Scheme 3 Synthesis of 1,2-disubstituted quinolin-4(1*H*)-ones **15** via Sonogashira cross-coupling/Cu-catalysed cyclization [38]

prepared via a sequential process from α,β-ynones **13** and primary amines, omitting the isolation of the enaminone intermediates.

The mechanism proposed by the authors involves the initial coordination of nitrogen to copper, and the resulting complex undergoes an oxidative addition of the C–X bond to copper to afford the Cu(III) intermediate. Subsequent reductive elimination releases the product with concomitant regeneration of the Cu(I) species.

2.1.1 Carbonylative Sonogashira Reaction

Among the different methods described for the synthesis of 2-substituted quinolin-4(1*H*)-ones **19**, the Pd-catalysed carbonylative coupling of **2**-iodoanilines **16** with terminal arylacetylenes **17**, initially reported by Torii and Kalinin [39–41], appears to be the most versatile. The desired compounds were obtained in good yields using either $PdCl_2(PPh_3)_2$ or $PdCl_2$(dppf) in the presence of an excess Et_2NH which acts as solvent and base and plays a key role in the cyclization step (Scheme 4). 2-Iodoanilines are preferred to 2-bromoanilines and react both as a free base and in the hydrochloride form although the latter in lower yield [40]. The reaction of arylacetylenes generally gave better yields than aliphatic acetylenes, and functional groups such as thiophenyl, acetal, THP, ester, keto and ether tolerate the reaction conditions. Reaction with alkylated aniline under the same conditions led to the corresponding enamine **20** as main product (52%), and only 20% of quinolin-4(1*H*)-one **19** was obtained, although subsequent treatment of the enamine with sodium hydride in THF led to quinolin-4(1*H*)-one **19** quantitatively [39, 41]. Decrease of CO pressure or temperature drops the reaction yield.

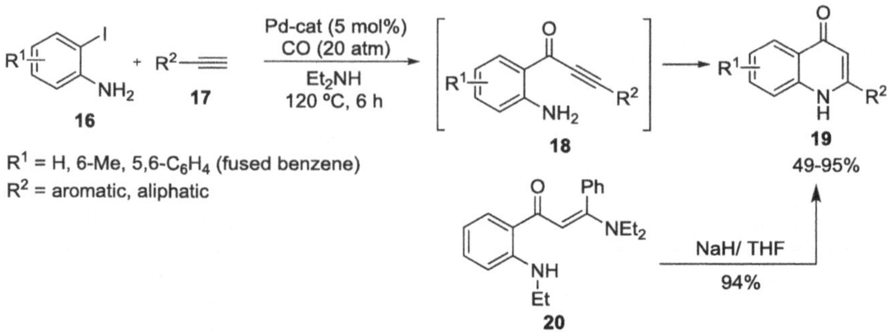

Scheme 4 Synthesis of 2-substituted quinolin-4(1*H*)-ones **19** by Pd-catalysed carbonylative coupling of 2-iodoanilines **16** with arylacetylenes **17** according to Torii [39, 41]

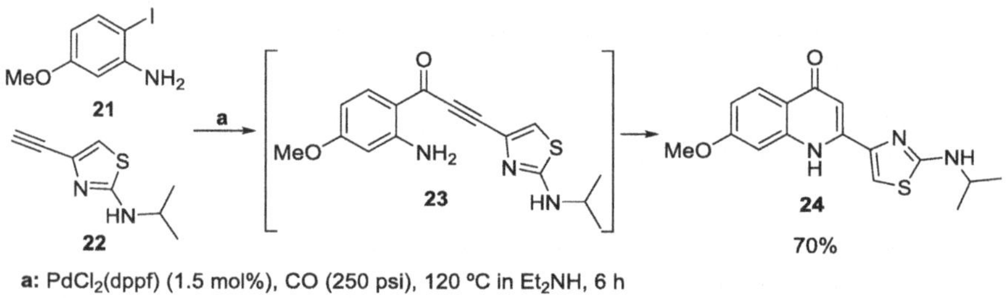

Scheme 5 A convergent synthesis of the quinolin-4(1*H*)-one **24** by Pd-catalysed carbonylative Sonogashira coupling of 2-iodo-5-methoxyaniline **21** with thiazolylacetylene **22** [42]

Quinolin-4(1*H*)-one **24**, a key substructure of the hepatitis C virus NS3 protease inhibitor BILN2061, was synthesized via a Pd-catalysed carbonylative Sonogashira coupling of 2-iodo-5-methoxyaniline **21** with thiazolylacetylene **22** (Scheme 5) [42].

Djacovitch and co-workers have found that 2-substituted quinolin-4(1*H*)-ones could be selectively obtained through a one-pot two-step multi-catalysis using sequentially $PdCl_2$(dppp) and Et_2NH as catalysts (Scheme 6, **a**) [43]. The intermediate **27** is selectively produced by a Pd-catalysed carbonylative Sonogashira coupling between 2-iodoanilines **25**, alkynes **26** and CO (5 bar), using Et_3N as the base in the presence of $PdCl_2$(dppp) as catalyst. In the second step, an organocatalysed cyclization occurs after the addition of Et_2NH to give the expected quinolin-4(1*H*)-ones **28** in high yields and selectivity. This method although successful still suffers from the need of homogeneous catalysts which are tedious to remove and could result in high Pd and ligand contamination of the final products that is not acceptable when dealing with animal and human health. More recently the same authors extended the procedure to the use of heterogeneous catalysts associating the [Pd(PNP)]@SBA-15 catalyst to a grafted amine catalyst as [NH_2]@SBA-3 in a one-pot tandem [Pd/amine] mode that allowed, for example, the selective synthesis of 2-phenylquinolin-4(1*H*)-one in a suitable 61% isolated yield (Scheme 6, **b**) [44]. Interestingly, such an approach resulted in a strong decrease of

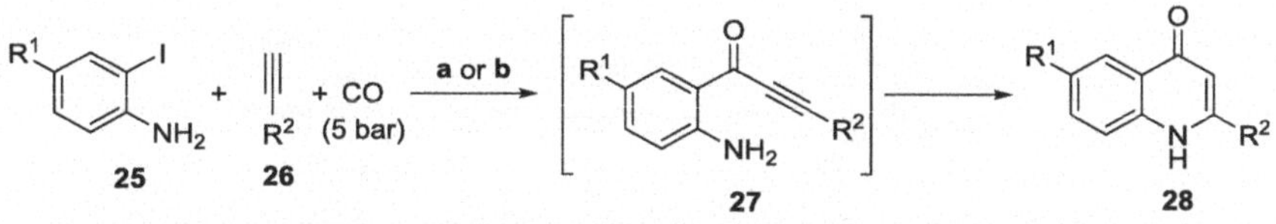

a: Homogeneous catalysis: R^1 = H, NO_2, Cl, F; R^2 = Ph, 4-OMePh, Bu; 26-98%
2-iodoaniline (3 mmol), alkyne (1.2 equiv), $PdCl_2(dppp)$ (1 mol%), Et_3N (4 equiv), toluene, 80 °C, 6 h; then Et_2NH (4 equiv), rt, 2 h.

b: Heterogeneous catalysis: R^1 = H; R^2 = Ph; 60-72%
run 1: 2-iodoaniline (3 mmol), phenylacetylene (1.2 equiv), Pd(PNP)@SBA-15 (0.1 mol%), $[NH_2]$@SBA-3 (1 mol%), Et_3N (2.5 equiv), anisole (5 mL), 80 °C. For the next runs 2-3: the same stoichiometry was used but the amount of reactants and solvent was adjusted according to the collected amount of recovered catalysts mixture.

Scheme 6 Synthesis of 2-substituted quinolin-4(1*H*)-ones **28** by a carbonylative Sonogashira coupling reaction under homogeneous and heterogeneous Pd catalysis [43–45]

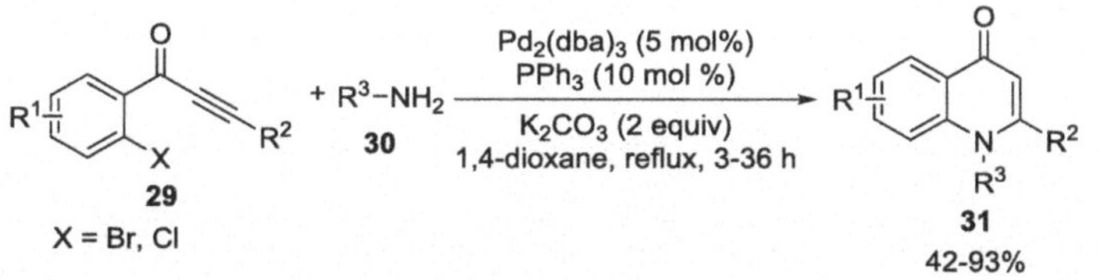

R^1 = H, Br
R^2 = 4-EtPh, 4-OMePh, 4-ClPh, 4-FPh, 3-ClPh, Ph, pyridin-3-yl
R^3 = Ph, 4-MePh, 3-MePh, 2-MePh, 4-OMePh, 2-OMePh, 4-NH_2Ph, 4-OHPh, 4-ClPh, 4-BrPh, 4-NO_2Ph, 4-CO_2EtPh, 4-AcPh, $BuNH_2$, naphtalen-1-yl, pyrimidin-5-yl

Scheme 7 Synthesis of functionalised quinolin-4(1*H*)-ones **31** through an efficient Pd-catalysed tandem amination approach [47]

Pd contamination in the final products since only 3–5 ppm of Pd was found in the crude quinolin-4(1*H*)-ones, while 40 ppm was measured when using homogeneous catalytic system. The overall reaction time was also reduced from 7 to 3 days (in the same reaction conditions). Recycling of the {[Pd(PNP)]@SBA-15/$[NH_2]$@SBA-3} catalyst mixture was successful for 3 runs [45, 46].

2.2 Buchwald–Hartwig Reaction

Functionalized quinolin-4(1*H*)-ones **31** were synthesized in good to excellent yields in one step through an efficient Pd-catalysed tandem amination approach, starting from easily accessible 2-haloaryl acetylenic ketones **29** and primary amines **30**, involving a sequential double C–N bond formation (Scheme 7) [47].

The reaction of **29** (X=Br, R^1=H) with aniline in the presence of $Pd(PPh_3)_4$ in 1,4-dioxane using K_2CO_3 as base afforded the corresponding quinolin-4(1*H*)-one **31** in 71% yield. A similar result was obtained using the $PdCl_2(dppf)–CH_2Cl_2$

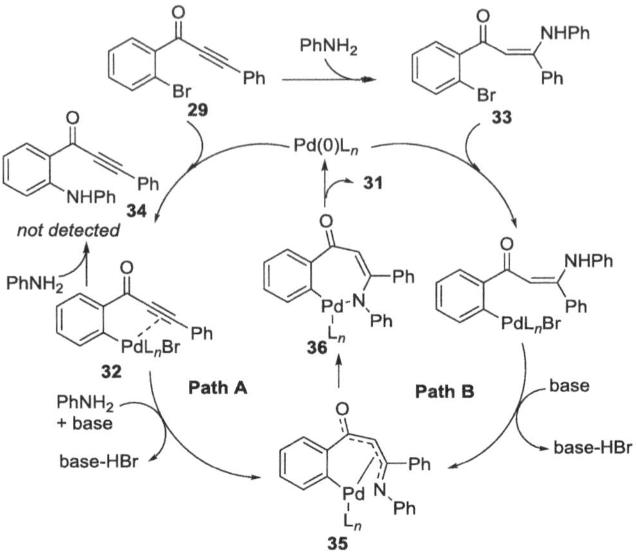

Scheme 8 Plausible mechanism for the synthesis of quinolin-4(1*H*)-ones **31** [47]

complex as catalyst. Improvements were made using $Pd_2(dba)_3$-$CHCl_3$ as catalyst combined with Xantphos or dppp, but PPh_3 proved to be the best ligand. A range of commercially available aryl amines can be employed to give the corresponding products in moderate to good yields (61–93%); however, with aliphatic amines such as butylamine, the product was obtained in moderate yield (42%). This reaction is also compatible with a variety of ynones substituted with aryl and pyridyl groups.

Two pathways are proposed as a possible mechanism for this reaction (Scheme 8), which may involve either oxidative addition of Pd(0) to the C–Br bond in **29** (intermediate **32** in Path A) or conjugate addition of aniline to **29** (intermediate **33** in path B) in the first step. The formed intermediate **32** presumably leads to **34** through Buchwald–Hartwig amination, or activation of a C≡C bond in **32** through coordination to the Pd and attack by aniline to form intermediate **35** can occur. Both pathways will go through intermediates **35** and **36**, followed by reductive elimination of Pd(0) to give the desired quinolin-4(1*H*)-one. On the basis of some experiments conducted by Xu and Zhao [47], path A could be the major pathway to afford the target quinolin-4(1*H*)-one.

More recently, Zhu and Shen reported an efficient Pd-catalysed tandem amination protocol for the synthesis of 1,2-disubstituted quinolin-4(1*H*)-ones **39** from easily accessible chalcones **37** and primary amines **38** in which the Pd catalyst [$Pd(OAc)_2$] plays a dual role, namely, in the Buchwald–Hartwig coupling and catalytic dehydrogenation (Scheme 9) [48]. $Pd_2(dba)_3$ that was a good catalyst in the Pd-catalysed tandem amination of 2-haloaryl acetylenic ketones and primary amines [47] proved to be less effective than $Pd(OAc)_2$ in this transformation (55% and 74%, respectively). A procedure using PPh_3 as a ligand in refluxing anhydrous

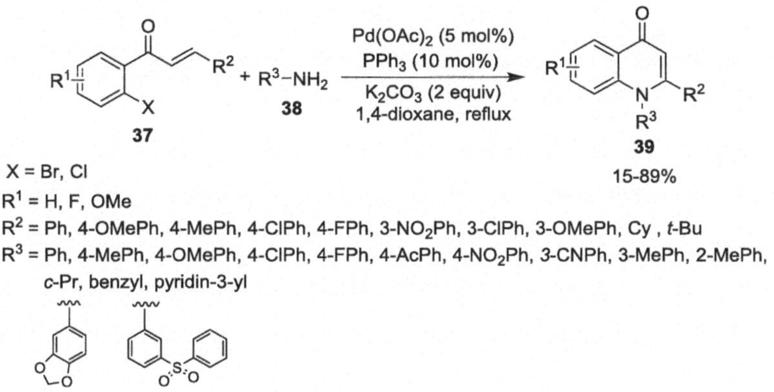

Scheme 9 Synthesis of 1,2-disubstituted quinolin-4(1*H*)-ones **39** through a Buchwald–Hartwig coupling/Michael addition sequence [48]

1,4-dioxane and K_2CO_3 as base was efficient with aromatic, heteroaromatic and aliphatic amines, the later in slightly lower yields. Due to the low oxidative addition reactivity of the C–Cl bond, 2-chloro-substituted chalcones gave quinolin-4(1*H*)-ones in lower yield than the correspondent bromo derivatives, even upon raised temperature.

2.3 Sequential Pd-Catalysed Amidation and Base-Promoted Cyclization

A variety of 2-substituted quinolin-4(1*H*)-ones **42** were obtained via sequential Pd-catalysed amidation of 2′-bromoacetophenones **40** followed by base-promoted intramolecular cyclization in a one-pot procedure under mild conditions (Scheme 10) [49]. The best solvent–base combination for this reaction was 1,4-dioxane/Cs_2CO_3. The addition of a strong base, either NaOH or NaO*t*-Bu, was found to be necessary to avoid the hydrolysis of the amide before cyclization. The scope of the reaction was quite general for both coupling partners although acyclic secondary amides were not successful.

2.4 Stille Reaction

Yamanaka and co-workers described the synthesis of 2-methylquinolin-4(1*H*)-one **46** in 57% overall yield from the Pd-catalysed cross-coupling reaction of 3-methyl-5-(tributylstannyl)isoxazole **44** with 2-bromonitrobenzene **43** followed by the catalytic hydrogenation of **45** over Raney nickel (Scheme 11) [50].

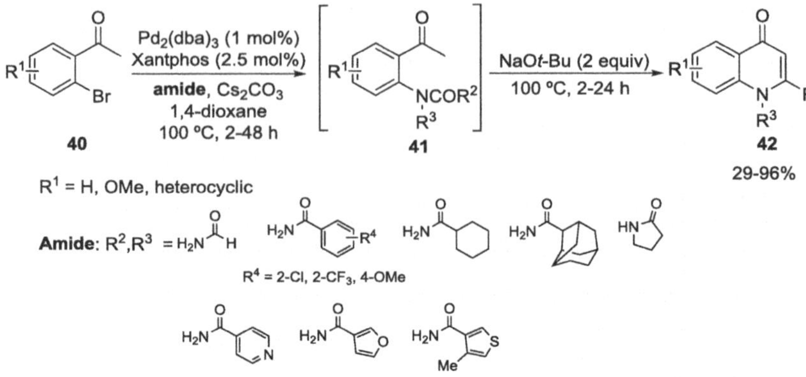

Scheme 10 Synthesis of 2-substituted quinolin-4(1*H*)-ones **42** via sequential Pd-catalysed amidation followed by base-promoted intramolecular cyclization [49]

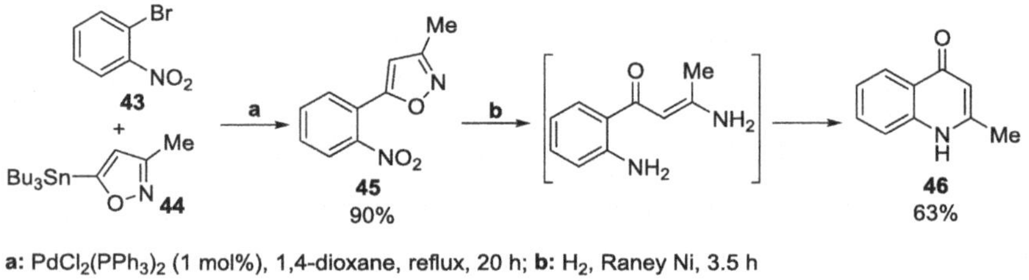

Scheme 11 Synthesis of 2-methylquinolin-4(1*H*)-one **46** via Stille reaction followed by catalytic hydrogenation [50]

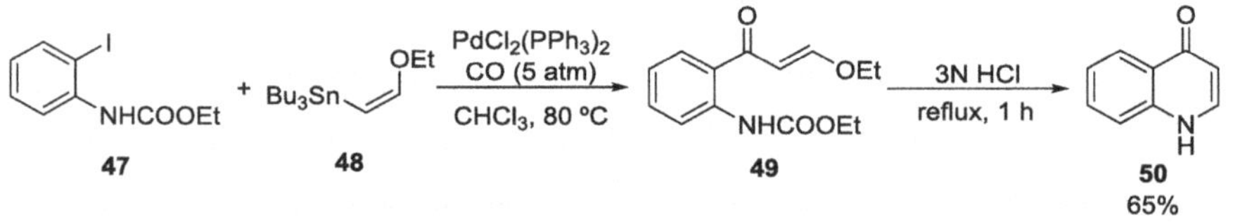

Scheme 12 Synthesis of quinolin-4(1*H*)-one **50** via carbonylative Stille reaction followed by cyclization [51]

Later, the same author reported the Pd-catalysed carbonylative coupling of ethyl 2-iodophenylcarbanylate **47** with (*Z*)-tributyl(2-ethoxyvinyl)stannane **48** under CO atmosphere [51]. The obtained ethyl (*E*)-2-(3-ethoxy-1-oxoprop-2-en-1-yl)phenyl carbanilate **49** underwent cyclization under acidic conditions to give quinolin-4 (1*H*)-one **50** (Scheme 12).

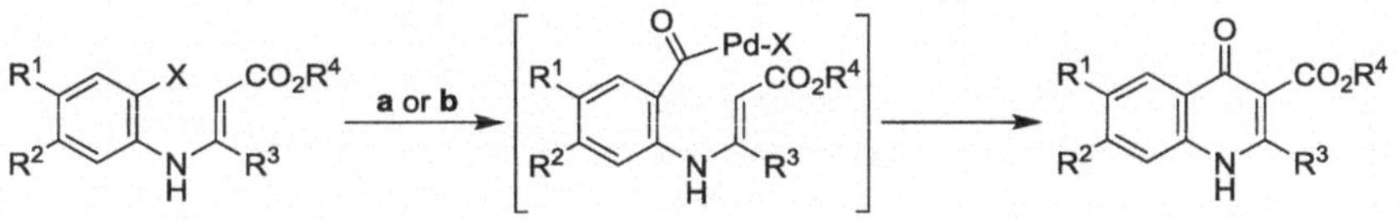

51 **52**

X = Br, I; R^1 = H, F; R^2 = H, Cl, F; R^3 = Me, CO_2Me, CH_2CO_2Me; R^4 = Me
a: 1) $Pd(OAc)_2$ (5 mol%), PPh_3 (20 mol%), CO (20 $kgcm^{-2}$ for X = I and 30 $kgcm^{-2}$ for X = Br), K_2CO_3, DMF, 120 °C, 20 h; 2) diazomethane; 24-82%

X = Br, I; R^1 = CN, CO_2Et, NO_2; R^2 = H; R^3 = CF_3; R^4 = Et
b: $Pd(OAc)_2$ (1 mol%), PPh_3 (19 mol%), CO, $NaHCO_3$, DMF, 120 °C, 2 h; 54-77%

Scheme 13 Synthesis of quinolin-4(1*H*)-ones **52** via Pd-catalysed carbonylative cyclization of 2-haloenamines **51** [52–54]

2.5 Cyclocarbonylation Reaction

Carbonylation of 3-substituted 3-(2-haloarylamino)prop-2-enoates **51** in the presence of Pd catalyst under CO atmosphere resulted in heterocyclization to form a variety of 2-substituted quinolin-4(1*H*)-one-3-carboxylates **52** [52]. Cyclocarbonylation of iodoenamine **51** was effective at 20 kg cm^{-2}, whereas 30 kg cm^{-2} was required for the bromoenamine **51**. After treatment with diazomethane, to avoid the presence of quinolin-4(1*H*)-one-3-carboxylic acid, quinolin-4(1*H*)-ones **52** were obtained in moderate to good yields (55–82%) although derivatives involving 2-methoxycarbonyl group or 6,7-difluoride gave lower yields (24% and 37%, respectively) (Scheme 13, **a**). Under the same conditions, but using CO at atmospheric pressure, Stanforth et al. reported the synthesis of three 2-trifluoromethylated quinolin-4(1*H*)-ones (Scheme 13, **b**) [53, 54].

Gabriele and co-workers isolated quinolin-4(1*H*)-one **55** from Pd-catalysed cyclization–alkoxycarbonylation of 1-[2-(trimethylsilylethynyl)phenyl]urea **53** using Pd/C-Bu_4NI as catalyst in the presence of KF for the in situ deprotection (Scheme 14) [55, 56]. Formation of quinolin-4(1*H*)-one **55** is quite intriguing, and the authors described it from some sort of rearrangement, probably via intermediate formation of the benzoxazine derivative **54**. A reasonable mechanistic hypothesis, proposed by the authors, that requires further investigation is depicted in Scheme 15. The nature of the substituent is crucial for the product formation. In the case of an unsubstituted urea **53**, quinolin-4(1*H*)-one **55** was the only isolated reaction product, but when it is substituted, no quinolin-4(1*H*)-one was obtained [55].

Alper and co-workers reported a Pd-catalysed intermolecular cyclocarbonylation of 2-iodoanilines **56** with diethyl ethoxycarbonylbutendienoate **57** to produce highly functionalized 2,3-dihydroquinolin-4(1*H*)-ones **58** in one step with moderate to good yields (Scheme 16) [57]. The method involves a Michael addition and subsequent carbonylation reactions using the catalytic system of $Pd_2(dba)_3$/2-(di-*tert*-butylphosphino)biphenyl in MeCN at 80°C under 500 psi of CO. The use of CH_2Cl_2 or THF as solvent favours the formation of the Michael

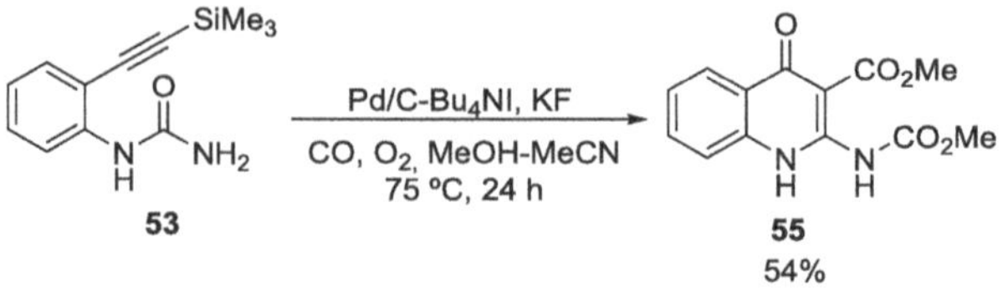

Scheme 14 Synthesis of quinolin-4(1*H*)-one **55** via Pd-catalysed oxidative carbonylation of 1-[2-(trimethylsilylethynyl)phenyl]urea **53** [55, 56]

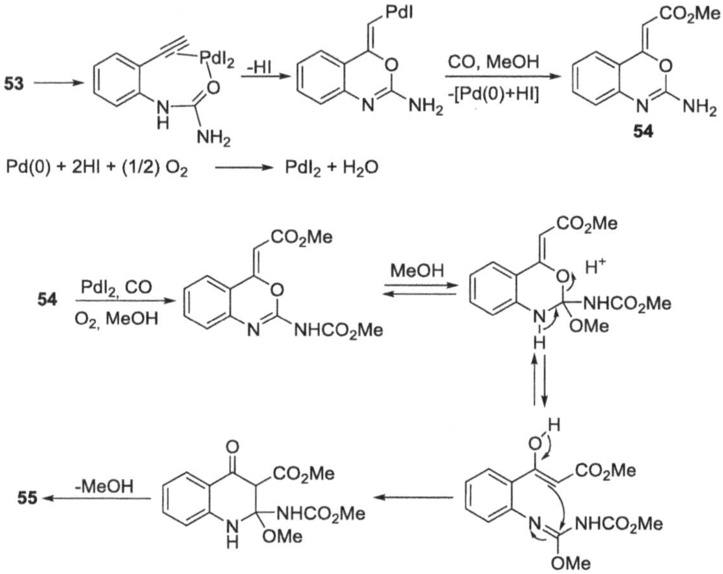

Scheme 15 Proposed reaction mechanism for the formation of quinolin-4(1*H*)-one **55** via Pd-catalysed oxidative carbonylation of 1-[2-(trimethylsilylethynyl)phenyl]urea **53** [55, 56]

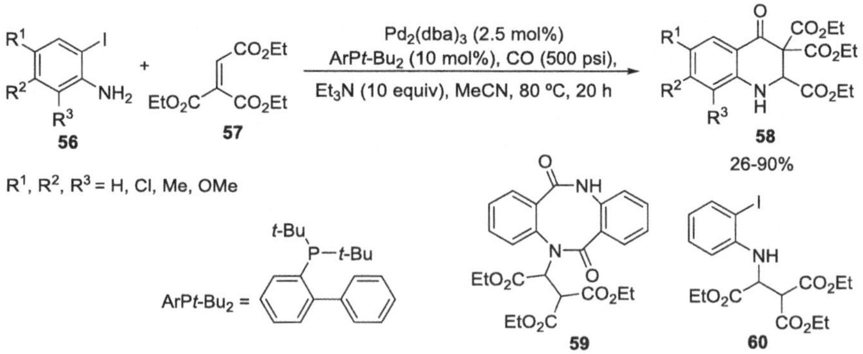

Scheme 16 Synthesis of 2,3-dihydroquinolin-4(1*H*)-ones **58** by Pd-catalysed intermolecular cyclocarbonylation of 2-iodoanilines **56** and diethyl ethoxycarbonyl butendienoate **57** [57]

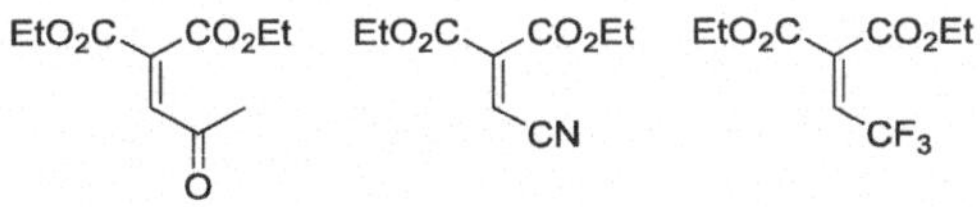

Fig. 2 Michael acceptors [57]

addition product **60** in more than 80% yield being the unique reaction product with no carbonylation taking place. Electron-donating phosphines, such as trialkylphosphines and dialkylarylphosphines, tend to give better yields; however, the authors found that 2-(di-*tert*-butylphosphino)biphenyl was the better ligand. Xantphos behaved differently and led to the formation of **59** much more selectively than when using any other ligands. The reaction is sensitive to the electronic nature of the substituents at the *para*-position relatively to the iodide group. Highly electron-donating or electron-withdrawing groups, such as methoxyl and chlorine, afforded **58** in lower yields (39% and 26%, respectively). The substituents influence the rate of the carbonylation and Michael addition steps and both can take place independently. Thus a favourable balance of the rate between these two reactions led to the successful results. No product was isolated in the reaction of **56** with a tetrasubstituted olefin indicating that initial Michael addition cannot take place probably due to steric effects. Other types of Michael acceptors (Fig. 2) were examined leading to unsatisfactory yields even after a brief screening (phosphines and bases) to optimize the reaction conditions (14–35%).

The formation of **58** involves the Michael addition of 2-iodoaniline **56** to diethyl ethoxycarbonylbutendienoate **57** as the first step to produce the Michael adduct **60**. Then the phosphine-ligated Pd(0) species undergoes oxidative addition to the C–I bond of **60**, followed by insertion of CO giving the aroyl–Pd intermediate **61**. Nucleophilic attack of the internal malonate anion on the aroyl-Pd intermediate **61** completes the catalytic cycle affording quinolin-4(1*H*)-ones **58** regenerating the Pd(0) species. Formation of **59** was explained by intermolecular double carbonylation of **56** with **60** formed in situ. First intermolecular carbonylation takes place between **56** and **60** affording an acyclic amide **62** and/or **63**, which can then undergo a second intramolecular carbonylation to give the final product **59** (Scheme 17).

A Pd(0)-catalysed termolecular queuing process involving oxidative addition of an aryl iodide **64** followed by carbonylation, allene insertion and capture of the resulting π-allyl-Pd(II) species by an internal *N*-nucleophile was described [58]. CO at atmospheric pressure was employed in contrast to Alper's reported cascades of a similar type [59] which were carried out at high pressures of CO (20 atm). Under optimal conditions, a (3+1+2)-cycloaddition reaction of 2-iodo-1-tosylaniline was undertaken using allene (1 atm) and CO (1 atm) to afford quinolin-4(1*H*)-ones **65** in good yields (55–99%) (Scheme 18). This reaction tolerates substituents on both allene and aryl iodide allowing access to an array of heterocycles with *s-cis* enone moieties.

In the first step of the catalytic cycle, the Pd(0) catalyst undergoes oxidative addition to the aryl iodide **64** bond followed by coordination and insertion of CO. The acyl–Pd(II) intermediate **66** adds to the allene at the central carbon atom

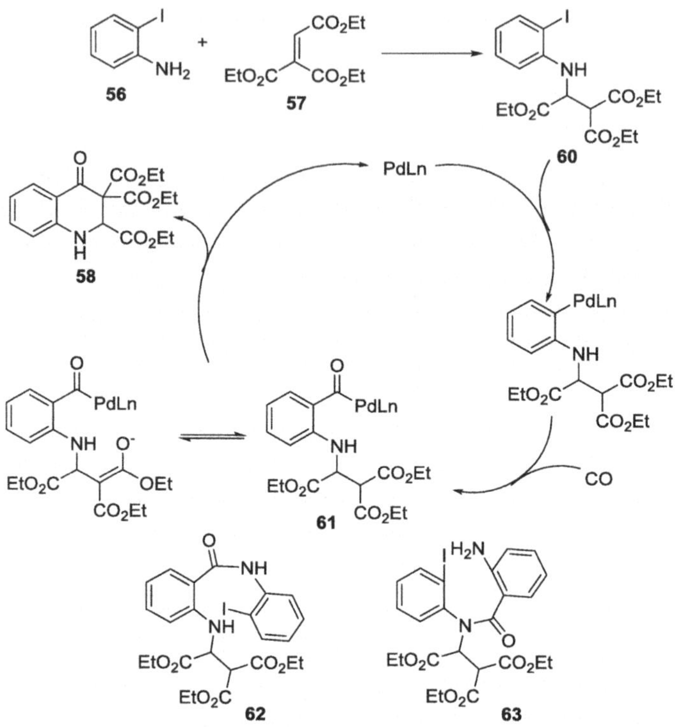

Scheme 17 Probable reaction mechanism for the formation of 2,3-dihydroquinolin-4(1*H*)-ones **58** by Pd-catalysed intermolecular cyclocarbonylation of 2-iodoanilines **56** and diethyl ethoxycarbonyl butendienoate **57** [57]

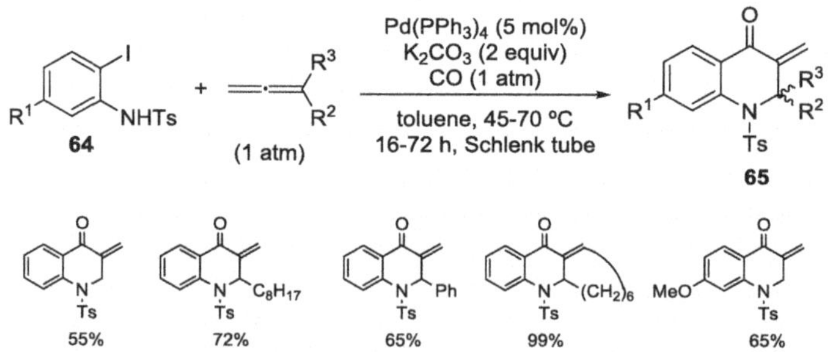

Scheme 18 Synthesis of quinolin-4(1*H*)-ones **65** via a Pd-catalysed cascade carbonylation–allene insertion [58]

to furnish a π-allyl-Pd(II) species **67** which undergoes nucleophilic attack by the internal nucleophile to give the enone product **65** (Scheme 19) [58].

These authors also described a one-pot quinolin-4(1*H*)-one synthesis–Michael addition starting from 2-iodo-1-tosylaniline **64**, CO (1 atm) and allene (1 atm) in toluene at 45°C during 48 h (Scheme 20) [58]. Both aliphatic and heteroaromatic

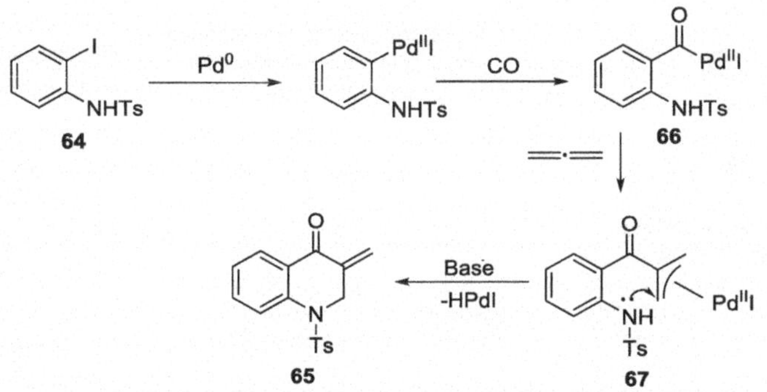

Scheme 19 Proposed reaction mechanism for the formation of quinolin-4(1*H*)-ones **65** via a Pd-catalysed cascade carbonylation–allene insertion [58]

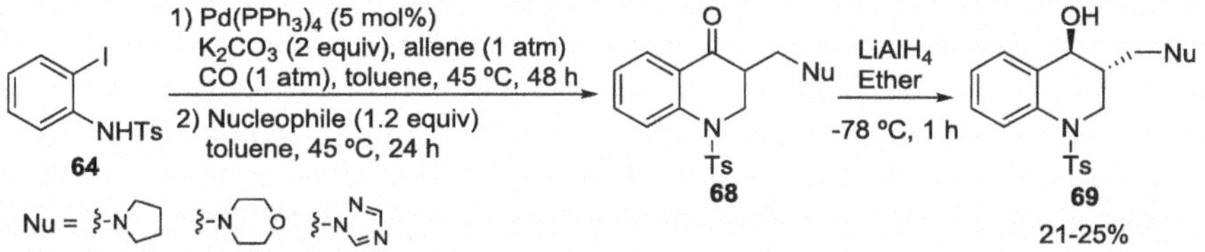

Scheme 20 Synthesis of γ-amino alcohols **69** via Pd-catalysed cascade carbonylation–allene insertion–Michael addition followed by reduction of the corresponding quinolin-4(1*H*)-ones **68** [58]

N-nucleophiles were effective, and in the case of 1,2,4-triazole, only the 1-substituted triazole was observed. As the obtained Michael adducts **68** were sensitive to retro-Michael addition, the authors performed the reduction with $LiAlH_4$, without prior isolation, obtaining the corresponding γ-amino alcohols **69** as single stereoisomers.

2.6 *Others*

An efficient metal-catalysed decarboxylative cross-coupling reaction of quinolin-4(1*H*)-one-3-carboxylic acids **70** with various (hetero)aryl halides **71** has been described (Scheme 21) [60]. An extensive screening of various reaction parameters (Pd, ligand, solvent, base and temperature) showed that PdI_2, $Pd(OAc)_2$ and $PdCl_2$ were less effective than $PdBr_2$. The nature of phosphine ligand has an important influence on selectivity and optimal conditions on the reaction of **70** (R^1=Ph; R^2=H) with 4-iodoanisole **71** involved the combination of the bidentate phosphine DPEphos with $PdBr_2$ in toluene/DMA at 150°C. The use of microwave irradiation provided shortening of reaction time and increase of quinolin-4(1*H*)-one **72** yield (MW, 1 h, 81%; CH, 8 h, 77% and 1 h, 60%). The bimetallic system $PdBr_2/Ag_2CO_3$ is necessary for the coupling to occur; no product could be formed in the absence of

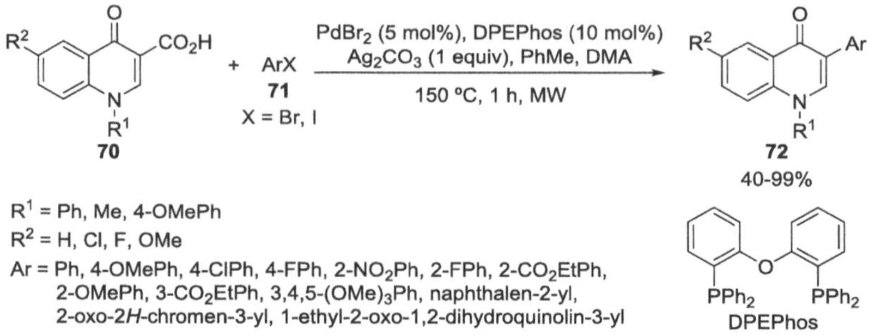

Scheme 21 Metal-catalysed decarboxylative coupling reaction of quinolin-4(1*H*)-one-3-carboxylic acids **70** with (hetero)aryl halides **71** [60]

$PdBr_2$ or when Ag_2CO_3 was replaced by other bases. Using optimal conditions under microwave irradiation, electron-rich and electron-deficient *ortho*-, *meta*- and *para*-substituted aryl iodides and bromides all efficiently underwent decarboxylative coupling in good yields (40–99%), and the coupling with heterocyclic halides was also successful [40% for 3-bromocoumarin and 57% for 3-bromoquinolin-2(1*H*)-one]. Both *N*-alkyl- and *N*-arylquinolin-4(1*H*)-one-3-carboxylic acids **70** having electron-donating or electron-withdrawing groups on the aromatic nucleus led to the formation of the corresponding coupled products **72** in good yields (60–90%). Excellent chemical selectivity was observed for **70** (R^2=Cl), preserving the C–Cl bond, which could undergo further metal-catalysed functionalization reactions. This protocol is an attractive alternative to the existing methods for the synthesis of 3-(hetero)arylquinolin-4(1*H*)-ones **72** and was also applied to the synthesis of 1-methyl-3-phenylquinolin-2(1*H*)-one (44%).

Y. Hamada and co-workers reported an interesting method for the synthesis of 3-substituted 2,3-dihydroquinolin-4(1*H*)-ones starting from 2-aminobenzaldehydes and allylic acetates, using a one-pot multi-catalytic cascade process: Pd-catalysed allylic amination–thiazolium salt-catalysed Stetter reaction (Scheme 22) [61, 62]. Under optimized conditions the one-pot sequential two-step multi-catalytic cascade process afforded 2,3-dihydroquinolin-4(1*H*)-ones **76** in excellent yields (94–99%) and is applicable for 2-aminobenzaldehydes **73** bearing electron-donating and electron-withdrawing substituents on the aromatic ring, although, in contrast to that of γ-acetoxy α,β-unsaturated esters **74** (R^2=CO_2Et, CO_2*t*-Bu), the reaction did not proceeded for γ-acetoxy α,β-unsaturated nitrile **74** (R^2=CN) (Scheme 22, **a**). When performing the one-pot sequential multi-catalytic cascade process in the presence of both catalysts (one-step procedure), the reaction proceeded well with the three different allylic acetates **74** giving 2,3-dihydroquinolin-4(1*H*)-ones **76** in excellent yields (98–99%), but, on the other hand, the allylic amination occurred only for the unsubstituted 2-aminobenzaldehyde **73**: R=H (98%) (Scheme 22, **b**).

A range of 2,3-dihydroquinolin-4(1*H*)-ones were synthesized by Pd(0)-catalysed intramolecular coupling of β-(2-iodoanilino)esters in moderate to good

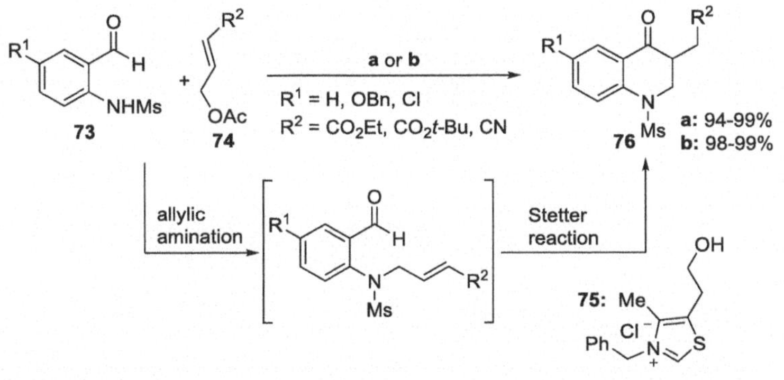

a: *Two-step procedure:* 1) $Pd(OAc)_2$ (5 mol%), PPh_3 (12 mol%), *i*-Pr_2NEt (5 equiv), *t*-BuOH (0.1 M), rt, 6-16 h; 2) **75** (20-30 mol%), 50 °C, 12-24 h;
b: *One-step procedure:* **75** (20-30 mol%), $Pd(OAc)_2$ (5 mol%), PPh_3 (12 mol%), *i*-Pr_2NEt (5 equiv), *t*-BuOH (0.1 M), 50 °C, 12-24 h.

Scheme 22 Synthesis of 3-substituted 2,3-dihydroquinolin-4(1*H*)-ones **76** by Pd-catalysed allylic amination/thiazolium salt-catalysed Stetter cascade reaction [61]

yields (Scheme 23) [63]. The use of $Pd(PPh_3)_4$ as the catalyst and Cs_2CO_3 as base in THF resulted exclusively in the reduction product **79** (40%) although using K_3PO_4 as base in combination with Et_3N and toluene led to an increase in the yield of ketone **78** to 65%. These compounds resulted from the nucleophilic substitution at the alkoxycarbonyl group. Substituents on the aromatic ring have little effect on the carbopalladation reaction evidencing that the nucleophilicity of the aryl–Pd species does not appear to be affected by the electronic properties of the substituent. The low yield for the product derived from **77** (R^1=6-CO_2Me) is mainly a consequence of the competitive retro-Michael fragmentation of the β-amino ester. The benzyl ester counterpart of **77** (R^4=Bn) can also be used as substrate; however, the desired product is formed in better yields when a methyl ester, R^4=Me, is used (50% and 65%, respectively). The reaction of amino esters **77** without hydrogen atoms α to the carbonyl group proceeded smoothly to give exclusively the corresponding ketones **78** in high yields (79–88%). On the other hand, no competition between nucleophilic attack at the carbonyl group and α-arylation was observed in the reactions of amino esters **77** which contain hydrogen atoms α to the carbonyl group; however, ketones **78** (35–67%) were obtained together with the reduction products **79** (25–45%).

This Pd(0)-catalysed cyclization involves the oxidative addition of the aryl iodide to a Pd(0) species affording a four-membered azapalladacycle **80**. A carbopalladation between the α-aryl–Pd moiety and the alkoxycarbonyl group would then lead to the alkoxide-Pd(II) chelate **82**. The coordination of the N atom to the Pd centre in **80** brings the carbonyl group nearer to the metal to facilitate the formation of a transient chelated intermediate **81** in which the carbonyl group is coordinated to the Pd centre and increases the electron density on the Pd centre to enable the otherwise unfavourable carbopalladation reaction to occur. β-Alkoxide elimination from **82** would afford ketone **78** and a Pd(II) alkoxide, which would finally undergo β-hydride elimination to regenerate the Pd(0) catalyst. The

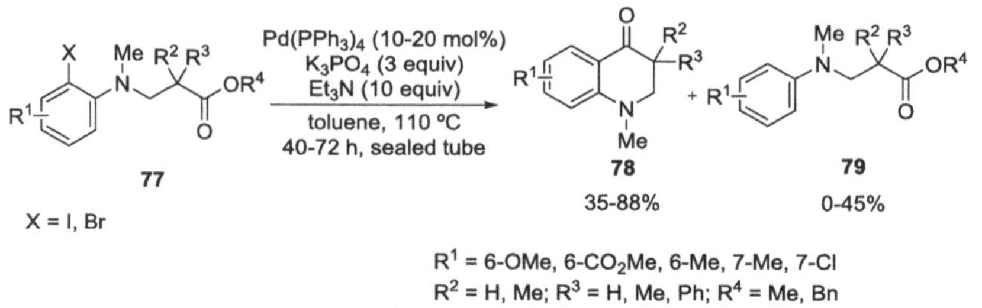

Scheme 23 Synthesis of 2,3-dihydroquinolin-4(1*H*)-ones **78** by Pd(0)-catalysed intramolecular coupling of β-(2-iodoanilino)esters **77** [63]

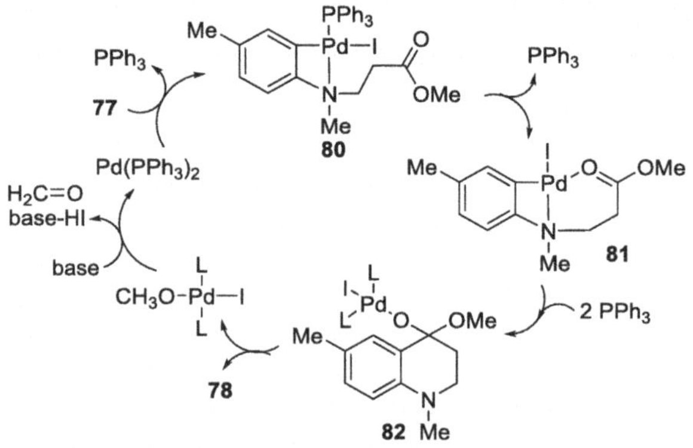

Scheme 24 Proposed Pd-catalysed acylation reaction mechanism in the formation of 2,3-dihydroquinolin-4(1*H*)-ones **78** [63]

reduction sequence is supported by the isolation of significant amounts of benzaldehyde in the reaction of benzyl ester **77** (R^4=Bn) (Scheme 24).

3 Synthesis of Quinolin-2(1*H*)-ones

3.1 Pd-Catalysed Cross-Coupling Reactions

3.1.1 Heck Reaction

In 1978, Heck and co-workers reported for the first time the synthesis of quinolin-2(1*H*)-ones from vinylic substitution of 1,2-disubstituted olefins with 2-iodoanilines [64]. Reaction of 2-iodoanilines **83** with dimethyl maleate **84** using $Pd(OAc)_2$ as catalyst and Et_3N as base in acetonitrile at 100°C afforded, as expected from previous reaction with 4-iodoaniline, the appropriate intermediate amino esters that in situ cyclize to the quinolin-2(1*H*)-ones **85**, in moderate to good yields (71%, 55% and 30% for R=H, Br and OH, respectively), because the β-carboxymethyl group was very close to the amino group (Scheme 25). If diethyl

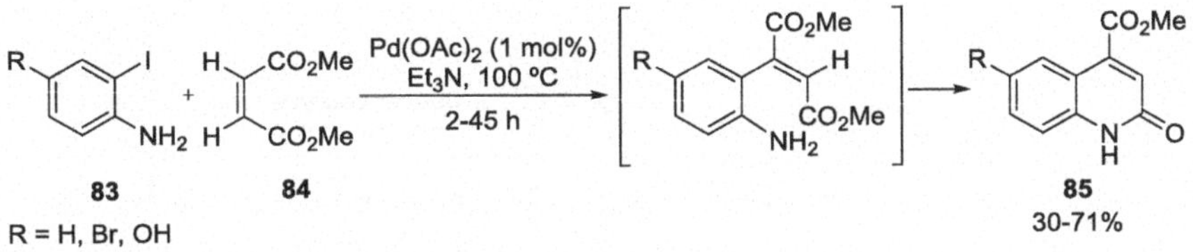

Scheme 25 Synthesis of quinolin-2(1*H*)-ones **85** from the Heck reaction of 2-iodoanilines **83** with dimethyl maleate **84** [64]

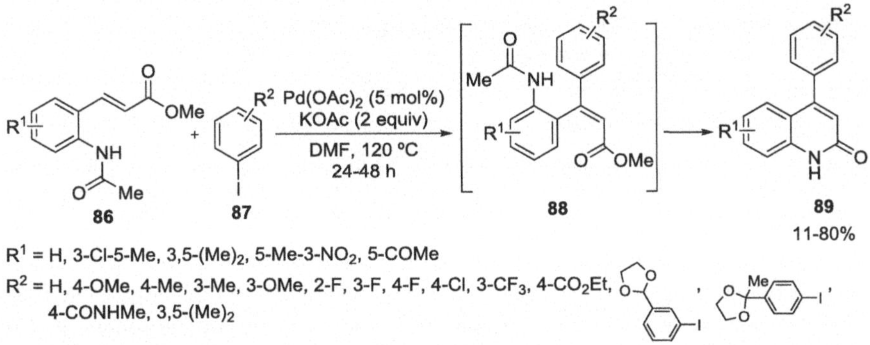

Scheme 26 Synthesis of 4-arylquinolin-2(1*H*)-ones **89** from methyl β-(2-acetamidoaryl)acrylates **86** through a domino Heck cyclization with aryl iodides **87** [65]

fumarate is used, instead of dimethyl maleate, the stereochemistry of the intermediate amino ester did not allow cyclization, because the β-carboxymethyl group was far from the amino group, without first isomerizing, although quinolin-2(1*H*)-one **85** was isolated in 47% yield along with 20% of aniline. In this case, isomerization of the intermediate should occur fairly easily or the σ-bonded Pd intermediate cyclizes readily in this reaction.

4-Phenylquinolin-2(1*H*)-one was also prepared using this methodology. First, reaction of 2-iodoaniline with (*Z*)-*N*-phenylcinnamamide led to 4-phenylquinolin-2(1*H*)-one in good yield (66%), and, as described above, the use of (*E*)-*N*-phenylcinnamamide afforded the 4-phenylquinolin-2(1*H*)-one but only in 15% yield. Another way to prepare 4-phenylquinolin-2(1*H*)-one was the reaction of (*E*)-2-aminocinnamic acid with iodobenzene, which gives the expected quinolin-2(1*H*)-one in good yield (71%).

Cacchi and co-workers described a straightforward approach to free NH 4-arylquinolin-2(1*H*)-ones **89** (30–80%) from readily available methyl β-(2-acetamidophenyl)acrylates **86** through a domino Heck cyclization reaction with aryl iodides **87** in the presence of $Pd(OAc)_2$ and KOAc in DMF at 120°C (Scheme 26) [65]. The vinylic substitution intermediate **88** was only observed when the reaction was performed at lower temperatures (25% at 80°C and 30% at 60°C).

The same authors have also shown that β-(2-bromophenyl)acrylamide **90** can also be used in the synthesis of 4-arylquinolin-2(1*H*)-ones **92** through a sequential Heck reaction–Cu-catalysed cyclization, using neutral, electron-rich

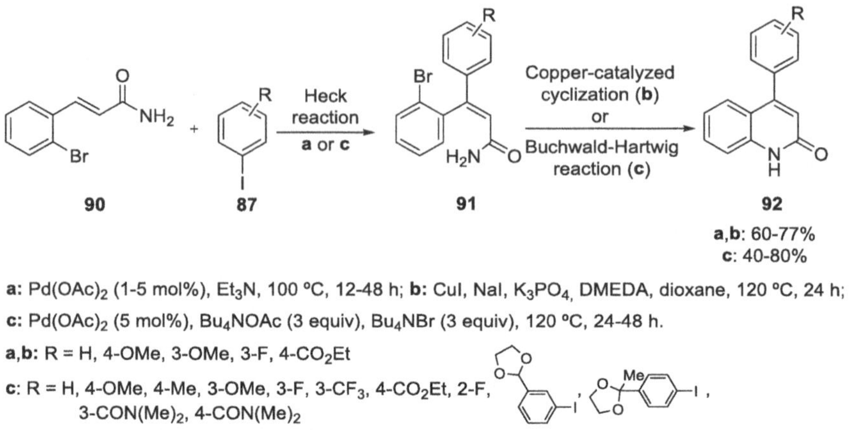

Scheme 27 Synthesis of 4-arylquinolin-2(1*H*)-ones **92** through a sequential Heck–Cu-catalysed cyclization process (**a,b**) or through a pseudo-domino Heck–Buchwald–Hartwig process (**c**) [66, 67]

and electron-poor aryl iodides (Scheme 27, **a,b**). The optimized cyclization conditions involve 20 mol% CuI, 2 equiv. NaI, 2 equiv. K_3PO_4 and 0.4 equiv. DMEDA in 1,4-dioxane at 120°C; the reaction did not work in the absence of CuI [66]. Later they have shown that the reaction sequence can also be done using a pseudo-domino process that involves two mechanistically independent catalytic cycles, a Heck reaction of 2-bromocinnamamide with aryl iodides followed by an intramolecular Buchwald–Hartwig C–N bond-forming reaction (Scheme 27, **c**) [67]. Phosphine-free $Pd(OAc)_2$ is used as the pre-catalyst and a molten tetrabutylammonium acetate/tetrabutylammonium bromide as the reaction medium, at 120°C, in the Heck reaction.

Methoxyquinolin-2(1*H*)-ones were prepared from methoxylated pivaloylaminobenzenes via metalation, Heck coupling reaction and cyclization [68]. The quinolin-2(1*H*)-one ring was obtained in two steps from the iodo derivatives **93**, which were synthesized by metalation of the corresponding polymethoxypivaloylaminobenzenes with BuLi. Then the methyl methoxypivaloylaminocinamates **94** were prepared via Heck coupling reaction of **93** with methyl acrylate using Pd $(OAc)_2$ as catalyst and Et_3N as base in acetonitrile at 100°C. Cyclization to quinolin-2(1*H*)-one **95** was then performed in acidic medium in fairly good overall yields (19–62%) (Scheme 28).

The coupling–cyclization of 2-iodoaniline **96** with α,β-unsaturated carbonyl compounds **97** in DMF at 100°C in the presence of a catalytic amount of a Pd catalyst along with a base afforded 3-substituted quinolin-2(1*H*)-ones **98** in moderate to good yields (67–76%) (Scheme 29) [69]. The best conditions found were the use of 5 mol% $Pd(OAc)_2$ as catalyst, 10 mol% PPh_3 as ligand and 6 equiv. NaOAc as base. In the same conditions, $PdCl_2(PPh_3)_2$ was also effective. If a ketone group is linked to the vinyl carbon of the starting alkenes instead of a carboalkoxyl group, 2,3- or 2,4-disubstituted quinolines were obtained, depending if the α- or the β-position of the alkene is substituted.

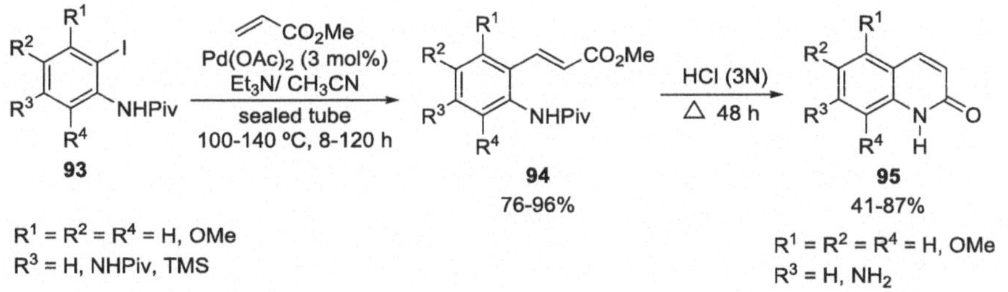

Scheme 28 Synthesis of methoxyquinolin-2(1*H*)-ones **95** from methoxylated iodopivaloylaminobenzenes **93** via Heck coupling reaction and cyclization [68]

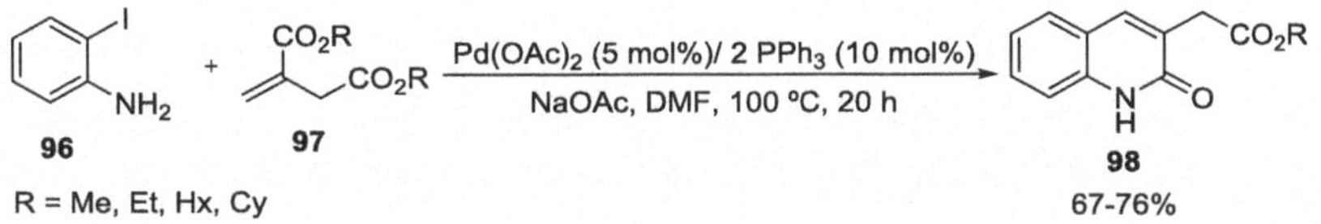

Scheme 29 Synthesis of 3-substituted quinolin-2(1*H*)-ones **98** via Heck coupling–cyclization of 2-iodoaniline **96** with α,β-unsaturated carbonyl compounds **97** [69]

4-Methoxyphenylquinolin-2(1*H*)-one **102** was prepared via a ligand-free Pd-catalysed Heck–Matsuda reaction of cinnamate ester **99** with arene diazonium-tetrafluoroborate **100** followed by a Cu-catalysed asymmetric 1,4-reduction–cyclization of the obtained *β*,*β*-diaryl acrylate **101** (Scheme 30, **a**) [70]. The presence of the base (in this case 2,6-di-*tert*-butyl-4-methylpyridine) is important for the excellent stereoselectivity (>95:5 *Z*/*E*) due to the scavenging of PdH, preventing its reinsertion which is responsible for isomerization of the Heck adducts. Based on the enantioselective synthesis of *β*,*β*-diaryl propanoates by Cu-catalysed 1,4-reduction of β,β-diaryl acrylate using $Cu(OAc)_2$ and *R*-JOSIPHOS as ligand in the presence of an excess of PMHS, the cyclization of intermediates **101** gave almost quantitatively quinolin-2(1*H*)-one **102** using 4 equiv. of PHMS/*t*-BuOH (an intramolecular amidation reaction took place in preference to the 1,4-reduction) (Scheme 30, **c**). By using 10 equiv. of PHMS/*t*-BuOH, the expected dihydroquinolin-2(1*H*)-one **103** is obtained in good yield (Scheme 30, **b**). Dihydroquinolin-2(1*H*)-one **103** can also be prepared by subjecting quinolin-2(1*H*)-one **102** to a second Cu-catalysed 1,4-reduction (Scheme 30, **d**).

A Pd-catalysed one-pot sequential Heck–reduction–cyclization (HRC) methodology led to a selective synthesis of 3,4-dihydroquinolin-2(1*H*)-ones **106** and **109** using either heterogeneous catalysts (Scheme 31) or mixed homogeneous–heterogeneous catalysts (Scheme 32) with in situ generated Pd/C [46, 71]. The overall reaction sequence proceeds under mild conditions providing good to high isolated yields starting either from 2-(2-nitrobenzyl)acrylates **104** and aryldiazonium salts **105** or 2-nitrobenzenediazonium salts **107** and acrylates **108**. Recycling experiments showed that the reused heterogeneous Pd/C catalyst was not able to promote another HRC sequence, but was still highly active for hydrogenation reactions.

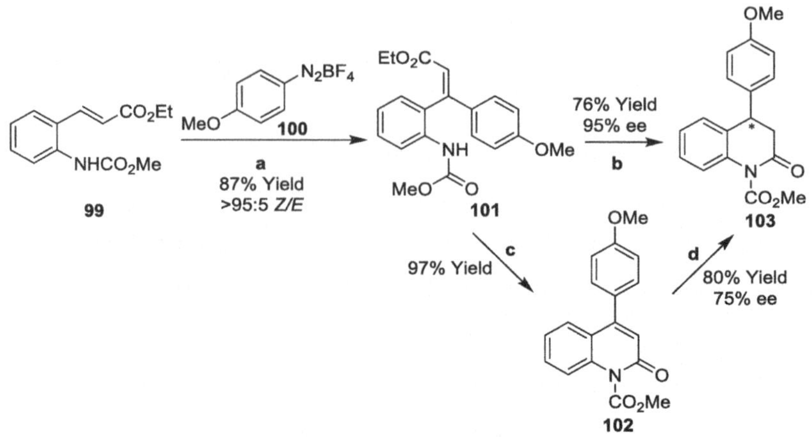

a: $Pd(OAc)_2$ (5 mol%), DTBMP (1.2 equiv), EtOH:BnCN (1:1), 60 °C, 14 h; **b:** $Cu(OAc)_2$ (3 mol%), *R*-JOSIPHOS (4 mol%), PHMS (10 equiv), *t*-BuOH (10 equiv), toluene, rt, 16 h; **c:** $Cu(OAc)_2$ (3 mol%), *R*-JOSIPHOS (4 mol%), PHMS (4 equiv), *t*-BuOH (4 equiv), toluene, rt, 16 h; **d:** $Cu(OAc)_2$ (3 mol%), *R*-JOSIPHOS (4 mol%), PHMS (10 equiv), *t*-BuOH (10 equiv), toluene, rt, 16 h.

Scheme 30 Synthesis of 4-methoxyphenylquinolin-2(1*H*)-one **102** via a Pd-catalysed Heck–Matsuda reaction of cinnamate ester **99** with arene diazoniumtetrafluoroborate **100** followed by a Cu-catalysed asymmetric 1,4-reduction-cyclization [70]

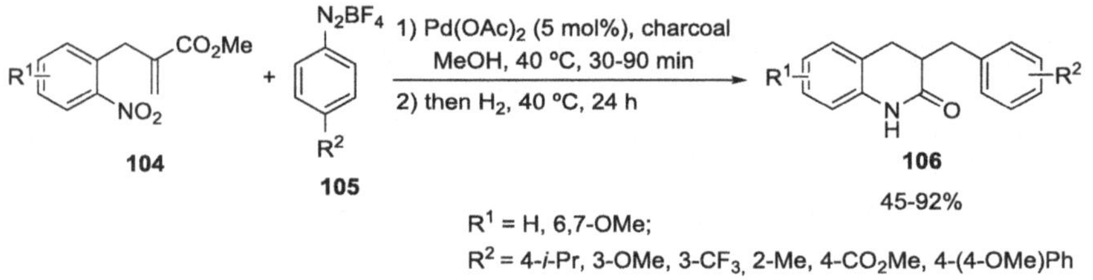

Scheme 31 Heterogeneous tandem HRC synthesis of 3,4-dihydroquinolin-2(1*H*)-ones **106** from 2-(2-nitrobenzyl)acrylates **104** and aryldiazonium salts **105** [71]

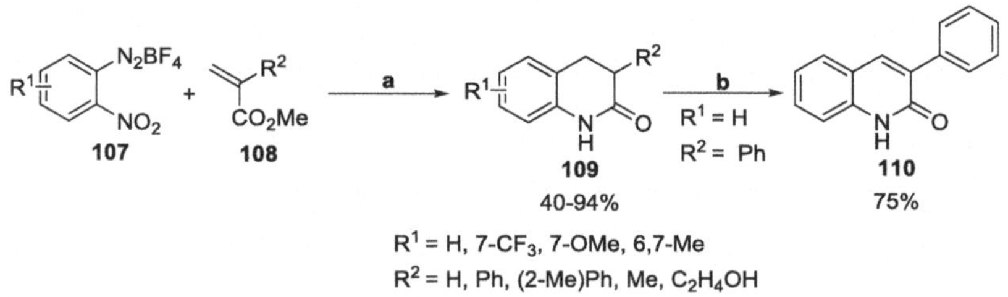

a: 1) $Pd(OAc)_2$ (5 mol%), 1,4-dioxane, 40 °C, 30-90 min; 2) then charcoal, H_2, 40 °C, 24 h
b: *t*-BuOH, KO*t*-Bu, DMSO, 25 °C, 24 h

Scheme 32 Homogeneous–heterogeneous sequential HRC synthesis of 3,4-dihydroquinolin-2(1*H*)-one **109** from 2-nitrobenzenediazonium salts **107** and acrylates **108** [71]

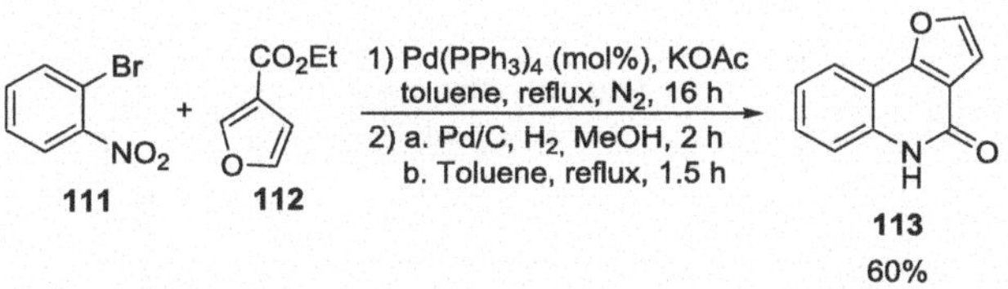

Scheme 33 Synthesis of furo[3,2-*c*]quinolin-4(5*H*)-one **113** through two sequential Pd-catalysed reactions of ethyl 3-furoate **112** with 1-bromo-2-nitrobenzene **111** [72]

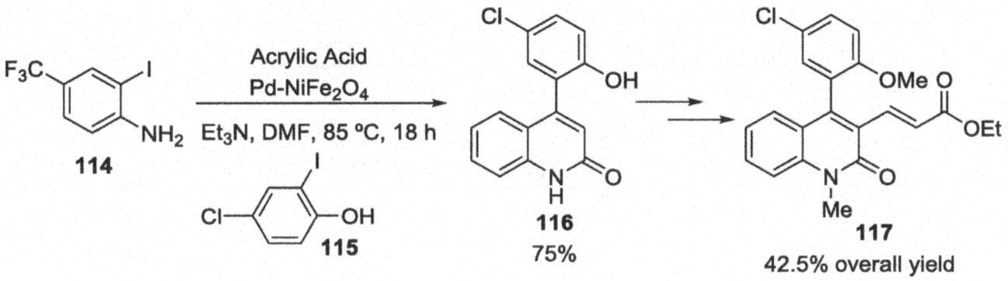

Scheme 34 Synthesis of 4-arylquinolin-2(1*H*)-one **116** by Heck reaction of 2-iodoanilines **114** with acrylic acid and 2-iodophenol **115** followed by in situ cyclization [73]

3,4-Dihydroquinolin-2(1*H*)-ones could be easily dehydrogenated under mild oxidative conditions (Scheme 32).

Two efficient sequential Pd-catalysed reactions for synthesis of furo[3,2-*c*] quinolin-4(5*H*)-one **113** in 60% overall yield were reported as shown in Scheme 33 [72]. Arylation of ethyl 3-furoate **112** with 1-bromo-2-nitrobenzene **111** in the presence of $Pd(PPh_3)_4$ afforded 2-aryl furoate (80%). Hydrogenation of the nitro group with Pd/C as catalyst and subsequent cyclization occurred uneventfully to provide furo[3,2-*c*]quinolin-4(5*H*)-one **113** (75%).

A convenient one-pot method for the synthesis of 4-arylquinolin-2(1*H*)-one **116** has been described [73]. The successive Heck reaction on substituted 2-iodoaniline **114** with acrylic acid and 2-iodophenol **115** catalysed by a Pd/nickel ferrite catalyst followed by in situ cyclization was the key step in the synthesis of 4-arylquinolin-2 (1*H*)-one **116**. The scope of this methodology was extended to the synthesis of bioactive 3-alkenyl derivatives of 4-arylquinolin-2(1*H*)-ones **117** (Scheme 34).

5-Butyl-1-methyl-1*H*-imidazo[4,5-*c*]quinolin-4(5*H*)-one **119** was obtained by the Pd-catalysed intramolecular Heck cyclization of *N*-phenyl-1*H*-imidazole-4-carboxamide **118** (Scheme 35) [74].

Tricyclic fused quinolone derivatives were prepared in three steps, an intramolecular Heck cyclization of *N*-(hetero)arylcarboxamides **123** being the key transformation [75]. These compounds were prepared by treatment of the acyl chlorides, generated in situ from the commercially available pyrrole-, thiophene- and furan-2- and 3-carboxylic acids **120**, with the appropriated 2-iodoanilines **121**, followed by *N*-methylation to avoid N–Pd complexation. Heck cyclization at the position 2 or 3 of the heterocyclic nucleus was performed with $Pd(PPh_3)_4$ as catalyst and KOAc as base affording the correspondent tricyclic fused quinolones **124** in moderate to high yields (Scheme 36).

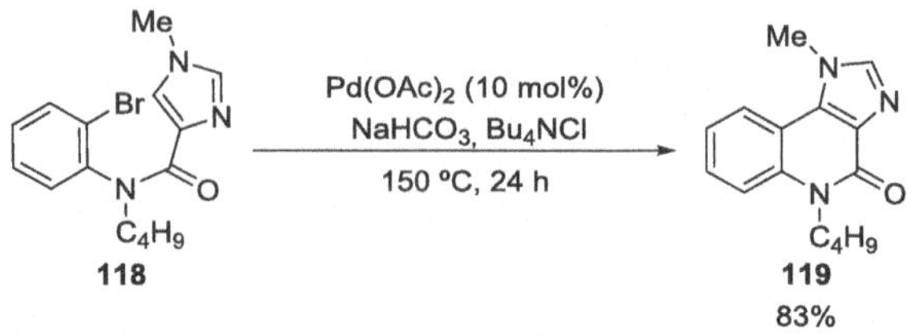

Scheme 35 Synthesis of 5-butyl-1-methyl-1*H*-imidazo[4,5-*c*]quinolin-4(5*H*)-one **119** by an intramolecular Heck cyclization reaction [74]

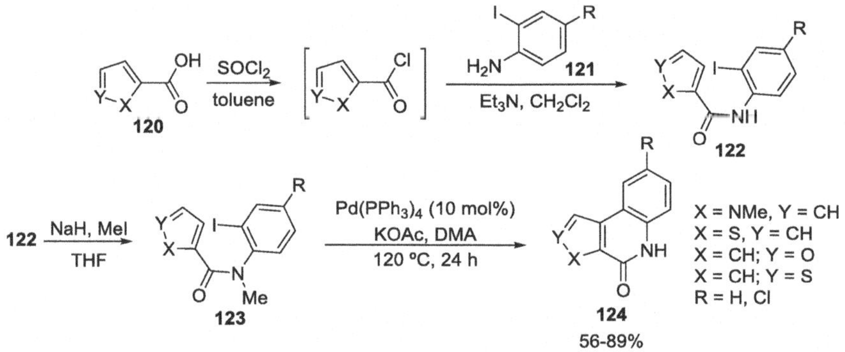

Scheme 36 Synthesis of tricyclic fused quinolones **124** by an intramolecular Heck cyclization of *N*-(hetero)arylcarboxamides **123** [75]

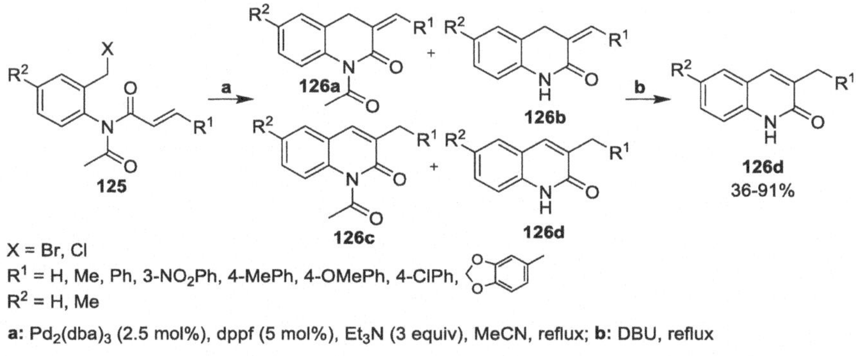

Scheme 37 Synthesis of 3-alkylquinolin-2(1*H*)-ones **126d** by intramolecular Heck cyclization of *N*-acetyl-*N*-[2-(halomethyl)aryl]acrylamides **125** [76]

3-Alkylquinolin-2(1*H*)-ones **126d** were synthesized by the Pd-catalysed intramolecular Heck cyclization of *N*-acetyl-*N*-[2-(halomethyl)aryl]acrylamides **125** [76]. Besides the expected and obtained 3-alkylquinolin-2(1*H*)-one **126d**, another three cyclization products **126a–126c** were also isolated. Compounds **126a** and **126c** are the acetylated products of **126b** and **126d**, respectively. Treatment of the reaction mixture with DBU successfully converted **126a–126c** into 3-alkylquinolin-2(1*H*)-one **126d** (Scheme 37). The combination of $Pd_2(dba)_3$ and

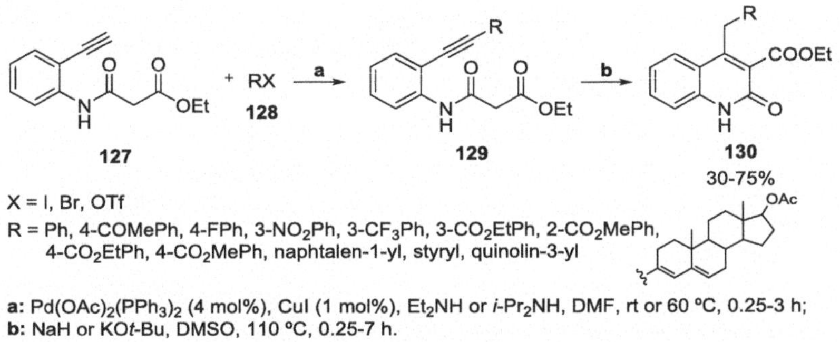

Scheme 38 Synthesis of 3,4-disubstituted quinolin-2(1*H*)-ones **130** by a Pd-catalysed reaction of ethyl *N*-(2-ethynyl)malonanilide **127** followed by an intramolecular carbocyclization [77]

dppf in the presence of Et_3N in refluxing MeCN followed by treatment of the reaction mixture with DBU leads to 3-alkylquinolin-2(1*H*)-ones **126d** in good to excellent yields.

3.1.2 Sonogashira Reaction

3,4-Disubstituted quinolin-2(1*H*)-ones **130** have been prepared through a Pd-catalysed reaction of the easily available ethyl *N*-(2-ethynyl)malonanilide **127** with aryl, heteroaryl and vinyl halides or vinyl triflates **128** followed by an intramolecular cyclization of the resulting coupling derivatives **129** under basic conditions [77]. This carbocyclization involves an intramolecular nucleophilic attack of the carbanion, generated from **129**, on the carbon–carbon triple bond affording, after protonation, a six-membered ring methylidene intermediate that isomerizes to the quinolin-2(1*H*)-one **130** (Scheme 38). The nature of the substituent in the acetylenic moiety is crucial for the success of the cyclization step. Best results were obtained in the presence of aromatic rings bearing electron-withdrawing substituents (60–75%), and no quinolin-2(1*H*)-ones were obtained in the presence of the electron-donating *p*-methoxyphenyl or butyl groups.

3.1.3 Carbonylative Annulation

A Pd-catalysed carbonylative annulation of terminal alkynes **132** by 2-iodoaniline derivatives **131** under CO in the presence of pyridine and $Pd(OAc)_2$ afforded 3- and 4-substituted quinolin-2(1*H*)-ones **133** and **134** (Scheme 39) [78]. Terminal alkynes bearing alkyl, phenyl, silyl, hydroxyl, ester and cyano substituents can be used in this process affording quinolin-2(1*H*)-ones in moderate yields. Reaction with phenylacetylene and triethylsilylacetylene afforded only the 3-substituted quinolin-2(1*H*)-one **133**. Removal of the carbamate protecting group by treating the crude reaction with 1 M of ethanolic NaOH is necessary; otherwise, mixtures of

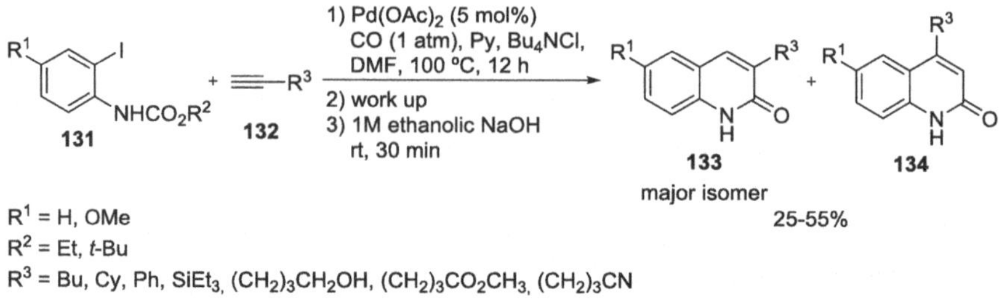

Scheme 39 Synthesis of 3- and 4-substituted quinolin-2(1*H*)-ones **133** and **134** via Pd-catalysed carbonylative annulation of terminal alkynes **132** by 2-iodoanilines **131** [78]

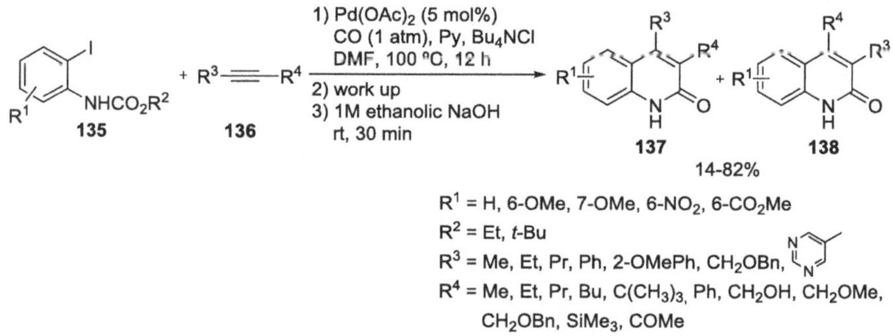

Scheme 40 Synthesis of quinolin-2(1*H*)-ones **137** and **138** via Pd-catalysed carbonylative annulation of internal alkynes **136** by 2-iodoanilines **135** [79]

unprotected and protected quinolin-2(1*H*)-ones were obtained. Both 3- and 4-substituted quinolin-2(1*H*)-ones **133** and **134** were isolated in the reactions with terminal alkynes bearing long alkyl chains. Such unusual behaviour for terminal alkynes indicates that the key step in this process is the insertion of the terminal alkyne into the carbon–Pd bond and not the Sonogashira-type coupling [39, 41] leading to quinolin-2(1*H*)-ones and not quinolin-4(1*H*)-ones. This reactivity pattern was unambiguously proven by an isotope-labelling experiment.

Later, the same authors described the synthesis of 3,4-disubstituted quinolin-2(1*H*)-ones **137** and **138** from ligand-free Pd-catalysed carbonylative annulation of internal alkynes **136** by 2-iodoanilines **135** under CO (Scheme 40) [79]. Best results were obtained using mild electron-withdrawing substituents as *p*-toluenesulfonyl, trifluoroacetyl and alkoxycarbonyl. The nitrogen substituent is lost during the course of the reaction leading to the formation of *N*-unsubstituted quinolin-2(1*H*)-ones, save for aminocarbonyl and ethoxycarbonyl groups being the correspondent protected quinolin-2(1*H*)-one isolated in 7% and 11% yields, respectively. Ethoxycarbonyl group was removed with 1 M ethanolic NaOH. A wide variety of internal alkynes were effective in this process although the use of unsymmetrical alkynes leads to the formation of mixtures of regioisomers **137** and **138** with low regioselectivity, which arise from the two possible modes of alkyne insertion into the aryl–Pd bond. Electron-deficient alkynes are very poor

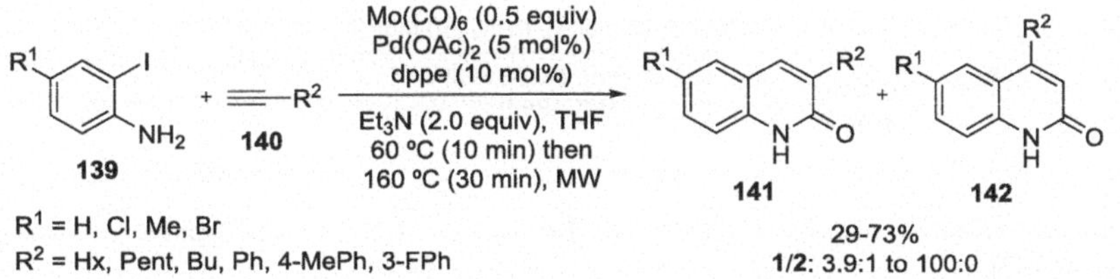

Scheme 41 Synthesis of 3- and 4-substituted quinolin-2(1*H*)-ones **141** and **142** via microwave-assisted Pd-catalysed carbonylative annulation of 2-iodoanilines **139** with terminal alkynes **140** [80]

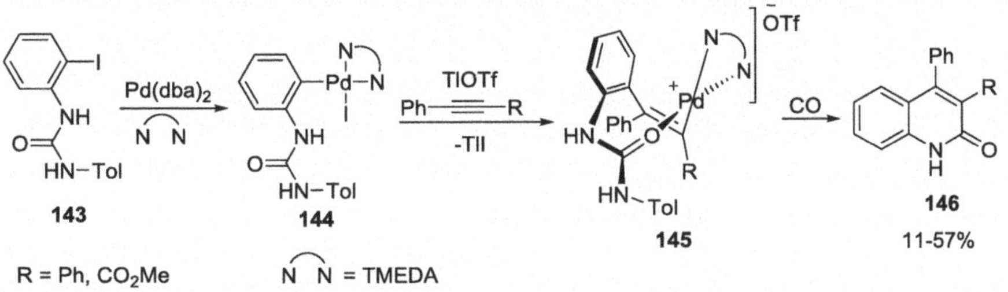

Scheme 42 Synthesis of 3-substituted 4-phenylquinolin-2(1*H*)-ones **146** through reaction of vinyl–Pd complexes **145** with CO [81]

substrates for the carbonylative annulation. Electron-rich 2-iodoanilines can be employed as annulating agents, but if the substitution is *para* to the iodine, the yield is lower. Carbonylative annulation of electron-poor 2-iodoanilines afforded the corresponding quinolin-2(1*H*)-ones in lower yields than the parent system.

The synthesis of 3- and 4-substituted quinolin-2(1*H*)-ones **141** and **142**, the latter ones as minor products, was achieved from Pd-catalysed carbonylative annulation of unprotected 2-iodoanilines **139** and terminal alkynes **140** using the commercially available molybdenum hexacarbonyl as a convenient and solid CO source (Scheme 41) [80]. The reactions were conducted at 160°C for 30 min under microwave irradiation. Et_3N proved to be the best base in terms of yield and regioselectivity. $Pd(OAc)_2$ exhibited good catalytic activity, and dppe and THF were the best ligand and solvent choice for regioselectivity and yields. Different substituted 2-iodoanilines and alkyl alkynes were used without significant loss in reaction yield or efficiency. Aryl-substituted alkynes gave only the 3-substituted quinolin-2(1*H*)-one **141**. Neither internal alkynes nor *N*-protected 2-iodoanilines can be used in this carbonylative annulation.

Vinyl–Pd complexes **145**, obtained from the oxidative addition of 1-(2-iodophenyl)-3-*p*-tolylurea **143** to $Pd(dba)_2$ in the presence of 1 equiv. of TMEDA and subsequent reaction of **144** with internal alkynes, react with CO to give Pd and two 3-substituted 4-phenylquinolin-2(1*H*)-ones **146** (R=Ph, 11% and R=CO_2Me, 57%) (Scheme 42) [81]. The reaction mechanism should follow CO insertion into the Pd–C–vinyl bond of **145** to give an acyl–Pd derivative, which undergoes a C–N coupling with loss of Pd and *p*-tolyl isocyanate.

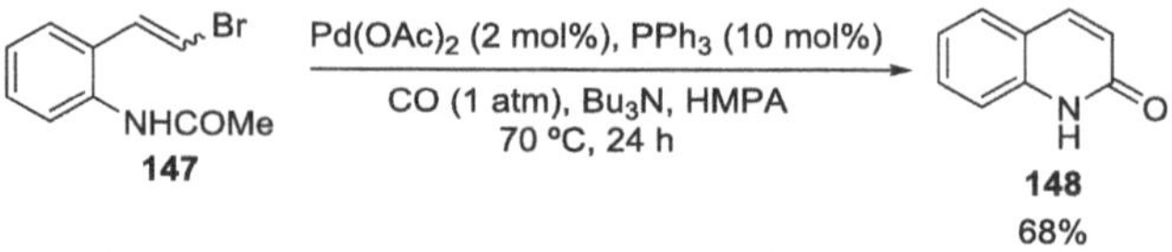

Scheme 43 Synthesis of quinolin-2(1*H*)-one **148** by Pd-catalysed carbonylation of vinyl bromides **147** [82]

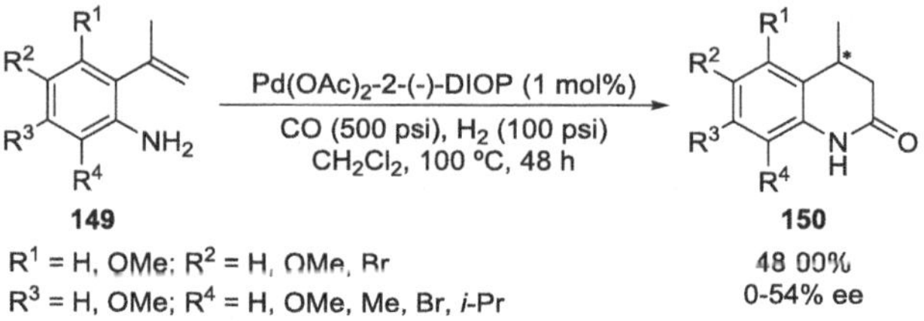

Scheme 44 Synthesis of 3,4-dihydro-4-methylquinolin-2(1*H*)-ones **150** by asymmetric cyclocarbonylation of 2-(1-methylvinyl)anilines **149** [83]

3.1.4 Cyclocarbonylation Reaction

In 1979 Ban and co-workers described the Pd-catalysed carbonylation of vinyl bromides bearing an internal amide group [82]. The (*Z*)-isomer **147** of the vinyl bromide smoothly gave, in the presence of the catalytic system $Pd(OAc)_2$ and PPh_3, quinolin-2(1*H*)-one **148** with loss of the *N*-acetyl group (Scheme 43). The corresponding (*E*)-isomer in the same reaction conditions gave only 7.7% yield of the same quinolin-2(1*H*)-one **148** because palladation occurs at the position of the halogen atom of the vinyl halide.

Chiral 3,4-dihydro-4-methylquinolin-2(1*H*)-ones **150** were prepared by asymmetric cyclocarbonylation of 2-(1-methylvinyl)anilines **149** (up to 54% ee) using a catalyst system consisting of $Pd(OAc)_2$-2(−)-DIOP [83] [$Pd(OAc)_2$ (1 mol%), (−)-DIOP (2 mol%)], in CH_2Cl_2 at 100°C for 48 h, under CO (500 psi) and hydrogen pressure (100 psi) (Scheme 44). Other chiral ligands [(−)-bppm, (+)-BINAP and *R*, *R*-BDPP] proved to be inferior to (−)-DIOP in terms of enantioselectivity. Enantioselectivity was not influenced by the amount of (−)-DIOP (1, 2, 6 equiv.) or Pd precursors [$Pd_2(dba)_3 \cdot CHCl_3$, $Pd(acac)_2$, $Pd(CF_3COO)_2$, π-allyl-Pd chloride dimer] employed, while it was slightly sensitive to the ratio of CO and hydrogen pressure. In the absence of hydrogen, the yield is very low (37%) although optical purity was still maintained (33% ee). Substituents at the 4- and 5- position of 2-(1-methylvinyl)anilines **149** showed some effect on enantioselectivity (**150**, R^2=R^3=OMe, 20% ee; R^2=Br, 31% ee). Stronger effects were found in the case of **149** (R^1=OMe); the reaction was not enantioselective affording the correspondent racemic mixture, while the carbonylation of **149** (R^4=OMe) proceeded in a more enantioselective manner affording the correspondent quinolone **150** in nearly quantitative yield (99%, 54% ee). For 6-substituted anilines **149**, excellent yields of

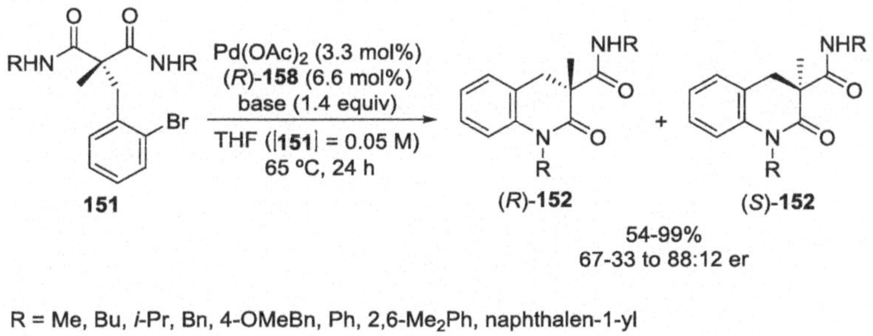

Scheme 45 Synthesis of enantiopure 3,4-dihydroquinolin-2(1*H*)-ones **152** by an enantioselective Buchwald–Hartwig reaction [84]

150 (R^4=Me, 98%) and **150** (R^4=*i*-Pr, 99%) were obtained, whereas the carbonylation of **149** (R^2=R^4=Br) afforded the correspondent 3,4-dihydro-4-methylquinolin-2(1*H*)-one **150** in moderate yield (48%).

3.1.5 Buchwald–Hartwig Reaction

Desymmetrized 3,4-dihydroquinolin-2(1*H*)-ones **152** were synthesized by an unusual enantioselective Buchwald–Hartwig reaction involving symmetrical α-(2-bromobenzyl)malonamides **151**, in nearly quantitative yields and enantiomeric ratios up to 88:12 (Scheme 45) [84].

Using enantiopure (*R*)-**153** as ligand, **152** (R=Bn) was isolated in 85% yield but in a 57:43 er. Other ligands [(*R*)-xylyl BINAP **154**, (*R*)-H8-BINAP **157**, (*S*)-**155** and (*R*)-**156** (Fig. 3)] and more polar solvents (MeCN and DMSO) demonstrated detrimental effects or inferior results in the yield of the reaction without any enhancement in the enantiomeric ratios. Different ratios of Pd to (*R*)-**158** were investigated from 1:1 to 1:3 because this monophosphine may complex Pd differently from the bisphosphines represented in Fig. 3. The yield was slightly lower for 1:3 ratio, but enantioselectivity was nearly the same for all ratios studied. A series of bases were screened using refluxing THF, but the best results were obtained with K_3PO_4 and Cs_2CO_3 affording the product in nearly quantitative yield with good enantioselectivity. While most of quinolin-2(1*H*)-ones **152** were obtained in 99%, more sterically demanding amide nitrogen substituents led to lower reaction yields (*i*-Pr, 90%; 2,6-Me_2Ph, 54%; naphthalen-1-yl, 79%), although enantioselectivity increased with the steric bulk of the substituent (ranging from 69:31 er for Me up to 85:15 er and 88:12 er for *i*-Pr and 4-OMeBn, respectively). In the case of *N*-aryl substrates, the authors observed the same enantiomeric ratios of *N*-phenyl and *N*-benzylquinolin-2(1*H*)-ones (79:21 er).

Enantioselective intramolecular double *N*-arylation of malonamides **159** bearing 2-bromoarylmethyl groups catalysed by a Pd–BINAP complex furnished C2-symmetric spirobis[3,4-dihydroquinolin-2(1*H*)-ones] **160** in up to 70% ee (Table 1) [85]. The scope of this reaction was investigated using a variety of

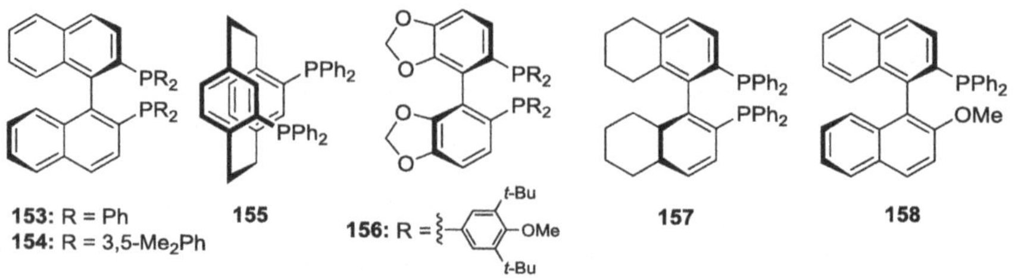

Fig. 3 Phosphine ligands discussed in Sect. 3.1.5 [84]

malonamides **159**, and it was demonstrated that substrates having a 2-chlorobenzyl groups were less reactive, although the reaction with the iodide and bromine analogues took place smoothly to give quinolin-2(1*H*)-ones **160** in high yield and moderate enantioselectivity (Table 1, entries 1–3). The reaction enantioselectivity is also affected by the substituent on the aromatic ring, moderate optical purity being obtained for 5-Cl substituents, whereas nearly racemic products were produced for 5-OMe- and 3-NO_2-substituted malonamides (Table 1, entries 6–8). It was also possible to introduce a 1,3-benzodioxole ring leading to quantitative formation of the corresponding spirobis[3,4-dihydroquinolin-2(1*H*)-one], in 57% ee, and no selectivity was observed for sterically more demanding substrates (Table 1, entries 9 and 10).

According to Sasai and co-workers, the formation of **160** seems to be initiated by the oxidative addition of one of the bromoarenes in **159** to a catalytically active Pd (0) complex. Then desymmetrization of the amide groups in the resulting Pd (II) complex **161** proceeds affording palladacycle **162**. The following C–N bond-forming reductive elimination yields monocyclized compound **163** and regenerates the Pd catalyst. The subsequent intramolecular *N*-arylation of **163** furnishes the desired spirobis[3,4-dihydroquinolin-2(1*H*)-one] **160**, enantioselectivity being determined in the first cyclization. The authors also confirmed that a kinetic resolution process is involved in the second cyclization (Scheme 46).

3.1.6 Ullmann Reaction

A simple, economical and effective two-step procedure for the synthesis of quinolin-2(1*H*)-ones was described [86]. The Pd(0)-mediated Ullmann cross-coupling reaction of ester **164** with 1-bromo-2-nitrobenzene gives the nitro ester **165** which engages in reductive cyclization on exposure to hydrogen in the presence of Pd/C affording the quinolin-2(1*H*)-one **166** (Scheme 47).

3.1.7 Others

Benzothieno[2,3-*c*]quinolin-6(5*H*)-ones **169** were synthesized starting from 2-haloanilines **167** and alkyl 3-bromobenzo[*b*]thiophene-2-carboxylates **168** by

Table 1 Synthesis of spirobis[3,4-dihydroquinolin-2(1*H*)-ones] **160** by intramolecular double *N*-arylation of malonamides **159** [85]

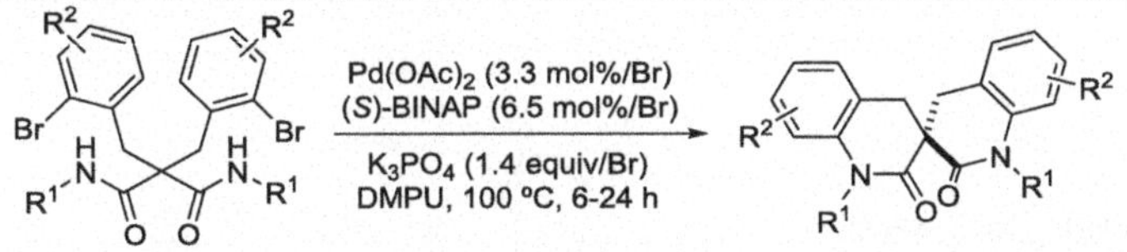

159 **160**

Entry	Substrate **159**	Product **160**	
		Yield%[a]	ee%[b]
	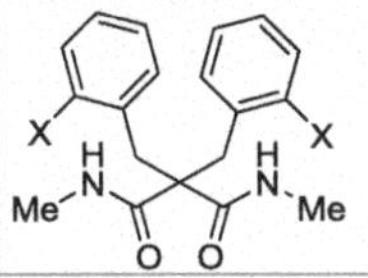	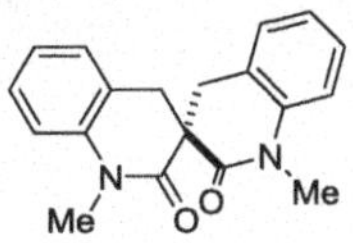	
1	X=Br	99	70
2	X=Cl	8	6
3	X=I	94	38
	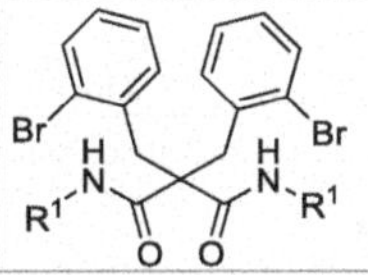	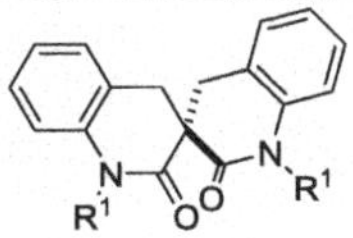	
4	R^1=Et	99	52
5	R^1=Bn	99	49
	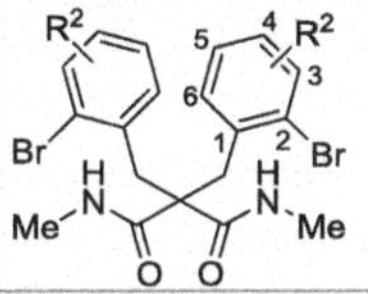	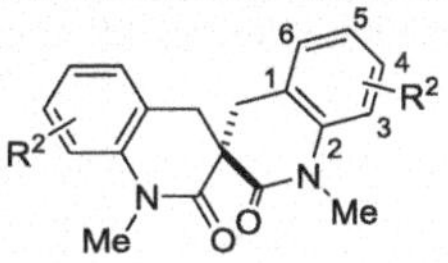	
6[c]	R^2=5-Cl	85	48
7[c]	R^2=5-OMe	90	6
8	R^2=3-NO_2	42	5
	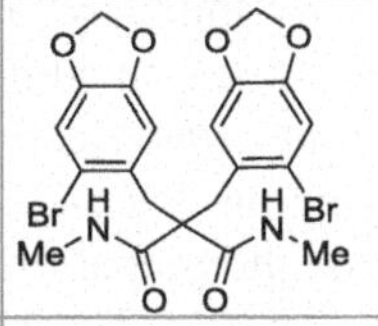	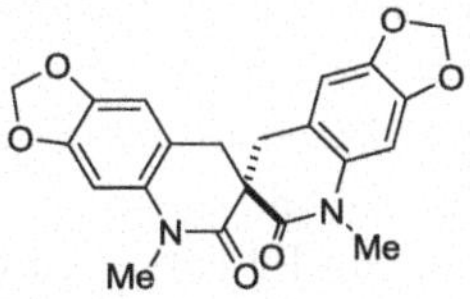	
9		99	57
	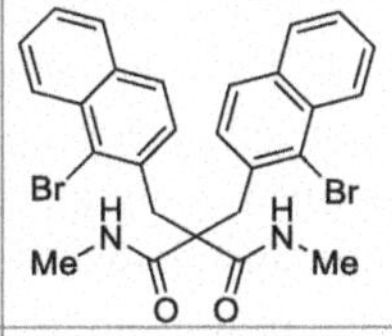	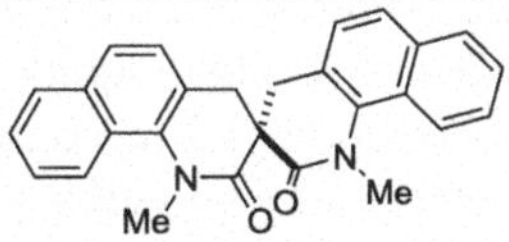	
10[c]		95	*rac*

[a]Isolated yield
[b]Determined by HPLC analysis
[c]The reactions were performed in toluene using Cs_2CO_3 as the base for 24–72 h

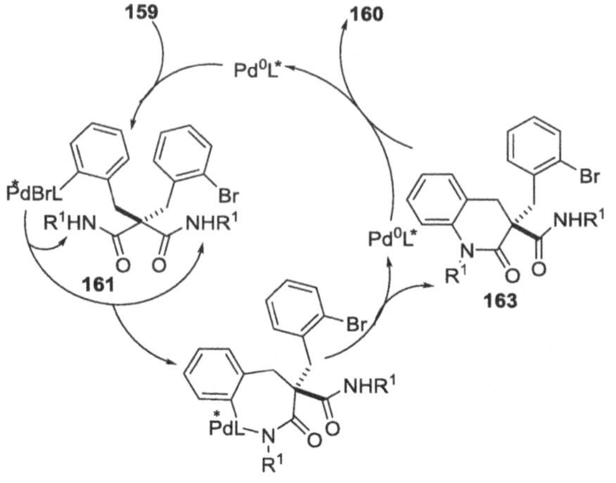

Scheme 46 Plausible catalytic cycle to give quinolin-2(1*H*)-ones **160** [85]

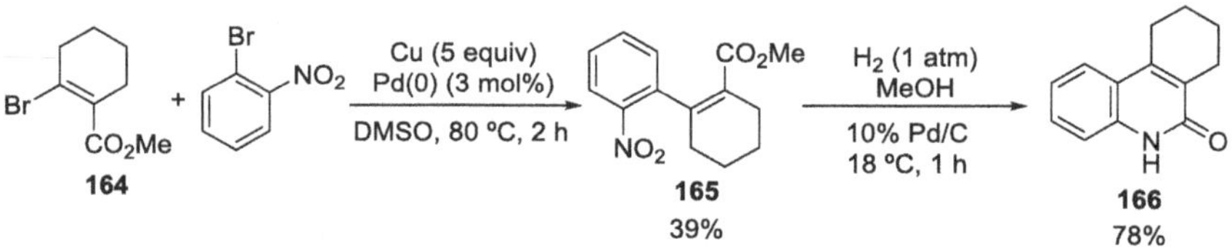

Scheme 47 Synthesis of quinolin-2(1*H*)-one **166** via Pd(0)-mediated Ullmann cross-coupling reductive cyclization sequence [86]

one-pot three-step Pd-catalysed borylation, Suzuki coupling and amidation (lactamization) (Scheme 48) [87]. The amidation reaction probably occurs in the Suzuki coupling intermediate through a nucleophilic attack of the nitrogen atom of the amine on the carbonyl of the ester group with loss of methanol or ethanol giving the corresponding tetracyclic quinolone. The use of an electron-rich sterically hindered ligand, such as 2-(dicyclohexylphosphanyl)biphenyl and $Ba(OH)_2{\cdot}8H_2O$ as base, is convenient for sterically hindered substrates. In addition the borylation should be performed in the component bearing an *ortho*-electron-donating group with the other Suzuki coupling component having an *ortho*-electron-withdrawing group. The in situ Pd-catalysed borylation avoids the preparation and isolation of boronic acids or esters and occurs with atom economy. The borylation occurred either using 2-bromo or 2-chloroanilines enhancing the scope of the reaction.

Synthesis of 3,4-dihydroquinolin-2(1*H*)-ones **172** was achieved starting from *N*-(1′-alkoxy)cyclopropyl-2-haloanilines via Pd-catalysed cyclopropane ring expansion (Scheme 49) [88]. Good yields of **171** were obtained using electron-rich biphenyl-based phosphine ligands; however, the more sterically demanding XPhos provides the better yield. The use of K_2CO_3 as base led to the best result, while KO*t*-Bu did not give the desired compound, and DMF provided better yields than toluene or 1,4-dioxane. Bromo- and iodoaniline derivatives gave better yields with a shorter reaction

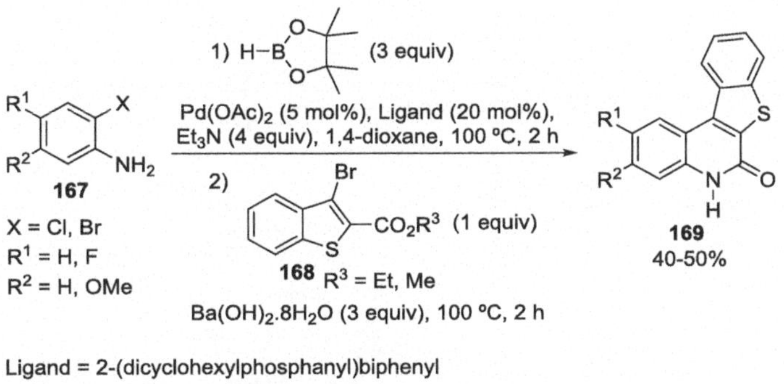

Scheme 48 Synthesis of benzothieno[2,3-*c*]quinolin-6(5*H*)-ones **169** by one-pot three-step Pd-catalysed borylation, Suzuki coupling and amidation [87]

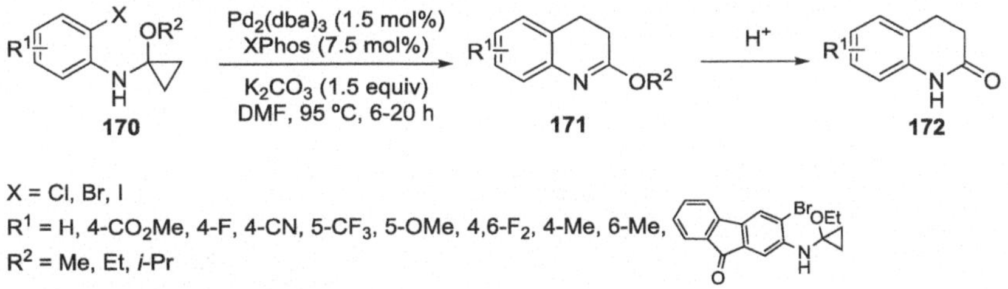

Scheme 49 Synthesis of 3,4-dihydroquinolin-2(1*H*)-ones **172** via Pd-catalysed cyclopropane ring expansion [88]

time than chloroaniline. Hydrolysis of 2-alkoxy-3,4-dihydroquinoline **171** resulted in the formation of 3,4-dihydroquinolin-2(1*H*)-ones **172** in good yields. The reaction tolerates a variety of functional groups such as ester, nitrile, ether and ketone. The substrate bearing a nitro substituent did not give the desired compound.

The formation of **171** starts with the oxidative addition of aryl halide **170** to Pd (0) **173** followed by intramolecular ligand exchange to give the four-membered azapalladacycle **174**. Further Pd rearrangement accompanied with cyclopropane ring opening furnishes the energetically favourable seven-membered azapalladacycle **175**. Finally, reductive elimination produces 2-methoxy-3,4-dihydroquinoline **171** and the regeneration of the reactive Pd(0) species **173** (Scheme 50).

3.2 Ir- and Rh-Catalysed Cross-Coupling Reactions

3,4-Disubstituted quinolin-2(1*H*)-ones **178** and **179** were prepared through an efficient annulation of *N*-arylcarbamoyl chlorides **176** with internal alkynes **177** in the presence of an iridium complex [89]. The reaction scope was investigated and showed that various aliphatic and aromatic internal alkynes can be employed affording the corresponding quinolin-2(1*H*)-ones **178** and **179** in 57–95% yield

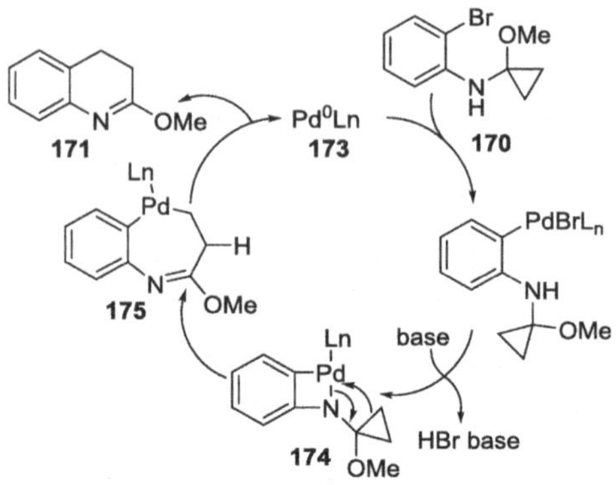

Scheme 50 Plausible reaction mechanism for the formation of 2-methoxy-3,4-dihydroquinolines **171** [88]

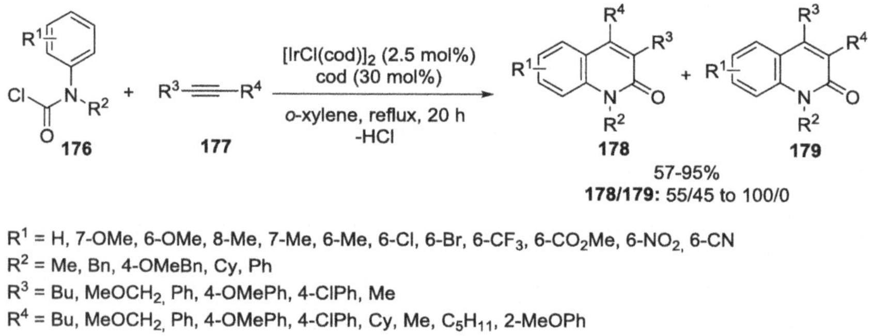

Scheme 51 Synthesis of quinolin-2(1*H*)-ones **178/179** by iridium(I)-catalysed annulation of *N*-arylcarbamoyl chlorides **176** with internal alkynes **177** [89]

(Scheme 51). When using unsymmetrical alkynes such as 1-cyclohexylpropyne and 1-phenylpropyne, the corresponding quinolin-2(1*H*)-ones were obtained in high yields albeit with low regioselectivities (**178/179**: 55/45 and 58/42, respectively). The use of unsymmetrical alkynes bearing an ether functionality improved the regioselectivity of the products (**178/179**: 72/28 and 82/18), possibly as a result of the directing effect of the oxygen atom. Terminal alkynes such as 1-decyne and phenylacetylene did not afford neither **178** nor **179**. Various carbamoyl chlorides can be employed in the reaction. Electron-rich and electron-poor aryl moieties on the nitrogen smoothly participated in the cyclization to afford the corresponding quinolin-2(1*H*)-ones in good to excellent yields (57–93%).

According to the authors, the reaction mechanism involves the oxidative addition of **176** to the iridium(I) species (Scheme 52, step 1), generating carbamoyl-chloroiridium(III) intermediate **180**. Next, an intramolecular cyclization (Scheme 52, step 2) affords five-membered iridacycle **181**. The authors demonstrated that cyclization must be electrophilic. The construction of the iridacycle

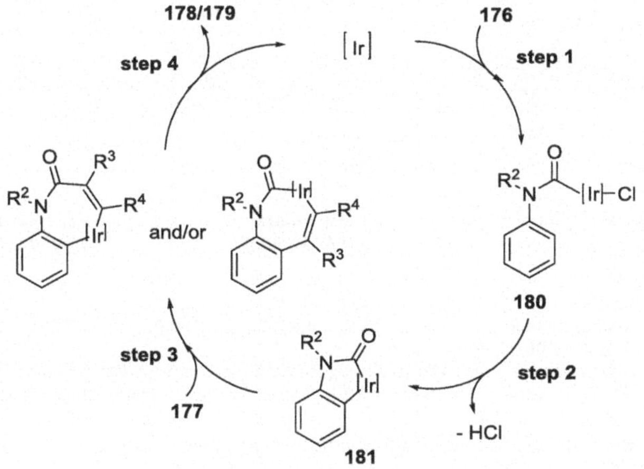

Scheme 52 Plausible reaction mechanism for the formation of 3,4-disubstituted quinolin-2(1*H*)-ones **178/179** [89]

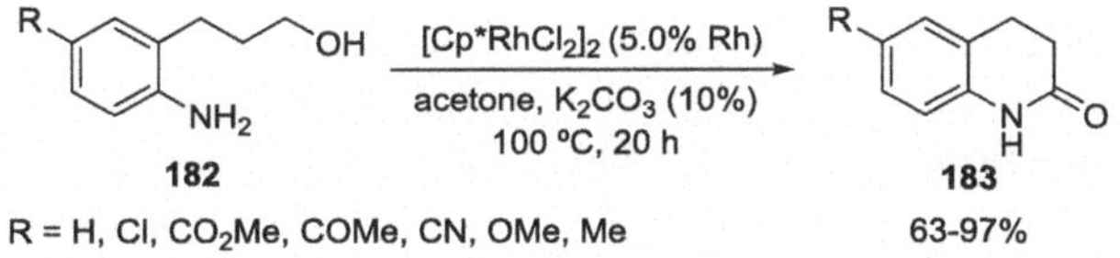

Scheme 53 Synthesis of 3,4-dihydroquinolin-2(1*H*)-ones **183** from 3-(2-aminophenyl)-1-propanols **182** catalysed by $[Cp^*RhCl_2]_2/K_2CO_3$ system [90]

181, the key intermediate in the catalytic reaction, plays a crucial role in suppressing the decarbonylation. Subsequent insertion of **177** (Scheme 52, step 3) followed by reductive elimination (Scheme 52, step 4) affords the quinolin-2(1*H*)-one and regenerates the iridium(I) species.

Transition metal-catalysed hydrogen transfer reactions of amino alcohols using a Cp*Rh complex as catalyst afforded 3,4-dihydroquinolin-2(1*H*)-ones [90]. Oxidative *N*-heterocyclization of 3-(2-aminophenyl)-1-propanol **182** (R=H) under the optimized conditions, in the presence of $[Cp^*RhCl_2]_2$ (0.0125 mmol, 5.0% Rh) and acetone as solvent, gave selectively 3,4-dihydroquinolin-2(1*H*)-one **183** (R=H) in 81% yield (Scheme 53). Decreasing reaction temperature (80°C) as well as the amount of acetone resulted in a lower yield. The authors concluded that acetone plays a key role as hydrogen acceptor and that other rhodium catalysts {$Cp^*Rh(OAc)_2{\cdot}H_2O$, $RhCl(PPh_3)_3$, $[RhCl(CO)_2]_2$} showed lower activity than $[Cp^*RhCl_2]_2$. Electron-donating substituent (OMe) afforded **183** in moderate yield (63%), while electron-withdrawing substituents gave excellent yields (96–97%) with the exception of CN group (71%), probably due to deactivation of the catalyst by coordination of CN substituent to the rhodium centre.

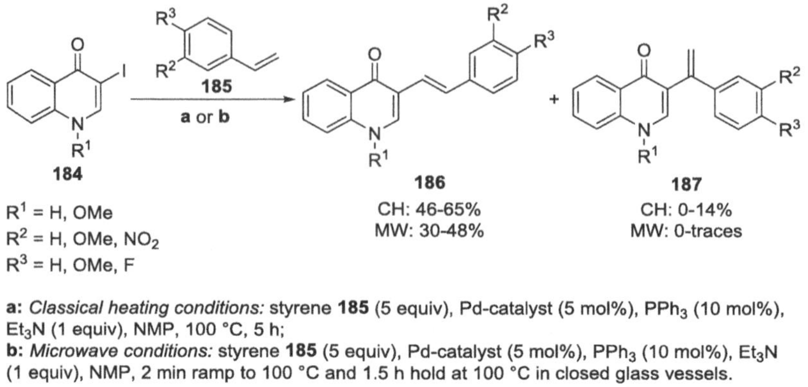

Scheme 54 Synthesis of (*E*)-3-styrylquinolin-4(1*H*)-ones **186** by Heck reaction of 3-iodoquinolin-4(1*H*)-ones **184** with styrene derivatives **185** [91]

4 Transformation of Quinolin-4(1*H*)-ones

4.1 Pd-Catalysed Cross-Coupling Reactions

4.1.1 Heck Reaction

(*E*)-3-Styrylquinolin-4(1*H*)-ones **186** were prepared from the Heck reaction of 3-iodoquinolin-4(1*H*)-ones **184** with styrene derivatives **185** (Scheme 54) [91]. The reaction of 3-iodoquinolin-4(1*H*)-one **184** (R^1=H) with styrene afforded the (*E*)-3-styrylquinolin-4(1*H*)-one **186** (R^1=R^2=R^3=H) in moderate yield (46%) using $Pd(PPh_3)_4$ as catalyst, PPh_3 as ligand and Et_3N as base in NMP at 100°C. Change of the Pd source [$Pd(OAc)_2$ or Pd/C], the ligand [$P(o\text{-tol})_3$] or the base (NaOAc or K_2CO_3) did not improve the yields; instead when $P(o\text{-tol})_3$ was used, the branched regioisomer 3-(1-phenylethenyl)quinolin-4(1*H*)-one **187** was obtained as the main product (**186**, 10%; **187**, 16%). This Heck procedure revealed to be efficient only when 3-iodoquinolin-4(1*H*)-one was *N*-methylated. The Heck reaction of 3-iodo-1-methylquinolin-4(1*H*)-one **184** led to the (*E*)-1-methyl-3-styrylquinolin-4(1*H*)-one **186** in better yield (55%) although the branched regioisomer **187** was also isolated (14%). When the reaction was performed under microwave irradiation, shortening of the reaction time occurred but lower yields were obtained (40% in 1.5 h). The best catalyst found for the unsubstituted styrene [$Pd(PPh_3)_4$] did not work well with substituted styrenes with the exception of styrene **185** (R^2=NO_2; R^3=H) (CH, 65%; MW, 45%). For other styrenes **185**, $PdCl_2$ proved to be more efficient (R^2=H; R^3=OMe, CH 59%, MW, 36%; R^2=R^3=OMe, CH 55%, MW 30%; R^2=H; R^3=F, CH 56%; MW 48%).

Later, the same authors described the Heck reaction of (*E*)-3-iodo-2-styrylquinolin-4(1*H*)-ones **188** with styrene **185**, leading to (*E*,*E*)-2,3-distyryl-quinolin-4(1*H*)-ones **189** in good yields (58–65%) (Scheme 55) [92]. $Pd(PPh_3)_4$

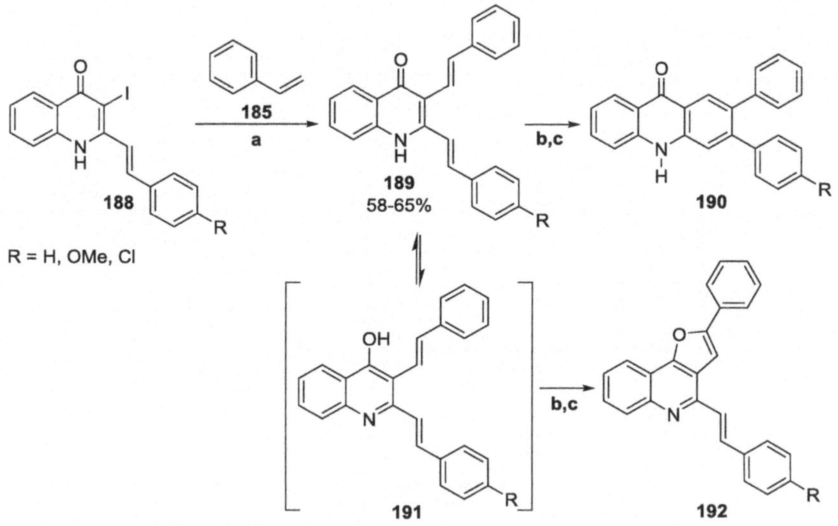

Scheme 55 Synthesis of 2,3-diarylacridin-9(10*H*)-ones **190** and (*E*)-2-aryl-4-styrylfuro[3,2-*c*] quinolines **192** via Heck reaction of (*E*)-3-iodo-2-styrylquinolin-4(1*H*)-ones **188** with styrene **185** followed by cyclization [92]

was the best catalyst [$Pd(OAc)_2$ and $PdCl_2$ proved to be unsuccessful], MeCN the most appropriated solvent and Et_3N the most suitable base.

The obtained (*E*,*E*)-2,3-distyrylquinolin-4(1*H*)-ones **189**, when heated at high temperatures, cyclize in two different ways. Electrocyclization and further in situ oxidation lead to 2,3-diarylacridin-9(10*H*)-ones **190**, while tautomerization and cyclization by nucleophilic attack of the hydroxyl oxygen atom to the β-position of the 3-styryl group and further in situ oxidation produce (*E*)-2-phenyl-4-styrylfuro[3,2-*c*]quinolines **192**. When the reaction was performed in refluxing 1,2,4-trichlorobenzene, acridin-9(10*H*)-ones **190** were obtained in low yields (16–37%) and (*E*)-2-phenyl-4-styrylfuro[3,2-*c*]quinolines **192** as the main products (35–66%). The use of an acidic medium (PTSA), in order to displace the tautomerism of 4-hydroxyquinoline **191** to the quinolone **189**, and the presence of iodine to favour the isomerization of the double bonds improved the yield of acridin-9(10*H*)-ones **190** (35–40%) although (*E*)-2-phenyl-4-styrylfuro[3,2-*c*]quinolines **192** were again obtained as the main products (41–56%).

The Heck reaction of 6-iodoquinolin-4(1*H*)-ones **194** with macrolides (from azithromycin) **193** was studied, aiming the preparation of compounds with high antibacterial activity [93]. Macrolones (macrolide + quinolone) **195** were isolated in good to excellent yields when using the catalytic system $Pd(OAc)_2/P(o\text{-tol})_3$. The optimized protocol required 2.5 mol excess of quinolone in DMF as the solvent and two different temperatures (65°C and 75°C). The desirable macrolones **196** were

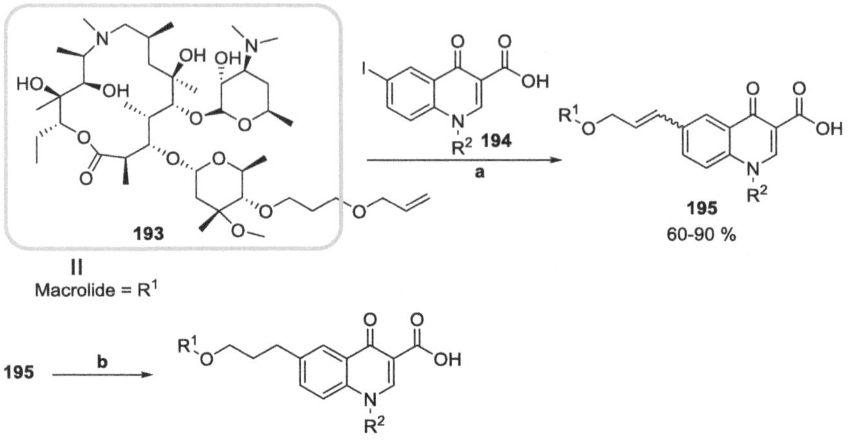

196

R^2 = c-Pr, Et, $(CH_2)_2OMe$, NMe_2, t-Bu, i-Pr, $(CH_2)_2NMe_2$

a: $Pd(OAc)_2$ (20 mol%), $P(o\text{-tol})_3$ (40 mol%), **194** (2.5 equiv), Et_3N (4 equiv), DMF, 65 °C, 2 h, then 75 °C, 18 h;

b: Pd/C 10 wt % (10 mol%), H_2 (3 bar), MeOH, 15 h (Complete conversion of **195** into **196**, isolated with > 92% purity.

Scheme 56 Synthesis of macrolones **196** via Heck reaction of macrolide **193** with 6--iodoquinolin-4(1*H*)-ones **194** followed by Pd/C hydrogenation [93]

isolated with $>92\%$ purity after complete hydrogenation of **195** catalysed by Pd/C (Scheme 56).

4.1.2 Suzuki–Miyaura Reaction

3-(4-Methoxyphenyl)-5-trifluoromethanesulfonatequinolin-4(1*H*)-ones **197** were transformed into derivatives **198** through a modified Suzuki methodology, the desired quinolin-4(1*H*)-ones **198** being obtained in moderate to good yields (Scheme 57) [94]. This protocol was applied successfully to the synthesis of 2-(3-methoxyphenyl)quinolin-4(1*H*)-ones **199** analogues [95].

The synthesis of novel and medicinally important quinolin-4(1*H*)-ones via mono and/or sequential Suzuki–Miyaura cross-coupling reactions was reported [96]. Coupling of quinolin-4(1*H*)-one **200** using a Pd/SPHOS catalytic system led to 3-substituted quinolin-4(1*H*)-ones **201** in high yields (Scheme 58), better yields being obtained with 3-iodo- than with 3-bromoquinolin-4(1*H*)-ones. For the less soluble quinolin-4(1*H*)-one, DMF was required as solvent. This reaction was effective for coupling quinolin-4(1*H*)-ones with all major subclasses of substrates with moderate to excellent yields.

To highlight the divergent nature of this synthetic route, the authors used 6-chloro-3-iodoquinolin-4(1*H*)-ones **200**, taking advantage of the reactivity differences between the iodo- and chloro-substituents. Previously the 3-position was subjected to Suzuki–Miyaura coupling (Scheme 58), and then a subsequent

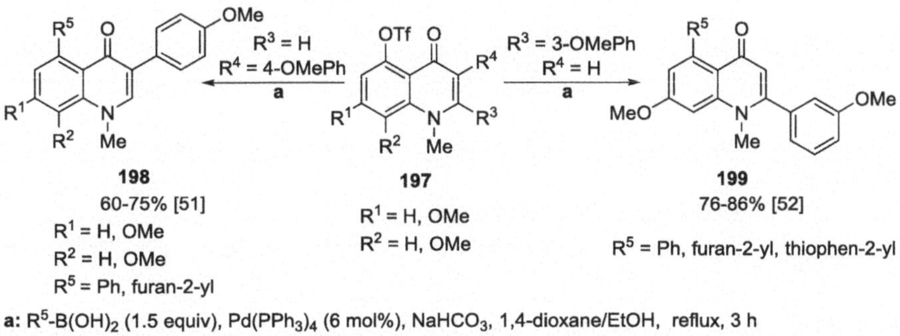

Scheme 57 Synthesis of quinolin-4(1*H*)-ones **198** and **199** by Suzuki coupling reaction of 5-trifluoromethanesulfonatequinolin-4(1*H*)-ones **197** with boronic acids [94, 95]

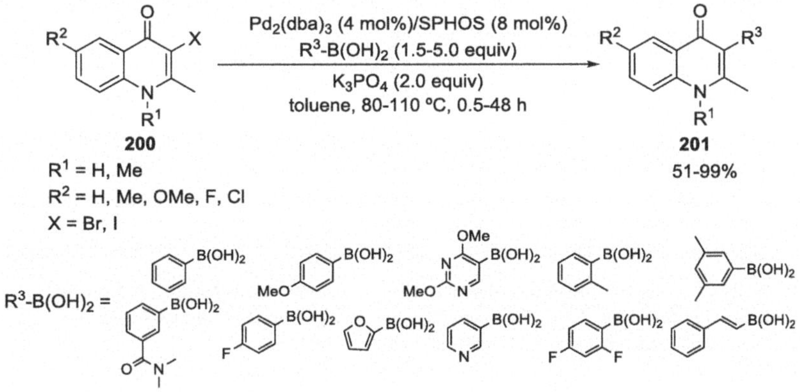

Scheme 58 Synthesis of 3-substituted quinolin-4(1*H*)-ones **201** via Suzuki–Miyaura cross-coupling reaction of quinolin-4(1*H*)-ones **200** with boronic acids [96]

coupling with different boronic acids at the 6-position was performed, generating complex quinolin-4(1*H*)-ones **202** (Scheme 59). Very good yields were obtained for 1-methylquinolin-4(1*H*)-ones (70–94%), while the NH containing 6-chloro-3-(2-furanyl)quinolin-4(1*H*)-ones only produces modest yields of the coupling product (41–47%) [96].

Sequential Suzuki–Miyaura arylation at the 3- and 6-positions of 1-substituted 6-bromo-3-iodoquinolin-4(1*H*)-ones **203** can be carried out regioselectively under standard conditions and controlled reaction temperature [97]. The chemoselective functionalization of C-3 position of the 6-bromo-3-iodoquinolin-4(1*H*)-one **203** with arylboronic acids afforded 3-aryl-6-bromoquinolin-4(1*H*)-ones **204** in good yields (80–84%), except when 2-furanboronic acid (49%) and 4-methoxyphenylvinylboronic acid (47%) were employed, without concurrent formation of regioisomeric and/or bis-coupling products (Scheme 60, **a**).

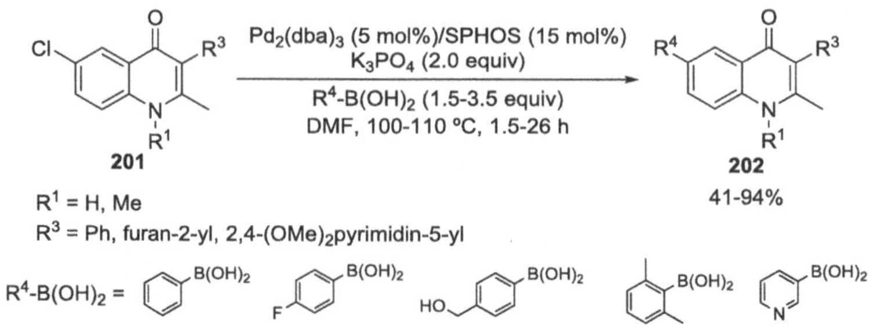

Scheme 59 Synthesis of quinolin-4(1*H*)-ones **202** via Suzuki–Miyaura cross-coupling reaction of 3-aryl-6-cloroquinolin-4(1*H*)-ones **201** with boronic acids [96]

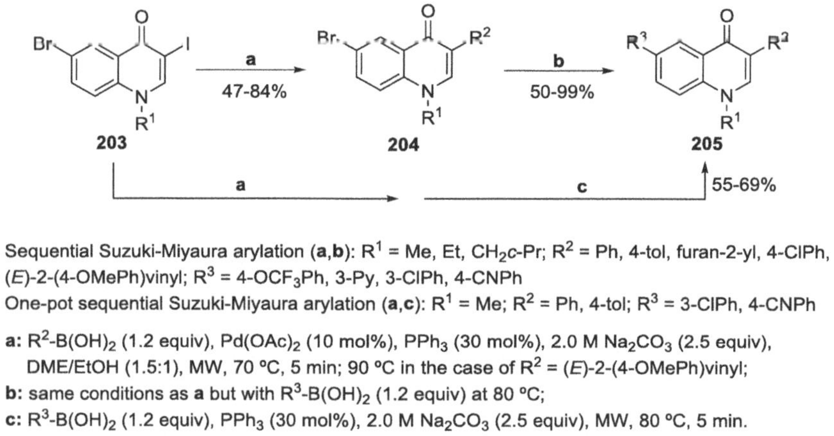

Scheme 60 Synthesis of trisubstituted quinolin-4(1*H*)-ones **205** by sequential Suzuki–Miyaura (**a**,**b**) or one-pot double Suzuki–Miyaura (**a**,**c**) reactions of 1-substituted 6-bromo-3-iodoquinolin-4(1*H*)-ones **203** with arylboronic acids [97]

Further elaboration of compounds **204** into trisubstituted quinolin-4(1*H*)-ones **205** was carried out through sequential Suzuki–Miyaura cross-coupling reaction by exploiting the reactivity of a 6-bromine substituent (Scheme 60, **b**). The authors performed the reaction under the same conditions used to generate compounds **204** by increasing the reaction temperature to 80°C. The trisubstituted quinolin-4(1*H*)-ones **205** were obtained in good yields (67–99%) with complete conversion of the substrate **204** with the exception of the reaction employing 4-(trifluoromethoxy) phenylboronic acid (50%) [97]. These authors also investigated the one-pot sequential Suzuki–Miyaura reaction and obtained the desired trisubstituted derivatives **205** in overall yields comparable or slightly lower than those obtained in the stepwise synthesis [Scheme 60, **a**,**b**; 70% for **205** (R^1=Me, R^2=Ph, R^3=3-ClPh) and 72% for **205** (R^1=Me, R^2=4-tol, R^3=3-CNPh)] in a total reaction time of 10 min, without the isolation of the monoarylated intermediate (Scheme 60, **a**,**c**).

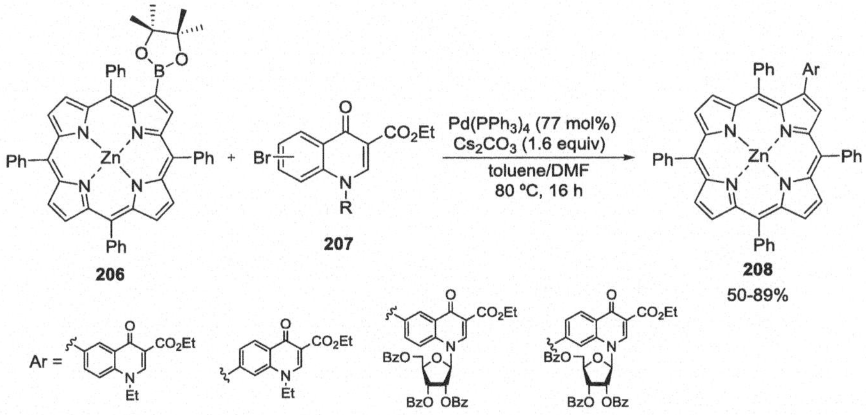

Scheme 61 Synthesis of porphyrin-quinolin-4(1*H*)-one conjugates **208** by Suzuki–Miyaura reaction of β-borylated porphyrin **206** with 6- or 7-bromoquinolin-4(1*H*)-ones **207** [98]

Porphyrin-quinolin-4(1*H*)-one conjugates were synthesized from the Suzuki–Miyaura coupling reaction of a β-borylated porphyrin **206**, prepared by borylation of the correspondent 3-bromotetraphenylporphyrinatozinc(II) with pinacolborane in the presence of $PdCl_2(PPh_3)_2$, with 6- or 7-bromoquinolin-4(1*H*)-ones **207** containing *N*-ethyl and *N*-D-ribofuranosyl substituents (Scheme 61) [98]. Quinolones **207** bearing an *N*-ethyl substituent were more reactive than those bearing *N*-ribonucleosides providing the corresponding porphyrin-quinolin-4(1*H*)-one conjugates **208** in very good yields (82–89%). The other conjugates **208** were isolated in lower yields (50–51%), but nearly 50% of the starting porphyrin **206** was recovered in both cases. Attempts to improve the outcome of the coupling process between **206** and *N*-ribonucleosides quinolone derivatives **207** were not successful [e.g. by increasing both the reaction time and the amount of bromoquinolin-4(1*H*)-ones] probably due to steric effects caused by the ribofuranosyl group. Basic hydrolysis and deprotection of the ester and benzoyl groups of ribose moieties in conjugates **208** followed by the acid demetalation afforded the correspondent conjugates that were evaluated as singlet oxygen generators.

4.1.3 Sonogashira Reaction

Diverse 2-substituted furo[3,2-*c*]quinolines **210** were prepared by a one-pot process involving Sonogashira-type coupling followed by the electrophilic or transition metal-mediated cyclization of the resulting alkynes possessing a suitable nucleophilic group in the proximity of the triple bond (Scheme 62) [99]. Better yields were obtained when using $Pd(PPh_3)_4$ (85%) or $PdCl_2(PPh_3)_2$ (80%) in place of Pd/C-PPh_3 (70%), although the latter is cheaper. DMF was found to be a better solvent when compared to THF or MeCN, and CuI is crucial in this reaction; otherwise,

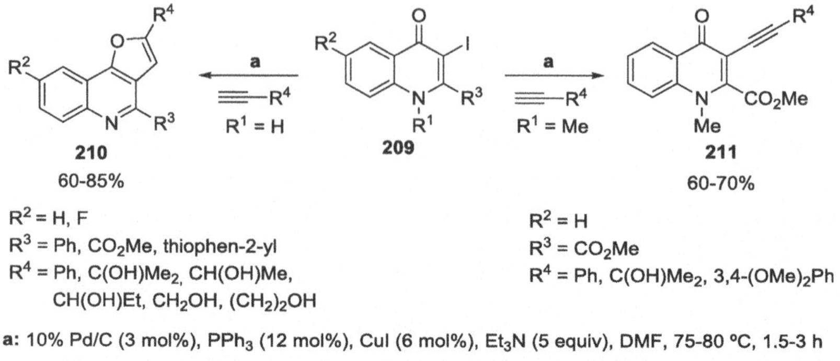

Scheme 62 Pd/C-PPh_3-mediated synthesis of 2-substituted furo[3,2-*c*]quinolines **210** and 3--alkynylquinolin-4(1*H*)-ones **211** [99]

only deiodinated product was formed. Absence of PPh_3 when using Pd/C resulted in a poor yield (22%). Good yields of the desired furo[3,2-*c*]quinolines **210** were obtained regardless the nature of terminal alkynes used.

The key features of the present tandem coupling–cyclization process are the transition metal-mediated activation of the triple bond of the 3-alkynylquinoline generated in situ followed by an intramolecular attack of the oxygen on the activated triple bond with subsequent proton transfer and release of the metal ion to give the desired furoquinoline **210**. The NH of the quinolin-4(1*H*)-one ring has a critical role in facilitating the participation of C-4 quinoline oxygen in the cyclization step. Indeed when the authors performed the reaction of methyl 3-iodo-1-methyl-4-oxo-1,4-dihydroquinoline-2-carboxylate **209** with terminal alkynes under the same conditions, only 3-alkynylquinolin-4(1*H*)-ones **211** were isolated as a result of a normal Sonogashira coupling reaction, and formation of furoquinolines **210** was not observed, even in trace amounts (Scheme 62).

Sonogashira protocol was used to synthesize different macrolone derivatives with high antibacterial activity, starting from 6-iodoquinolin-4(1*H*)-ones [93, 100, 101]. An example is shown in Scheme 63 where Sonogashira reaction of macrolides **212** with 6-iodoquinolin-4(1*H*)-ones **194** leads to macrolones **213** in moderate yields. Pd/C-catalysed hydrogenation was performed to obtain the desirable macrolones **214**.

Another group of macrolones (with tricyclic quinolone moiety) **217** (R^3=H) and **218** was also synthesized by Sonogashira approach although, only when ethyl esters **215** of the parent acids were used, the reaction proceed in shorter reaction times and better yields (Scheme 64, **a**: R^3=H, **a***: R^3=Et). After hydrogenation (Scheme 64, **b**), esters **217** (R^3=Et) were hydrolysed to the corresponding carboxylic acids **218** (Scheme 64, **c***).

The Sonogashira reaction of 6-bromo-3-iodoquinolin-4(1*H*)-one **219** with TMSA yielded the expected coupling products in very satisfactory yields, with no further conversion to furo[3,2-*c*]quinoline derivatives (Scheme 65) [97]. Subsequently, Suzuki and Sonogashira cross-coupling reactions at the 6-position of

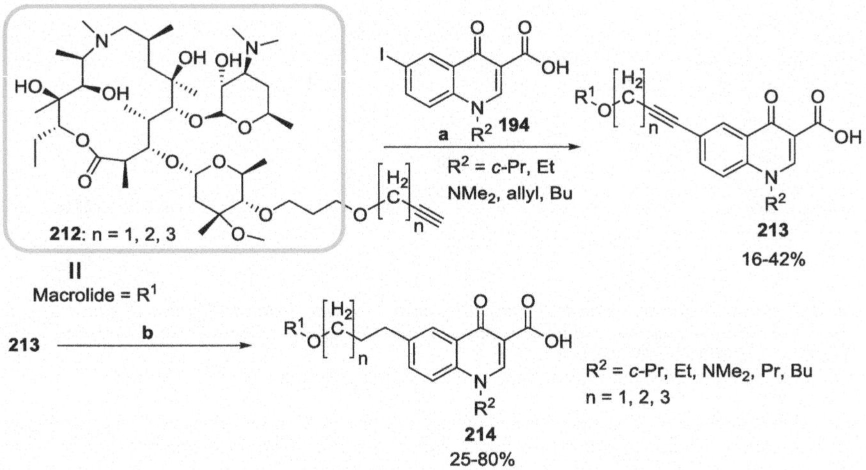

a: $PdCl_2(PPh_3)_2$ (5 mol%), CuI (20 mol%), **194** (1.8 equiv), Et_3N (10 equiv), MeCN, 50 °C, 16 h;
b: Pd/C 20 wt % (10 mol%), H_2 (2.2 bar), MeOH, 15 h.

Scheme 63 Synthesis of macrolones **214** via Sonogashira reaction of macrolide **212** with 6-iodoquinolin-4(1*H*)-ones **194** followed by Pd/C hydrogenation [93]

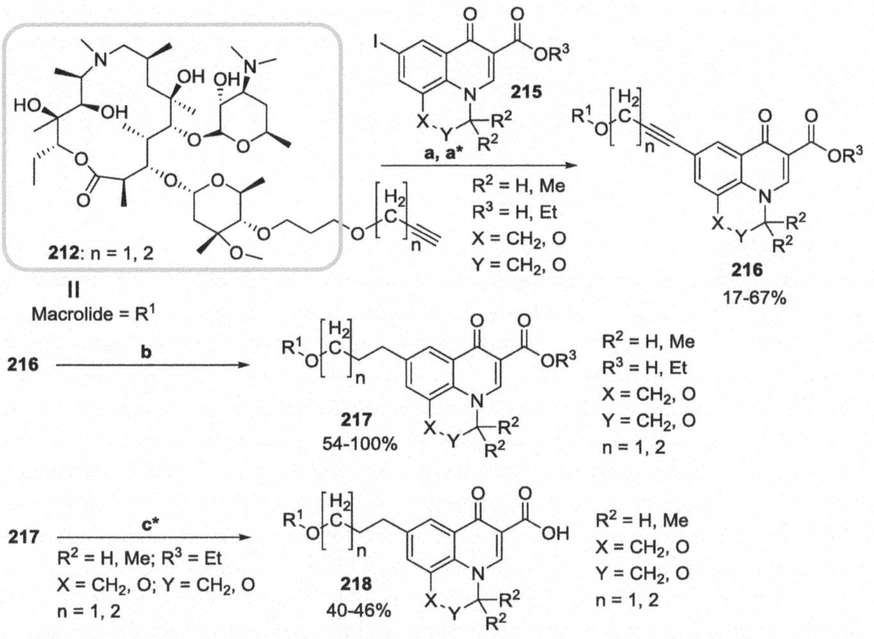

a: $PdCl_2(PPh_3)_2$ (5 mol%), CuI (20 mol%), **215**: R^3 = H (2.5 equiv), Et_3N (10 equiv), EtOH, 50 °C, 16 h; **a*:** $PdCl_2(PPh_3)_2$ (5 mol%), CuI (20 mol%), **215**: R^3 = Et (2.0 equiv), Et_3N (10 equiv), MeCN, 50 °C, 1 h; **b:** Pd/C 10 wt % (10 mol%), H_2 (2 bar), MeOH, 16 h; **c*:** THF/H_2O 2:1, 0.5 M LiOH 2.5 equiv, 3 h.

Scheme 64 Synthesis of macrolones **217** and **218** via Sonogashira reaction followed by Pd/C hydrogenation and hydrolysis of the corresponding esters **217** [93]

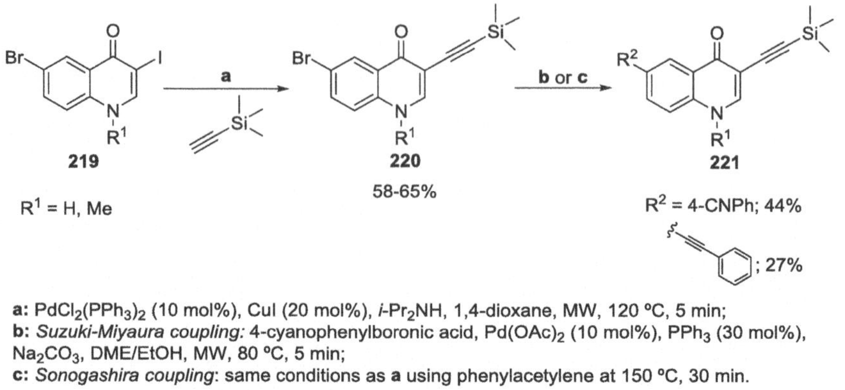

Scheme 65 Sonogashira reaction of 6-bromo-3-iodoquinolin-4(1*H*)-one **219** and subsequent Suzuki and Sonogashira reactions at the 6-position of quinolin-4(1*H*)-ones **220** [97]

substrates **220** were performed to afford the trisubstituted quinolin-4(1*H*)-ones **221**. Generally, harder conditions were necessary to functionalize the 6-position due to the lower reactivity of bromine compared to iodine, while, on the other hand, the increased structural complexity made substrates **220** more prone to decomposition and side products' formation. Therefore, cross-coupling reactions at the 6-position proceeded with slightly lower efficiency, leading to products **221** in moderate to low yields.

4.1.4 Stille Reaction

In a programme to develop new antibacterial agents, Laborde and co-workers described a series of 1-cyclopropylquinolin-4(1*H*)-ones **224** (R^5=*c*-Pr), bearing at C-7 a vinyl, a 1-cyclopentenyl or a 1,2,3,6-tetrahydropyridin-4-yl group, synthesized by Pd cross-coupling of 7-quinolyltriflate **222** (X=OTf) with an appropriately functionalized vinylstannane **223** (Scheme 66, **a**) [102, 103]. The reaction with a range of vinylstannanes **223** bearing different functional groups afforded the corresponding quinolin-4(1*H*)-ones **224** in moderate to good yields (22–88%) with complete chemo- and regioselectivity; the coupling takes place exclusively at the C-7 position of the quinolone nucleus even in the presence of the C-6 fluorine or the C-2 α,β-unsaturated keto-ester moiety. In some cases, additional amounts of the tin reagent and/or the catalyst [either $PdCl_2(PPh_3)_2$ or $Pd(PPh_3)_4$] were necessary to improve the yield of the coupled product. The low yield obtained when using the cyclopentenylstannane possessing a bulky *tert*-butoxycarbonyl group at the 3-position (31%) was attributed to unfavourable steric demands. Later, Reuman

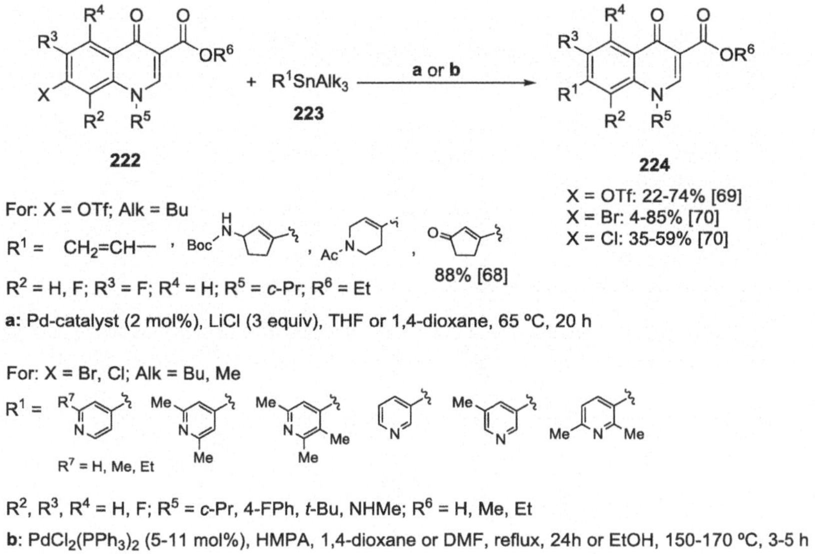

Scheme 66 Pd-catalysed Stille cross-coupling reaction of quinolin-4(1*H*)-ones **222** with functionalized stannanes **223** [102–104]

and co-workers reported a new series of 7-pyridinylquinolin-4(1*H*)-ones applying the same methodology but starting from 7-bromo- or 7-chloroquinolin-4(1*H*)-ones **222** (X=Br or Cl) (Scheme 66, **b**) [104]. The quinolin-4(1*H*)-one esters **224** were hydrolysed to the corresponding acids (50–73%) with sodium hydroxide or hydrochloric acid with concomitant removal of the *N*-protecting groups.

5-Substituted 3-(4-methoxyphenyl)quinolin-4(1*H*)-ones **225** have been synthesized in good yields via Stille cross-coupling reaction of the corresponding 3-(4-methoxyphenyl)-5-trifluoromethanesulfonatequinolin-4(1*H*)-ones **197** with tributyl(vinyl)stannane (67–85%) or with tributyl(thiophen-2-yl)stannane (64%) (Scheme 67) [94]. This protocol was successfully applied to the synthesis of 2-(3-methoxyphenyl)quinolin-4(1*H*)-one **226** [95].

4.1.5 Aminocarbonylation Reaction

A Pd-catalysed carbonylation of 6-bromo-3-iodoquinolin-4(1*H*)-ones **227** with amines, using $Pd(OAc)_2$ as catalyst and $Mo(CO)_6$ as CO source, provides compounds **228** in low yields (25–38%) (Scheme 68, **a**) [97]. Slight modifications of the experimental conditions did not improve the reaction yield. Even when gaseous CO and Pd/C as the catalyst were used (Scheme 68, **b**), the aminocarbonylation product was obtained in similar yield (35%).

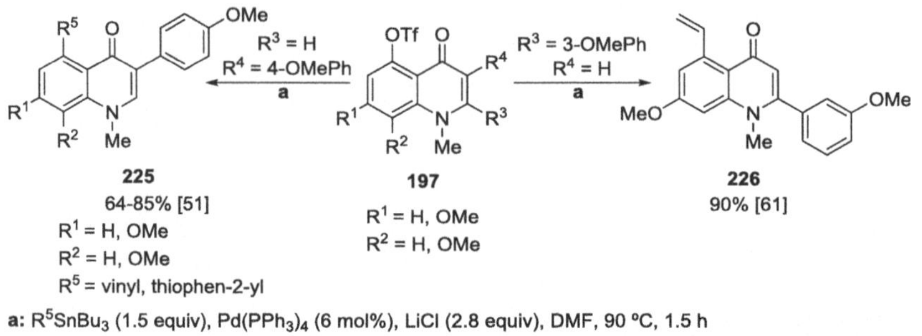

Scheme 67 Synthesis of quinolin-4(1*H*)-ones **225** and **226** by Stille coupling reaction of 5--trifluoromethanesulfonatequinolin-4(1*H*)-ones **197** with tributyl(vinyl)stannanes [94, 95]

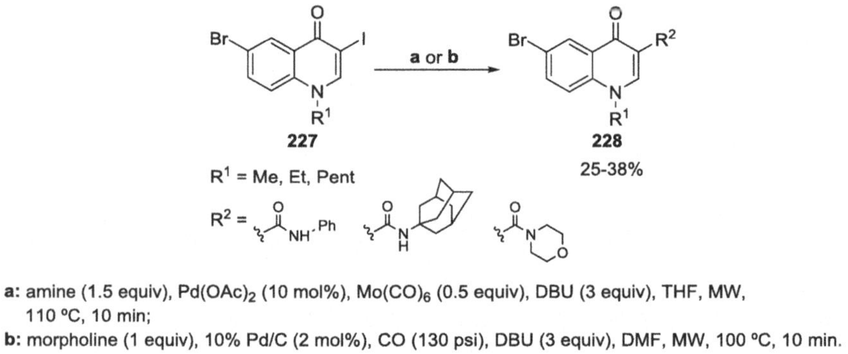

Scheme 68 Pd-catalysed aminocarbonylation of 6-bromo-3-iodoquinolin-4(1*H*)-ones **227** [97]

4.2 Cu-Catalysed Ullmann-Type Reaction

Ullmann-type reactions of 3-haloquinolin-4(1*H*)-ones **229** with various *N*-containing nucleophiles using an inexpensive copper catalyst system proceed rapidly under relatively mild conditions providing direct access to various 3-(1-substituted)-quinolin-4(1*H*)-ones **230** and **231** in good to excellent yields (Scheme 69) [105]. The reaction was dramatically facilitated when using K_2CO_3 or Cs_2CO_3 as bases, using Cu powder as catalyst and using DMEDA as the ligand in toluene at 135°C. The use of other ligands [e.g. (±)-*trans*-cyclohexane-1,2-diamine-based ligands, TMEDA, 1,10-phenanthroline, L-proline or ethyl 2-oxocyclohexanecarboxylate] induced a lowering of the conversion rate. The authors also found that in the coupling of 3-iodo-1-methylquinolin-4(1*H*)-one **229** (X=I) with 4-methoxybenzamide, the source of copper used has no influence on the reaction rate, since $CuSO_4$, CuI, CuBr and CuTC gave similar results than that of Cu powder. This Ullmann-type reaction proved to be general for the coupling with a large variety of nucleophiles, being compatible with primary

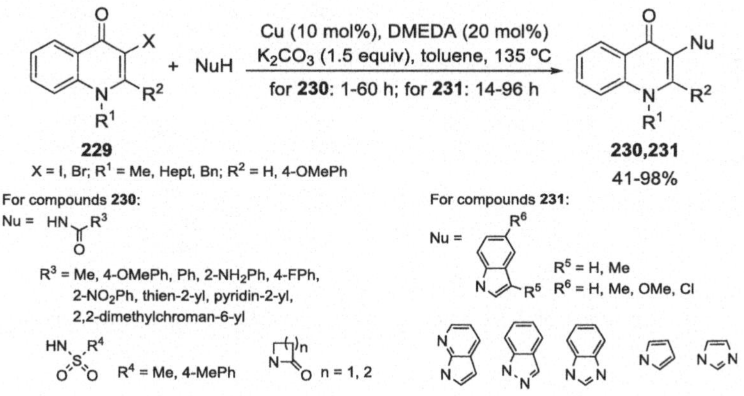

Scheme 69 Synthesis of quinolin-4(1*H*)-ones **230** and **231** by Cu-catalysed Ullmann-type reactions of 3-haloquinolin-4(1*H*)-ones **229** with various *N*-containing nucleophiles [105]

substituted (hetero)aromatic, aliphatic and cyclic amides, alkyl or arylsulfonamides, indoles, pyrroles, azaindoles, imidazoles and indazoles although in the case of less nucleophilic reagents, a longer reaction time was required. The efficiency of this catalytic system was also tested in the coupling of various nitrogen nucleophiles with other electrophilic coupling partners, such as 3-bromoquinolin-2(1*H*)-ones, and these electrophilic substrates efficiently undergo the coupling reaction giving the corresponding coupling products moderate to excellent yields (41–98%) [105]. When using 3,6-dibrominated quinolin-2(1*H*)-ones, the amination proceeded at the more activated C-3 position and yielded the monoaminated products in satisfactory yields (42–68%) under the conditions indicated in Scheme 69.

5 Transformation of Quinolin-2(1*H*)-ones

5.1 *Heck Reaction*

The precursor **234** of quinolone dimers paraensidimerins was synthesized by a Pd-catalysed Heck reaction of 3-iodoquinolin-2(1*H*)-one **232** with 2-methyl-3-buten-2-ol **233** (Scheme 70) [106]. The reaction of 4-acetoxyquinolin-2(1*H*)-one **232** (R=Ac), obtained by acetylation of 4-hydroxy-3-iodo-1-methylquinolin-2(1*H*)-one **232** (R=H), gave three products: the expected quinolone allylic alcohol **234** (30%); the diene **236** (11%), probably obtained by dehydration of allylic alcohol **234** by the Et_3NHI salt by-product formed in the reaction mixture; and the known alkaloid *N*-methylflindersine **235**, probably formed by deacetylation and subsequent intramolecular cyclization during the course of the reaction.

4-Acetylquinolin-2(1*H*)-one **239** was prepared via a very high α-regioselective Heck coupling of 4-tosylquinolin-2(1*H*)-one **237** with butyl vinyl ether (Scheme 71) [107]. The α-product **238** was obtained in very high α/β

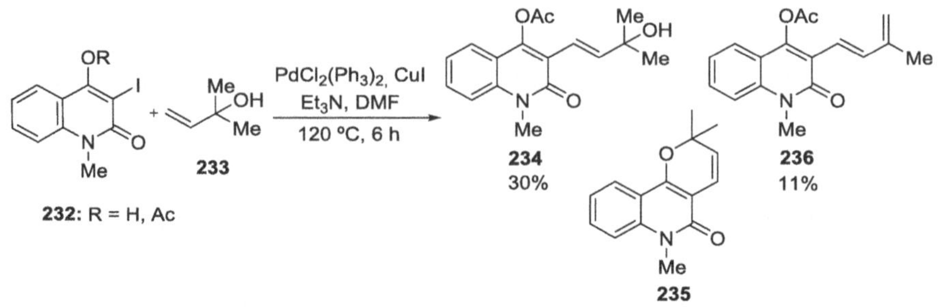

Scheme 70 Heck reaction of 3-iodoquinolin-2(1*H*)-one **232** with 2-methyl-3-buten-2-ol **233** [106]

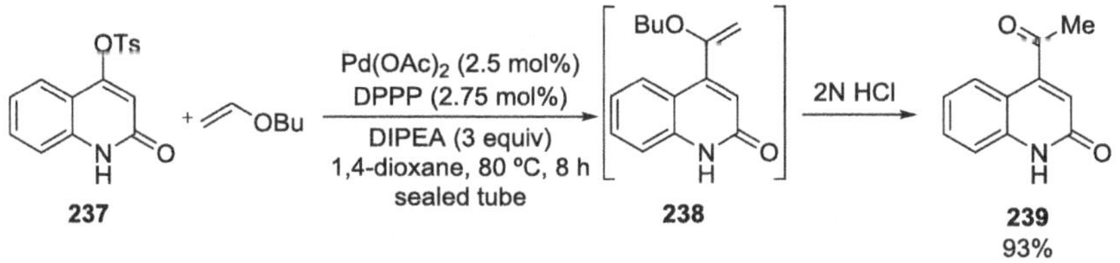

Scheme 71 Synthesis of 4-acetylquinolin-2(1*H*)-one **239** via Heck reaction of 4-tosylquinolin-2(1*H*)-one **237** with butyl vinyl ether followed by acidic treatment [107]

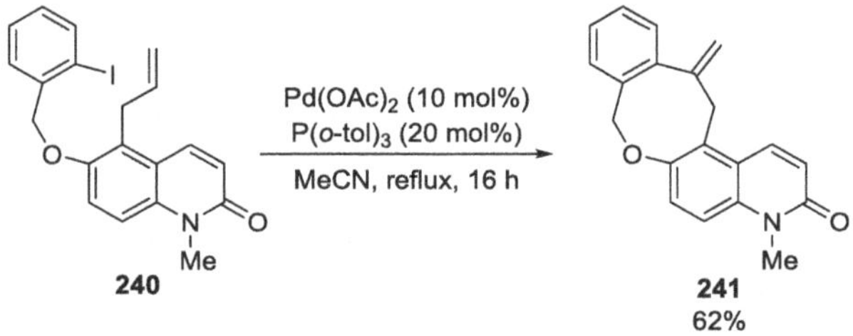

Scheme 72 Synthesis of benzoxocine-fused quinolin-2(1*H*)-one **241** from intramolecular Heck reaction of benzylallylquinolin-2(1*H*)-one **240** [108]

regioselectivity (>99/1) and after acidic treatment led to 4-acetylquinolin-2(1*H*)-one **239**.

One benzoxocine-fused quinolin-2(1*H*)-one **241** was prepared from the intramolecular Heck reaction of the benzylallylquinolin-2(1*H*)-one precursor **240** (Scheme 72) [108]. The reaction led to the regioselective formation of the 8-*exo* cyclization product **241**, and no 9-*endo* product was observed in this case.

Quinolin-2(1*H*)-one annulated benzazocines **243** were obtained from the Pd-catalysed intramolecular Heck reaction of unactivated allylquinolin-2(1*H*)-ones **242** (Scheme 73) [109]. The intramolecular Heck reaction afforded exclusively the *endo*cyclic product **243** in good yields. Once the exclusively 8-*exo* mode of cyclization is unusual, in this case, the authors believe that the formation of the *endo*cyclic products may occur via two possible pathways, namely, proceeding by

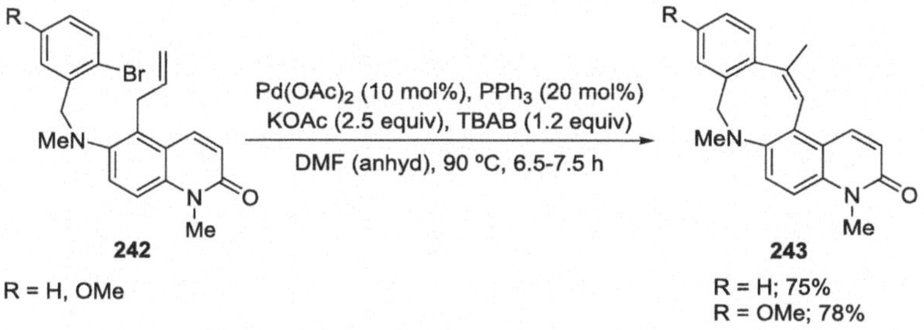

Scheme 73 Synthesis of quinolin-2(1*H*)-one annulated benzazocines **243** by intramolecular Heck reaction of allylquinolin-2(1*H*)-ones **242** [109]

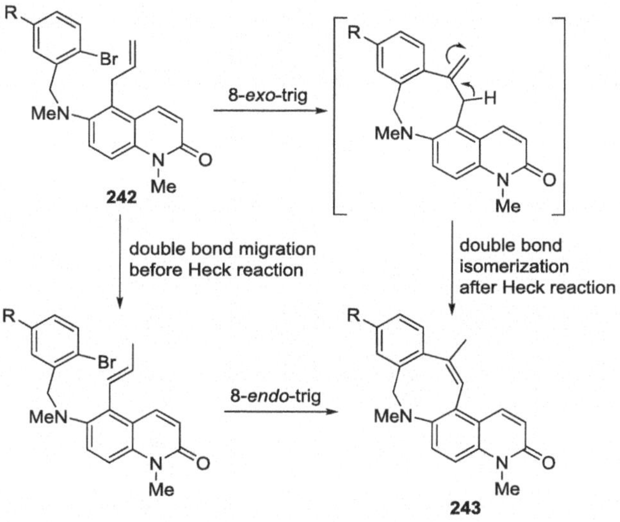

Scheme 74 Probable mechanistic path of the intramolecular Heck reaction of allylquinolin-2 (1*H*)-ones **242** [109]

the 8-*exo* mode of cyclization followed by double-bond isomerization and, alternatively, a double-bond isomerization prior to the Heck reaction leading to the subsequent 8-*endo*-Heck cyclization (Scheme 74).

Later, the same authors described the synthesis of quinolin-2(1*H*)-one annulated benzazocinones **245** applying the same methodology at unactivated allylquinolin-2 (1*H*)-ones **244** (Scheme 75) [110]. In this case, the optimal reaction conditions were identified using $Pd(PPh_3)_4$ as the catalyst. This ligand-free Heck coupling reaction leads exclusively to the expected 8-*exo*-Heck product **245**, in good yields, without any contamination of the 9-*endo*-Heck or 8-*exo*-isomerized product.

Using the methodology described above, two quinolin-2(1*H*)-one annulated benzazoninones **247** were synthesized starting from unactivated allylquinolin-2 (1*H*)-ones **246** (Scheme 76) [111]. The Heck reaction proceeded in reasonable

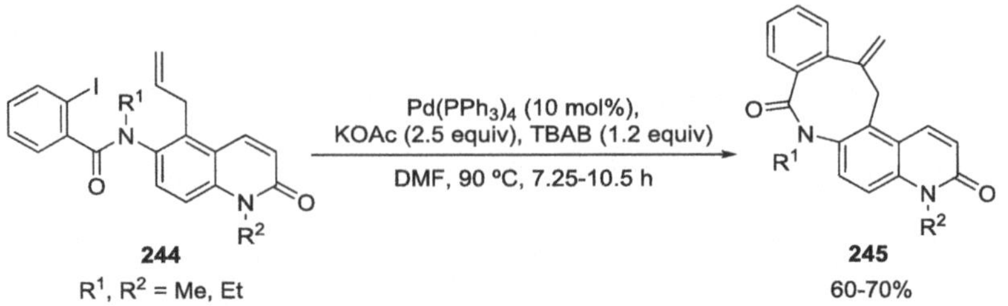

Scheme 75 Synthesis of quinolin-2(1*H*)-one annulated benzazocinones **245** by intramolecular Heck reaction of allylquinolin-2(1*H*)-ones **244** [110]

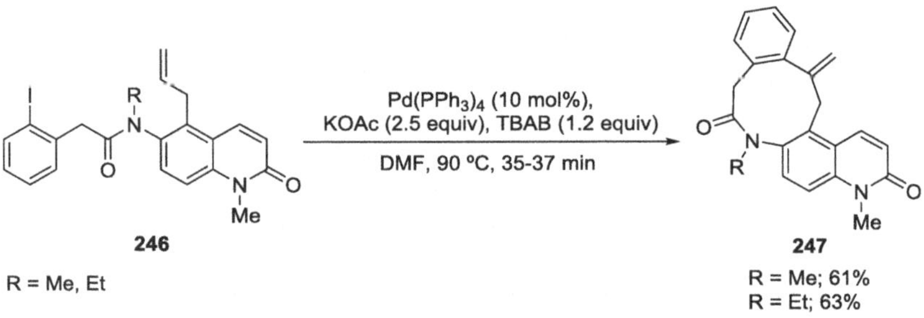

Scheme 76 Synthesis of quinolin-2(1*H*)-one annulated benzazoninones **247** by intramolecular Heck reaction of allylquinolin-2(1*H*)-ones **246** [111]

good yields with shorter reaction times, and only the product corresponding to the 9-*exo* mode of cyclization was obtained.

Pd-catalysed intramolecular Heck reaction of 3- and 4-(2-bromobenzyloxy) quinolin-2(1*H*)-ones **248** and **250**, previously synthesized [112, 113], under Jeffery's two-phase protocol afforded four tetracyclic quinolones **249** and **251** in very good yields (Scheme 77) [114].

A ligand-free intramolecular Heck reaction of 6-amino-5-bromoquinolin-2(1*H*)-ones **252** catalysed by $Pd(OAc)_2$ gave new pyrrole-fused quinolin-2(1*H*)-ones **253** (Scheme 78) [115]. This methodology is applicable to secondary and tertiary amines. Under similar reaction conditions, the intramolecular Heck reaction of quinolin-2(1*H*)-one **254** afforded the corresponding pyrrol-2(3*H*)-one fused quinolin-2(1*H*)-one **255**.

Later, another synthesis of pyrrolo[3,2-*f*]quinolin-7(6*H*)-ones **257** in very good yields was reported, by a ligand-free $Pd(OAc)_2$-catalysed intramolecular Heck reaction of enamines, prepared in situ from the condensation of 6-amino-5-bromo-1-methylquinolin-2(1*H*)-one **256** with cyclic and acyclic ketones (Scheme 79) [116]. The reaction is only effective when DMF is used as solvent and other Pd sources [$PdCl_2$, $PdCl_2(PPh_3)_2$, $Pd(PPh_3)_4$] were less efficient than Pd $(OAc)_2$.

First, substrates **256** condense with the ketone to generate **258** and produce the enamines **259** in the presence of a base. The formed enamines subsequently

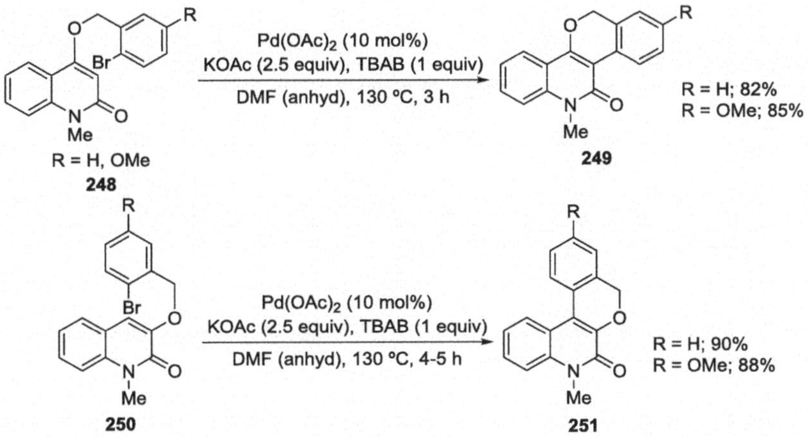

Scheme 77 Synthesis of tetracyclic quinolones **249** and **251** by intramolecular Heck reaction of 3- and 4-(2-bromobenzyloxy)quinolin-2(1*H*)-ones **248** and **250** [114]

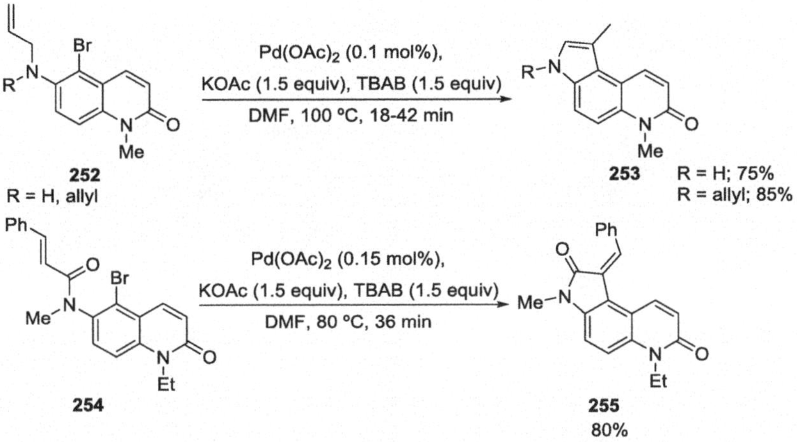

Scheme 78 Synthesis of pyrrolo[3,2-*f*]quinolin-7(6*H*)-ones **253** and **255** by intramolecular Heck reaction of 5-bromoquinolin-2(1*H*)-ones **252** and **254** [115]

underwent intramolecular palladium-catalysed Heck reaction to afford the products **257** (Scheme 80).

5.2 Suzuki–Miyaura Reaction

The synthesis of 4- and 3,4-substituted quinolin-2(1*H*)-ones **261** and **262** via Pd-catalysed regioselective cross-coupling reactions of 3-bromo-4-trifloxy-quinolin-2(1*H*)-ones **260** with arylboronic acids was described (Scheme 81)

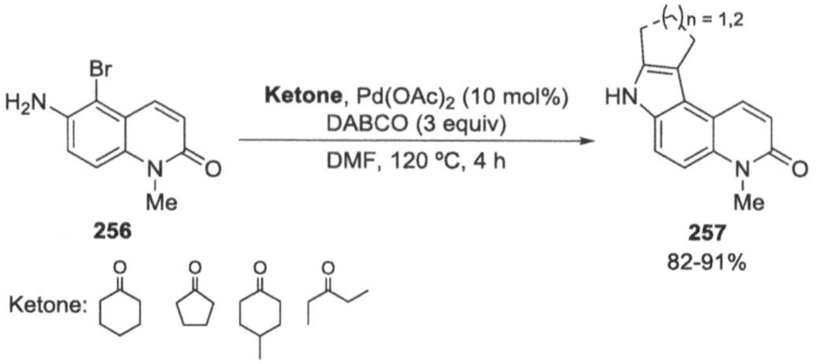

Scheme 79 Synthesis of pyrrolo[3,2-*f*]quinolin-7(6*H*)-ones **257** from an intramolecular Heck reaction of the in situ prepared enamines [116]

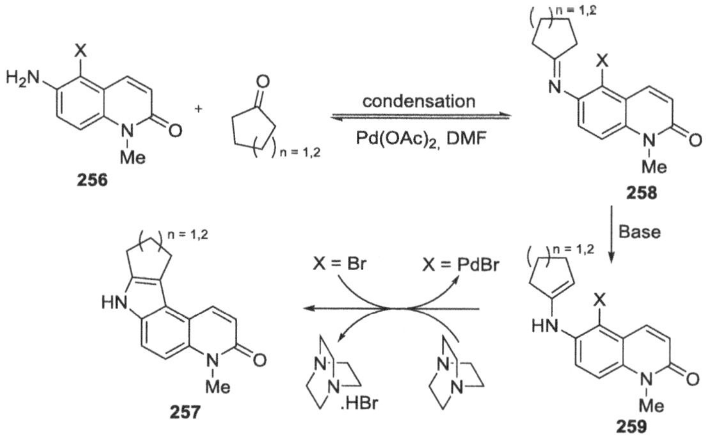

Scheme 80 Proposed mechanism for the synthesis of pyrrolo[3,2-*f*]quinolin-7(6*H*)-ones **257** [116]

[117]. Regiocontrolled cross-coupling of 3-bromo-4-trifloxyquinolin-2(1*H*)-one **260** could be achieved by tuning the temperature and the amount of arylboronic acid. When using $PdCl_2(PPh_3)_2$ as catalyst in the reaction with 4-methoxyphenylboronic acid (1.5 equiv.) at 50°C, the corresponding product **261** was obtained in 20% yield together with disubstituted compound **262** (60%). At room temperature **261** was obtained as the major product (78%) and only traces of **262** were detected. Reducing the amount of 4-methoxyphenylboronic acid to 1.1 equiv., **261** was the only product (81%). At 60°C with a higher excess of 4-methoxyphenylboronic acid (2.5–3.0 equiv.), only **262** was generated (87%). Both electron-withdrawing and electron-donating-substituted arylboronic acids are suitable coupling partners, giving similar reaction yields.

Compound **261** (R^1=3-CF_3Ph) was further elaborated under the conditions shown in Scheme 82, and when it was employed as substrate in the Suzuki–Miyaura

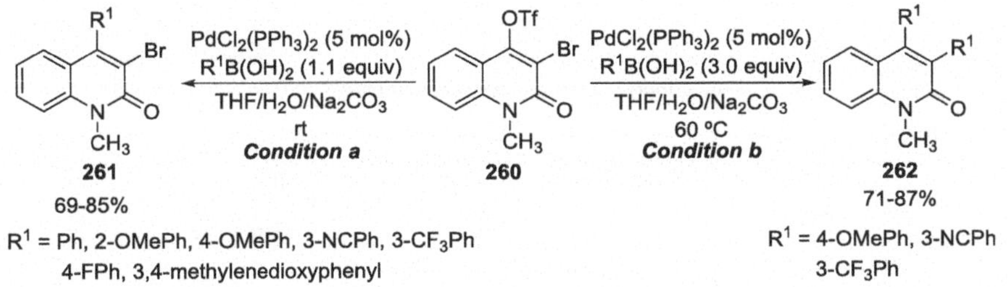

Scheme 81 Synthesis of 4- and 3,4-substituted quinolin-2(1*H*)-ones **261** and **262** via regioselective Suzuki–Miyaura reactions of 3-bromo-4-trifloxyquinolin-2(1*H*)-ones **260** [117]

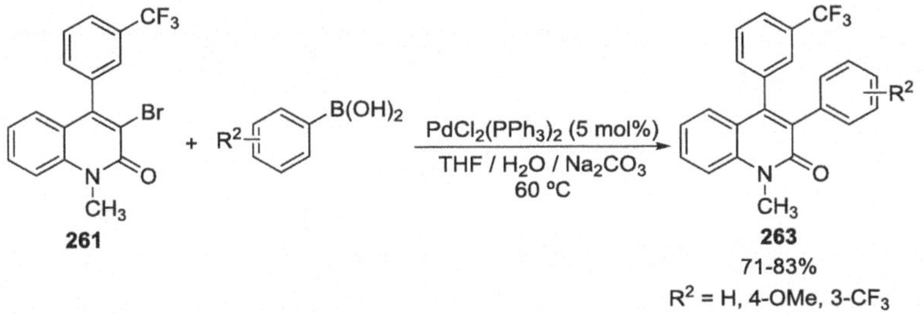

Scheme 82 Synthesis of 3,4-disubstituted quinolin-2(1*H*)-ones **263** by Suzuki–Miyaura coupling reaction of **261** with arylboronic acids [117]

reaction with different arylboronic acids, good yields of the corresponding products **263** were obtained [117].

Functionalized 4,4′-bisquinolones **265** were prepared in good to excellent yields through Pd-catalysed one-pot borylation–Suzuki cross-coupling reaction of 4-chloroquinolin-2(1*H*)-one precursors **264** employing controlled MW irradiation (Scheme 83) [118]. A strong base such as KOH in combination with $PdCl_2$(dppf) as catalyst is needed to achieve excellent conversions. Solvents such as DMSO, DMF and toluene promoted the formation of the dehalogenated product, while 1,4-dioxane, dichloromethane and 1,2-dichloroethane minimized dehalogenation. 1-Chlorobutane proved to be the best solvent in this particular case.

5.3 Sonogashira Reaction

Pyrrolo[3,2-*f*]quinolin-7(6*H*)-ones **269** were synthesized in excellent yields by sequential coupling and cyclization reactions of aryl halides with (trimethylsilyl) acetylene with concurrent elimination of the TMS substituent (Scheme 84) [119]. Acetylenic amines possessing an electron-donating group on the nitrogen atom also underwent Cu(I)-catalysed cyclization. Quinolones **266** were prepared

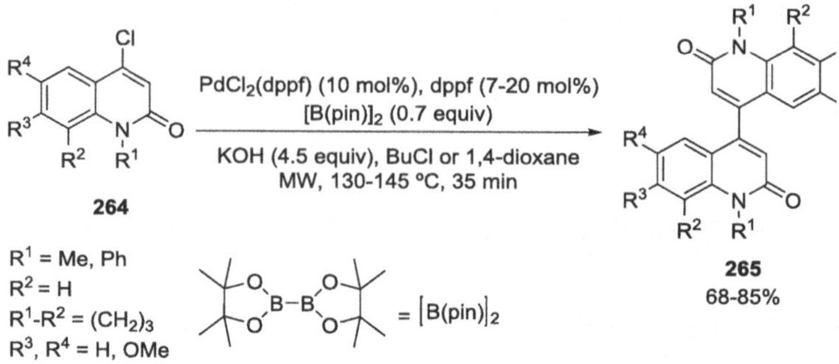

Scheme 83 Synthesis of 4,4′-bisquinolones **265** by one-pot borylation–Suzuki cross-coupling of 4-chloroquinolin-2(1*H*)-one **264** [118]

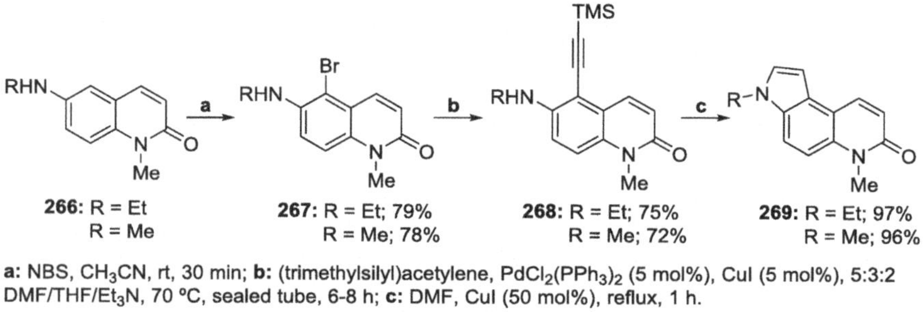

Scheme 84 Synthesis of pyrrolo[3,2-*f*]quinolin-7(6*H*)-ones **269** by sequential Sonogashira coupling and cyclization reactions of 5-bromoquinolin-2(1*H*)-ones **267** with (trimethylsilyl)acetylene [119]

from commercially available quinolines [120, 121] and then brominated, and the resulting bromo derivatives **267** transformed into the required heteroannulation precursors **268** by a Sonogashira coupling with (trimethylsilyl)acetylene using $PdCl_2(PPh_3)_2$ as catalyst and CuI as the cocatalyst. The reactions were optimized by smoothly heating the reaction in a sealed tube. Heteroannulation of the acetylenic amines to give **269** was achieved by refluxing the precursors **268** in DMF in the presence of CuI.

Acetylenic amines **268** and **270** were obtained by Sonogashira coupling of the corresponding 6-amino-5-bromoquinolin-2(1*H*)-one derivatives **267** with (trimethylsilyl)acetylene or phenylacetylene, respectively. Both reactions were performed in the presence of $PdCl_2(PPh_3)_2$ as catalyst and CuI as cocatalyst, although in slightly different experimental conditions (Scheme 85) [122].

Pyrrolo[3,2-*f*]quinolin-7(6*H*)-ones were also synthesized in high yield by cycloisomerization of previously described acetylenic amines **270** using $AuCl_3$ in the absence of any silver salts or any other bases. Under these conditions acetylenic amines with electron-donating groups readily undergo cycloisomerization (Scheme 86, **a**). No cycloisomerization product was obtained with trimethylsilyl-substituted acetylenic amines **268**, even when using other solvents (toluene,

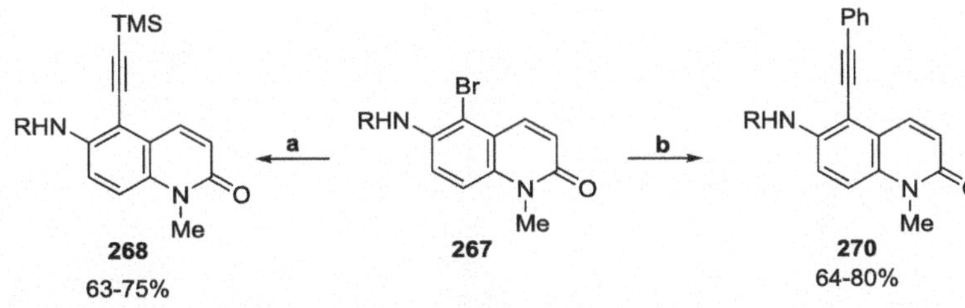

R = H, Me, Et

a: (trimethylsilyl)acetylene, $PdCl_2(PPh_3)_2$ (4.8 mol%), CuI (11 mol%), 2:2:5 Et_3N/THF/DMF, sealed tube, 80 °C, 9 h; **b:** phenylacetylene, $PdCl_2(PPh_3)_2$ (5.7 mol%), CuI (11 mol%), 2:5 Et_3N/DMF, 120 °C, 1.5 h.

Scheme 85 Sonogashira coupling of 6-amino-5-bromoquinolin-2(1*H*)-ones **267** with phenylacetylene and (trimethylsilyl)acetylene [122]

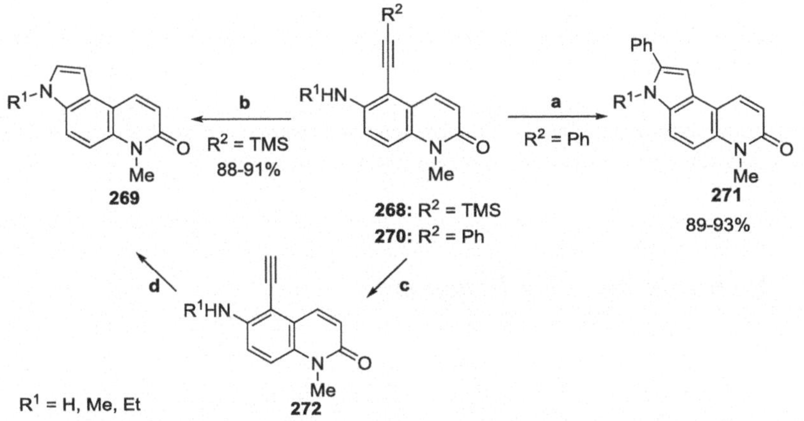

R¹ = H, Me, Et

a: $AuCl_3$ (1 mol%), EtOH, 80 °C, 4-7 h; **b:** $AuCl_3$ (1 mol%), EtOH, K_2CO_3, 70 °C, sealed tube, 8.5–10 h; **c:** K_2CO_3, MeOH, 30 °C, 4 h; **d:** $AuCl_3$ (1 mol%), EtOH, 80 °C.

Scheme 86 Synthesis of pyrrolo[3,2-*f*]quinolin-7(6*H*)-ones **269** and **271** by gold-catalysed cycloisomerization of acetylenic amines **268** and **270** [122]

1,4-dioxane, DMF or MeCN) and increasing the amount of catalyst to 5 mol% or the temperature, probably due to the presence of bulky trimethylsilyl group. Cyclized products **269** were only obtained in very good yield when **268** was treated with $AuCl_3$, using K_2CO_3 as base (Scheme 86, **b**). The trimethylsilyl group was initially eliminated by the reaction with K_2CO_3, and then cyclization proceeded in a one-pot transformation with the unsubstituted alkyne moiety, which was verified by the reaction of the unsubstituted alkyne **272** (Scheme 86, **c** and **d**). The time required for the cyclization of compounds **268** in these conditions is relatively long because the triple bond at the terminal position is not activated [122].

Another series of pyrrolo[3,2-*f*]quinolin-7(6*H*)-ones **275** were synthesized from the intramolecular hydroamination reaction of acetylenic amines **274**, obtained from **273**, catalysed by $PdCl_2/FeCl_3$ (Scheme 87) [123]. $PdCl_2(PPh_3)_2$ and $Pd(OAc)_2$ were

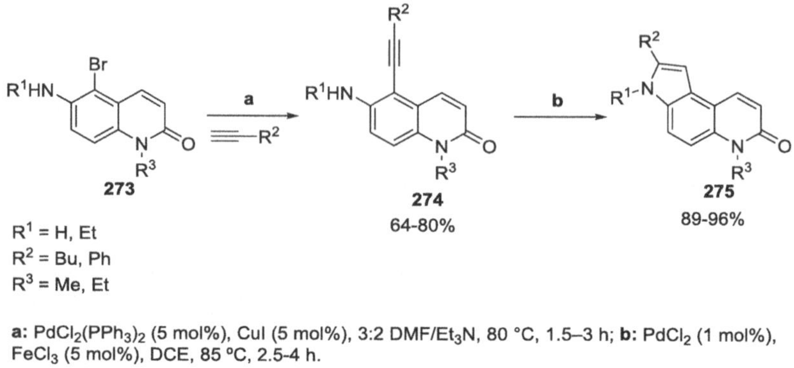

Scheme 87 Synthesis of pyrrolo[3,2-*f*]quinolin-7(6*H*)-ones **275** by $PdCl_2/FeCl_3$-catalysed intramolecular hydroamination of acetylenic amines **274** [123]

also tested as Pd(II) sources but found to be ineffective, unlike $PdCl_2$ that is efficient in only 1 mol%. The presence of $FeCl_3$ is necessary for this hydroamination reaction to occur and may facilitate the reoxidation of Pd(0) to Pd(II) in the catalytic cycle. This cyclization proceeded well in the presence of aromatic- and aliphatic-substituted alkynes and with both protected and unprotected amines.

5.4 *Buchwald–Hartwig Reaction*

A series of 3-(*N*-substituted-amino)quinolin-2(1*H*)-ones **277–280** were synthesized in good to excellent yields by the Pd-catalysed C–N coupling reaction starting from 3-bromoquinolin-2-(1*H*)-ones **276**. Several nucleophiles, including amines, amides and benzylcarbamate, as well as the less nucleophilic alkyl- or arylsulfonamides, were effective in this transformation (Scheme 88) [124]. The C–N bond-forming reaction was also studied with urea leading to the double heteroarylated coupling product **280** in good yield (61%). In the presence of two carbon–bromine atoms of quinolin-2-(1*H*)-one **276** (R^2=Br), the coupling proceeded at the more activated C-3 position and yielded the mono-substituted product (44–48%) together with the disubstituted one (15–27%).

Later, the same authors described a Pd-catalysed coupling reaction of 3-bromo-1-methylquinolin-2(1*H*)-one **276** with indole and 5-methoxyindole affording the coupling products **282** in excellent yields (Scheme 89) [125].

5.5 *Aminocarbonylation Reaction*

Quinolin-2(1*H*)-one-6-carboxamides **288**, potential gonadotropin-releasing hormone antagonists, were obtained via the Pd-catalysed aminocarbonylation reaction

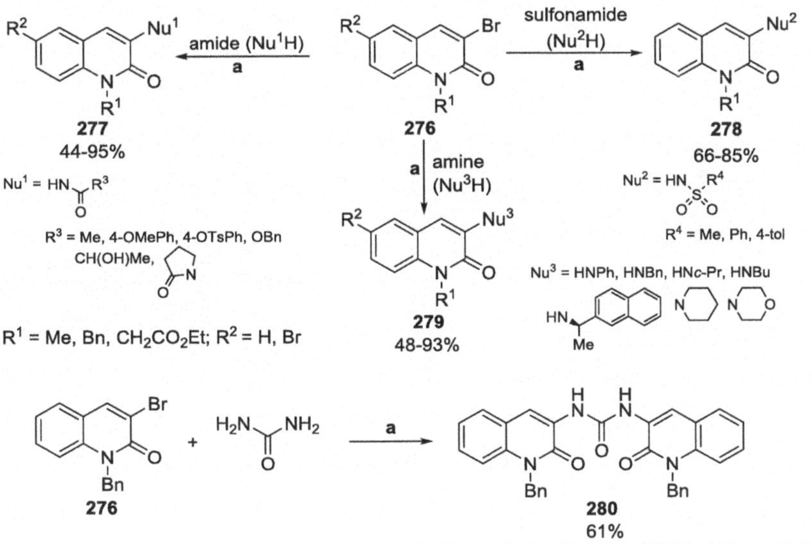

a: $Pd(OAc)_2$ (5 mol%), Xantphos (5 mol%), Cs_2CO_3 (2 mmol), 1,4-dioxane, 100 °C, 10 h, sealed Schlenk tube.

Scheme 88 Synthesis of 3-(*N*-substituted-amino)quinolin-2(1*H*)-ones **277–280** by Pd-catalysed C–N coupling reaction of 3-bromoquinolin-2-(1*H*)-ones **276** [124]

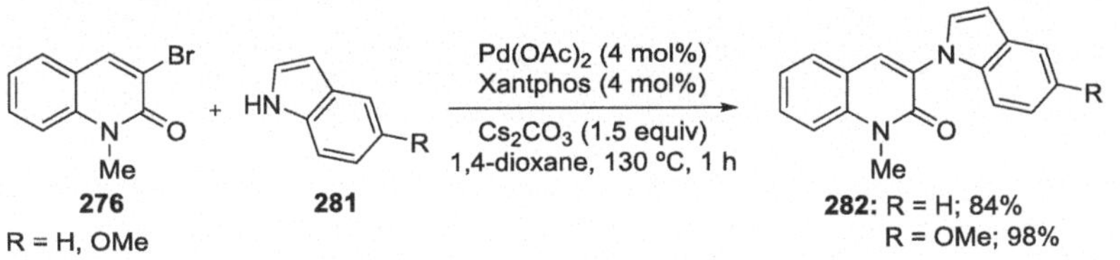

Scheme 89 Synthesis of 3-aminoquinolin-2(1*H*)-ones **282** by Pd-catalysed coupling reaction of 3-bromo-1-methylquinolin-2(1*H*)-one **276** with indoles **281** [125]

of 4-hydroxyquinolin-2(1*H*)-one **283** [126]. *O*-Alkylation of **283** with either (±)-**284** or (*S*)-**284** using the Mitsunobu protocol afforded the *O*-alkyl ethers (±)-**285** and (*S*)-**285**. Then Pd-catalysed aminocarbonylation of the iodo compound **285** with a variety of primary and secondary amines **286** led to the protected quinolin-2(1*H*)-one-6-carboxamides **287**. Removal of Boc protecting group of piperidines **287** using trifluoroacetic acid afforded the targeted amides **288** in quantitative yields (Scheme 90).

Quinolin-2(1*H*)-one-6-carboxamide **291** was also prepared by alkylation of 4-hydroxyquinolin-2(1*H*)-one **283** with (*S*)-2-(2-chloroethyl)-1-methylpiperidine **289** followed by aminocarbonylation of intermediate **290** with 4-aminopyrimidine (Scheme 91).

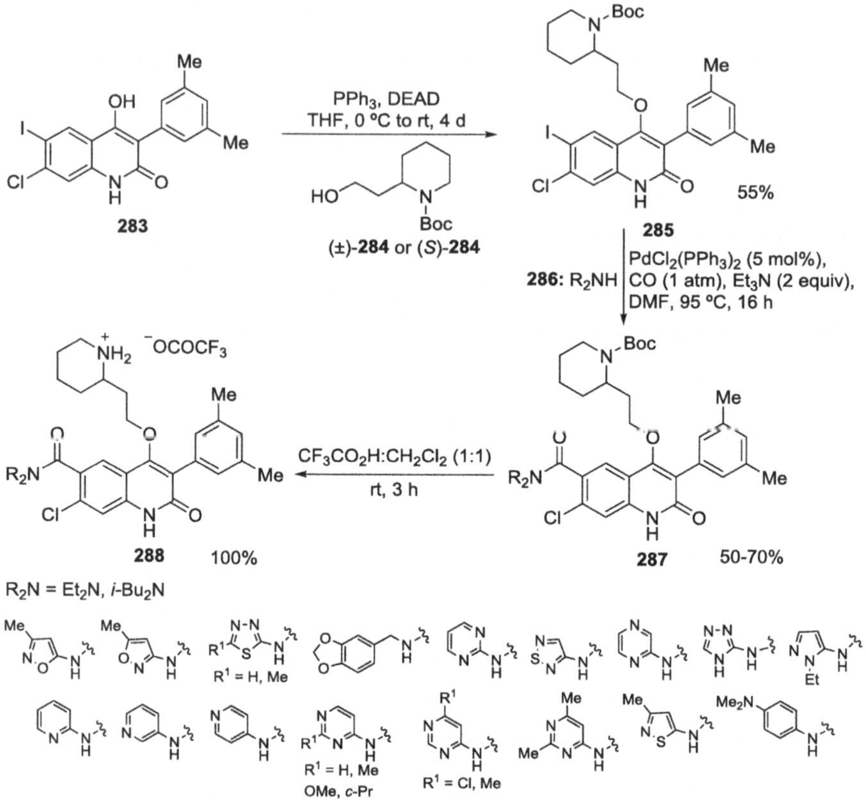

Scheme 90 Synthesis of quinolin-2(1*H*)-one-6-carboxamides **288** by Pd-catalysed aminocarbonylation of 4-hydroxyquinolin-2(1*H*)-one **283** with several amines **286** [126]

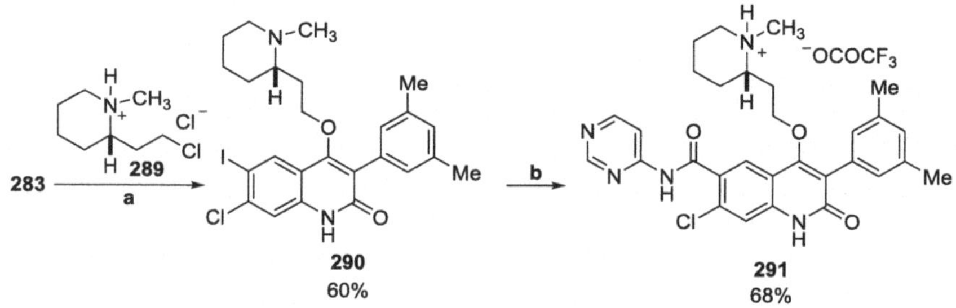

a: K_2CO_3, DMF, 80 °C, 3 h; **b:** 4-aminopyrimidine, $PdCl_2(PPh_3)_2$ (5 mol%), CO (1 atm), Et_3N (2 equiv), DMF, 95 °C, 16 h.

Scheme 91 Synthesis of quinolin-2(1*H*)-one-6-carboxamide **291** by Pd-catalysed aminocarbonylation of 4-hydroxyquinolin-2(1*H*)-one **283** with 4-aminopyrimidine **289** [126]

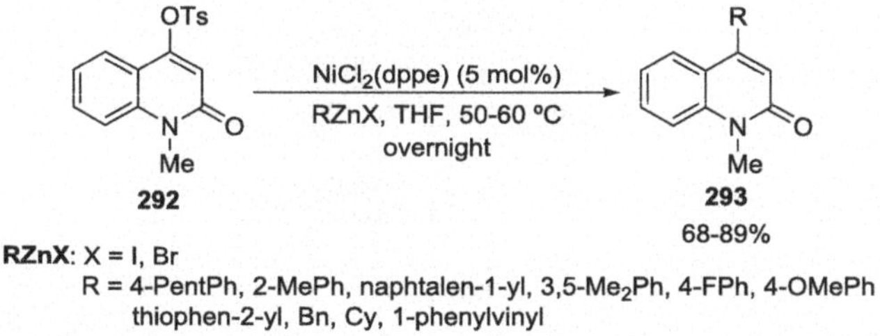

Scheme 92 Synthesis of 4-substituted quinolin-2(1*H*)-ones **293** via Ni-catalysed cross-coupling of 1-methyl-4-tosyloxyquinolin-2(1*H*)-one **292** with organozinc reagents [127]

5.6 Ni-Catalysed Cross-Coupling Reaction

4-Substituted quinolin-2(1*H*)-ones **293** were synthesized via Ni-catalysed cross-coupling of 4-tosylate **292** with organozinc reagents as an alternative to the Suzuki–Miyaura cross-coupling reaction of 1-methyl-4-tosyloxyquinolin-2(1*H*)-one **292** and arylboronic acids that were not well succeeded and no product was detected [127]. The use of zinc reagent as substrate under Pd-catalysed conditions using different Pd catalysts [$Pd(PPh_3)_4$, $PdCl_2(PPh_3)_2$, $Pd(OAc)_2$, $PdCl_2$, $PdCl_2$-$(MeCN)_2$, $PdCl_2(PhCN)_2$, $Pd_2(dba)_3$] in the reaction of 4-tosylate **292** with 4-pentylphenylzinc iodide afforded quinolin-2(1*H*)-ones **293** in low yields (10–30%) even with the addition of ligand, with different solvents or temperatures. When $NiCl_2$(dppp) (5 mol%) was used as the catalyst, 71% yield of the desired product **293** (R=4-PentPh) was afforded in 12 h at 50°C. After optimization of the reaction conditions, $NiCl_2$(dppe) was identified as the best catalyst (**293**, R=4-PentPh; 84%). The method is simple, and air-stable, inexpensive tosylates are used under extremely mild conditions. Alkylzinc and electron-rich and electron-poor arylzinc reagents were suitable for this reaction and gave the desired products in good yields (Scheme 92).

6 Synthesis of Acridin-9(10*H*)-ones

An efficient and regioselective route to acridin-9(10*H*)-ones through application of a combined Buchwald–Hartwig aryl amination-directed remote metalation (DreM) protocol has been disclosed [128, 129]. Diarylamines **296** were prepared by Buchwald–Hartwig C–N cross-coupling of 2-halobenzamides **294** with anilines **295**, in good yields, as an alternative to the harsh conditions, tedious workup and purification procedures typical of Ullmann chemistry. In light of the inability to effect LDA-mediated cyclization of the unprotected diarylamines, these derivatives were *N*-methylated in excellent yields. Upon treatment with LDA, *N*-methyldiarylamines were converted to acridin-9(10*H*)-ones **297** or dibenzo[*b*,*f*]azepinones **298**, in good to excellent yields with regioselectivity depending upon the presence or absence of directed metalation groups (Scheme 93). When R^4=H, only the

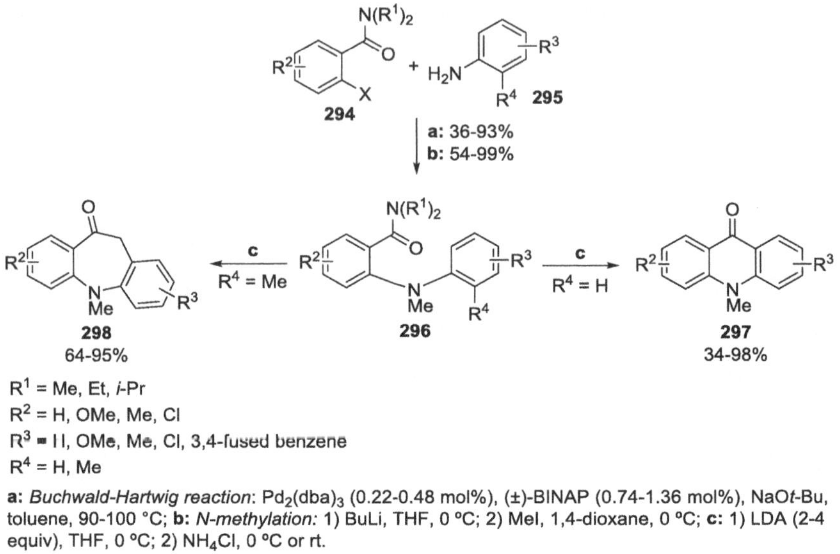

Scheme 93 Synthesis of acridin-9(10*H*)-ones **297** by a combined Buchwald–Hartwig aryl amination (DreM) [128, 129]

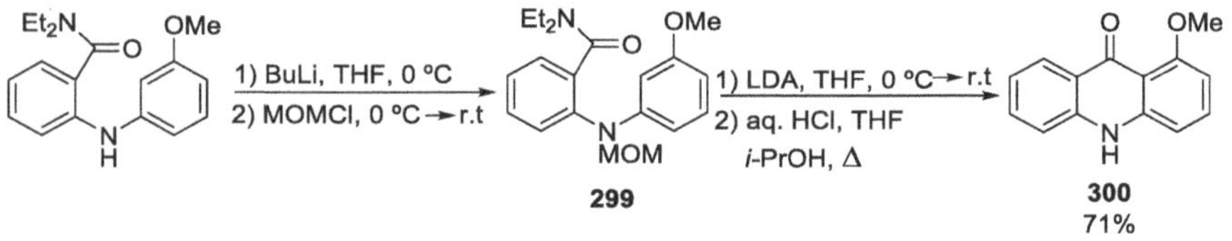

Scheme 94 Synthesis of *N*-unsubstituted acridin-9(10*H*)-one **300** from a MOM-protected diarylamine **299** through a DreM protocol [129]

acridin-9(10*H*)-ones **297** were obtained, while when R^4=Me, regioselectivity for dibenzo[*b*,*f*]azepinones **298** over acridin-9(10*H*)-ones **297** was quite remarkable, except when a methoxyl or a chlorine group is present in the *para*-position relatively to the methyl group (R^4=Me). This synthetic route represents a vast improvement over classical methods for acridin-9(10*H*)-one synthesis, offering significant advantages when compared to the classical Friedel–Crafts reactions.

This synthetic strategy requires the *N*-alkylation of substrates prior to LDA-mediated cyclization. To overcome the limitation of *N*-functionalization post-cyclization, the authors used the MOM-protected diarylamine **299** and applied the standard LDA conditions (Scheme 94). A smooth cyclization ensued to give the product which, upon mild acidic hydrolysis, afforded acridin-9(10*H*)-one **300** in good overall yield, thus extending the DreM approach for the construction of *N*-unsubstituted acridin-9(10*H*)-ones.

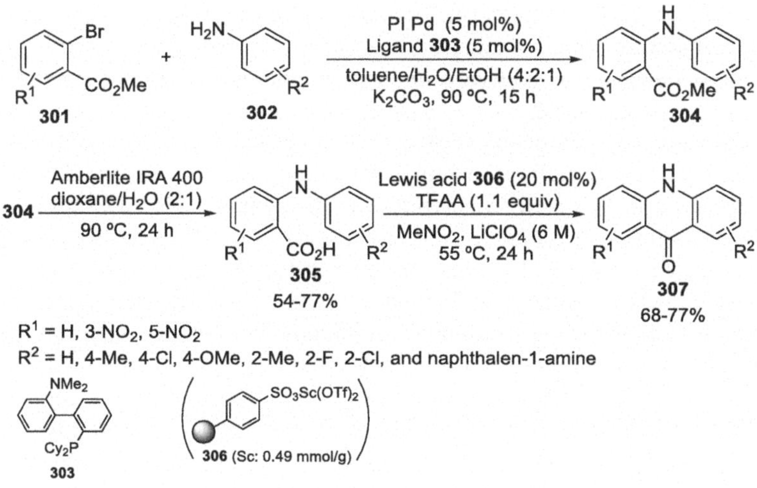

Scheme 95 Synthesis of acridin-9(10*H*)-ones **307** using polymer-supported Pd and Sc catalysts [130]

Another methodology to the synthesis of *N*-unsubstituted acridin-9(10*H*)-ones is based on a combined use of polymer-supported Pd and Sc catalysts in arylamination and intramolecular Friedel–Crafts acylation reactions, respectively (Scheme 95) [130]. The best conditions for the amination step involved the use of PI Pd [Pd $(PPh_3)_4$ immobilized onto polystyrene-based copolymers using polymer-incarcerated (PI) method] and ligand **303** in a toluene–water–ethanol solvent system and K_2CO_3 as a stoichiometric base. The obtained compounds **304** after hydrolysis underwent intramolecular Friedel–Crafts reaction in the presence of TFAA and a polymer-supported Sc catalyst in a $MeNO_2$-$LiClO_4$ solution and afforded acridin-9(10*H*)-ones **307** in good yields. The supported catalysts used can be recovered quantitatively by simple filtration and reused several times without loss of activity.

More recently the synthesis of 2,3-diarylacridin-9(10*H*)-ones **190** by electrocyclization of the Heck products (*E*,*E*)-2,3-distyrylquinolin-4(1*H*)-ones **189** was described and was already discussed in Sect. 4.1.1 [92].

7 Conclusion

We have described the extensive work in the syntheses and transformations of quinolin-2 and quinolin-4(1*H*)-ones by using metal-catalysed cross-coupling reactions. These transition metal-catalysed procedures have been developed providing increased tolerance towards functional groups, leading generally to higher reaction yields. A great number of these transformations involved Pd-catalysed reactions, although the use of copper, zinc, nickel, gold, tin, iridium, rhodium and scandium can also be found. The Sonogashira cross-coupling transformation is one of the most typical and used metal-catalysed reactions.

Although there are a relatively interesting number of studies on the metal-catalysed cross-coupling synthesis and transformations of quinolones, this field still needs further development. For instance, homoeopathic Pd-catalysed transformations are rare (usually 1–5 mol% Pd is employed), thus opening space to further research in this area, concerning the development of Pd-catalysed cross-coupling reactions using milder conditions with lower Pd loadings using more efficient catalytic systems. Likewise the search for heterogeneous catalysed procedures that include either the formation of the desired heterocycles or their transformation to highly substituted compounds is another important research topic. Developments in these two new fields can be important to overcome some drawbacks of metal-catalysed procedures, as metal contamination of bioactive products, generally over accepted limits expressed in medicinal regulations that often preclude further industrial development of these methods, towards more environmentally friendly and economically cheaper methods.

The study on the metal-catalysed cross-coupling reactions for the synthesis of acridin-9(10*H*)-ones is almost inexistent, since only three studies have been disclosed. This means that this field presents an interesting issue for future research.

The asymmetric catalysis to obtain enantiopure quinolones is another underdeveloped topic since only four examples were found concerning the synthesis of 3,4-dihydroquinolin-2(1*H*)-ones by Pd-catalysed cyclocarbonylation or Buchwald–Hartwig reaction and Cu-catalysed cyclization.

References

1. Michael JP (1991) Quinoline, quinazoline, and acridone alkaloids. Nat Prod Rep 8:53–68
2. Michael JP (2008) Quinoline, quinazoline, and acridone alkaloids. Nat Prod Rep 25:166–187
3. Huse H, Whiteley M (2011) 4-Quinolones: smart phones of the microbial world. Chem Rev 111:152–159
4. Winter R, Kelly JX, Smilkstein MJ, Hinrichs D, Koop DR, Riscoe MK (2011) Optimization of endochin-like quinolones for antimalarial activity. Exp Parasitol 127:545–551
5. Xia Y, Yang Z-Y, Xia P, Bastow KF, Nakanishi Y, Nampoothiri P, Hamel E, Brossi A, Lee K-H (2003) Antitumor agents. Part 226: synthesis and cytotoxicity of 2-phenyl-4-quinolone acetic acids and their esters. Bioorg Med Chem Lett 13:2891–2893
6. Wang S-W, Pan S-L, Huang Y-C, Guh J-H, Chiang P-C, Huang D-Y, Kuo S-C, Lee K-H, Teng C-M (2008) CHM-1, a novel synthetic quinolone with potent and selective antimitotic antitumor activity against human hepatocellular carcinoma in vitro and in vivo. Mol Cancer Ther 7:350–360
7. Uchida M, Tabusa F, Komatsu M, Morita S, Kanbe T, Nakagawa K (1987) Studies on 2(1*H*)-quinolinone derivatives as gastric antiulcer active agents. Synthesis and antiulcer activities of optically active alpha-amino acid derivatives of 2(1*H*)-quinolinone and oxindole. Chem Pharm Bull 35:853–856
8. Norman P (2002) Tipifarnib (Janssen Pharmaceutica). Curr Opin Investig Drugs 3:313
9. Li Q, Woods KW, Wang W, Lin N-H, Claiborne A, Gu W-z, Cohen J, Stoll VS, Hutchins C, Frost D, Rosenberg SH, Sham HL (2005) Design, synthesis, and activity of achiral analogs of

2-quinolones and indoles as non-thiol farnesyltransferase inhibitors. Bioorg Med Chem Lett 15:2033–2039
10. Hopkins AL, Ren J, Milton J, Hazen RJ, Chan JH, Stuart DI, Stammers DK (2004) Design of non-nucleoside inhibitors of HIV-1 reverse transcriptase with improved drug resistance properties. 1. J Med Chem 47:5912–5922
11. Freeman GA, Andrews CW III, Hopkins AL, Lowell GS, Schaller LT, Cowan JR, Gonzales SS, Koszalka GW, Hazen RJ, Boone LR, Rob G, Ferris RG, Creech KL, Roberts GB, Short SA, Weaver K, David J, Reynolds DJ, Milton J, Ren J, Stuart DI, Stammers DK, Chan JH (2004) Design of non-nucleoside inhibitors of HIV-1 reverse transcriptase with improved drug resistance properties. 2. J Med Chem 47:5923–5936
12. Tedesco R, Shaw AN, Bambal R, Chai D, Concha NO, Darcy MG, Dhanak D, Fitch DM, Gates A, Gerhardt WG, Halegoua DL, Han C, Hofmann GA, Johnston VK, Kaura AC, Liu N, Keenan RM, Lin-Goerke J, Sarisky RT, Wiggall KJ, Zimmerman MN, Duffy KJ (2006) 3-(1,1-Dioxo-2*H*-(1,2,4)-benzothiadiazin-3-yl)-4-hydroxy-2(1*H*)-quinolinones, potent inhibitors of hepatitis C virus RNA-dependent RNA polymerase. J Med Chem 49:971–983
13. Joule JA, Mills K (2000) Quinolines and isoquinolines: reaction and synthesis. In: Heterocyclic chemistry, 4th edn. Blackwell Science, Oxford
14. Fuson RC, Burness DN (1946) A new synthesis of 2-aryl-4-hydroxyquinolines. J Am Chem Soc 68:1270–1272
15. Ogata Y, Kawasaki A, Tsujimura K (1971) Kinetics and mechanism of the formation 4-hydroxyquinoline from methyl anthranilate. Tetrahedron 27:2765–2770
16. Tabarrini O, Manfroni G, Fravolini A, Cecchetti V, Sabatini S, De Clercq E, Rozenski J, Canard B, Dutartre H, Paeshuyse J, Neyts J (2006) Synthesis and anti-BVDV activity of acridones as new potential antiviral agents. J Med Chem 49:2621–2627
17. Kelly JX, Smilkstein MJ, Brun R, Wittlin S, Cooper RA, Lane KD, Janowsky A, Johnson RA, Dodean RA, Winter R, Hinrichs DJ, Riscoe MK (2009) Discovery of dual function acridones as a new antimalarial chemotype. Nature 459:270–273
18. Nguyen HT, Lallemand M-C, Boutefnouchet S, Michel S, Tillequin F (2009) Antitumor *Psoropermum* xanthones and *Sarcomelicope* acridones: privileged structures implied in DNA alkylation. J Nat Prod 72:527–539
19. Wallstab A, Koester M, Böhme M, Keppler D (1999) Selective inhibition of MDR1 P-glycoprotein-mediated transport by the acridone carboxamide derivative GG918. Br J Cancer 79:1053–1060
20. You J, Zhang W, Zhang Q, Zhang L, Yan C, Zhang Y (2002) Development of a precolumn derivatization method for the determination of free amines in wastewater by high-performance liquid chromatography via fluorescent detection with 9-(2-hydroxyethyl) acridone. Anal Chem 74:261–269
21. Faller T, Hutton K, Okafo G, Gribble A, Camilleri P, Games DE (1997) A novel acridone derivative for the fluorescence tagging and mass spectrometric sequencing of peptides. Chem Commun 1529–1530
22. Hagiwara Y, Hasegawa T, Shoji A, Kuwahara M, Ozaki H, Sawai H (2008) Acridone-tagged DNA as a new probe for DNA detection by fluorescence resonance energy transfer and for mismatch DNA recognition. Bioorg Med Chem 16:7013–7020
23. Qiu B, Guo L, Chen Z, Chi Y, Zhang L, Chen G (2009) Synthesis of N-4-butylamine acridone and its use as fluorescent probe for ctDNA. Biosens Bioelectron 24:1281–1285
24. García-Garrido SE, Caltagirone C, Light ME, Gale PA (2007) Acridinone-based anion receptors and sensors. Chem Commun 1450–1452
25. Singh P, Kaur J, Holzer W (2010) Acridone based Cu^{2+}-F^{-}/F^{-}-Cu^{2+} responsive ON/OFF key pad. Sens Actuators B 150:50–56
26. Huang C, Yan S-J, Li Y-M, Huang R, Lin J (2010) Synthesis of polyhalo acridones as pH-sensitive fluorescence probes. Bioorg Med Chem Lett 20:4665–4669

27. Delmas F, Avellaneda A, Di Giogio C, Robin M, De Clercq E, Timon-David P, Galy JP (2004) Synthesis and antileishmanial activity of (1,3-benzothiazol-2-yl)amino-9-(10*H*)-acridinone derivatives. Eur J Med Chem 39:685–690
28. Goodell JR, Madhok AA, Hiasa H, Ferguson DM (2006) Synthesis and evaluation of acridine- and acridone-based anti-herpes agents with topoisomerase activity. Bioorg Med Chem 14:5467–5480
29. Winter RW, Kelly JX, Smilkstein MJ, Dodean R, Bagby GC, Rathbun RK, Levin JI, Hinrichs D, Riscoe MK (2006) Evaluation and lead optimization of anti-malarial acridones. Exp Parasitol 114:47–56
30. Boumendjel A, Macalou S, Ahmed-Belkacem A, Blanc M, Di Pietro A (2007) Acridone derivatives: design, synthesis, and inhibition of breast cancer resistance protein ABCG2. Bioorg Med Chem 15:2892–2897
31. Gopinath VS, Thimmaiah P, Thimmaiah KN (2008) Acridones circumvent P-glycoprotein-associated multidrug resistance (MDR) in cancer cells. Bioorg Med Chem 16:474–487
32. Rudas M, Nyerges M, Toke L, Pete B, Groundwater PW (1999) A convenient regioselective synthesis of pyrano[3,2-*b*]acridones involving nucleophilic addition to benzyne. Tetrahedron Lett 40:7003–7006
33. Zhao J, Larock RC (2007) Synthesis of xanthones, thioxanthones, and acridones by the coupling of arynes and substituted benzoates. J Org Chem 72:583–588
34. MacNeil SL, Wilson BJ, Snieckus V (2006) Anionic N-fries rearrangement of N-carbamoyl diarylamines to anthranilamides. Methodology and application to acridone and pyranoacridone alkaloids. Org Lett 8:1133–1136
35. Kundu NG, Mahanty JS, Das P, Das B (1993) Synthesis of quinolines and 2,3-dihydro-4(1*H*)-quinolones. Palladium catalysed reaction of *o*-iodoanilides with acetylenic carbinols. Tetrahedron Lett 34:1625–1628
36. Mahanty JS, De M, Das P, Kundu NG (1997) Palladium catalysed heteroannulation with acetylenic carbinols as synthons – synthesis of quinolines and 2,3-dihydro-4(1*H*)-quinolones. Tetrahedron 53:13397–13418
37. Pisaneschi F, Sejberg JJP, Blain C, Hei NW, Aboagye EO, Spivey AC (2011) 2-Substituted-2,3-dihydro-1*H*-quinolin-4-ones via acid-catalyzed tandem Rupe rearrangement-Donnelly-Farrell ring closure of 2-(3′-hydroxy-propynyl)anilines. Synlett 241–244
38. Bernini R, Cacchi S, Fabrizi G, Sferrazza A (2009) 1,2-Disubstituted 4-quinolones via copper-catalyzed cyclization of 1-(2-halophenyl)-2-en-3-amin-1-ones. Synthesis 1209–1219
39. Torii S, Okumoto H, Xu LH (1991) Palladium-catalyzed carbonylation to form 2-substituted 1,4-dihydro-4-oxo-quinoline. Tetrahedron Lett 32:237–240
40. Kalinin VN, Shostakovsky MV, Ponomaryov AB (1992) A new route to 2-aryl-4-quinolones via palladium catalyzed carbonylative coupling of *o*-iodoanilines with terminal arylacetylenes. Tetrahedron Lett 33:373–376
41. Torii S, Okumoto H, Xu LH, Sadakane M, Shostakovsky MV, Ponomaryov AB, Kalinin VN (1993) Synthesis of chromones and quinolones via Pd-catalyzed carbonylation of *o*-iodophenols and anilines in the presence of acetylenes. Tetrahedron 49:6773–6784
42. Haddad N, Tan J, Farina V (2006) Convergent synthesis of the quinolone substructure of BILN 2061 via carbonylative Sonogashira coupling/cyclization. J Org Chem 71:5031–5034
43. Genelot M, Bendjeriou A, Dufaud V, Djakovitch L (2009) Optimised procedures for the one-pot selective syntheses of indoxyls and 4-quinolones by a carbonylative Sonogashira/cyclisation sequence. Appl Catal A 369:125–132
44. Genelot M, Dufaud V, Djakovitch L (2011) Heterogeneous metallo-organocatalysis for the selective one-pot synthesis of 2-benzylidene-indoxyl and 2-phenyl-4-quinolone. Tetrahedron 67:976–981
45. Batail N, Genelot M, Dufaud V, Joucla L, Djakovitch L (2011) Palladium-based innovative catalytic procedures: designing new homogeneous and heterogeneous catalysts for the synthesis and functionalisation of N-containing heteroaromatic compounds. Catal Today 173:2–14

46. Djakovitch L, Batail N, Genelot M (2011) Recent advances in the synthesis of *N*-containing heteroaromatics via heterogeneously transition metal catalysed cross-coupling reactions. Molecules 16:5241–5267
47. Zhao T, Xu B (2010) Palladium-catalyzed tandem amination reaction for the synthesis of 4-quinolones. Org Lett 12:212–215
48. Fei X-D, Zhou Z, Li W, Zhu Y-M, Shen J-K (2012) Buchwald–Hartwig coupling/Michael addition reactions: one-pot synthesis of 1,2-disubstituted 4-quinolones from chalcones and primary amines. Eur J Org Chem 3001–3008
49. Huang J, Chen Y, King AO, Dilmeghani M, Larsen RD, Faul MM (2008) A mild, one-pot synthesis of 4-quinolones via sequential Pd-catalyzed amidation and base-promoted cyclization. Org Lett 10:2609–2612
50. Sakamoto T, Kondo Y, Uchiyama D, Yamanaka H (1991) Condensed heteroaromatic ring systems. XIX. Synthesis and reactions of 5-(tributylstannyl)isoxazoles. Tetrahedron 47:5111–5118
51. Sakamoto T, Yasuhara A, Kondo Y, Yamanaka H (1992) Condensed heteroaromatic ring systems. XX. Palladium-catalyzed carbonylative coupling of iodobenzenes with (*Z*)-1-ethoxy-2-(tributylstannyl)ethane. Chem Pharm Bull 40:1137–1139
52. Torii S, Okumoto H, Xu LH (1990) A direct approach to 2-substituted 1,4-dihydro-4-oxo-quinoline-3-carboxylates by palladium-catalyzed carbonylative cyclization. Tetrahedron Lett 31:7175–7178
53. Latham EJ, Stanforth SP (1996) Synthesis of indoles and quinolones by sequential Wittig and Heck reactions. Chem Commun 2253–2254
54. Latham EJ, Stanforth SP (1997) Synthesis of indoles and quinolones by sequential Wittig and Heck reactions. J Chem Soc Perkin Trans 1 2059–2063
55. Costa M, Cà ND, Gabriele B, Massera C, Salerno G, Soliani M (2004) Synthesis of 4*H*-3,1-benzoxazines, quinazolin-2-ones, and quinoline-4-ones by palladium-catalyzed oxidative carbonylation of 2-ethynylaniline derivatives. J Org Chem 69:2469–2477
56. Gabriele B, Salerno G, Costa M (2004) PdI_2-catalyzed synthesis of heterocycles. Synlett 2468–2483
57. Okuro K, Alper H (2012) Palladium-catalyzed intermolecular cyclocarbonylation of 2-iodoanilines with the Michael acceptor, diethyl ethoxycarbonylbutendienoate. J Org Chem 77:4420–4424
58. Grigg R, Liu A, Shaw D, Suganthan S, Woodall DE, Yoganathan G (2000) Synthesis of quinol-4-ones and chroman-4-ones via a palladium-catalysed cascade carbonylation-allene insertion. Tetrahedron Lett 41:7125–7128
59. Okuro K, Alper H (1997) Palladium-catalyzed carbonylation of *o*-iodophenols with allenes. J Org Chem 62:1566–1567
60. Messaoudi S, Brion J-D, Alami M (2012) Palladium-catalyzed decarboxylative coupling of quinolone-3-carboxylic acids and related heterocyclic carboxylic acids with (hetero)aryl halides. Org Lett 14:1496–1499
61. Nemoto T, Fukuda T, Hamada Y (2006) Efficient synthesis of 3-substituted 2,3-dihydroquinolin-4-ones using a one-pot sequential multi-catalytic process: Pd-catalyzed allylic amination-thiazolium salt-catalyzed Stetter reaction cascade. Tetrahedron Lett 47:4365–4368
62. Barluenga J, Rodríguez F, Fañanás FJ (2009) Recent advances in the synthesis of indole and quinoline derivatives through cascade reactions. Chem Asian J 4:1036–1048
63. Solé D, Serrano O (2007) Palladium-catalyzed intramolecular nucleophilic substitution at the alkoxycarbonyl group. Angew Chem Int Ed 46:7270–7272
64. Cortese NA, Ziegler CB Jr, Hrnjez BJ, Heck RF (1978) Palladium-catalyzed synthesis of 2-quinolones derivatives from 2-iodoanilines. J Org Chem 43:2952–2958
65. Bernini R, Cacchi S, Fabrizi G, Sferrazza A (2006) 3-Aryl-2-quinolones via a domino Heck reaction/cyclization process. Heterocycles 69:99–105

66. Bernini R, Cacchi S, De Salve I, Fabrizi G (2006) The Heck reaction of β-arylacrylamides: an approach to 4-aryl-2-quinolones. Synlett 2947–2952
67. Battistuzzi G, Bernini R, Cacchi S, De Salve I, Fabrizi G (2007) 4-Aryl-2-quinolones through a pseudo-domino Heck/Buchwald–Hartwig reaction in a molten tetrabutylammonium acetate/tetrabutylammonium bromide mixture. Adv Synth Catal 349:297–302
68. Fourquez JM, Godard A, Marsais F, Quéquiner G (1995) Regioselectivity of the metalation of polymethoxylated pivaloylaminobenzenes. Synthesis of methoxy-2(1*H*)-quinolones. J Heterocycl Chem 32:1165–1170
69. Cho CS, Kim JU (2007) An approach for quinolines via palladium-catalyzed Heck coupling followed by cyclization. Tetrahedron Lett 48:3775–3778
70. Taylor JG, Correia CRD (2011) Stereoselective synthesis of unsymmetrical β, β-diarylacrylates by a Heck–Matsuda reaction: versatile building blocks for asymmetric synthesis of β, β-diphenylpropanoates, 3-aryl-indole, and 4-aryl-3,4-dihydroquinolin-2-one and formal synthesis of (−)-indatraline. J Org Chem 76:857–869
71. Felpin F-X, Coste J, Zakri C, Fouquet E (2009) Preparation of 2-quinolones by Heck reduction-cyclization (HRC) reactions by using a multitask palladium catalyst. Chem Eur J 15:7238–7245
72. Glover B, Harvey KA, Liu B, Sharp MJ, Tymoschenko MF (2003) Regioselective palladium-catalyzed arylation of 3-carboalkoxy furan and thiophene. Org Lett 5:301–304
73. Borhade SR, Waghmode SB (2011) An efficient synthesis of 4-arylquinolin-2(1*H*)-ones and 3-alkenyl-4-arylquinolin-2(1*H*)-one using a Pd/$NiFe_2O_4$-catalyzed consecutive Heck reaction. Can J Chem 89:1355–1363
74. Kuroda T, Suzuki F (1991) Synthesis of 1*H*-imidazo[4,5-*c*]quinolin-4(5*H*)-one via palladium-catalyzed cyclization of N-(2-bromophenyl)-1*H*-imidazole-4-carboxamide. Tetrahedron Lett 32:6915–6918
75. Beccalli EM, Broggini G, Martinelli M, Paladino G, Zoni C (2005) Synthesis of tricyclic quinolones and naphthyridones by intramolecular Heck cyclization of functionalized electron-rich heterocycles. Eur J Org Chem 2091–2096
76. Liu Z, Shi C, Chen Y (2008) Synthesis of 3-alkyl-1*H*-quinolin-2-ones via palladium-catalyzed intramolecular cyclization of benzyl halides and α, β-unsaturated amides. Synlett 1734–1736
77. Arcadi A, Cacchi S, Fabrizi G, Manna F, Pace P (1998) Ethyl *N*-(*o*-ethynyl)malonanilide as a useful building block for the preparation of 3,4-disubstituted-2(1*H*)-quinolones, 3,4-disubstituted- and 2,3,4-trisubstituted quinolones. Synlett 446–448
78. Kadnikov DV, Larock RC (2003) Palladium-catalyzed carbonylative annulation of terminal alkynes: synthesis of coumarins and 2-quinolones. J Organomet Chem 687:425–435
79. Kadnikov DV, Larock RC (2004) Synthesis of 2-quinolones via palladium-catalyzed carbonylative annulation of internal alkynes by *N*-substituted *o*-iodoanilines. J Org Chem 69:6772–6780
80. Chen J-R, Liao J, Xiao W-J (2010) Microwave-assisted, palladium-catalyzed carbonylative cyclization – rapid synthesis of 2-quinolones from unprotected 2-iodoanilines and terminal alkynes. Can J Chem 88:331–337
81. Vicente J, Abad J-A, López-Serrano J, Jones PG, Nájera C, Botella-Segura L (2005) Synthesis and reactivity of *ortho*-palladated arylureas. Synthesis and catalytic activity of a C,N,C pincer complex. Stoichiometric syntheses of some N-heterocycles. Organometallics 24:5044–5057
82. Mori M, Chiba K, Ohta N, Ban Y (1979) A novel synthesis of cyclic imides and quinolone by use of palladium catalyzed carbonylation. Heterocycles 13:329–332
83. Okuro K, Kai H, Alper H (1997) Palladium-catalyzed asymmetric cyclocarbonylation of 2-(1-methylvinyl)anilines. Tetrahedron: Asymmetry 8:2307–2309
84. Porosa L, Viirre RD (2009) Desymmetrization of malonamides via an enantioselective intramolecular Buchwald–Hartwig reaction. Tetrahedron Lett 50:4170–4173

85. Takenaka K, Itoh N, Sasai H (2009) Enantioselective synthesis of C2-symmetric spirobilactams via Pd-catalyzed intramolecular double *N*-arylation. Org Lett 11:1483–1486
86. Banwell MG, Lupton DW, Ma X, Renner J, Sydnes MO (2004) Synthesis of quinolines, 2-quinolones, phenanthridines, and 6(5*H*)-phenanthridinones via palladium[0]-mediated Ullmann cross-coupling of 1-bromo-2-nitroarenes with β-halo-enals, -enones, or -esters. Org Lett 6:2741–2744
87. Queiroz M-JRP, Castanheira EMS, Lopes TCT, Cruz YK, Kirsch G (2007) Synthesis of fluorescent tetracyclic lactams by a "one pot" three steps palladium-catalyzed borylation, Suzuki coupling (BSC) and lactamization DNA and polynucleotides binding studies. J Photochem Photobiol A Chem 190:45–52
88. Tsuritani T, Yamamoto Y, Kawasaki M, Mase T (2009) Novel approach to 3,4-dihydro-2 (1*H*)-quinolinone derivatives via cyclopropane ring expansion. Org Lett 11:1043–1045
89. Iwai T, Fujihara T, Terao J, Tsuji Y (2010) Iridium-catalyzed annulation of *N*-arylcarbamoyl chlorides with internal alkynes. J Am Chem Soc 132:9602–9603
90. Fugita K-I, Takahashi Y, Owaki M, Yamamoto K, Yamaguchi R (2004) Synthesis of five-, six-, and seven-membered ring lactams by Cp*Rh complex-catalyzed oxidative *N*-heterocyclization of amino alcohols. Org Lett 6:2785–2788
91. Almeida AIS, Silva AMS, Cavaleiro JAS (2010) Reactivity of 3-iodo-4-quinolones in Heck reactions: Synthesis of novel (*E*)-3-styryl-4-quinolones. Synlett 462–466
92. Silva VLM, Silva AMS, Cavaleiro JAS (2010) New synthesis of 2,3-diarylacridin-9(10*H*)-ones and (*E*)-2-phenyl-4-styrylfuro[3,2-*c*]quinolones. Synlett 2565–2570
93. Jakopović IP, Kragol G, Forrest AK, Frydrych CSV, Štimac V, Kapić S, Škugor MM, Ilijaš M, Paljetak HC, Jelić D, Holmes DJ, Hickey DMB, Verbanac D, Haber VE, Alihodžić S (2010) Synthesis and properties of macrolones characterized by two ether bonds in the linker. Bioorg Med Chem 18:6578–6588
94. Joseph B, Béhard A, Lesur B, Guillaumet G (2003) Convenient synthetic routes to 5-substituted 3-(4-methoxyphenyl)-4-(1*H*)-quinolones. Synlett 1542–1544
95. Pain C, Célanire S, Guillaumet G, Joseph B (2003) Synthesis of 5-substituted 2-(4- or 3-methoxyphenyl)-4(1*H*)-quinolones. Tetrahedron 59:9627–9633
96. Cross RM, Manetsch R (2010) Divergent route to access structurally diverse 4-quinolones via mono or sequential cross-couplings. J Org Chem 75:8654–8657
97. Mugnaini C, Falciani C, De Rosa M, Brizzi A, Pasquini S, Corelli F (2011) Regioselective functionalization of quinolin-4(1*H*)-ones via sequential palladium-catalyzed reactions. Tetrahedron 67:5776–5783
98. Gomes ATPC, Cunha AC, Domingues MRM, Neves MGPMS, Tomé AC, Silva AMS, Santos FC, Souza MCBV, Ferreira VF, Cavaleiro JAS (2011) Synthesis and characterization of new porphyrin/4-quinolone conjugates. Tetrahedron 67:7336–7342
99. Venkataraman S, Barange DK, Pal M (2006) One-pot synthesis of 2-substituted furo[3,2-*c*] quinolines via tandem coupling-cyclization under Pd/C-copper catalysis. Tetrahedron Lett 47:7317–7322
100. Škugor MM, Štimac SS, Palej I, Lugarić D, Paljetak HC, Filić D, Modrić M, Dilović I, Gembarovski D, Mutak S, Haber VE, Holmes DJ, Ivezić-Schoenfeld Z, Alihodžić S (2010) Synthesis and biological activity of 4″-*O*-acyl derivatives of 14- and 15-membered macrolides linked to ω-quinolone-carboxylic unit. Bioorg Med Chem 18:6547–6558
101. Kapić S, Paljetak HC, Alihodžić S, Antolović R, Haber VE, Jarvest RL, Holmes DJ, Broskey JP, Hunt E (2010) 6-Alkylquinolone-3-carboxylic acid tethered to macrolides synthesis and antimicrobial profile. Bioorg Med Chem 18:6569–6577
102. Laborde E, Leshaski LE, Kiely JS (1990) Palladium-catalyzed intermolecular vinylic arylation of cycloalkenes. Applications to the synthesis of quinolone antibacterials. Tetrahedron Lett 31:1837–1840
103. Kiely JS, Laborde E, Lesheski LE, Bucsh RA (1991) Synthesis of 7-(alkenyl, cycloalkenyl, and 1,2,3,6-tetrahydro-4-pyridinyl)quinolones. J Heterocycl Chem 28:1581–1585

104. Reuman M, Daum SJ, Singh B, Wentland MP, Perni RB, Pennock P, Carabateas PM, Gruett MD, Saindane MT, Dorff PH, Coughlin SA, Sedlock DM, Rake JB, Lesher GY (1995) Synthesis and antibacterial activity of some novel l-substituted 1,4-dihydro-4-oxo-7-pyridinyl-3-quinolinecarboxylic acids. Potent antistaphylococcal agents. J Med Chem 38:2531–2540
105. Audisio D, Messaoudi S, Peyrat J-F, Brion J-D, Alami M (2011) A general copper powder-catalyzed Ullmann-type reaction of 3-halo-4(1*H*)-quinolones with various nitrogen-containing nucleophiles. J Org Chem 76:4995–5005
106. Neville CF, Barr SA, Grundon MF (1992) Approaches to the syntheses of dimeric quinolinone alkaloids. Tetrahedron Lett 33:5995–5998
107. Valente S, Kirsch G (2011) Facile synthesis of 4-acetyl-coumarins, -thiocoumarin and -quinolin-2(1*H*)-one via very high α-regioselective Heck coupling on tosylates. Tetrahedron Lett 52:3429–3432
108. Chattopadhyay SK, Neogi K, Singha SK, Dey R (2008) Sequential Claisen rearrangement and intramolecular Heck reaction as a route to medium-ring oxacycle-fused heterocycles. Synlett 1137–1140
109. Majumdar KC, Nandi RK, Samanta S, Chattopadhyay B (2010) Novel synthesis of heterocycle-annulated azocine derivatives of biological relevance by aromatic aza-Claisen rearrangement and intramolecular Heck reaction. Synthesis 985–990
110. Majumdar KC, Samanta S, Ghosh T (2012) Synthesis of coumarin- and quinolone-annulated benzazocinone frameworks by a palladium-catalyzed intramolecular Heck reaction. Synthesis 44:1711–1717
111. Majumdar KC, Ghosh T, Samanta S (2011) Facile synthesis of coumarin- and quinolone-annulated benzazoninone derivatives by an intramolecular Heck reaction strategy via 9-*exo*-trig cyclization. Synthesis 1569–1574
112. Majumdar KC, Mukhopadhyay PP (2003) Regioselective synthesis of 2*H*-benzopyrano [3,2-*c*]quinolin-7(8*H*)-ones by radical cyclization. Synthesis 97–100
113. Majumdar KC, Mukhopadhyay PP, Basu PK (2005) Regioselective synthesis of coumarin and quinolone-annulated spiro heterocycles via aryl radical cyclization. Synthetic Commun 35:1291–1299
114. Majumdar KC, Pal AK, Taher A, Debnath P (2007) Highly effective regioselective method for the synthesis of substituted coumarin- and quinolone-annulated heterocycles using a palladium(0)-catalyzed reaction. Synthesis 1707–1711
115. Majumdar KC, Chakravorty S, Shyam PK, Taher A (2009) Palladium-mediated Heck reaction to the synthesis of 3-substituted indoles and indolones. Synthesis 403–408
116. Majumdar KC, Ganai S, Chattopadhyay B, Ray K (2011) Palladium-catalyzed one-pot synthesis of pyrrole-annulated coumarin, quinolone, and 7-aza-indole derivatives. Synlett 2369–2373
117. Wu J, Zhang L, Sun X (2005) Synthesis of 3,4-disubstituted quinolin-2(1*H*)-ones via palladium-catalyzed regioselective cross-coupling reactions of 3-bromo-4-trifloxyquinolin-2(1*H*)-one with arylboronic acids. Chem Lett 34:550–551
118. Hashim J, Glasnov TN, Kremsner JM, Kappe CO (2006) Symmetrical bisquinolones via metal-catalyzed cross-coupling and homocoupling reactions. J Org Chem 71:1707–1710
119. Majumdar KC, Mondal S (2008) A new strategy for the synthesis of coumarin- and quinolone-annulated pyrroles via Pd(0) mediated cross-coupling followed by Cu (I) catalyzed heteroannulation. Tetrahedron Lett 49:2418–2420
120. Majumdar KC, Ghosh SK (1994) Studies on amine oxide rearrangements: regioselective synthesis of pyrano[3,2-*e*]indol-7-one. J Chem Soc Perkin Trans 1 2889–2894
121. Majumdar KC, Biswas P, Jana GH (1997) Studies on amine oxide rearrangements: regioselective synthesis of pyrrolo[3,2-*f*]quinolin-7-ones. J Chem Res (S) 310–311
122. Majumdar KC, Chatoppadhyay B, Samanta S (2009) A short route to the synthesis of pyrrolocoumarin and pyrroloquinolone derivatives by Sonogashira cross-coupling and gold-catalyzed cycloisomerization of acetylenic amines. Synthesis 311–317

123. Majumdar KC, De N, Roy B (2010) Iron/palladium-catalyzed intramolecular hydroamination: an expedient synthesis of pyrrole-annulated coumarin and quinolone derivatives. Synthesis 4207–4212
124. Messaoudi S, Audisio D, Brion J-D, Alami M (2007) Rapid access to 3-(*N*-substituted)-aminoquinolin-2(1*H*)-ones using palladium-catalyzed C–N bond coupling reaction. Tetrahedron 63:10202–10210
125. Soussi MA, Audisio D, Messaoudi S, Provot O, Brion J-D, Alami M (2011) Palladium-catalyzed coupling of 3-halo-substituted coumarins, chromenes, and quinolones with various nitrogen-containing nucleophiles. Eur J Org Chem 5077–5088
126. Walsh TF, Toupence RB, Young JR, Huang SX, Ujjainwalla F, DeVita RJ, Goulet MT, Wyvratt MJ Jr, Fisher MH, Lo J-L, Ren N, Yudkovitz JB, Yang YT, Cheng K, Smith RG (2000) Potent antagonists of gonadotropin releasing hormone receptors derived from quinolone-6-carboxamides. Bioorg Med Chem Lett 10:443–447
127. Wu J, Sun X, Zhang L (2005) Efficient route to 4-substituted-2(5*H*)-furanones, 2(1*H*)-quinolones, and pyrones by nickel-catalyzed cross-coupling of arenesulfonates with organozinc reagents. Chem Lett 34:796–797
128. MacNeil SL, Gray M, Briggs LE, Li JJ, Snieckus V (1998) Direct *ortho* and remote metalation – cross coupling connections. Buchwald–Hartwig synthesis of 2-carbamoyl diarylamines. Regioselective anionic routes to acridones, oxindoles, dibenzo-[*b*,*f*]azepinones, and anthranilate esters. Synlett 419–421
129. MacNeil SL, Gray M, Gusev DG, Briggs LE, Snieckus V (2008) Carbanionic Friedel–Crafts equivalents. Regioselective directed *ortho* and remote metalation-C-N cross coupling routes to acridones and dibenzo[b, f]azepinones. J Org Chem 73:9710–9719
130. Nishio R, Wessely S, Sugiura M, Kobayashi S (2006) Synthesis of acridone derivatives using polymer-supported palladium and scandium catalysts. J Comb Chem 8:459–461

Top Heterocycl Chem (2016) 45: 231–304
DOI: 10.1007/7081_2014_136

Published online: 13 December 2014

Synthesis and Transformations of Oxygen Heterocycles

Zoltán Novák and András Kotschy

Abstract The recent developments in the transition metal-catalyzed synthesis and transformations of such oxygen-containing heteroaromatic systems are reviewed, where the oxygen is part of a five-membered ring.

Keywords Benzofuran · Benzoxazole · Catalysis · Furan · Isoxazole · Oxadiazole · Oxazole · Transition metal

Contents

Abbreviations

Ac	Acetyl
acac	Acetylacetonate

Z. Novák (✉)
MTA-ELTE "Lendület" Catalysis and Organic Synthesis Research Group, Institute of Chemistry, Eötvös Loránd University, Pázmány P. s. 1/A, H-1117 Budapest, Hungary

A. Kotschy (✉)
Servier Research Institute of Medicinal Chemistry, Záhony u. 7., H-1031 Budapest, Hungary

Ad	1-Adamantyl
BAr^F	Tetrakis[3,5-bis(trifluoromethyl)phenyl]borate
BINAM	2,2′-Bis(diphenylphosphinoamino)-1,1′-binaphthyl
BINAP	2,2′-Bis(diphenylphosphino)-1,1′-binaphthyl
BmimBF4	1-Butyl-3-methylimidazolium tetrafluoroborate
bpy	2,2′-Bipyridyl
BQ	1,4-Benzoquinone
cod	1,5-Cyclooctadiene
Cp	Cyclopentadienyl
Cp*	Pentamethylcyclopentadienyl
CuTC	Copper(I) thiophene-2-carboxylate
Cy	Cyclohexyl
Cy-DHTP	Cyclohexyl-dihydroxy-terphenylphosphine
DABCO	1,4-Diazabicyclo[2.2.2]octane
DBU	1,8-Diazabicycloundec-7-ene
DCE	1,2-Dichloroethane
DCM	Dichloromethane
dcpe	1,2-Bis(dicyclohexylphosphino)ethane
DEA	Diethylamine
diglyme	1-Methoxy-2-(2-methoxyethoxy)ethane
DIPA	Diisopropylamine
DMA	*N*,*N*-dimethylacetamide
DMAP	4-Dimethylaminopyridine
DME	1,2-Dimethoxyethane
DMEDA	*N*,*N*′-dimethylethylenediamine
DMF	*N*,*N*-dimethylformamide
DMSO	Dimethyl sulfoxide
DPE-Phos	2,2′-Bis(diphenylphosphino)-diphenyl ether
dppe	1,2-Bis(diphenylphoshino)ethane
dppf	1,1′-Bis(diphenylphoshino)ferrocene
dppp	1,3-Bis(diphenylphoshino)propane
DTAC	Dodecyl trimethylammonium chloride
EDIPA	Ethyldiisopropylamine
hfacac	Hexafluoroacetylacetonate
Im	Imidazol-1-yl
IMes	1,3-Bis(2,4,6-trimethylphenyl)imidazol-2-ylidene
IPr	1,3-Bis(2′,6′-diisopropylphenyl)imidazol-2-ylidene
MonoPhos	3,4-A′(dinaphthalen-4-yl)dimethylamine
NBS	*N*-bromosuccinimide
NHC	*N*-heterocyclic carbine
NIS	*N*-iodosuccinimide
NMI	*N*-methylimidazole
NMP	*N*-methyl-2-pyrrolidone
Ph	Phenyl

Phen	1,10-Phenanthroline
PIFA	(Bis(trifluoroacetoxy)iodo)benzene
Piv	2,2-Dimethyl-propanoyl
PMP	4-Methoxyphenyl
pTSA	*p*-Toluenesulfonic acid
PyBroP	Bromotripyrrolidinophosphonium hexafluorophosphate
SDS	Sodium dodecyl sulfate
Selectfluor	1-Chloromethyl-4-fluoro-1,4-diazoniabicyclo[2.2.2]octane bis (tetrafluoroborate)
SPhos	2-Dicyclohexylphosphino-2′,6′-dimethoxybiphenyl
Sylphos	1,1′-Bis(1-diphenylphosphino-1-methylethyl)ferrocene
TBAB	Tetrabutylammonium bromide
TEA	Triethylamine
Tf	Trifluoromethylsulfonyl
THF	Tetrahydrofuran
THP	Tetrahydropyranyl
TMEDA	Tetramethylethylenediamine
TMG	1,1,3,3-Tetramethylguanidine
TMHD	2,2,6,6-Tetramethyl-3,5-heptanedionate
tmpp	Tris(2,4,6-trimethoxyphenyl)phosphine
Ts	Tosyl, 4-toluenesulfonyl
Xphos	2-Dicyclohexylphosphino-2′,4′,6′-triisopropylbiphenyl

1 Introduction

Oxygen-containing heterocycles are a class amply represented both in natural products and in biologically active compounds; therefore, it is not surprising that their research has been always in the forefront of interest both in basic and applied science [1, 2]. Due to the vastness of the area, this chapter is limited to the last decade's (2004–2014) developments in the transition metal-catalyzed synthesis and transformations of mono- and bicyclic compounds that contain at least a five-membered aromatic ring with an oxygen atom. For those looking for a broader coverage of the area, a significant number of monographs and review articles are available.

The discussion of the subject is dived into chapters on the basis of the number of heteroatoms in the given ring systems; thus, the first block describes the synthesis and functionalization of furans and benzofurans, which is followed by the description of the synthesis and transformations of such five-membered oxygen heterocycles that also contain another hetero atom.

2 Synthesis of Furans and Benzofurans

This chapter summarizes the transition metal-catalyzed approaches for the creation of the furan ring. Reactions leading to furans and benzofurans are treated separately. Within the chapters describing the synthesis of these compound classes, a further classification was made on the basis of the type of chemical transformation that results in the ring closure.

2.1 Transition Metal-Catalyzed Synthesis of Benzofurans

The benzofuran ring is widely abundant both in nature and in synthetic compounds; therefore, it is not surprising that several approaches have been established for its preparation. The following chapter provides an overview of the recent developments in the transition metal-catalyzed construction of the furan ring when condensed to a benzene moiety.

2.1.1 Direct Ring Closure of Halophenols Including Sonogashira-Type Reaction: The Coupling/Ring Closure Approach

The benzofuran ring can be constructed via the utilization of Sonogashira reaction of *ortho*-halophenols and terminal acetylenes. In this chapter new variants of the transition metal-catalyzed coupling of terminal acetylenes and aryl halides and the subsequent ring closure are collected.

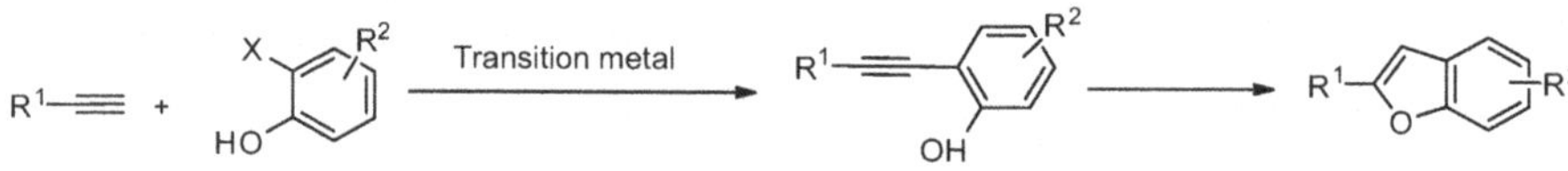

A new iron-catalyzed version of the Sonogashira reaction was developed by Bolm and coworkers [3]. The procedure utilizes simple $FeCl_3$ as catalyst in conjunction with DMEDA as ligand. The couplings of terminal acetylenes and aryl halides were achieved with 15 mol% $FeCl_3$ and 30 mol% DMEDA in toluene at 135°C in the presence of Cs_2CO_3 as base. Beyond the synthesis of simple internal acetylenes, the procedure allows the construction of benzofuran ring through the usual 5-endo-dig cyclization. With this strategy 2-phenyl benzofuran and 2-m-tolyl benzofuran were synthesized starting from the appropriate iodophenol and terminal acetylene in 51 and 50% yield, after the 72 h reaction time. Although the procedure offer a new possibility for the Sonogashira coupling by replacing palladium as catalyst, later it was demonstrated that the palladium contaminants in the other components had a significant role in the Sonogashira coupling [4].

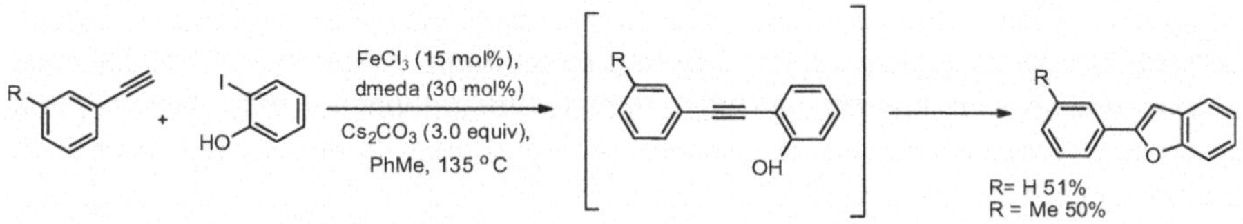

Later Shen and coworkers used 10 mol% CuI as catalyst for the construction of benzofuran and indole rings from 2-iodophenol or aniline in 1,4-dioxane at 100°C [5]. Several alkyl and aryl acetylenes were used as coupling partners and the substituted benzofurans were synthesized in 20–99% yield. The authors demonstrated that ultrapure CuI (99.999% purity) is less efficient in the coupling due to its low palladium content. However, addition of 100 ppb palladium to CuI significantly increased the conversion.

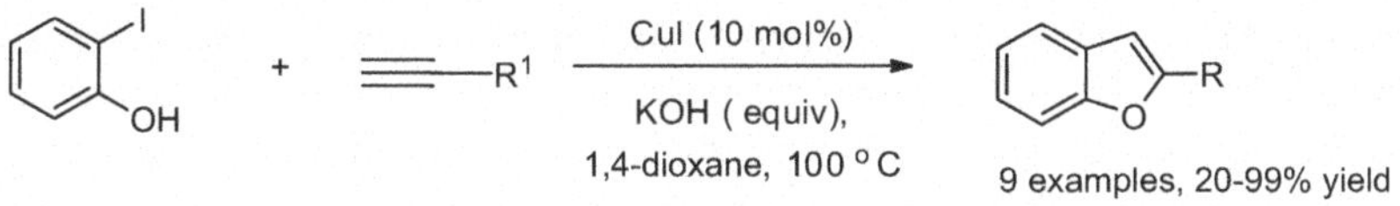

Copper triflate with *N,N*-disubstituted BINAM ligand is an efficient catalytic system for the construction of the benzofuran ring from 2-iodophenols and terminal acetylenes as it was demonstrated by Sekar [6]. In the presence of 20 mol% $Cu(OTf)_2$-BINAM catalyst in refluxing toluene, several 2-arylbenzofuran derivatives were obtained. With the utilization of substituted iodophenols, benzofurans could be isolated bearing alkyl group or halogen in position-5. Despite the relatively long reaction time (22–38 h), the desired heterocycles were obtained with good to excellent yield (48–91%). However, the possibility of palladium effect cannot be completely ignored in this case either since the procedure requires 3 equiv. of K_2CO_3, which can be a possible source of palladium contamination.

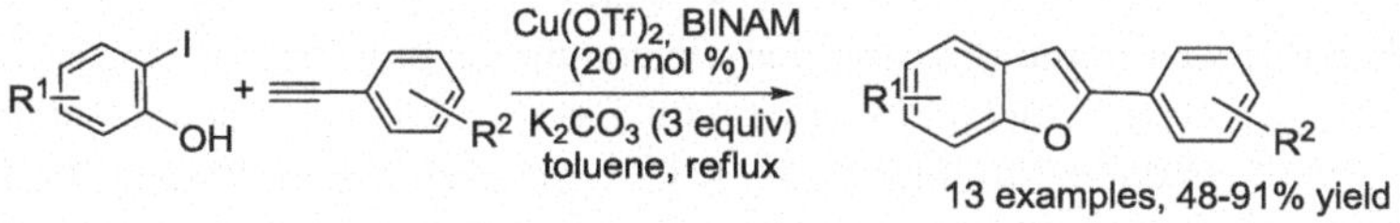

Besides the several homogeneous copper-catalyzed transformations, the benzofuran ring can also be constructed using a heterogeneous catalyst. An impregnated copper catalyst was developed and utilized by Ramon for the Sonogashira coupling of 2-iodophenols and different terminal alkynes [7]. After the coupling step the ring closure took place straightforward in a stepwise manner. The syntheses were performed in the presence of 1.3 mol% of $CuO\text{-}Fe_3O_4$ catalyst and 1.2 equiv. KOH in toluene at 130°C. After 24 h reaction time, the catalyst was simply removed with the aid of outer magnetic field and it was proved to be reusable ten times without the significant loss of activity. The methodology has good functional group tolerance considering the acetylene part. Several aryl acetylenes bearing electron-donating and electron-withdrawing groups were coupled with

2-iodophenol and the desired benzofurans were isolated with good yield. 2-Pyridylacetylene gave the 2-pyridylbenzofuran with 93% yield and alkylacetylenes were also successfully coupled with iodophenol, with the exception of ^{t}Bu-acetylene, which did not provide the appropriate product even after 72 h reaction time.

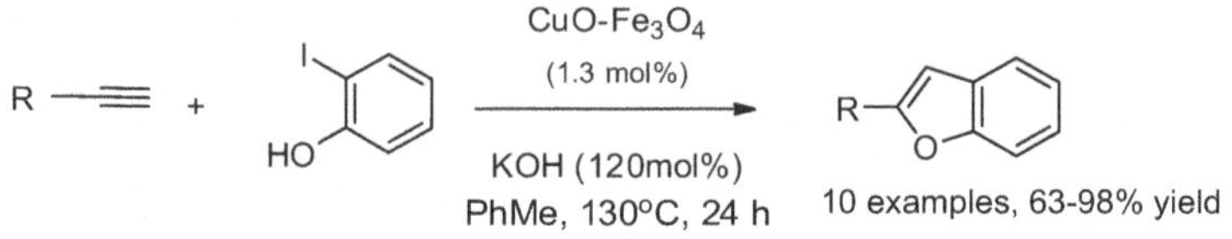

The palladium-catalyzed Sonogashira reaction can be achieved under both homogeneous and heterogeneous catalytic conditions. However, in situ generated palladium nanoparticles are also able to catalyze the coupling reaction, which was demonstrated by Ranu and coworkers [8]. Their methodology enables the coupling of aromatic iodides with terminal acetylenes under aqueous conditions in the presence of sodium dodecyl sulfate (SDS) at rt. For the successful coupling the nanoparticles were prepared in situ from Na_2PdCl_4. The Sonogashira conditions were also used for the synthesis of the benzofuran core using a general coupling-ring closure strategy. However, in this case the coupling required higher temperature (100°C) and longer reaction time (typically 12–18 h). The functionalized 2-aryl and 2-alkylbenzofurans were obtained in 75–86% yield, which demonstrates the robustness of the procedure.

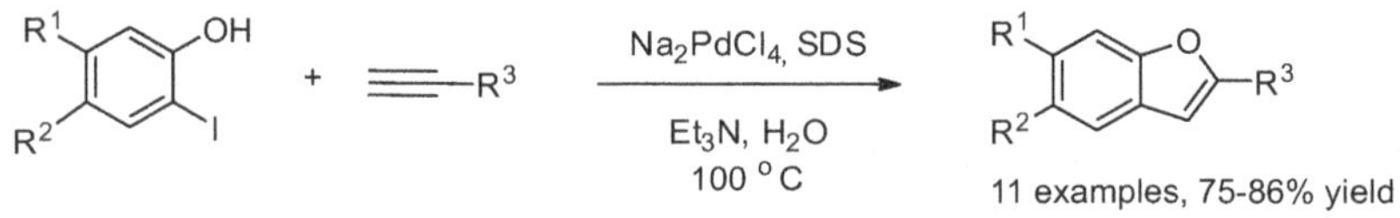

N-propargyl arylmethyl sulfoximines were also successfully coupled with 2-iodophenols under homogeneous catalytic conditions utilizing the frequently used $PdCl_2(PPh_3)_2$ catalyst (1 mol%) together with 5 mol% CuI. The reactions, developed by Bolm, were conducted in toluene at 70°C in the presence of 3 equiv. of 1,1,3,3-tetramethylguanidine (TMG) as base [9]. A versatile benzofuran library was built up with the use of differently substituted aryl sulfoximines and 2-iodophenols, and the procedure provided the products in excellent yield (64–95%).

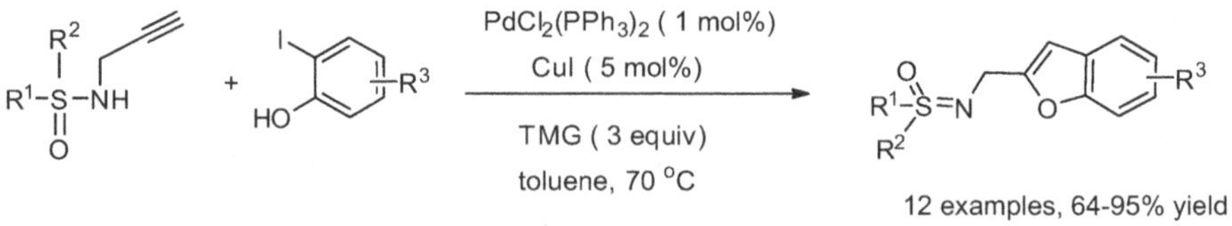

Most of the coupling-ring closure synthetic approaches for the construction of the benzofuran ring start from bromo- or iodophenols and terminal acetylenes.

Reactivity of chlorophenols is much lower due to the slower oxidative addition of the C–Cl bond to Pd. In general, aryl chlorides are less reactive in the Sonogashira reaction, and the presence of the hydroxyl group, which is required for the benzofuran formation, further deactivates the substrate for the coupling due to its strong electron donating property. Wang and Manabe described the first benzofuran synthesis from 2-chlorophenols following the common strategy [10]. The key point in their transformation is the choice of the appropriate ligand. The special bifunctional terphenyl ligand (Cy-HTP), which bears phosphano and hydroxyl groups proved to be an excellent ligand for the desired coupling and ring closure. In ^{t}BuOH at 110–120°C in the presence of ^{t}BuOLi, several benzofurans were synthesized with good yields. The additional advantage of the bifunctional ligand beyond its enhanced activity was exploited when dichlorophenols were selectively coupled with terminal acetylenes in position 2. The selectivity was ensured by the heteroaggregates of lithium phenoxides of the substrate and the ligand.

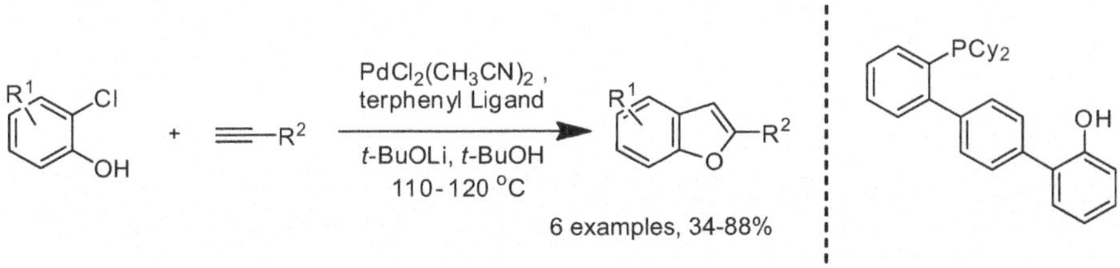

The catalytic system was improved later by Manabe with the design and utilization of dihydroxy-terphenylphosphine ligands bearing cyclohexyl groups on the phosphorous atom (Cy-DHTP) [11]. With this ligand the same transformations were performed, but the reaction time significantly decreased (from 22 h to 45 min) in the presence of Cy-DHTP. Dichlorophenols were also selectively coupled in position 2 with terminal acetylenes, and chlorobenzofurans were obtained. The chloro function offers the possibility of further functionalization via cross-coupling reactions. Taking advantage of this a one-pot sequential synthesis of disubstituted benzofurans was developed from dichlorophenols. The Cy-DHTP promoted coupling and ring closure was followed by a subsequent Suzuki coupling on the chloro function. The second coupling reaction required XPhos ligand. Utilizing this methodology several 2,4-, 2,5-, 2,6-, and 2,7-diarylated benzofurans were prepared in good to excellent yield (28–94%).

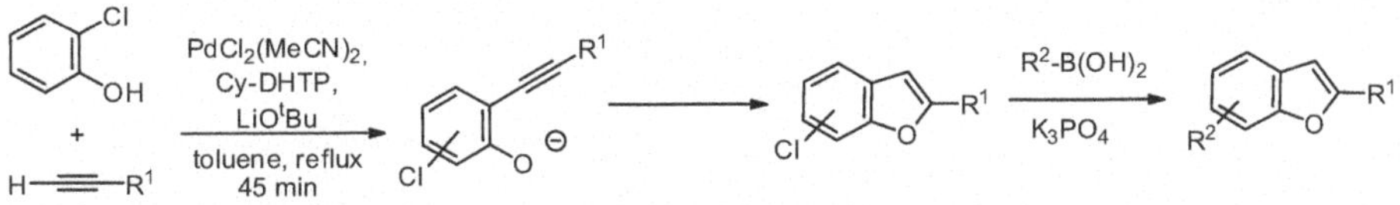

The similar sequential coupling – ring closure – coupling strategy was used by Arcadi et al. for the synthesis of 2,5,7-trisubstituted benzofurans from halophenols. Such 2-bromo- or 2-chloro-6-iodophenols were used as substrates that contained other substituents in position 4 (Me, F, Ph, $Me(CH_2)CO_2$) [12]. The Sonogashira reaction with terminal acetylenes was achieved with 10 mol% CuI catalyst and 20 mol% L-proline ligand in the presence of 2 equiv. K_2CO_3 in 1,4-dioxane at 120°C in 1–8 h. The appropriate 2-aryl or 2-alkyl 7-chloro or 7-bromobenzofurans were isolated in 60–85% yield. In a separate reaction the halo function in position 7 was successfully transformed via different cross-coupling reactions. In a palladium-catalyzed Suzuki reaction, 30 different 7-arylbenzofurans were prepared in 94–99% yield. 7-Alkynylbenzofurans were also synthesized via palladium-catalyzed Sonogashira coupling (yields varied between 50 and 99%), while the efficiency of the copper-catalyzed amination of benzofurans was demonstrated with 18 examples and typical yields of the C–N bond-forming reaction were 56–99%.

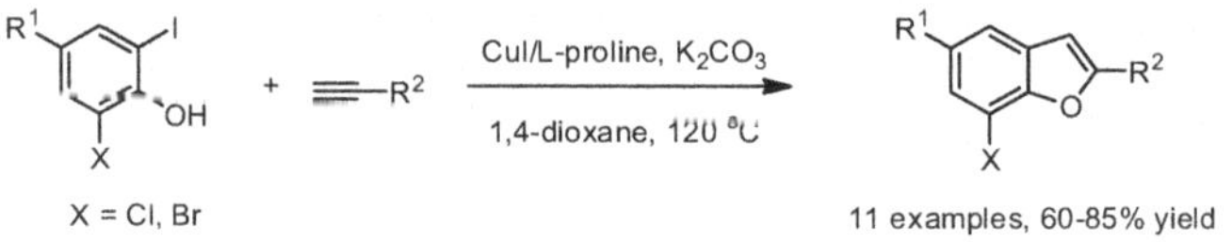

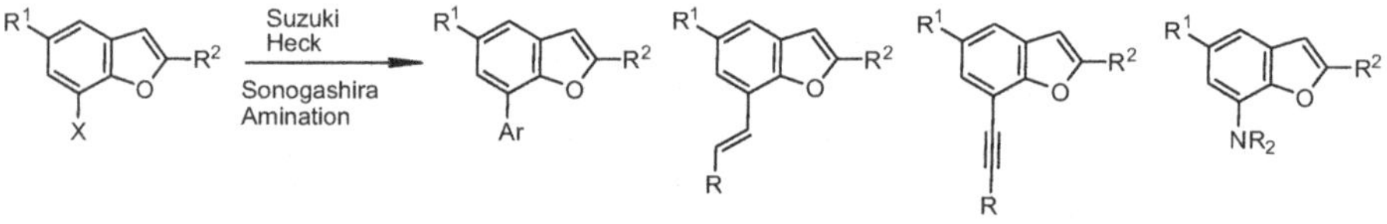

2,3-Disubstituted benzofurans were prepared by Larock from 2-iodophenols, terminal acetylenes, and aryl iodides utilizing palladium-catalyzed Sonogashira reactions with the aid of microwave irradiation [13]. The coupling between the iodophenol and the terminal acetylene was achieved with 3 mol% $PdCl_2(PPh_3)_2$ and 2 mol% CuI in TEA/THF at 25°C; then the appropriate aryl iodide and MeCN were added. The arylation of the preformed benzofuran ring in position 3 took place in 25 min at 100°C under microwave conditions. 27 examples demonstrate the usefulness of the procedure, and different aryl and heteroaryl groups were introduced into position 2 and 3 of the benzofuran ring via this sequential methodology.

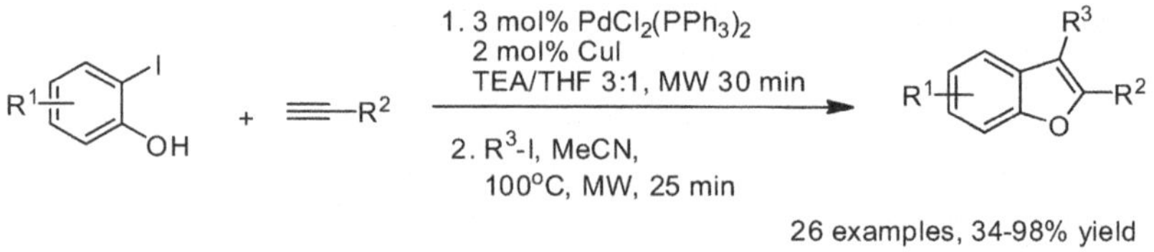

2-Aminomethyl-substituted benzofurans were prepared by Russo et al. in a one-pot three component reaction from propargyl halides, 2-iodophenols, and secondary amines [14]. The nucleophilic substitution on the propargyl halide propargyl was achieved by using the amine in excess (solvent); then the palladium-catalyzed Sonogashira reaction took place between the iodophenol and the terminal acetylene, followed by the ring closure affording the desired benzofuran derivatives. The reaction was achieved with 5 mol% $PdCl_2(PPh_3)_2$ and 10 mol% CuI at 80°C.

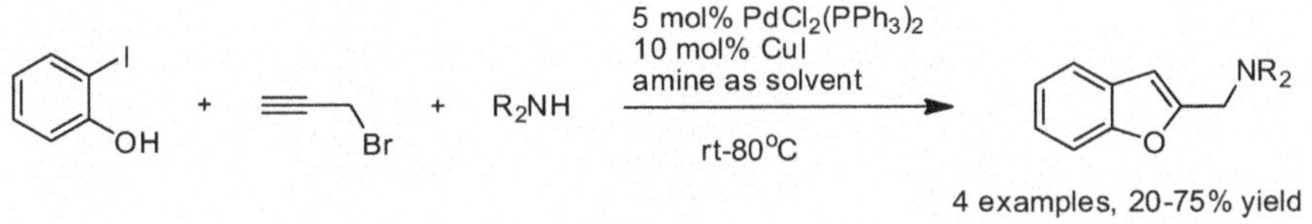

Based on Sonogashira coupling strategy, the benzofuran ring can also be constructed from three components: aryl iodides, *O*-protected 2-iodophenols, and an acetylene surrogate. For example, monoprotected acetylenes straightforwardly connect to aryl iodides under standard palladium-copper-catalyzed Sonogashira coupling conditions. Then the removal of the protecting group from the acetylene generates a new terminal acetylene ready to undergo another Sonogashira coupling. When this second Sonogashira coupling takes place with iodophenol, the benzofuran ring is formed in the traditional way. This approach was explored by Kotschy and coworkers utilizing ethynylcyclohexanol as an acetylene source [15]. The first coupling was carried out at rt in DIPA in the presence of 1 mol% $PdCl_2(PPh_3)_2$/CuI catalyst. After the completion of the coupling step, the silyl-protected iodophenol and KOH was added. The strong base removed the carbinol-protecting group from the acetylene and the second coupling took place with the iodophenol. After the addition of TBAF, the desilylation of the hydroxyl group and a subsequent ring closure occurred and the desired 2-arylbenzofurans were isolated in good yield (nine examples, 62–87%). The effectiveness of the procedure was demonstrated by the one-pot synthesis of the natural product Vignafuran.

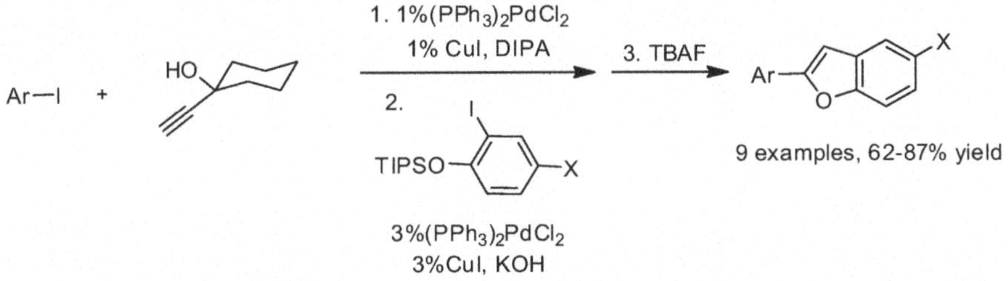

2.1.2 Electrophile-Induced Ring Closure

The cyclization of *ortho*-hydroxyl-aryl-acetylenes represents a powerful tool for the construction of the benzofuran core. With the aid of the coordination of transition metals to the triple bound, the nucleophilic attack of the oxygen is facilitated. Based on this phenomenon simple 2-aryl or alkyl benzofurans can be prepared after the ring closure. However, in the last couple of years, several transition metal-catalyzed methodologies were developed for the synthesis of 2,3-disubstituted benzofurans from *ortho*-alkynylphenols.

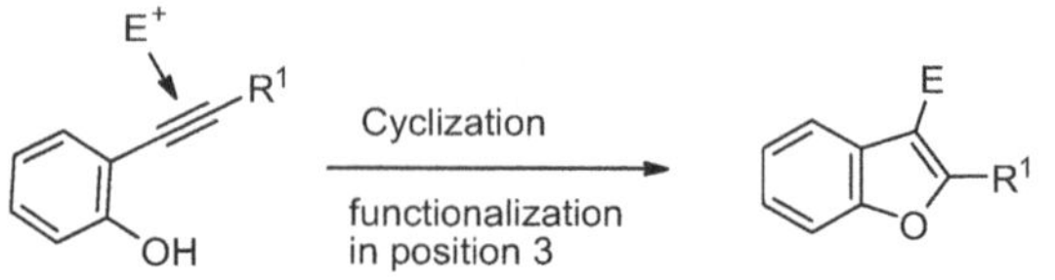

A palladium-catalyzed cascade heterocyclization-oxidative Heck coupling was developed by Álvarez et al. for the synthesis of alkenyl-substituted benzofurans from *ortho*-ethynylphenols [16]. In the presence of palladium(II) complexes and KI, the alkynes were reacted with olefins at 80°C and the appropriate 2,3-disubstituted benzofurans were isolated in 31–91% yield.

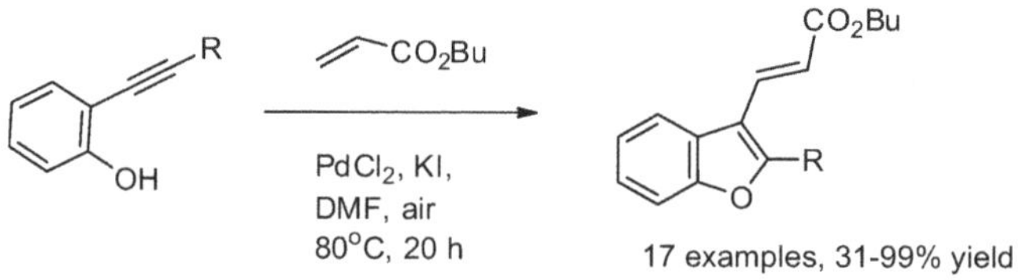

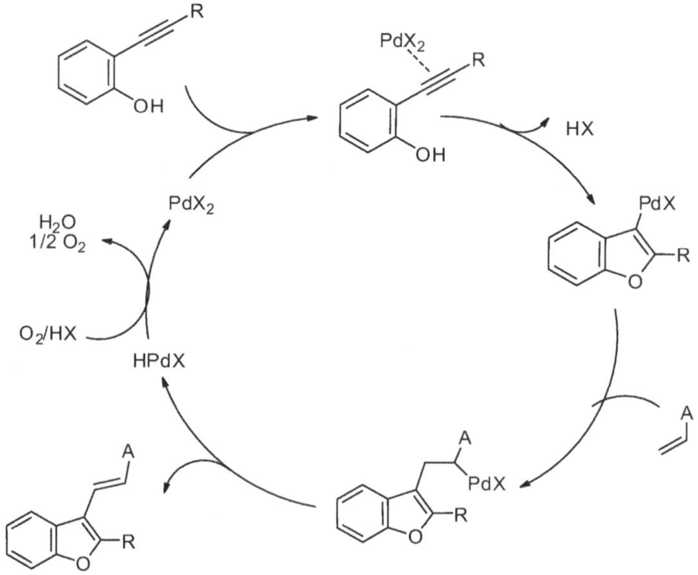

The first mechanistic step of the transformation is the coordination of the palladium(II) species to the triple bond, which induces the ring closure. The palladated benzofuran undergoes alkene insertion followed by β-hydride elimination providing the 2-aryl-3-alkenylbenzofuran and HPdX. The oxidative media is necessary to keep the palladium in the +2 oxidation state.

The palladium-catalyzed annulation strategy was used by Hu et al. for the synthesis of 2,3-diarylated benzofurans from *o*-alkynylphenols and aryl iodides [17]. The transformation was achieved with 5 mol% Pd_2dba_3 and 10 mol% bipyridyl ligand in MeCN at 50°C in 5 h, and the 2,3-disubstituted benzofurans were isolated in 52–87% yield. The synthetic methodology was used in the solid phase synthesis of benzofurans from on-bead iodophenols. However, for the successful annulation step 2.2 equiv. of palladium was required.

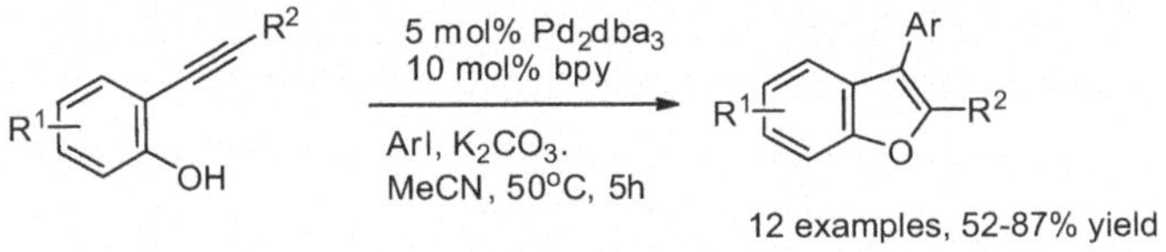

A different approach for the construction of the benzofuran ring was developed by Wang and coworkers [18]. In their copper-catalyzed transformation, the heterocyclic system was built up from tosylhydrazones and terminal acetylenes. In the presence of base, the tosylhydrazone provides diazo derivatives and the acetylene gives copper acetylide. The reaction of these two species enables the formation of an allene intermediate which undergoes cyclization by the attack of nucleophilic OH group. The protonolysis of the alkyl cuprate intermediate provided 2-benzylbenzofurans as the product. The reactions were performed with 10 mol% CuBr catalyst in the presence of Cs_2CO_3 as base in MeCN at 100°C and the desired products were isolated in 48–91% yield after 4 h reaction time. The functional group tolerance of the transformation was demonstrated on 16 examples.

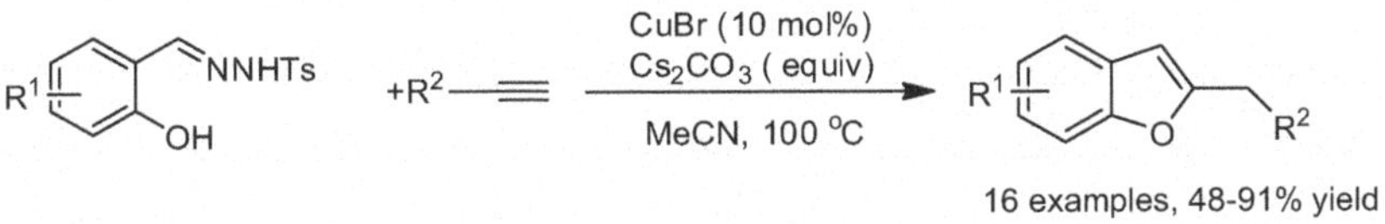

Later, the transformation was further developed by Zhou with the opposite approach regarding the key functional groups of the reactants [19]. The benzofuran synthesis was achieved efficiently with *O*-ethynyl-phenols and aromatic hydrazones with diverse substrates. In this latter approach ^{t}BuOLi was used as base instead of Cs_2CO_3. The formed copper acetylide reacts with the in situ generated diazo compound and afford ethynyl-copper carbene. Followed by the migratory insertion of carbene carbon and the protonation step, the allene intermediates undergo copper-promoted intramolecular cyclization to give the desired benzofuran molecules. The synthetic utility of the transformation was demonstrated with 20 examples and various benzofurans were isolated in 62–95% yield.

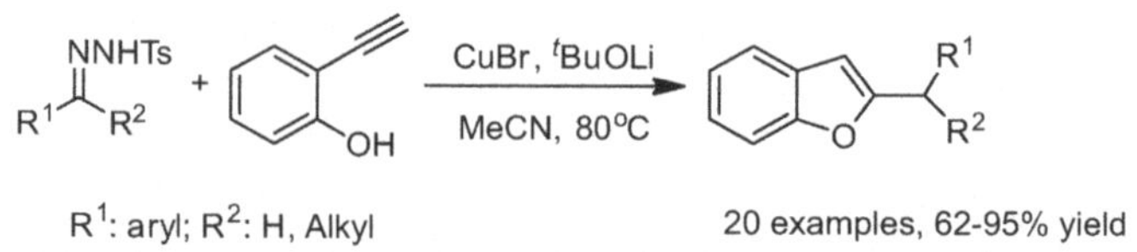

o-Ethynylphenols are suitable substrates for the synthesis of benzo[b]furan-3-carboxylic acids using the palladium-catalyzed carboxylative annulation strategy [20]. With $Pd(MeCN)_2Cl_2$ in the presence of AgOTf and $2\text{-}PyPPh_2$ under 1 atm CO in MeCN, the ring closure on the acetylene part takes place straightforwardly providing the desired 3-carboxylic acid derivatives in 1–2 h at 50°C. Utilizing the optimized reaction conditions, several 2-aryl- and alkyl-substituted benzofuran 3-carboxylic acid derivatives were isolated in 66–88% yield.

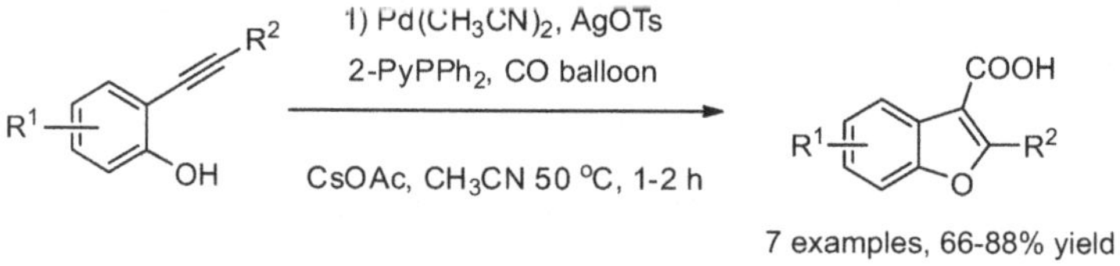

2,3-Disubstituted benzofurans can be prepared from *o*-ethynylphenols via $ZnCl_2$-assisted cyclization using BuLi as base [21]. The benzofuranyl zinc intermediates were transmetalated with copper(I) salts to form the corresponding cuprates. The organocuprates were readily reacted with several organic electrophiles such as allyl bromides, acid chlorides, enone systems, and aldehydes. Utilizing this transformation several 2,3-disubstituted benzofurans became accessible.

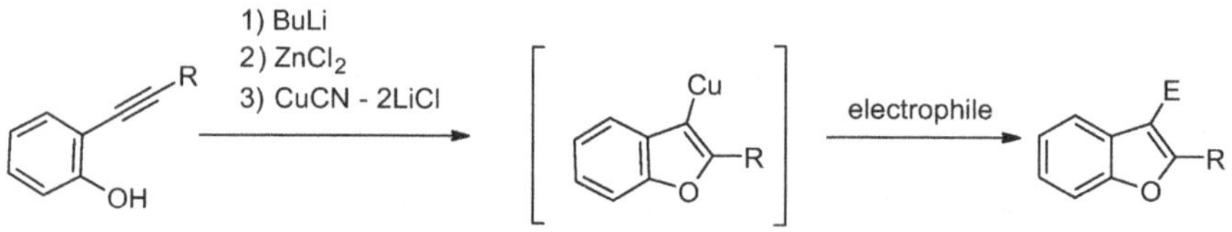

The palladium-catalyzed cyclization/coupling strategy was also utilized by Alvarez and Aurrecoechea for the synthesis of 3-ethenylbenzofurans from *o*-ethynylphenols [22]. The oxypalladation cyclization step starts with the coordination of the palladium catalyst to the triple bond, which induces the ring closure and results in the formation of a benzofuryl palladium species. Through the Heck-type coupling (insertion, β-elimination), 3-ethynylbenzofurans are formed. Utilizing the developed reaction conditions, several 2-aryl 3-alkenyl-benzufurans were prepared at 80–100°C in the presence of $PdCl_2$ and KI in DMF. The methodology provides the desired compounds in 31–91% yield.

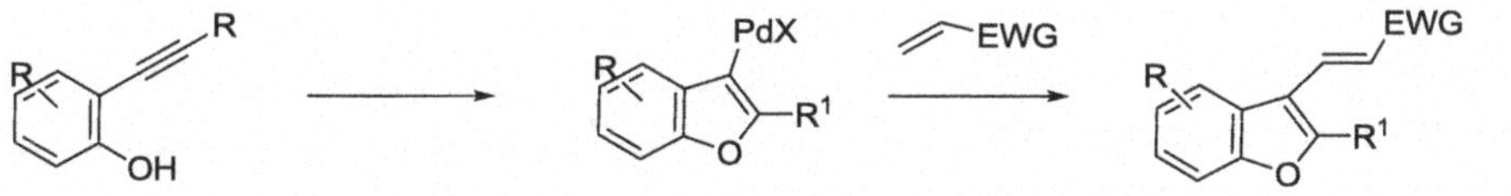

Change of the transition metal catalyst from palladium to rhodium provides 3-alkyl-substituted benzofurans in the reaction of *o*-ethynylphenols and electron deficient alkenes [23]. The cyclizations were conducted in dioxane/water (20:1) solvent mixtures at 90°C. In the presence of rhodium complexes the desired alkyl-substituted benzofurans were isolated with up to 96% yield, but in several cases the appropriate alkene derivatives also formed as side products (ca. 5%). A more detailed mechanistic study and expansion of this rhodium-catalyzed transformation was performed later by Lautens and coworkers [24].

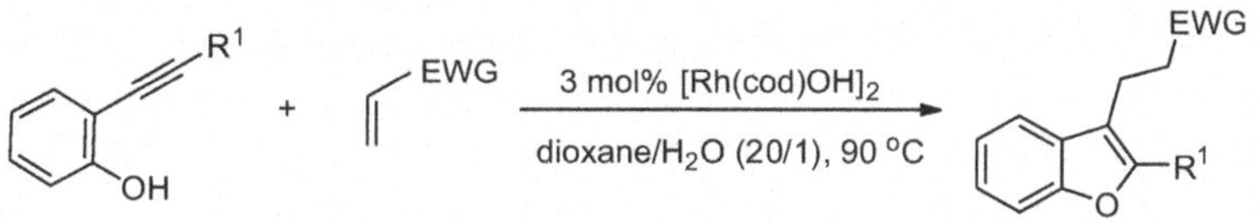

Han and Lu demonstrated that the utilization of cationic palladium species in the synthesis of 2-substituted-3-hydroxymethylbenzofurans is a powerful tool for the tandem annulation strategy [25]. The $[Pd(dppp)(H_2O)_2](BF_4)_2$ catalyst dissolved in THF enables the cyclization of 2-ethynylphenols and a subsequent functionalization in position 3 with aldehydes as electrophilic reagents. The synthetic utility of the methodology was demonstrated with 21 examples where the desired benzofurans were prepared in 37–100% yield in 2–5 h at 45°C.

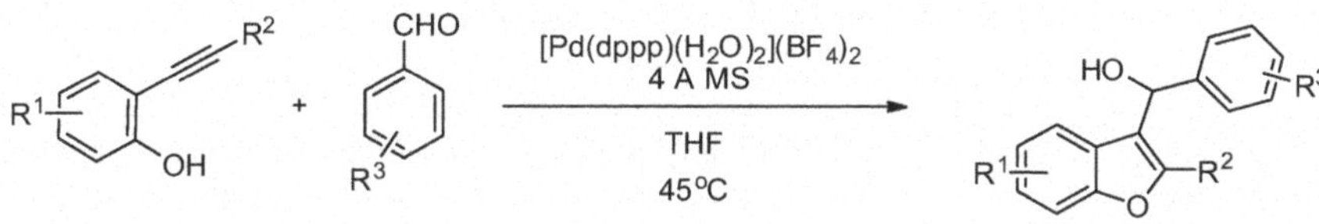

Alkynylphenyl acetals were transformed into 2,3-disubstituted benzofurans in a platinum-catalyzed ring closing reaction. The methodology developed by the Yamamoto group enables the synthesis of 2-alkyl and arylbenzofurans bearing alkyl ether function in position 3 in excellent yield (61–94%) [26]. The reaction takes place at 30°C in toluene. However, the catalyst loading is 2–100 mol% $PtCl_2$ depending on the substrates and the required reaction time varies between 1 h and 4 days. The ring closure is initiated by the coordination of the platinum to the triple bond which induces the nucleophilic attack of the oxygen. The cyclization product is the oxonium ion which is transformed trough the migration of the alkoxyalkyl group from position 2 to 3. The elimination of the platinum chloride closes the catalytic cycle and provides the desired benzofuran.

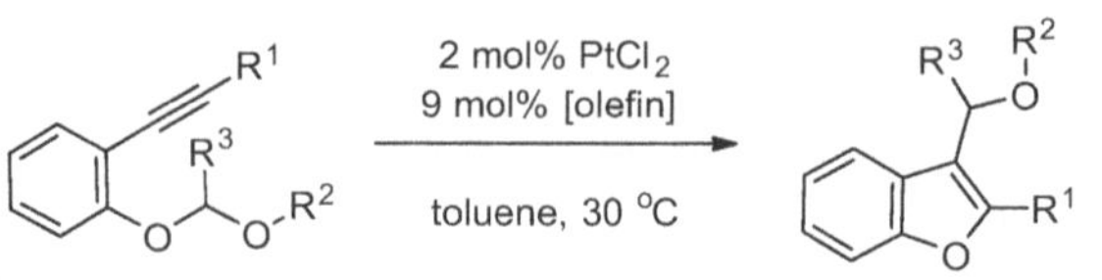

At the same time a similar strategy was used for the construction of the benzofuran ring from *ortho*-ethynylphenols and ethers by Fürstner and Davies [27]. The $PtCl_2$-catalyzed reactions of ethynylphenols provided the 2-alkyl and arylbenzofurans in good yield (88–98%) at 80°C in toluene in 1–5 h.

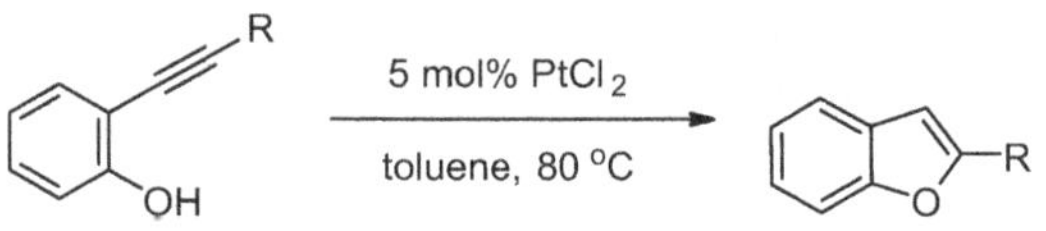

9 example, 88-98%

In case of the allyl ether derivatives, the ring closure is accompanied with the transfer of allyl group to position 3 from the oxygen providing 2,3-disubstituted benzofurans via intramolecular carboalkoxylation. The authors showed that the best results can be reached when the reaction was performed under CO atmosphere.

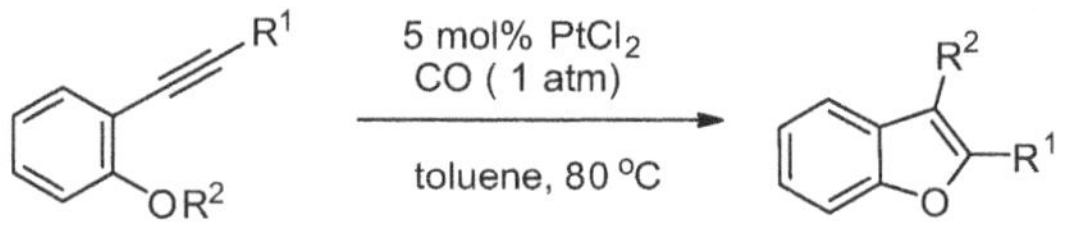

12 example, 54-98%

The platinum-catalyzed carboalkoxylation strategy was extended for the synthesis of 2,3-disubstituted benzofurans bearing a protected hydroxymethyl group in position 3. The alkyl migration takes place in case of MOM-, BOM-, and SEM-protected alcohols in 0.5–2 h and provides the benzofurans in excellent yield (62–95%).

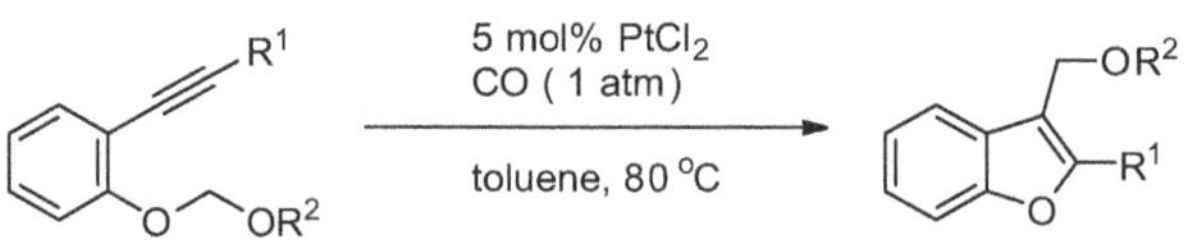

10 example, 62-95%

The palladium-catalyzed electrophilic annulation of 2-alkynylphenol derivatives with disulfides or diselenides enables the synthesis of 2,3-disubstituted benzofurans containing C–S or C–Se bond in position 3 [28]. The cyclization takes place in MeCN at 80°C in the presence of iodine in 36 h. The methodology provides the desired sulfides in good yield when substituted *ortho*-arylethynylanisole derivatives are subjected to ring closure, but alkylacetylenes cannot be transformed. Reactions with diselenides generally require shorter reaction time (6–18 h) for the preparation of selenobenzofurans in good yields (69–89%).

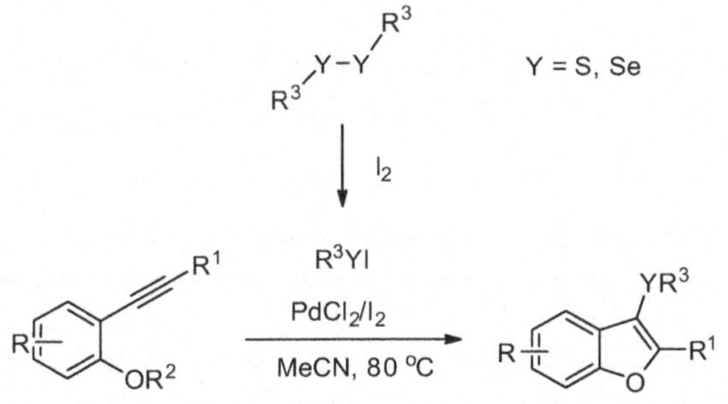

The reaction does not occur in the absence of iodine. It is suggested that the iodine reacts first with the disulfide and forms PhSI. Several synthetic methodologies exist for the iodine-mediated electrophilic cyclization for the synthesis of 3-iodobenzofurans, but in this case Li and coworkers ruled out the formation of iodobenzofuran intermediates through the absence of transformation between 3-iodobenzofuran and diphenylsulfide under the applied conditions.

Pyne developed a copper-mediated reaction for the synthesis of 3-iodo, 3-bromo, and 3-cyanobenzofurans from *ortho*-ethynylphenol derivatives [29]. The cyclization reaction was carried out in DMF at 135–140°C under oxygen atmosphere in 16 h in the presence of 2.2 equiv. of the appropriate Cu(I) salt. The copper salt delivered the incoming group to the 3 position of the benzofuran ring.

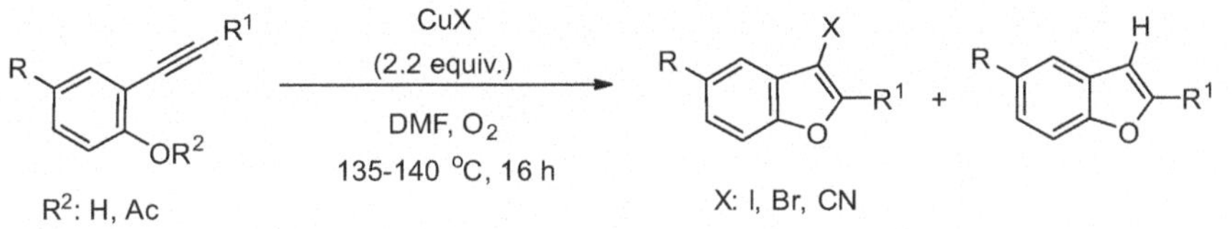

The similar cyclization-substitution strategy was developed by Zeni for the synthesis of 3-benzofuranyl-sulfides, selenides, and tellurides [30]. Although the procedure requires 1 equiv. of iron(III) salt, the desired compounds can be obtained in good yield under mild reaction conditions.

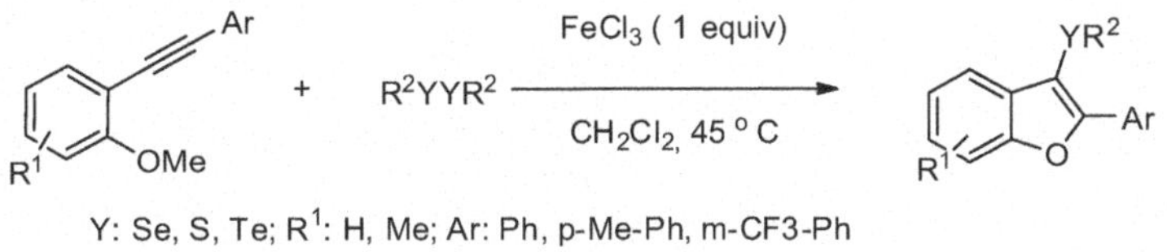

The benzofuran ring can also be constructed starting from aryl ethers such as α-aryloxy-ketones and allyl phenyl ethers in transition metal-catalyzed cyclization [31]. The cyclodehydration of α-aryloxy-ketones bearing an *ortho* directing group proceeds in an iridium-catalyzed transformation. In the presence of $Ir(cod)_2BARF$ catalyst and *rac*-BINAP ligand in chlorobenzene at 135°C, the cyclization is complete in 24 h providing the multisubstituted benzofurans in good to excellent yield (82–99%).

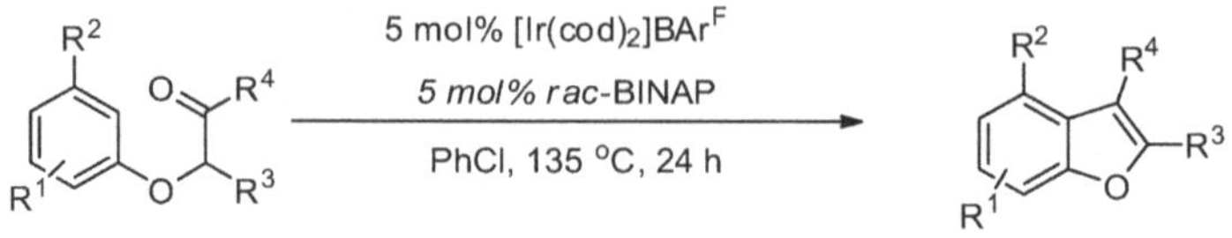

Shibata improved later his procedure and the cyclization was complete at rt in 8 h in DCE using a Cp*$IrCl_2$/$AgSbF_6$/$Cu(OAc)_2$ catalyst system, and the appropriate benzofuran products were obtained in the same excellent yields [32].

Using the oxidative cyclization strategy, several 2-methylbenzofurans were synthesized by Youn and Eom [33] from allyl aryl ethers in the presence of $Pd(MeCN)Cl_2$ catalyst and benzoquinone oxidant in dioxane.

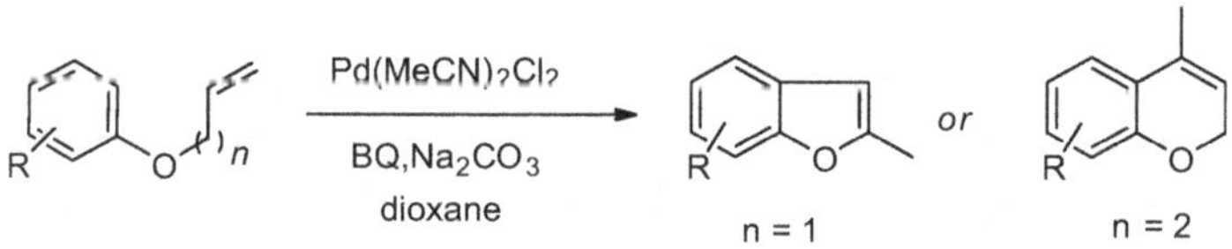

Allyl aryl ethers bearing an ynol motif in the *ortho*-position are excellent substrates for the synthesis of benzofuran 2-acetic acid esters [34] and amides [35] in deallylative carbonylation reactions. Under 30 atm CO pressure, the palladium-catalyzed coupling proceeds in the presence of alcohols or amines providing the appropriate esters or amides in good yield.

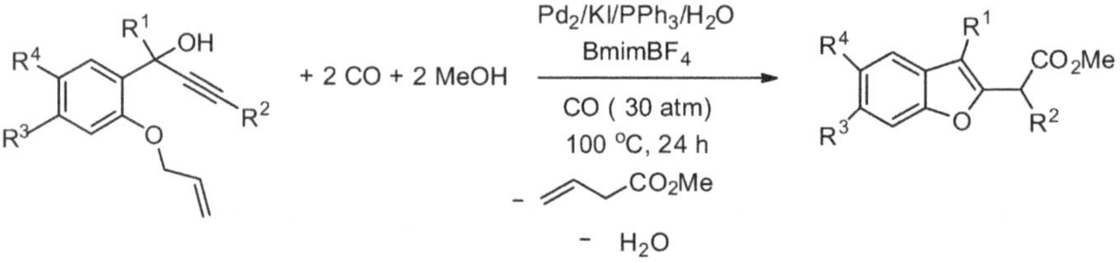

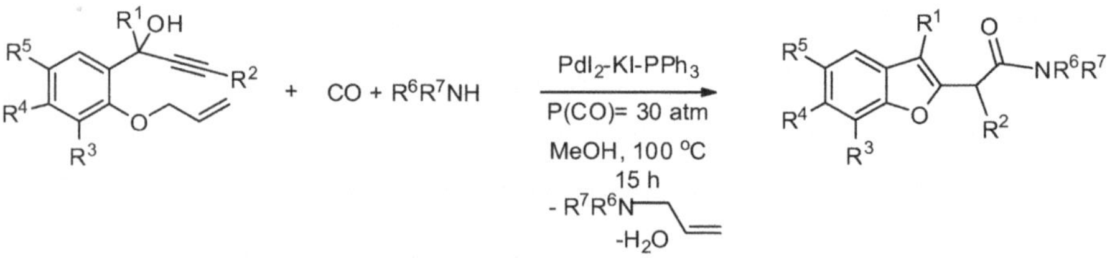

2.1.3 Synthesis of Benzofurans from Acetylene Alternatives

Construction of the benzofuran ring can also be achieved from 2-*gem*-dibromovinyl phenols [36]. The advantage of this strategy is that it enables the introduction of bromo, selenide, aryl, and ethynyl function into position 2 of the benzofuran ring utilizing palladium- or copper-catalyzed cross-coupling chemistry. In general the first step of the transformation is the ring closure through one of the bromines. The 2-bromobenzofuran intermediate offers further functionalization to obtain the desired products. Utilizing the procedure of Wang, 2-selenyl or sulfenylbenzofurans were prepared. Dibromovinyphenols were reacted with diaryl selenides or sulfides in the presence of t-BuLi and magnesium metal. The reaction proceeds with 10 mol% CuI in DMSO in 12 h at 110°C providing the appropriate products in 53–81% yields.

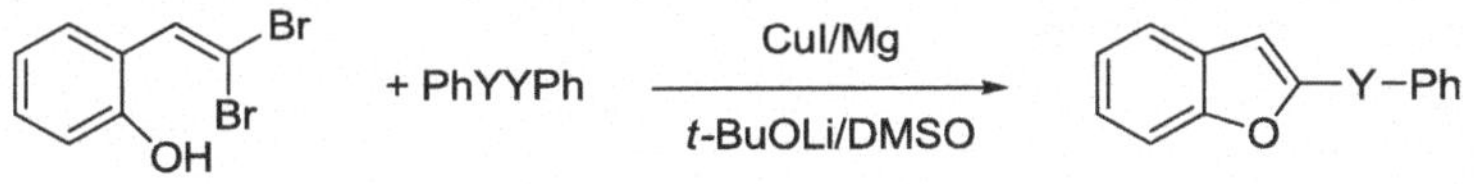

Under typical Sonogashira coupling conditions, in the presence of a Pd/C-CuI catalyst system, dibromovinylphenols smoothly react with terminal acetylenes in diisopropylamine providing 2-ethynylbenzofuran derivatives. In the procedure developed by Lautens and coworkers, both alkyl and aryl acetylenes are applicable for the coupling [37].

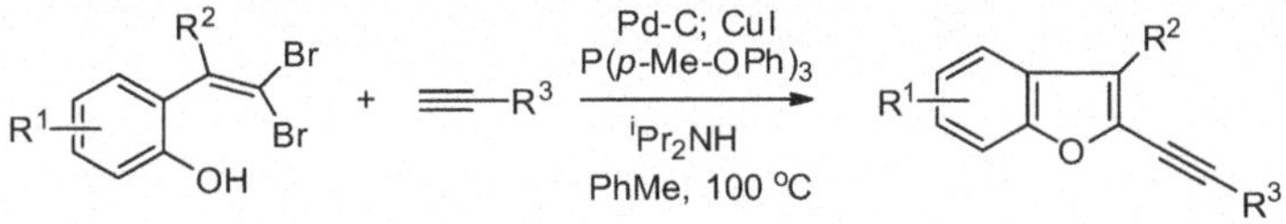

In the absence of any nucleophilic coupling partner, the transformation allows the isolation of 2-bromobenzofuran derivatives when the dibromovinyl compound is treated with 5 mol% CuI in the presence of K_3PO_4 in THF at 80°C for 6 h [38].

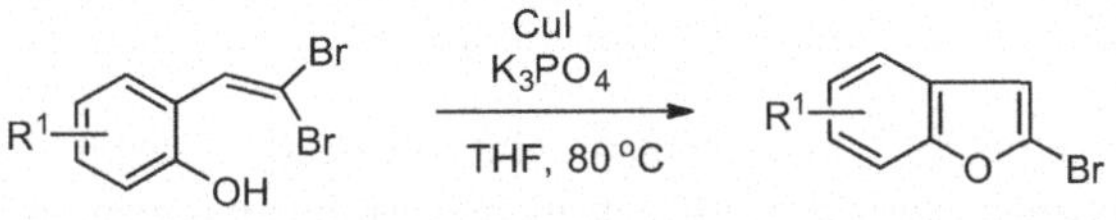

Beyond cross-coupling chemistry polyfluorinated aromatic compounds can be coupled directly via C–H functionalization to obtain 2-arylbenzofurans in the presence of CuI/phenanthroline catalyst system in dioxane at 125°C [39].

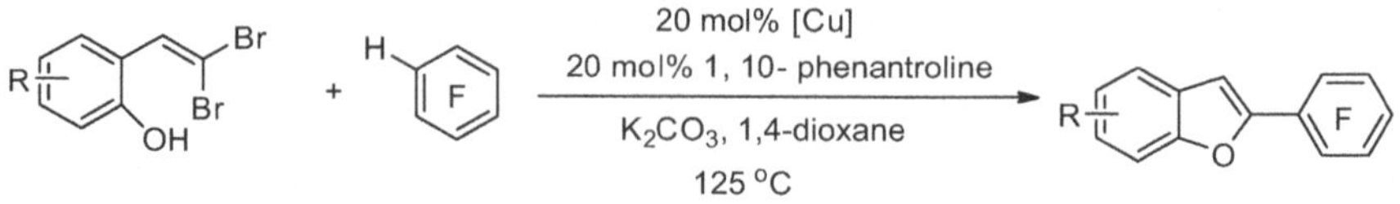

2,3-Disubstituted benzofurans were synthesized in a multicomponent reaction starting from salicylaldehyde, a secondary amine, and terminal or silyl-protected aryl acetylenes in a copper(II)-catalyzed reaction carried out in toluene or MeCN at reflux temperature [40]. The procedure enabled the synthesis of a versatile benzofuran library containing amino substituent in position 3.

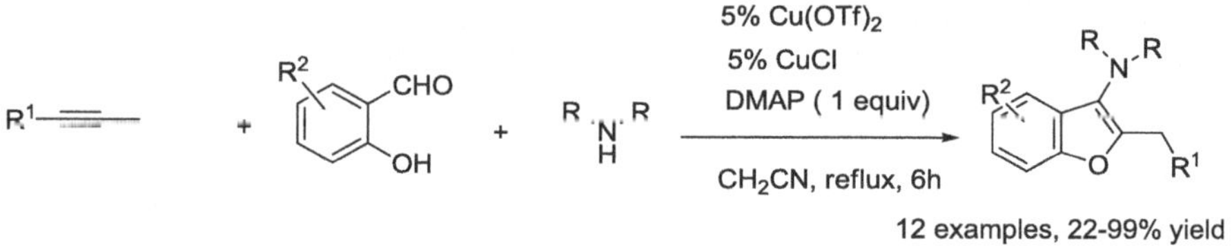

The benzofuran ring can be constructed from *o*-bromobenzyl ketones in copper- or palladium-catalyzed transformations. Domínguez and SanMartin utilized CuI/TMEDA catalyst for the ring closure under aqueous conditions [41].

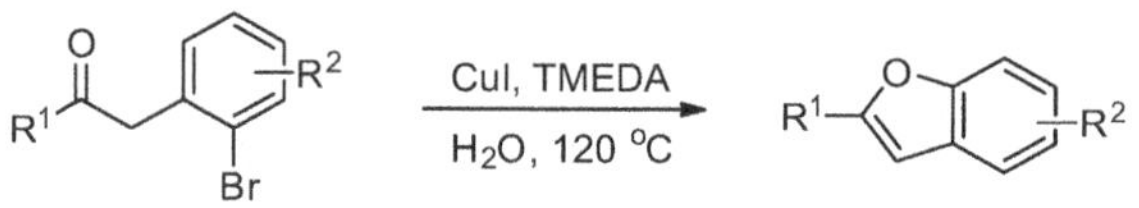

The similar transformation was developed by Bolm, using sub-mol% $CuCl_2$ beside $FeCl_3$ as catalyst in DMF. The desired 2-arylbenzofurans were isolated in good yields [42].

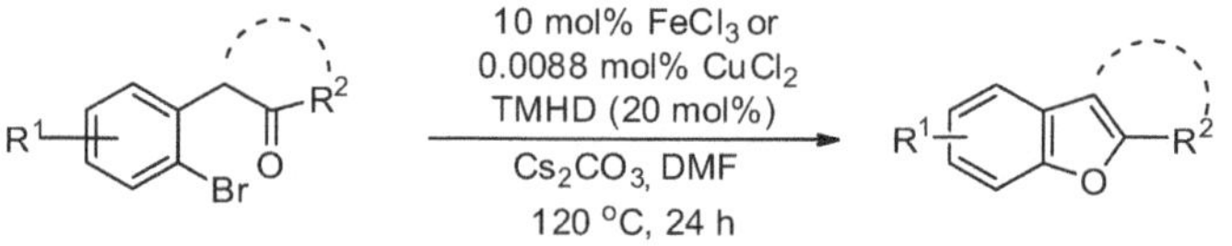

Faragó and Kotschy synthesized the desired 2-arylbenzofurans from 2-bromobenzylketones with the utilization of Pd_2dba_3 catalyst in the presence of NHC ligand [43].

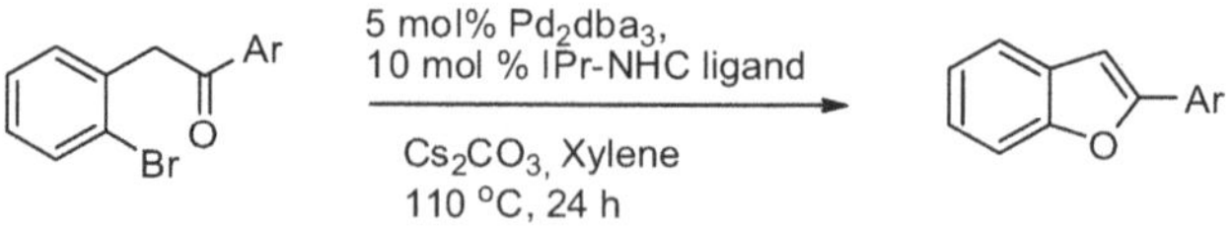

A novel methodology was developed for the construction of the benzofuran ring utilizing the transition metal-catalyzed oxidative coupling of aromatic alcohols. The coupling partner in most cases is an acetylene derivative. Jiang and coworkers developed a copper-catalyzed protocol for the construction of 2,3-disubstituted benzofurans from phenols and diaryl or aryl-alkyl acetylenes using molecular oxygen as oxidant [44]. The coupling reaction takes place in $MeNO_2$ at 120°C and the products were isolated in 72–93% yield.

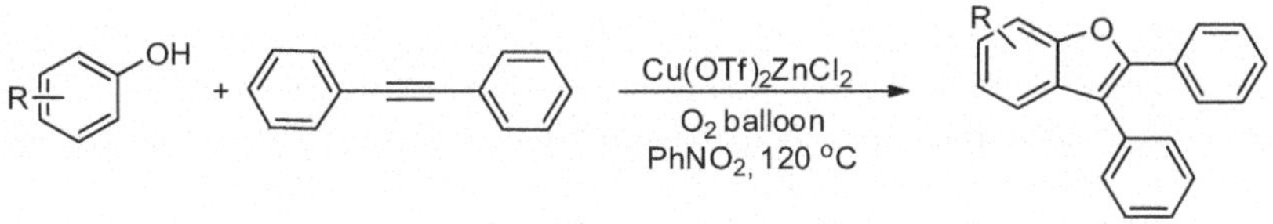

The same strategy was used by Shi and coworkers, but the transformation used a rhodium catalyst and the procedure required the addition of 1 equiv. of $Cu(OTf)_2$, 2.5 equiv. $AgPF_6$, 20 mol% acetanilide as ligand, and 50 mol% dodecyl trimethylammonium chloride as phase transfer catalyst [45].

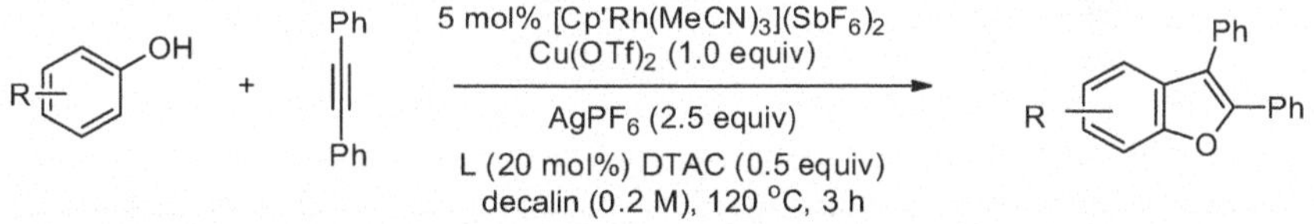

When 1-bromoacetylenes were used for this oxidative ring closure strategy, the palladium-catalyzed reaction afforded 2-aryl-substituted benzofurans from phenols at 130°C in DMF, as it was demonstrated by Wang and coworkers [46].

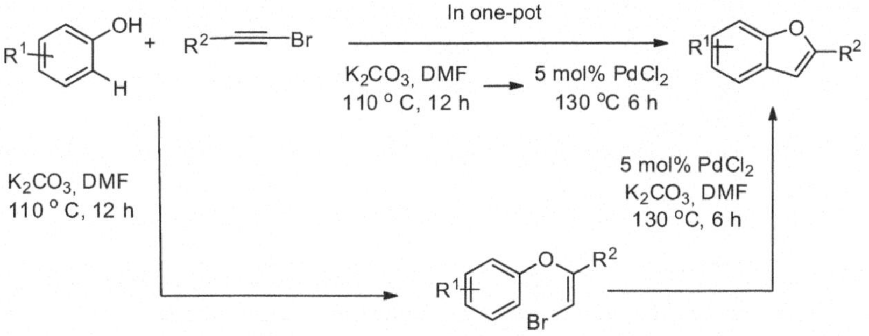

In the iron-catalyzed one-pot cascade reaction of propargylic alcohols and phenols, polysubstituted benzofurans or naphthopyrans can be prepared efficiently. The synthetic procedure developed by Han and Yuan enables the efficient synthesis of a diverse collection of oxygen heterocycles [47].

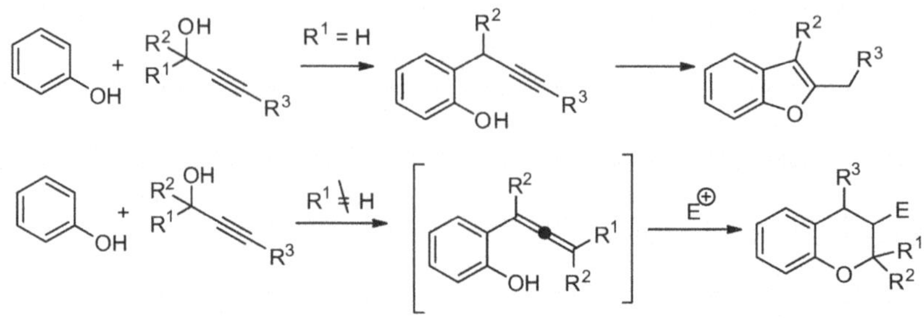

Iron(III) catalyst in the presence of di-*tert*-butylperoxide catalyzes the formation of benzofurans from phenol derivatives and β-ketoesters [48]. It was demonstrated that the reaction performed in DCE at 100°C was accelerated by water or various alcohols and Brönsted acids. The desired polysubstituted benzofuran derivatives were synthesized in 20–75% yield.

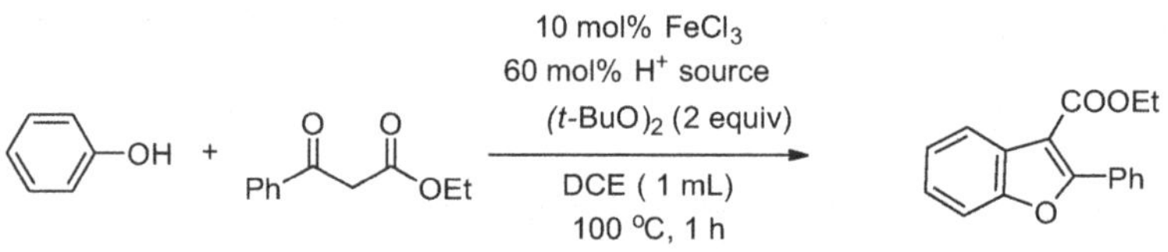

Another iron(III)-mediated oxidative approach was demonstrated by Zhao and coworkers [49]. Their procedure involves the oxidative C–O bond formation of α-aryl cyanoketones, which transformation provides 2-aryl-3-cyanobenzofurans in 50–94% yield.

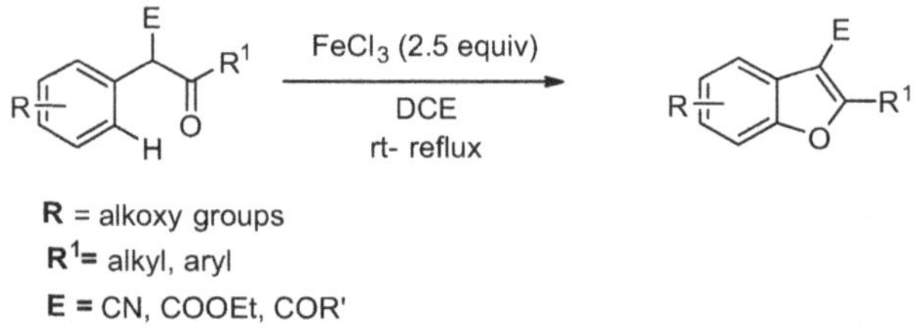

2.2 *Transition Metal-Catalyzed Synthesis of Furans*

The synthesis of furan derivatives from acetylene derivatives generally takes place via the formation of propargyl vinyl ether intermediates. Several transition metal-catalyzed transformations were developed in the last couple of years for the construction of the furan ring. This synthetic strategy enables the formation of polysubstituted furan rings. In this chapter the most important methodologies are summarized.

Polysubstituted furans can be synthesized in the palladium-catalyzed reaction of alkynoates and 2-yn-1-ols via the DABCO promoted propargyl vinyl ether formation in a two steps process [50]. Jiang and coworkers developed efficient conditions for the transformation and the desired tetrasubstituted furans were isolated on a broad scale in good yield (70–94%).

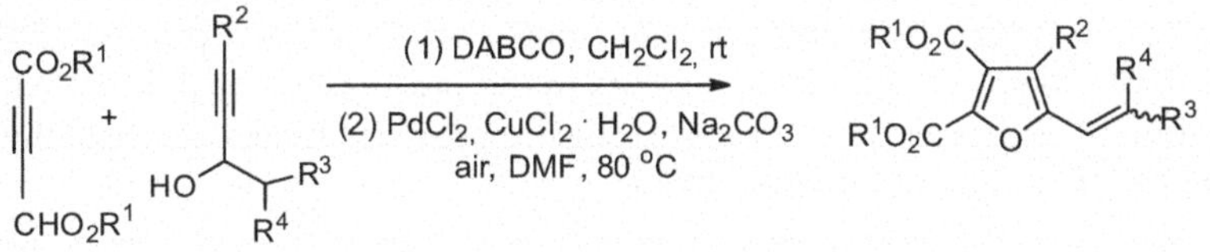

Jiang and coworkers also demonstrated that the same transformation could be achieved in a copper-catalyzed reaction. In the presence of Bu_3P or DABCO, internal acetylenes reacted smoothly with propargyl alcohols providing propargyl vinyl ethers, which underwent ring closure in the presence of nanosized Cu_2O in DMF at 50–80°C or with Ag(I) salts at 50°C in toluene [51–53].

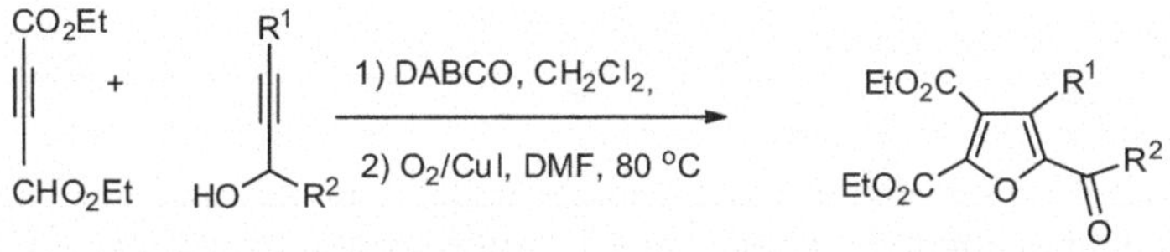

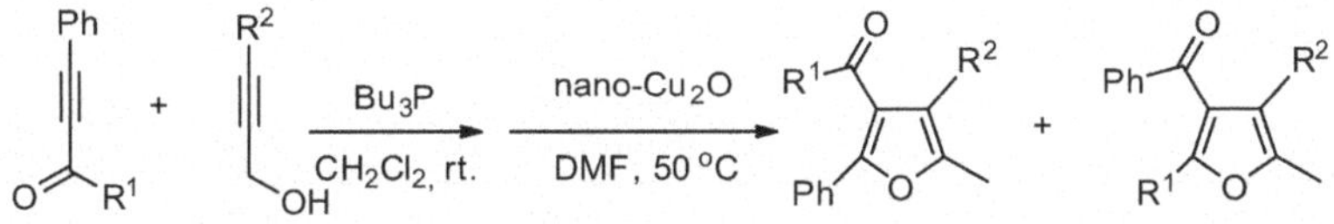

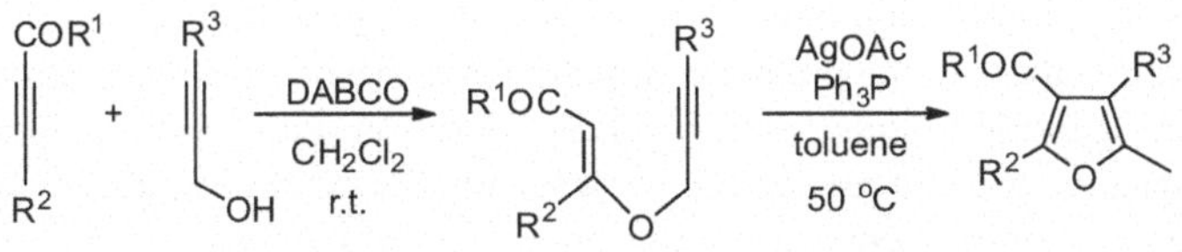

The synthesis of tetrasubstituted furans from propargyl vinyl ethers can be achieved in a gold-catalyzed transformation. Kirsch and coworkers demonstrated that the cyclization takes place straightforwardly at 23°C in DCM in the presence of 2 mol% $(PPh_3)_2AuCl$ catalyst [54].

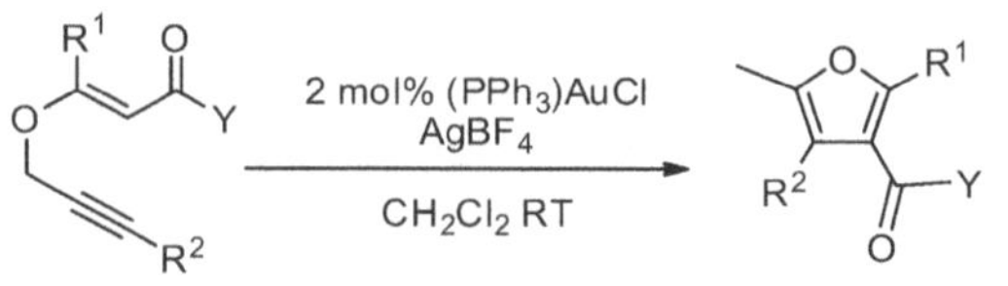

Besides noble metal catalysis, the synthesis of tetrasubstituted furans can also be achieved from propargylic alcohols and ynones in a two-step process [55]. The formation of propargyl vinyl ethers occurs in the presence of Bu_3P in DCM. After the addition of an iron(III) salt, the ring closure provides furan-2-carbaldehydes in DMSO at 80°C in good yield (53–83%).

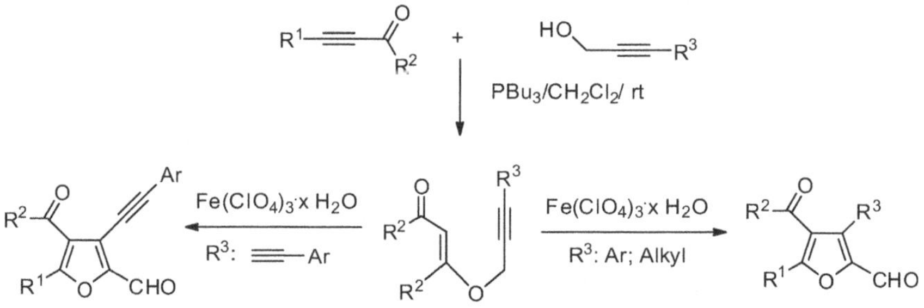

Jiang and coworkers demonstrated that the palladium-catalyzed ring closure also takes place with methyl vinyl ethers and 1,3-diketones in the presence of catalytic amount of $In(OTf)_3$ [56]. Applying the developed conditions several 2,4,5-trisubstituted furans were synthesized in DCE at 60–75°C under oxygen atmosphere.

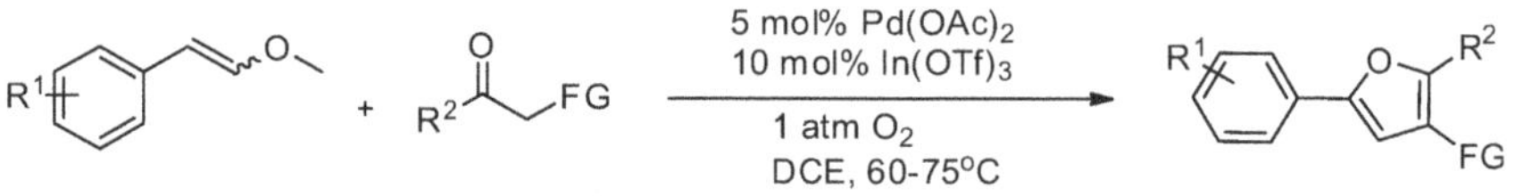

Another approach to the construction of the furan ring is based on enyne systems. Cyclization of enyneols or their equivalents (ynediols, alkynyl oxyranes), enyneones, and ethers in the presence of palladium, silver, or gold catalyst provides the furan ring generally under mild reaction condition with high efficiency.

Z-enynols are excellent precursors for tetrasubstituted furans. The cyclization of the conjugated system occurs in the presence of gold catalysts such as $AuCl_3$, Au $(PPh_3)Cl$ or $Au(PPh_3)OTf$ [57–59]. The latter form of phosphano gold catalyst can also be generated in situ from chlorides with AgOTf. The reactions take place in DCM or DMA at rt in 1–5 h providing the desired polysubstituted furan derivatives in typically good yield (up to 94%).

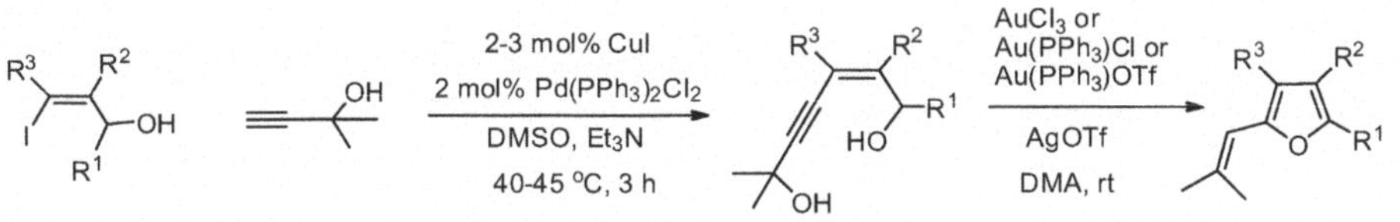

The cyclization of enyneols can also be achieved in the absence of transition metals as it was demonstrated by Liu and coworkers [60]. The synthesis of substituted furans was performed via PIFA (phenyliodine bis(trifluoroacetate))-I_2-mediated oxidative cycloisomerization of *Z*-enynols in THF. The furans were obtained 31–92% yield.

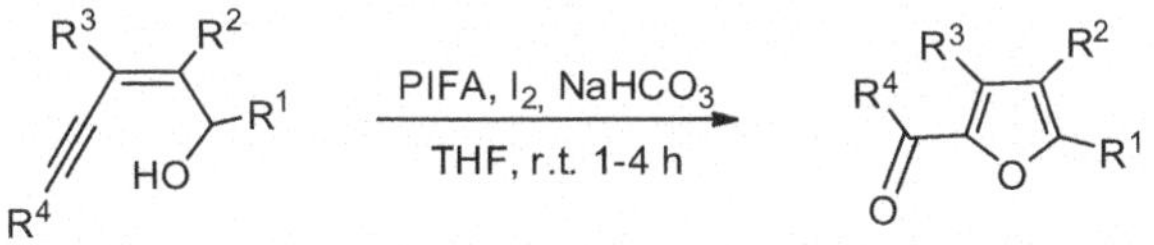

Using a similar approach trisubstituted furans can be prepared in the highly efficient palladium-catalyzed cascade reaction of aryloxy enynes. In the presence of $Pd(PPh_3)_4$ catalyst and Cs_2CO_3, the ring closure occurs at 70°C in DMF in 1.5–3 h. Trisubstituted furans [61] were isolated in 57–92% yield, while disubstituted furans [62] were obtained in 44–91% yield.

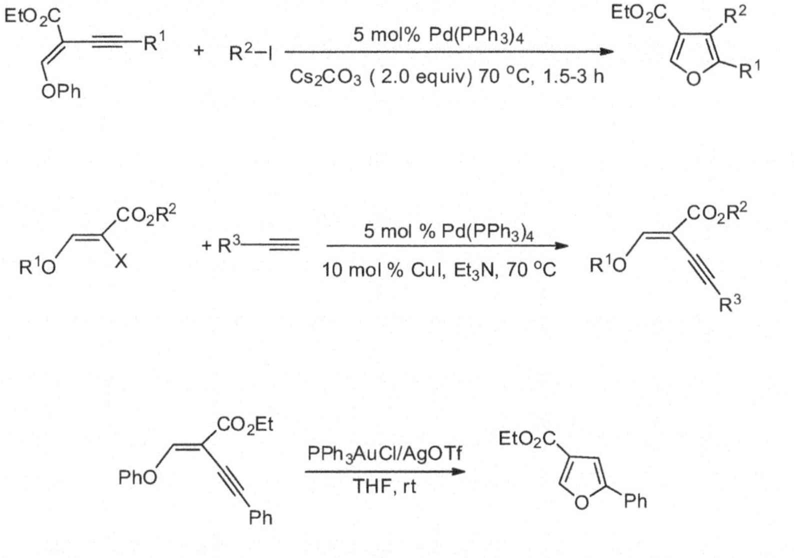

Besides *Z*-enynols 3-alkyne-1,2-diols can be easily transformed into substituted furans in transition metal-catalyzed cyclizations. In 2007, Knight and coworkers developed a heterogeneous $AgNO_3$/-SiO_2-catalyzed synthesis for trisubstituted furans [63]. The reaction takes place in DCM at 20°C in 3 h providing the appropriate furan derivatives in almost quantitative yields.

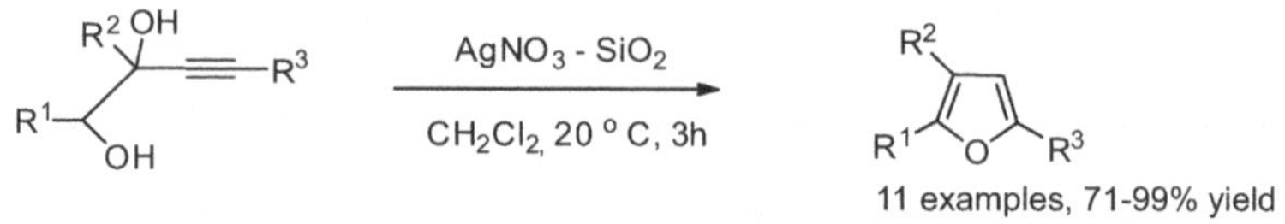

A similar gold-catalyzed cyclization strategy was developed by Aponick and coworkers [64]. For the open flask cyclization, 2 mol% Au[P(t-Bu)$_2$(*o*-biphenyl)]Cl and 2 mol% AgOTf were used as catalyst. The transformation was typically complete within 1 h at 0°C and provided the desired furans in excellent yield (75–99%).

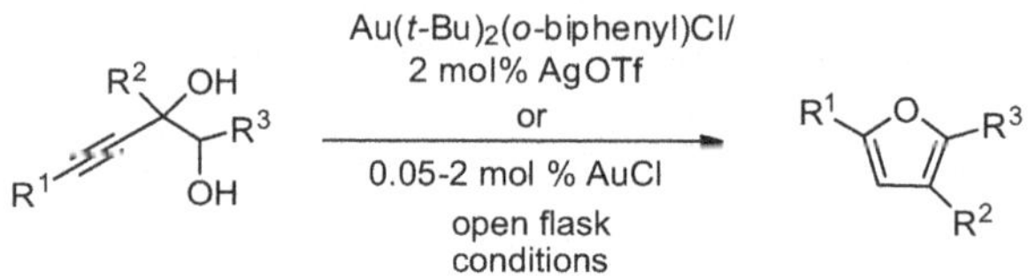

At the same time Akai and coworkers described the efficient cyclization of ethynyl-1,2-diols to furans using AuPPh$_3$Cl-AgNTf$_2$ or AuPPh$_3$Cl-AgOTf catalyst systems [65]. The advantage of the methodology is the utilization of a low amount (0.05–0.5 mol%) of catalyst, ambient temperature, and the high isolated yields.

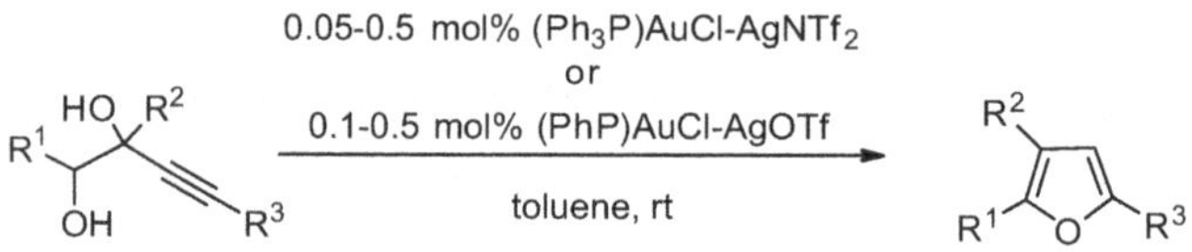

The previous examples showed that the gold-catalyzed cyclizations occur very efficiently under homogeneous catalytic conditions. The Akai research group developed a procedure for the gold-catalyzed cyclization utilizing an immobilized gold catalyst [66]. The gold center was attached to polystyrene support through a linked PPh$_3$ ligand. The polystyrene-supported phosphane ligand binds the gold catalyst quantitatively, and the subsequent anion exchange (chloride to OTf or NTf$_2$) provided efficient catalyst for the cyclization of alkynyl 1,2-diols. With the utilization of the methodology, the desired furans could be obtained with similar efficiency to the homogeneous version. The catalyst was reused several times without the loss of a significant amount of gold. The applicability of the polystyrene-supported gold catalyst in flow chemistry was also demonstrated.

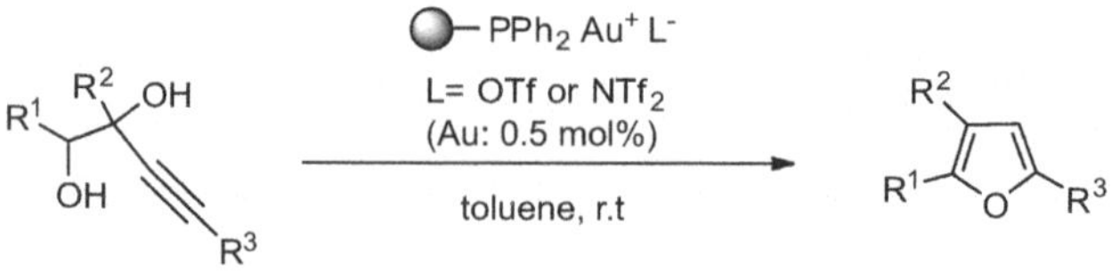

Similarly to alkynyl diols, alkynyl oxyranes can also be used in the cyclization. In the presence of silver or gold catalysts, the furan ring is constructed in an efficient manner as it was demonstrated by Pale [67, 68] and Aurrecoechea [69].

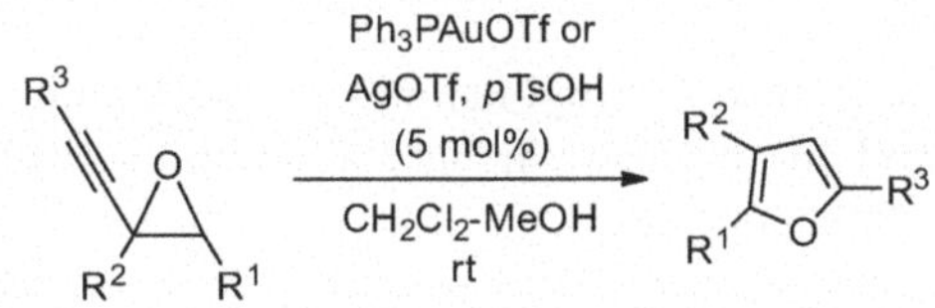

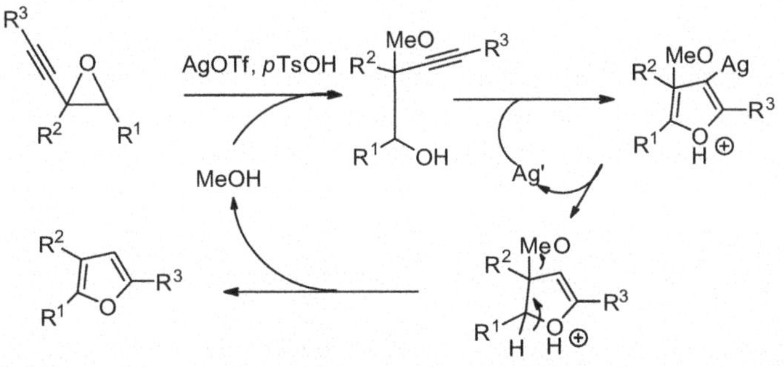

The gold-catalyzed cyclization strategy developed by Dembinski provides easy access to 2,5-substituted 3-fluorofurans. Starting from 1,4-disubstituted alkynons, silyl enol ethers were prepared for subsequent fluorination with Selectfluor. The formed fluorobutynone derivatives straightforwardly underwent gold-catalyzed cyclization in the presence of Ph_3PAuCl providing 3-fluorofurans [70, 71].

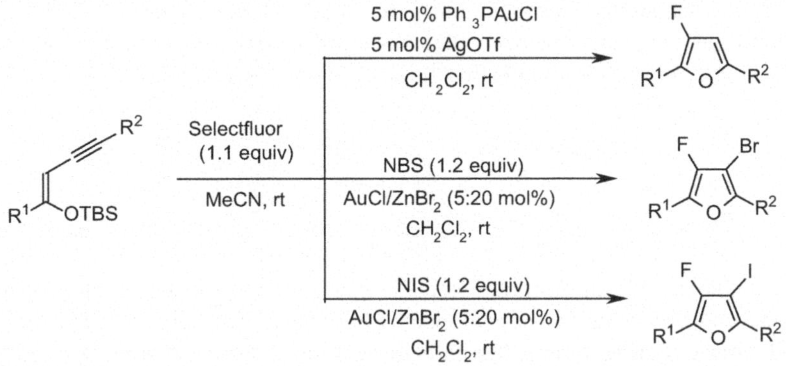

A palladium-catalyzed three component domino cyclization process was developed by Zhang and coworkers for the synthesis of tetrasubstituted furans from alkynylalkenones, vinyl ketones, and alcohols [72]. This transformation provides easy access to fully substituted furan derivatives with versatile substitution patterns.

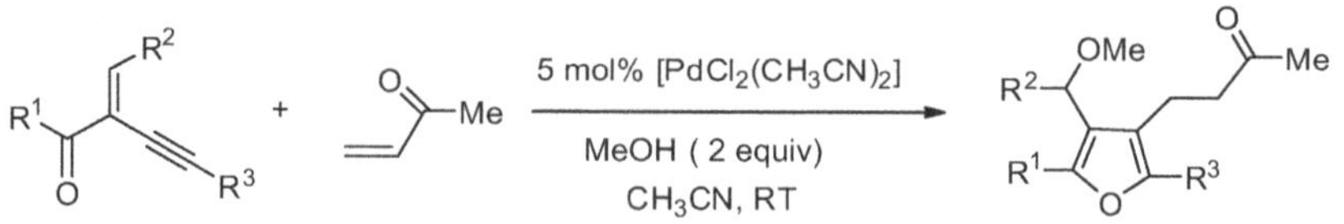

Using allyl chloride instead of vinyl ketones, the palladium-catalyzed transformation provides 3-allyl furans in a Michael addition-cyclization-cross-coupling sequence from alkynylalkenones [73].

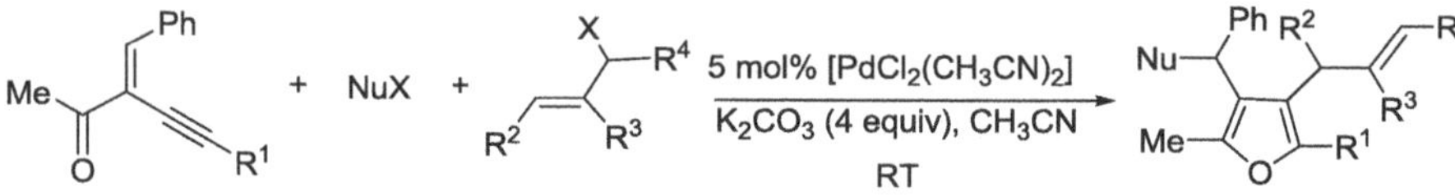

The three component approach enabled the introduction of aryl groups into position 3 when a diaryl iodonium salt was involved in the reaction as electrophile. The extension of the three component domino reaction was developed by Zhang and Li, and its efficiency was demonstrated with numerous examples [74].

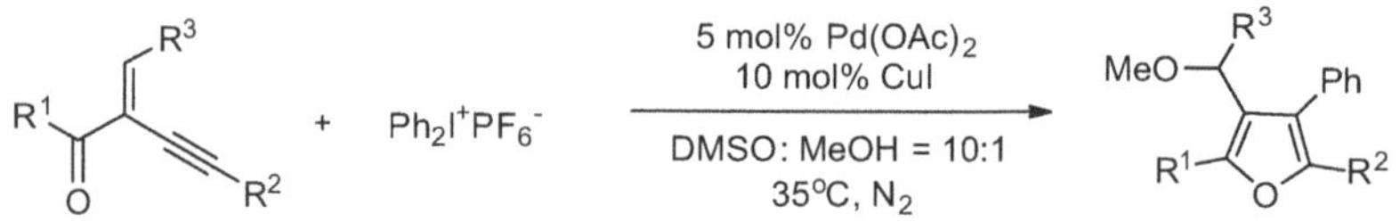

Acetylenes are widely utilized in furan syntheses. The following examples demonstrate their versatile applicability in the synthesis of tri or tetrasubstituted furan derivatives. In these transformations gold, copper, rhodium, silver, indium, and palladium catalyst are used.

The gold-catalyzed three component cascade cyclization of phenylglyoxal derivatives, secondary amines, and terminal alkynes provides 2,5-diaryl 3-aminofurans [75]. The reaction takes place in MeOH at 60°C in the presence of 5 mol% $AuBr_3$ providing the desired furans in good yield (65–93%).

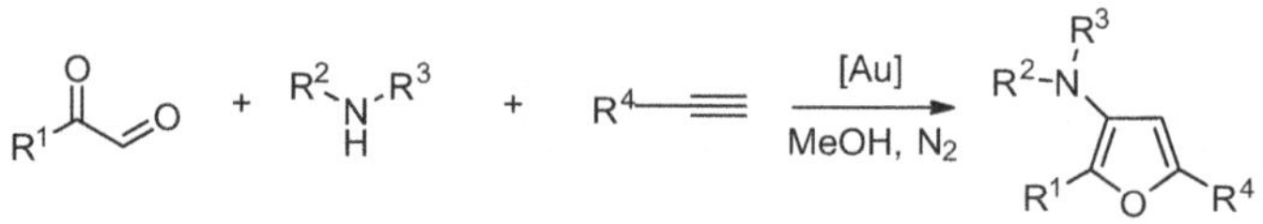

The synthesis of 2-aminofurans in the copper-catalyzed [3 + 2] cycloaddition of diazoesters and enamines was developed by Park and coworkers [76]. The tetrasubstituted furans bearing amino function in position 2 were isolated in good yield (16 examples, 52–76%) using $Cu(hfacac)_2$. The process also offers access to 2-unsubstituted furans (nine examples, 54–71%) through the elimination of amines from the 2,3-dihydrofuran intermediate in the presence of *p*-TsOH.

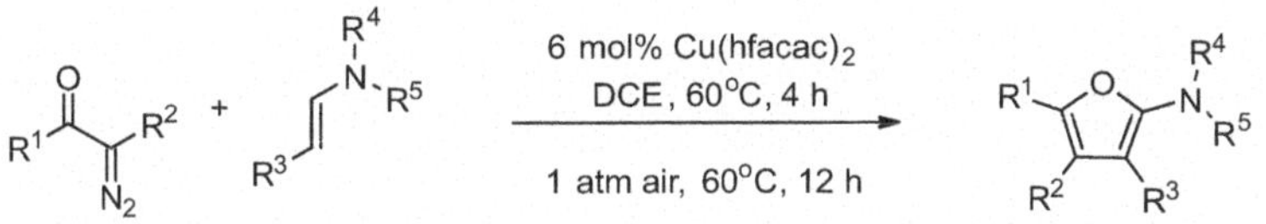

A silver-mediated oxidative C–C coupling of diketones and terminal acetylenes afforded trisubstituted benzofurans with high selectivity [77]. The transient presence of an alkynylalkenone is suggested as the result of C–C bond formation. The furan ring is formed through this intermediate in a silver-assisted cyclization step, and the diversely substituted furan derivatives (24 examples, 43–95%) were obtained in good yield.

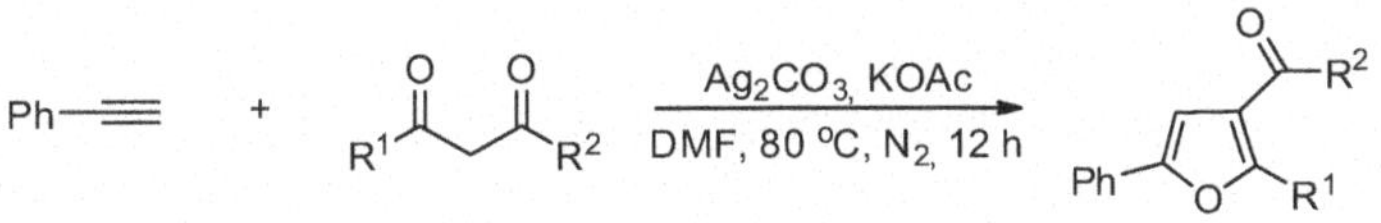

The rhodium-catalyzed hydroacylation-cyclization sequence enables the formation of furan ring from propargyl alcohols and aldehydes [78]. In the first step of the reaction, the hydroacylation occurs providing the hydroxyl enone intermediate, which cyclizes straightforwardly to the substituted furans. The methodology developed by Willis and coworkers enables the synthesis of diverse methylthioalkyl- or aryl-substituted furans in good yield (42–93%).

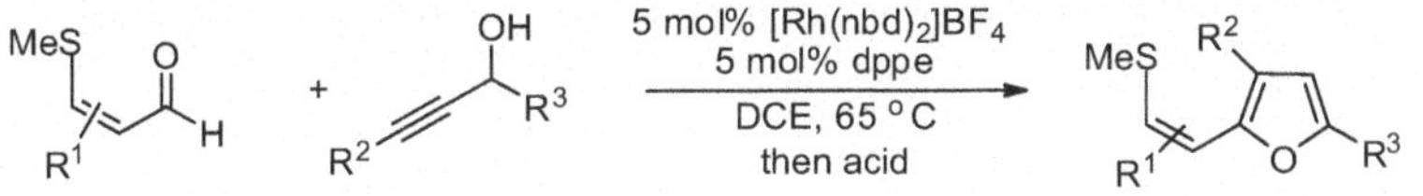

Propargylic alcohols are suitable starting materials for the construction of the furan ring in an $InCl_3$-catalyzed reaction [79]. The heterocycling ring is constructed utilizing 1,3-diketones or acetoacetates as reaction partners. The Lewis acid-catalyzed ring closure takes place in chlorobenzene at 110°C affording tetrasubstituted furans in 44–91% yield.

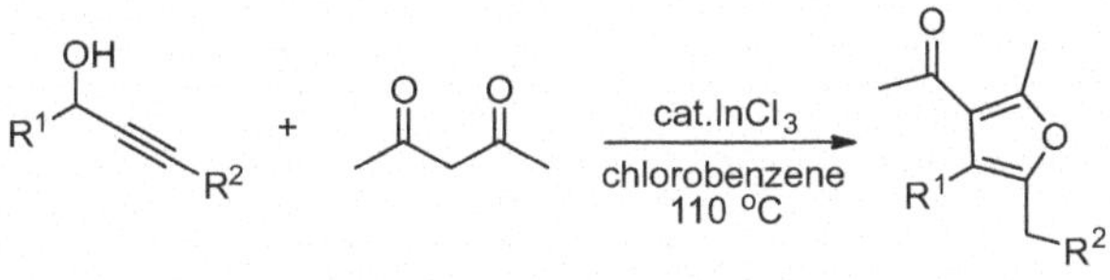

The same propargylation-cycloisomerization strategy was utilized by Zhan and coworkers in the presence of $Cu(OTf)_2$. In their work a high number of tetrasubstituted furans were synthesized from propargylic alcohols or acetates and 1,3-dicarbonyls [80] or silyl enol ethers [81].

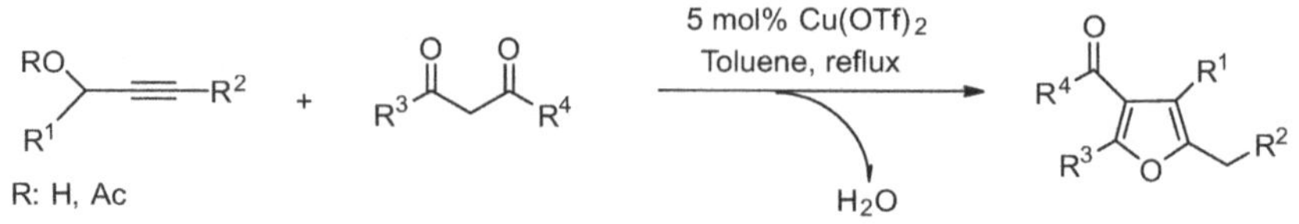

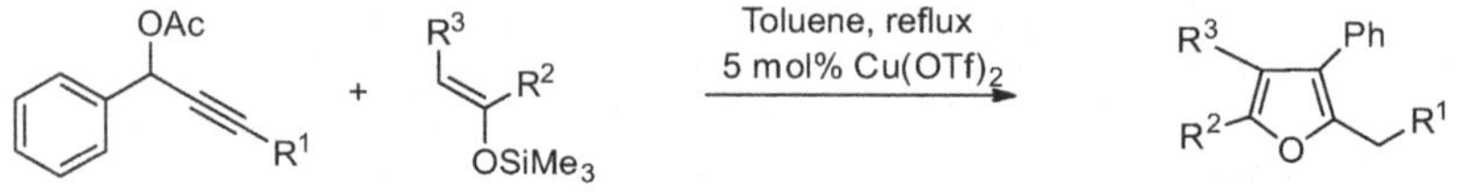

The reaction of THP-protected propargylic alcohols with acid chlorides enables the synthesis of 3-halofurans or 3-chloro-4-iodofurans in a palladium-catalyzed one-pot three component transformation [82]. This efficient reaction provides access to iodo and chloro functionalized furans with high diversity (23 examples, 31–73% yield).

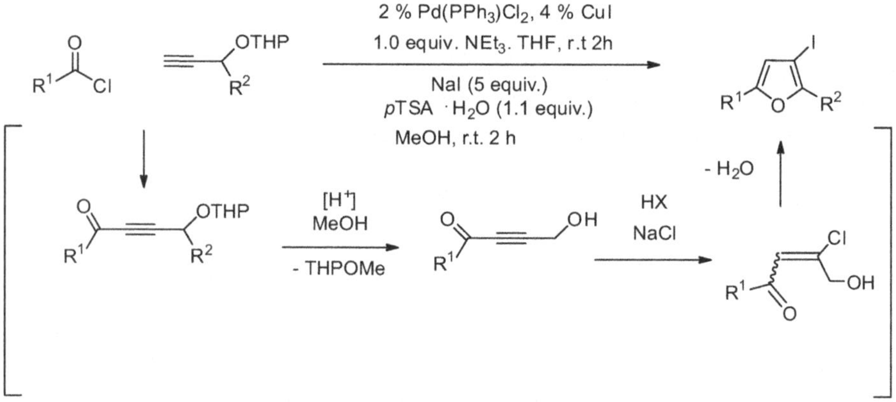

3-Yne-1,2-diol derivatives can be transformed to furan-3-carboxylic acids in alcoholic media via oxidative palladium-catalyzed cyclization under carbon monoxide atmosphere [83]. The catalytic transformation provides the tri- and tetrasubstituted furan carboxylic acid esters in good to excellent yields (21 examples, 56–93%).

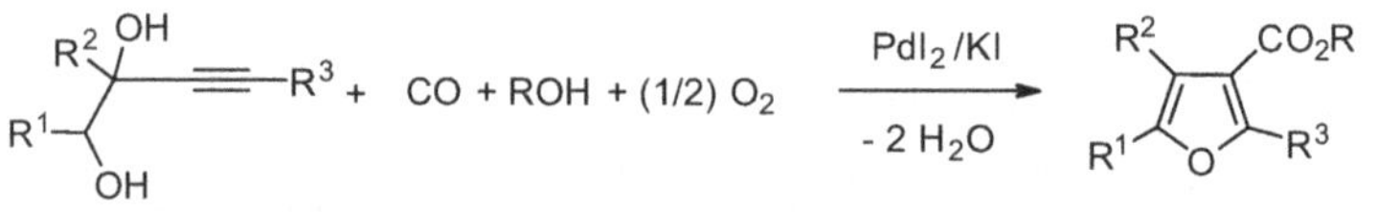

A double carbonylative process was used by Beller and coworkers for the synthesis of 3-ketofurans from aryl iodides and terminal acetylenes [84]. The carbon monoxide is the source of carbon and oxygen atoms both for the furan ring and the carbonyl function introduced into position 3 of the heterocycle. The synthesis was performed in the presence of 2 mol% of $Pd(OAc)_2$ and 4 mol % P(*o*-tolyl)$_3$ ligand under 10 bar CO. The synthetic utility of the methodology was demonstrated on 16 examples and the desired furans were isolated in 53–85% yield.

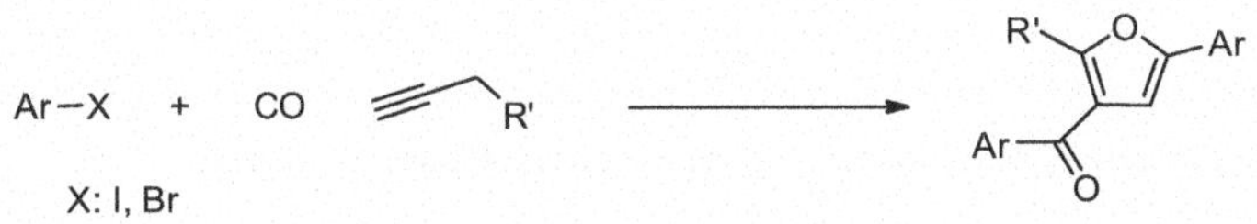

The ring closure of diynes provides 2,5-disubstituted furans where the heteroatom originates from water. Thus, 2,5-diarylfurans can be prepared from 1,4-butadiynes in copper- or gold-catalyzed ring closure. Jiang and coworkers developed a CuI-catalyzed synthesis of furans in a cyclization process [85]. It was also demonstrated that the furans (12 examples) could be obtained from 1-bromoacetylenes with high efficiency (68–93%).

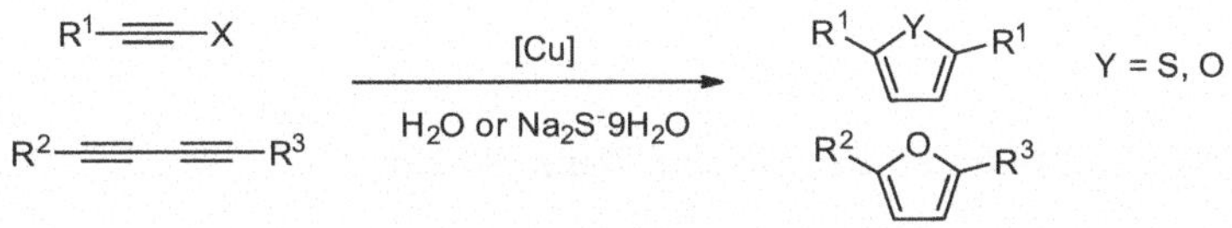

Nolan and coworkers utilized an Au-NHC complex to perform the same cyclization process to obtain 2,5-disubstituted furans (12 examples, 62–84%) [86].

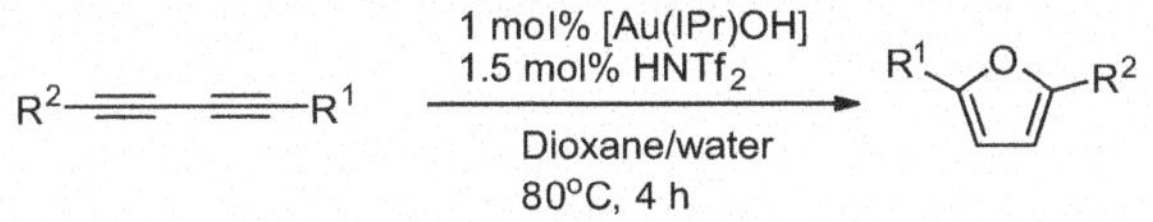

Gold-phosphano complexes also catalyze the cyclization of butadiynes as it was demonstrated by Skrydstrup's group [87]. This procedure not only enables the synthesis of 2,5-diarylfurans (seven examples, 59–84%) but also 2,5-diaminofurans were prepared (four examples, 51–85%) from diamino butadiynes.

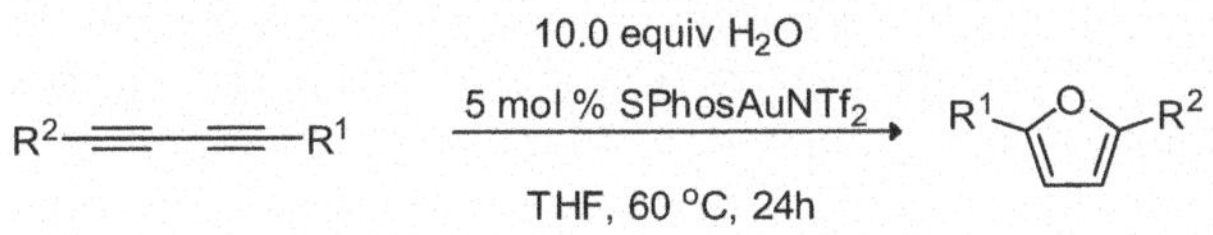

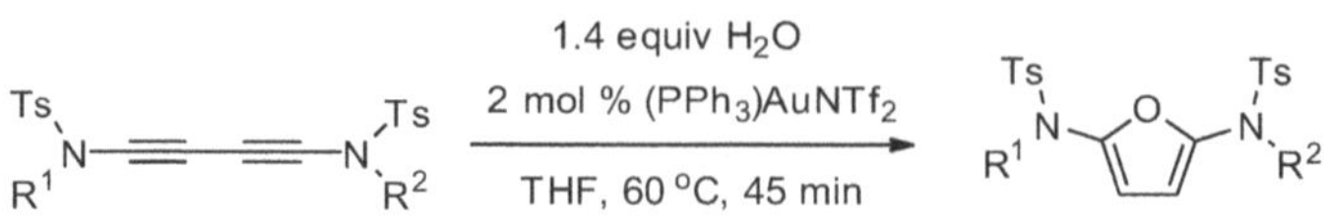

2.3 *Functionalization of Furans and Benzofurans*

In this chapter the latest methods for the functionalization of furans and benzofurans utilizing transition metal-catalyzed transformations are summarized. There are two major synthetic approaches for the construction of a new carbon–carbon bond. First one is the traditional cross-coupling chemistry and the second one is the direct C–H functionalization of the heterocyclic ring. These two types of transformation are discussed separately. Each section is organized by the position of the incoming substituent on the furan and the benzofuran ring.

2.3.1 Cross-Coupling on Furans and Benzofurans

There are several synthetic possibilities for the functionalization of the furan ring via cross-coupling reactions. Traditional couplings such as the Heck, Sonogashira, Suzuki, Hiyama, and Negishi reactions are all well known. Palladium is the most frequently used catalyst for these transformations, which usually gives excellent yields.

Functionalization of Furans in the 2-Position via Cross-Coupling

In the recent years several new palladium-catalyzed cross-coupling reaction conditions were developed which are based on the utilization of efficient new phosphane ligands. 2-Chloro, bromo, and iodo furans are appropriate substrates for cross-coupling (Entries 1–7). Several examples show the utility of furyl organometallic species such as boronic acids (Entries 8–12), tin reagent (Entry 13), silyl (Entries 14, 15), and zinc derivatives (Entry 16) in coupling with aryl halides and tosylates. Aluminum and indium derivatives were also used for the coupling (Entries 17, 18), while desulfinylative and decarboxylative couplings were also described.

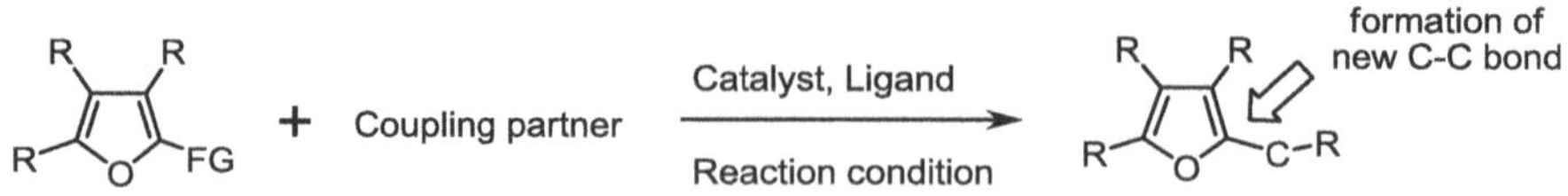

Entry	FG	Coupling partner	Catalyst, ligand	Reaction conditions	Yield	Examples	References
1	Cl	$PhB(OH)_2$	2% $Pd(OAc)_2$, 2.4% XPhos	$BuOH/H_2O$, CsOH, 25°C, 5 min–1 h	91–99%	3	Zhou [88]
2	Br	Ar-CCH	5% $PdCl_2(PPh_3)_2$, 10% CuI	THF, DEA, 20°C, 24 h	75–90%	3	Langer [89]
3	Br	$RCHCH_2$	5% $Pd(OAc)_2$, 10% SPhos or XPhos	DMF, TEA, 120°C, 36 h	78–94%	17	Langer [90]
4	Br	CH_2CH-BF_3K	2% $PdCl_2$, 6% PPh_3	THF/H_2O, Cs_2CO_3, 85°C, 22 h	70%	1	Brown [91]
5	Br	$RSnBu_3$	$PdCl_2dppf$	DMF, 105–110°C, 21 h	20–69%	13	Hocek [92]
6	Br	Het-H	$Pd(OAc)_2$,	DMA, KOAc, 120°C, 16 h	55–91	19	Doucet [93]
7	I	InX_2Li	4%, $Pd(dppf)Cl_2$,	NMP, 40°C, 4 h	70%	1	Knochel [94]
8	$B(OH)_2$	Ar-Cl	2% PdBiphenylamine-XPhos	THF, K_3PO_4, rt, 30 min	82–99%	3	Buchwald [95]
9	$B(OH)_2$	Ar-Cl, Ar-Br	3% $Pd(OAc)_2$, 6% DABCO,	DMF, Cs_2CO_3, 40°C, 19–21 h	90–98%	2	Xie [96]
10	$B(OH)_2$	3-Cl-Py, 2-OTs-Py, 2-Cl-thiophene, 2-OTs-thiophene	2% $Pd(OAc)_2$, 2.4% XPhos	$BuOH/H_2O$, CsOH, 25°C, 5 min–72 h	90–98%	8	Zhou [88]
11	$B(OH)_2$	Ar-Br	$PdCl_2(PPh_3)_2$,	EtOH-DME, Na_2CO_3, 60°C	29–86%	14	Debnath [97]
12	$B(OH)_2$	SO_2Cl	8% $Pd(PPh_3)_4$	THF, K_2CO_3, reflux	40	1	Vogel [98]
13	Sn(*n*-Bu)$_3$	Ar-Br, Ar-Cl or cyclopentenyl-Cl	Pd_2dba_3, azaphosphatrane	CsF, Dioxane, rt-110°C, 48 h	87–97%	5	Verkade [99]
14	$SiMe_2OH$	Ar-I, Ar-Br	Pd_2dba_3, $^tBu_3P$	NaH, Toluene, rt-50°C, 1–24 h	60–82%	10	Regens [100]
15	$SiMe_2OH$	Ar-I	Pd_2dba_3, $(2furyl)_3As$ or $Pd_2Cl(^tBu_3)_2$	NaH, Toluene, rt-50°C, 1–24 h	61–82%	8	Baird [101]
16	ZnBr	3-bromo-pyridine	$NiCl_2$, $(EtO)_2P(O)H$	110°C, 22 h	59	1	Knochel [102]

(continued)

Entry	FG	Coupling partner	Catalyst, ligand	Reaction conditions	Yield	Examples	References
17	$Al_{2/3}Cl$	Ar-Br, Ar-I	$Pd(tmpp)_2Cl_2$	THF:DMF 1:2, 50°C, 5–6 h	65–72%	2	Knochel [103]
18	InX_2LiX	ArI	4%, $Pd(dppf)Cl_2$,	NMP, 40°C, 4 h	73–78%	2	Knochel [94]
19	SO_2Li	Ar-Br	$PdCl_2$, PPh_3	DMF/water 170°C, 8 min, MW	65%	1	Forgione [104]
20	COOK	Ar-Br	CuI, $Pd(acac)_2$ 1,10-phenanthroline	NMP, quinoline, 190°C, 50 W, 5 min, MW	38%	1	Hartung [105]
21	COOK	Naphthyl-OTs	2.5–7.5% Cu_2O/phen, 5% $Pd(acac)_2$, 7.5% XPhos	NMP,180°C, 100 W, 2 min, MW	78	1	Goossen [106]

Functionalization of Furans in the 3-Position

Compared to the number of transformations in position 2 of the furan ring, there are fewer examples for the cross-coupling reactions in which the functional group is introduced into position 3. 3-Bromofurans can couple with alkenes in Heck reaction (Entry 1) and with arylboronic acid (Entry 2) in Suzuki coupling. However, 3-furylboronic acid derivatives also undergo palladium-catalyzed Suzuki coupling with aryl or hetaryl halides and tosylates (Entries 3–5).

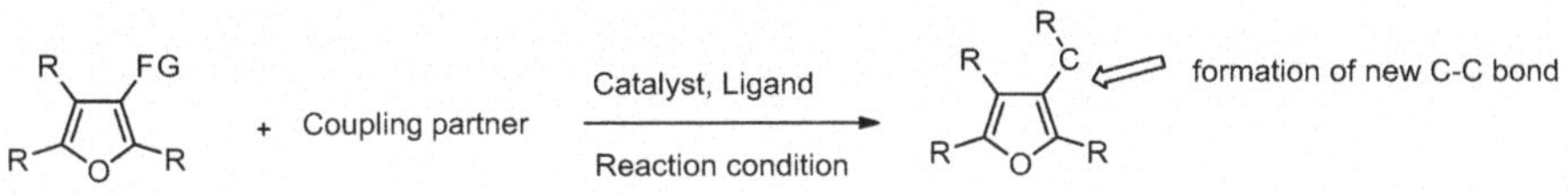

Entry	FG	Coupling partner	Catalyst	Conditions	Yield	Examples	References
1	Br	$RCHCH_2$	5% $Pd(OAc)_2$, 10% SPhos or Xphos	DMF, TEA, 120°C, 36 h	78–94%	17	Langer [90]
2	Br	$Ar\text{-}B(OH)_2$	2% PdBiphenylamine-XPhos	THF, K_3PO_4, rt, 30 min	45%	1	Buchwald [95]
3	$B(OH)_2$	Ar-Cl	2% PdBiphenylamine-XPhos	THF, K_3PO_4, rt, 30 min	80%	1	Buchwald [95]
4	$B(OH)_2$	3-Cl-Py, 2-OTs-Py, 2-Cl-thiophene, 2-OTs-thiophene	2% $Pd(OAc)_2$, 2.4% XPhos	BuOH/ H_2O, CsOH, 25°C, 5 min–1 h	90–93%	4	Zhou [88]
5	$B(OH)_2$	2-Cl, 6-methoxy-pyridine	2% Pd-NHC complex	tBuOH, 30°C, 24 h	70%	1	Lough [107]

Functionalization of Benzofurans in the 2-Position

2-Bromo- and iodobenzofurans are smoothly coupled with terminal acetylenes in the presence of palladium catalysts generally with high isolated yields. 2-Benzofurylboronic acids are also applicable reagents for the Suzuki coupling as several examples demonstrate their utility in this type of cross-coupling reaction (Entries 4–8). Silanes and sulfinyl groups on the heterocyclic ring can also be transformed to aryl function in their palladium-catalyzed reaction with aryl halides (Entries 9–11).

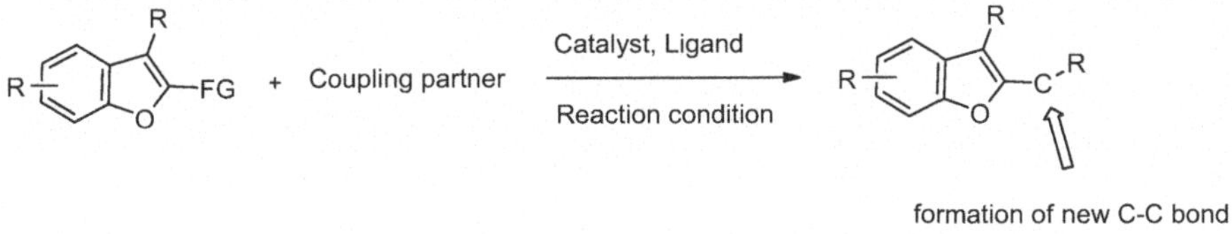

Entry	FG	Coupling partner	Catalyst	Conditions	Yield	Examples	References
1	Br	Ph-CCH	2 mol% $PdCl_2(PPh_3)_2$, 7 mol % CuI	DMF, TEA, rt4-6.5 h	50–98%	3	Mao [108]
2	Br	Ar-CCH	5 mol% $PdCl_2(PPh_3)_2$ 10 mol% CuI	THF, DEA, 20°C, 24 h	75–90%	3	Langer [89]
3	I	Ar-CCH	2 mol% $PdCl_2(PPh_3)_2$, 4 mol % CuI	THF, TEA, rt, 5 h	94%	1	Barret [109]
4	$B(OH)_2$	Ar-Cl	2 mol% PdBiphenylamine-XPhos	THF, K_3PO_4, rt, 30 min	97–99%	3	Buchwald [95]
5	$B(OH)_2$	3-Cl-Py, 2-OTs-Py, 2-Cl-thiophene, 2-OTs-thiophene	2 mol% $Pd(OAc)_2$, 2.4 mol% XPhos	BuOH/H_2O, CsOH, 25°C, 5 min–4 h	90–92%	4	Zhou [88]
6	$B(OH)_2$	Ar-Br	various Pd catalysts	DME, Na_2CO_3, 80°C, 8 h	21–90%	5	Antane [110]
7	$B(OH)_2$	ArCHCH-Br	5 mol% $Pd(PPh_3)_4$	DME/water, Na_2CO_3, 90°C, 20 h	57%	1	Barret [109]
8	$B(OH)_2$	3-Br thiophene	2 mol% Pd-NHC complex	^{i}PrOH, 30°C, 24 h	85%	1	Lough [107]
9	$SiMe_2ONa$	Ar-Br	2.5.5 mol% $^tBu_2Pd$	Toluene, THF or dioxane 60–70°C, 3.5–8 h	58–99%	7	Denmark [111]
10	TMS	Ar-I	$PdCl_2(PPh_3)_2$, $AgNO_3$, KF	DMSO, 100°C	43–86%	7	Mori [112]
11	SO_2Li	Ar-Br	$PdCl_2$, PPh_3	DMF/Water, 170°C, 8 min, MW	82%	1	Forgione [104]

Functionalization of Benzofurans in the 3-Position

There are two procedures reported for the cross-coupling of 3-iodobenzofuran. Terminal alkynes react straightforwardly with the iodo species under standard Sonogashira conditions (Entry 1). The introduction of an alkenyl group into this position was achieved with tin reagents, and the desired 3-alkenyl-benzofurans were isolated with good yields (Entry 2).

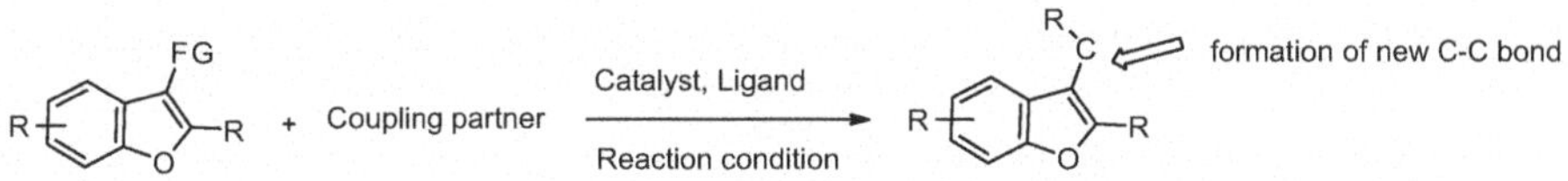

Entry	FG	Coupling partner	Catalyst	Conditions	Yield	Examples	References
1	I	R-CCH	1% $PdCl_2(PPh_3)_2$, 2% CuI	TEA, rt–60°C, 2–24 h	40–99%	22	Zeni [113]
2	I, OTf	Bu_3Sn-alkene	$Pd(PPh_3)_4$, CuI	DMF, CsF, 40°C	56–80%	6	Wada [114]

Functionalization of Benzofurans in the 5- and 7-Position

The functionalization of the benzofuran ring in positions 5 or 7 has also been described mostly in palladium-catalyzed Suzuki coupling. 5-Bromo, 7-chloro, and 7-bromo derivatives were all coupled with arylboronic acid (Entries 1–4). The opposite approach with benzofuran-7-ylboronic acid has also provided the arylated product in reaction with an aryl bromide (Entry 5).

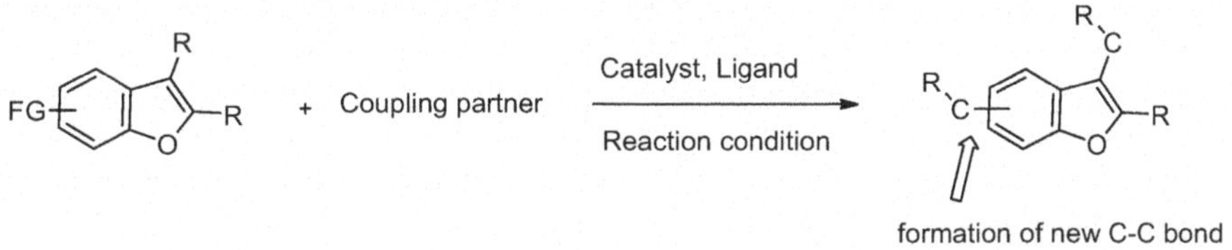

Entry	FG	Coupling partner	Catalyst	Conditions	Yield	Examples	References
1	5-Br	Ar-B $(OH)_2$	Benzothiazole based Pd complex, $PdCl_2PPh_3$	H_2O, TBAB, KOH,160°C, 250 W, 10–20 min	6.5–96 %	6	Dawood [115]
2	7-Cl	ArB $(OH)_2$	Pd_2dba_3, SPhos	1,4-dioxane, K_3PO_4, 100°C, 2–4 h	95–99%	15	Goggiamani [116]
3	7-Cl	NHR_2	Pd_2dba_3, XPhos	toluene, NaO^tBu, 80°C, 12–40 h	70–99%	4	Goggiamani [116]
4	7-Br	ArB $(OH)_2$	$Pd(PPh)_4$	DME (aq), Na_2CO_3,	81%	1	Goggiamani [116]
5	7-B $(OH)_2$	Ar-Br	$Pd(PPh)_4$		67%	1	Goggiamani [116]

2.3.2 Direct Functionalization of Furans and Benzofurans via C–H Activation

Beyond the traditional cross-coupling reactions, the direct functionalization of the furan C–H bonds is also possible. The most frequently used catalysts in this transformation are palladium based. However, there are some examples for the efficient application of rhodium, cobalt, or iridium catalysts too.

Functionalization of Furans via C–H Activation in Position 2

The direct arylation of the furan ring leads to the functionalization of the 2-position. These reactions can be performed with aryl chlorides (Entries 1–2), bromides (Entries 3–10), and iodides (Entries 11–15). The oxidative coupling with terminal olefins provides 2-alkenylated furans (Entries 16–21). The synthesis of 2-aryl or 2-hetarylfurans is also possible in palladium-catalyzed cross-dehydrogenative couplings with the utilization of aromatic and heteroaromatic systems (Entries 22–25).

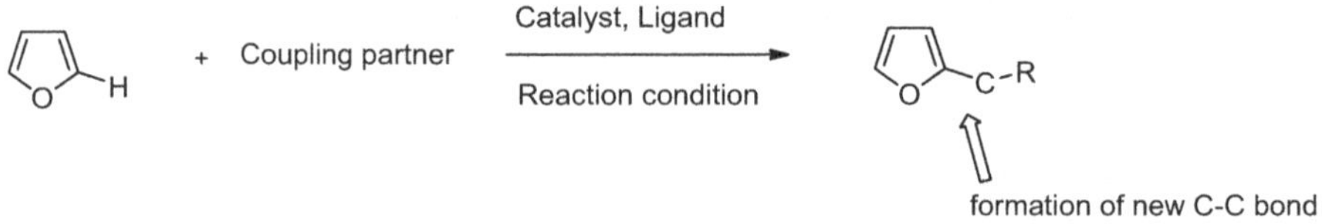

Entry	Coupling partner	Catalyst	Conditions	Yield	Examples	References
1	ArCl	$PdClNHCPCy_3$,	PivOH, DMA, K_2CO_3, 110°C, 12–15 h	41–89%	30	Lee [117]
2	ArCl	$Pd(OAc)_2$, Cy_2P-*o*-biphenyl	K_3PO_4, NMP, 100°C, 24 h	50–92%	11	Daugulis [118]
3	ArBr	$Pd(OAc)_2$, PCy_3	PivOH, K_2CO_3, DMF, 180°C, 10 min	85%	1	Kappe [119]
4	ArBr	$Pd(OAc)_2$,	DMA, KOAc, 130°C, 17 h	42–76%	13	Doucet [120]
5	ArBr	$Pd(OAc)_2$, ferrocene-based phosphane ligand	DMA, KOAc, 150°C, 16 h	38–93%	28	Doucet [121]
6	ArBr	$Pd(OAc)_2$,	DMA, KOAc, 150°C, 16 h	30–87%	24	Doucet [122]
7	ArBr	$Pd(OAc)_2$, PCy_3HBF_4,	PivOH, K_2CO_3, DMA, 100°C, 16 h	44–70%	3	Fagnou [123]
8	ArBr	$(PdCl\text{-allyl})_2$, Tedicyp	NaOAc, DMA, 150°C, 20 h	70–88%	9	Santell [124]
9	ArBr	$PdCl(C_3H_5)dppb$	KOAc, DMA, 150°C, 16 h	63–90%	16	Doucet [125]
10	ArBr	$Pd(OAc)_2$	KOAc, DMA, 130°C, 17 h	49–91%	35	Dixneuf [126]
11	ArI	$Pd(OAc)_2$	105°C, 24 h, K_2CO_3, DMF	62–71%	2	Catellani [127]
12	ArI	RhCl(CO){P[OCH(CF3)2]3}2	150–120°C, MW, m-xylene/DME, Ag_2CO_3,	64–66%	2	Itami [128]
13	ArI	$Pd(OAc)_2$, electron deficient phosphane ligand	PivOH, K_2CO_3, Ag_2CO_3,DMA, 100°C, 16 h	43–46%	2	Fagnou [129]
14	ArI	Co-porphyrin	furan, tBuOH, KOH, 200°C, 30–60 min	52–62%	2	Chan [130]
15	ArI	$[Ir(COD)(Py)PCy_3]PF_6$	Ag_2CO_3, m-xylene, 160°C, 18 h	54–72%	3	Itami [131]
16	*N*-butyl acrylate	$Pd(OAc)_2$,	AgOAc, pyridine, 120°C, 12 h	64–90%	4	Zhang [132]
17	*N*-butyl acrylate	$Pd(OAc)_2$,	$Cu(OAc)_2$, LiOAc, DMF, 120°C, 12 h	53–67%	2	Miura [133]
18	Styrenes	$Pd(OAc)_2$,	BQ, $Cu(OAc)_2$, $EtCO_2H$, Et_2O or THF 40–60°C	50–78%	23	Muzart [134]

(continued)

Entry	Coupling partner	Catalyst	Conditions	Yield	Examples	References
19	Styrenes	$Pd(OAc)_2$,	BQ, AcOH/DMSO, rt, 24–48 h	40–79%	6	Le Bras [135]
20	Allylacetate	$Pd(OAc)_2$,	Ag_2CO_3, DMSO/Dioxane, 110°C, 15 h	60%	1	Liu [136]
21	*N*-Boc-allylamines	$Pd(OAc)_2$,	Ag_2CO_3, $Cu(OAc)_2$, DMF/DCE, 120°C, 14 h	65–83%	7	Xiao [137]
22	Ar-H	$Pd(OAc)_2$, monophos	*N*-fluoropyridinium triflate, AcOH, 100°C, 2 h	61–78%	10	Seayad [138]
23	Xanthines-H, benzimidazoles, pyridine and quinoline *N*-oxides	$Pd(OAc)_2$	$Cu(OAc)_2$, pyridine, dioxane, 120°C, 20 h	60–95%	6	You [139]
24	F_5Ph-H	$Pd(OAc)_2$	Ag_2CO_3, HOAc, DMF/DMSO, 120°C, 3 h	54–73%	2	Zhang [140]
25	HetAr-H	$(RhCp^*Cl_2)_2$, $AgSBF_6$	CsOPiv, Cu(II) (2-ethylhexanoate), t-AmylOH, 120°C, 20 h	48–84%	12	Glorious [141]

Functionalization of Furans via C–H Activation in Position 3

The direct arylation of the furan ring in position 3 occurs when position 2 and 5 is occupied by another functional group. Both arylation and alkenylation of the furan ring can be achieved in palladium-catalyzed oxidative coupling using aryl bromides and iodides (Entries 1, 2) or terminal alkenes (Entries 3–6).

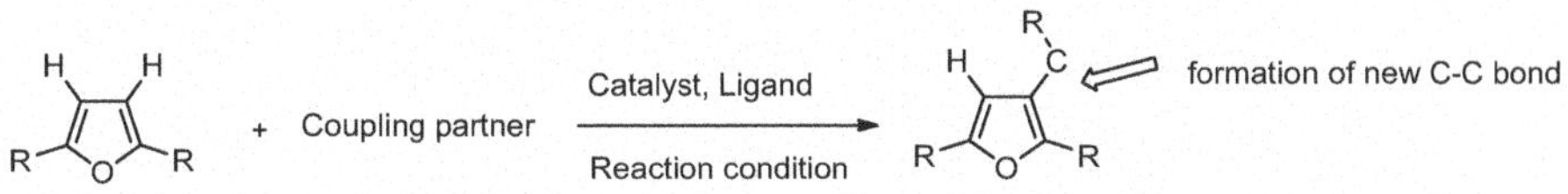

Entry	Coupling partner	Catalyst	Conditions	Yield	Examples	References
1	Ar-Br	$(PdCl\text{-}Allyl)_2$,	DMA, KOAc, 120°C, 12 h	36–76%	18	Doucet [142]
2	ArI, ArBr	$Pd(OAc)_2$, PPh_3,	DMA, Cs_2CO_3, 120°C	62–90%!	29	Yang [143]
3	Styrenes	$Pd(OAc)_2$,	BQ, $Cu(OAc)_2$, $EtCO_2H$, Et_2O or THF 40–60°C	52–66%	6	Le Bras [134]
4	Styrenes	$Pd(OAc)_2$,	BQ, AcOH/DMSO, rt, 24–48h	40–54%	2	Le Bras [135]
5	Acrylates, styrene	$Pd(OAc)_2$	$Cu(OAc)_2$, LiCl, DMF	44–91%	12	Zhu [144]
6	*N*-Boc-allylamines	$Pd(OAc)_2$,	Ag_2CO_3, $Cu(OAc)_2$, DMF/DCE, 120°C, 14 h	80–83%	2	Xiao [137]
7	Butyl acrylate	$[Cp^*RhCl_2]_2$	$AgSbF_6$, Ag_2CO_3, diglyme, 120°C	63	1	Miura [145]

Functionalization of Benzofurans via C–H Activation in Position 2

Similarly to the furan ring, direct functionalization of benzofurans is preferred in position 2. The C–H bond can be transformed in palladium-catalyzed reaction with the utilization of aryl halides (Entries 1–5), boronic acids (Entry 6), or diazonium salts (Entry 7). Alkenylation and oxidative arylation via double C–H activation are also possible on this heterocyclic ring (Entries 8–11).

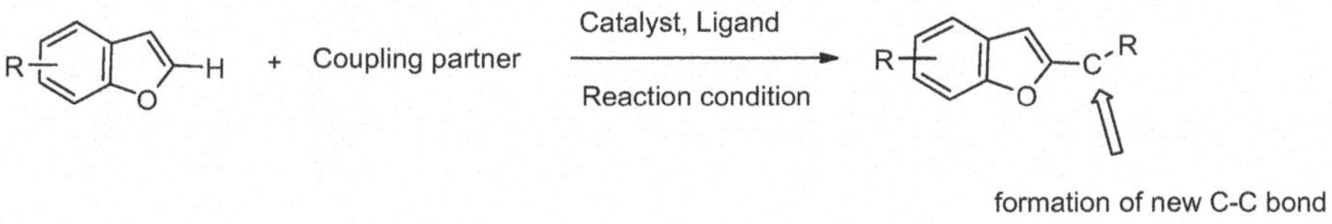

Entry	Coupling partner	Catalyst	Conditions	Yield	Examples	References
1	ArCl	$Pd(OAc)_2$, Sylphos	TBAB, KOAc, DMA, 150°C, 20h	38%	1	Doucet [146]
2	ArBr	$Pd(OAc)_2$, PCy_3	PivOH, K_2CO_3, DMF, 180°C, 60 min	50%	1	Kappe [119]
3	ArBr	$Pd(OAc)_2$, PCy_3HBF_4,	PivOH, K_2CO_3, DMA, 100°C, 16 h	29%	16	Fagnou [123]
4	ArBr	$Pd(P^tBu_3)_2$,	LiOtBu, DMF, 100°C	34–47%	2	Mori [147]
5	PhI	$PdBr_2BiPy$	Ag_2CO_3, Dioxane, 120°C, 13 h	48%	1	Itami [148]
6	$ArB(OH)_2$	$Pd(OAc)_2$, bis-oxazoline ligand,	TEMPO, TFA, DMF, 80°C, 12 h	67–87%	2	Itami [149]
7	ArN_2BF_4	Pd(OAc)2	MeOH, rt-65°C	33–61%	5	Correira [150]
8	Allylacetate	$Pd(OAc)_2$,	Ag_2CO_3, DMSO/ Dioxane, 110°C, 25 h	55%	1	Liu [136]
9	F_5Ph-H	$Pd(OAc)_2$	Ag_2O, O_2, PivOH, DMF/DMSO, 120°C	87%	1	Zhang [151]
10	F_5Ph-H	$Pd(OAc)_2$	Ag_2CO_3, HOAc, DMF/DMSO, 120°C, 3 h	78%	1	Zhang [140]
11	Xanthines-H	$Pd(OAc)_2$	$Cu(OAc)_2$, pyridine, dioxane, 120°C, 20h	66%	1	You [139]

Functionalization of Benzofurans via C–H Activation in Position 3

The preferred site of C–H functionalization on benzofurans is position 2; therefore, introduction of aryl or alkenyl groups in position 3 is achievable only when position 2 is occupied by another substituent. In these substrates the general C–H activation procedures enable the introduction of substituents into position 3. For this transformation aryl bromides (Entries 1, 2) and acrylates (Entries 3–5) were used.

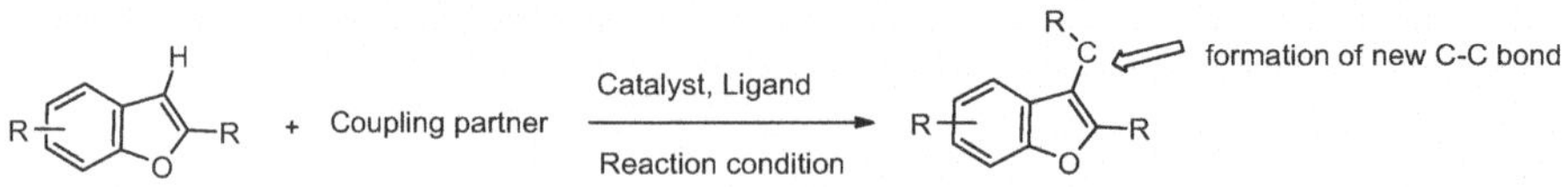

Entry	Coupling partner	Catalyst	Conditions	Yield	Examples	References
1	ArBr	$Pd(OAc)_2$, P(^tBu)$_2$MeHBF$_4$,	PivOH, Cs_2CO_3, mesitylene, 140°C, 24 h	40–61%	2	Fagnou [152]
2	ArBr	$Pd(OAc)_2$, P(^tBu)$_2$MeHBF$_4$	K_2CO_3, PicOH, mesitylene or DMA, 150°C, 16–63 h	18–98%	25	Bertounesque [153]
3	Allylacetate	$Pd(OAc)_2$,	Ag_2CO_3, DMSO-dioxane, 110°C, 25 h	55%	1	Liu [136]
4	Butyl acrylate	$[Cp^*RhCl_2]_2$	$AgSbF_6$, Ag_2CO_3, diglyme, 120°C	78	1	Miura [145]
5	Butyl acrylate	[Ru(p-cymene)Cl$_2$]$_2$	$Cu(OAc)_2$, LiOAc, DMF	66%	1	Miura [154]

3 Oxygen Heterocycles with Additional Heteroatom

Oxazole, isoxazole, [1,3,4]oxadiazole, and benzoxazole are ring systems whose derivatives are abundant both in nature and in the pool of synthetic compounds. The presence of the five-membered heteroaromatic ring in these compounds is the source of a collection of diverse transformations that result in the broad variability of their substitution pattern. In parallel the hydrogen bond acceptor nature of the heteroatoms in the ring in combination with the presence of several synthetically tractable vectors make them optimal building blocks in medicinal chemistry. Therefore, it is not surprising that the synthesis and functionalization of these molecular scaffolds has been widely studied and they are also a prime target for the development of novel synthetic methodologies. The following chapter doesn't attempt to cover this broad area in an exhaustive manner. It would rather like to give the reader a taste of the diversity of chemical approaches that were developed for and used on these ring systems in the last decade.

3.1 Synthesis of Oxygen Heterocycles with Additional Heteroatom

The synthesis of oxazoles, isoxazoles, oxadiazoles, and their benzologues can be achieved on several ways. This chapter is limited to those transformations, which result in the formation of the five-membered heterocyclic ring. Due to the broad range of reactions that can be applied, the selected examples are classified on the basis of the number of bonds formed.

3.1.1 Direct Ring Closure

The common feature of the transformations in this chapter is the fact that the ring formation is achieved by the formation of a single bond from the appropriate linear precursor.

The transition metal-catalyzed ring closure of linear compounds containing at least two unsaturated bonds has been well established. An extension of this protocol to the synthesis of oxazoles is the gold-catalyzed cycloisomerization of propargylic amides. The process reported by Hashmi [155] yields various 2-substituted oxazoles in the presence of gold(III) chloride. The authors were also able to identify some of the intermediates in the process supporting the gold(III)-promoted electrophilic activation of the triple bond as the opening step of the sequence.

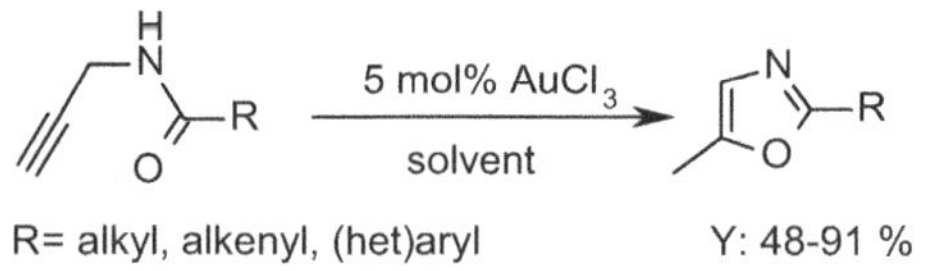

A similar process has been reported concomitantly by Nishibayashi and Uemura [156]. The formation of the oxazole ring follows the same protocol as above but this step is preceded by the ruthenium catalyzed synthesis of the propargyl amide from the appropriate amide and propargylic alcohol derivative. The one-pot protocol gave oxazoles bearing different substituents in the 2- and 4-position.

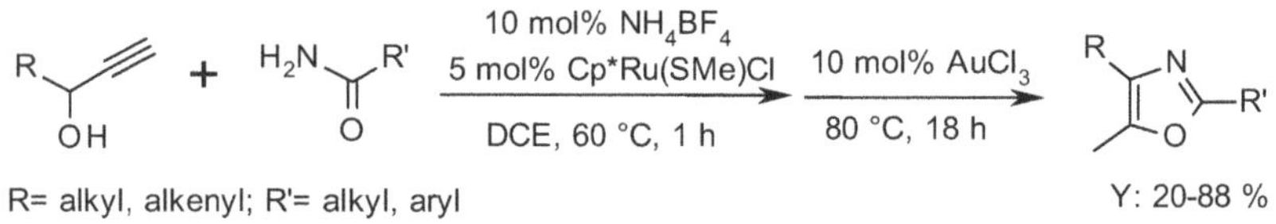

A recent three component transformation combining benzyl imines, acid chlorides, and acetylenes, reported by Strand [157], does also allow for the variation of the substituents in the 2-, 4-, and 5-position of the oxazole core. Although the process is modular, its efficiency varies with the selection of reagents, being more sensitive to the nature of the acetylene than of the acid chloride. The authors managed to bridge the gold catalyzed addition of the acetylides and imines, followed by the gold catalyzed cycloisomerization of the intermediate propargyl amides. The reaction was run at elevated temperature in a microwave reactor.

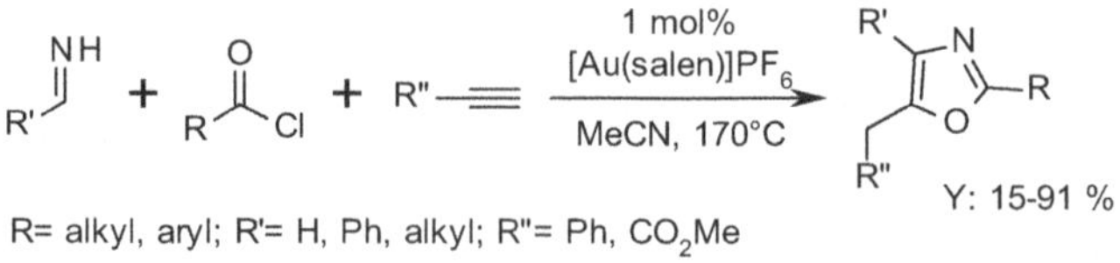

An extension of the propargyl amide cycloisomerization to enamides has been reported by Stahl recently [158]. The transformation requires the use of stoichiometric amounts of copper(II) chloride that acts both as the mediator of the ring closure and oxidizes the formed intermediate to oxazole. The addition of *N*-methylimidazole that might serve as a ligand to copper was found to be beneficial for the process. The incorporation of aromatic groups into the oxazole ring is usually more efficient than that of the aliphatic analogues. The reaction works equally well with *E*- and *Z*-enamides and is believed to follow a single electron transfer mechanism.

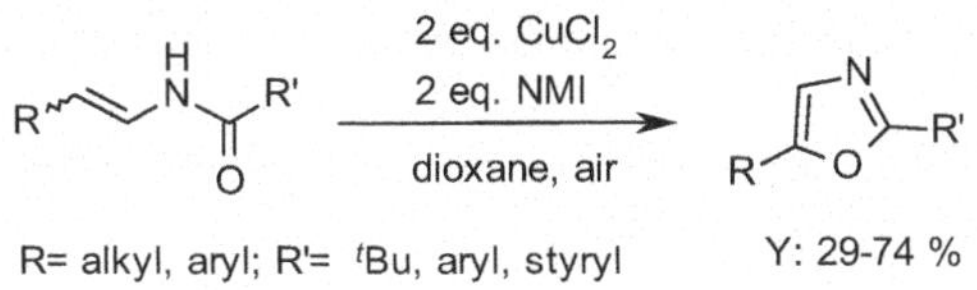

The transition metal-catalyzed synthesis of isoxazoles has also been explored. Miyata reported [159] a practical synthesis of 3,5-disubstituted isoxazoles via a silver catalyzed cyclization and subsequent protonation of alkynyl oxime ethers. The method tolerated the variation of the 3- and 5-substituents in a fairly broad range. The benzyl moiety attached to the oxime in the starting material is lost during the process.

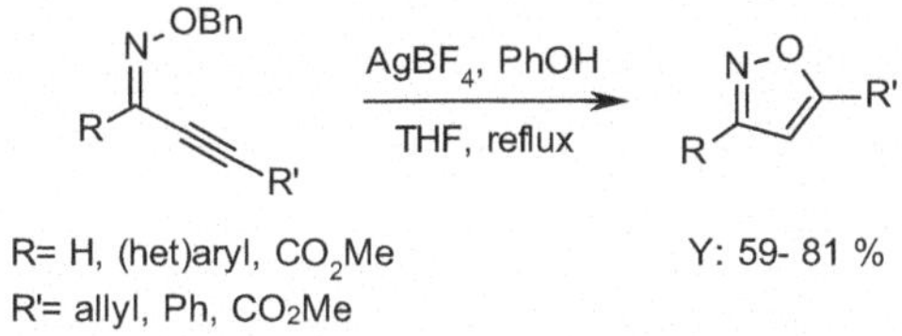

The same research group has reported the gold(III)-catalyzed rearrangement of the allyl ether analogue of the alkynyl oximes [160]. In this case the allyl group is not lost in the process, but the intermediate *O*-allylisoxazolium derivative undergoes a Claisen-type rearrangement to deliver the allyl group into the 4-position of the product.

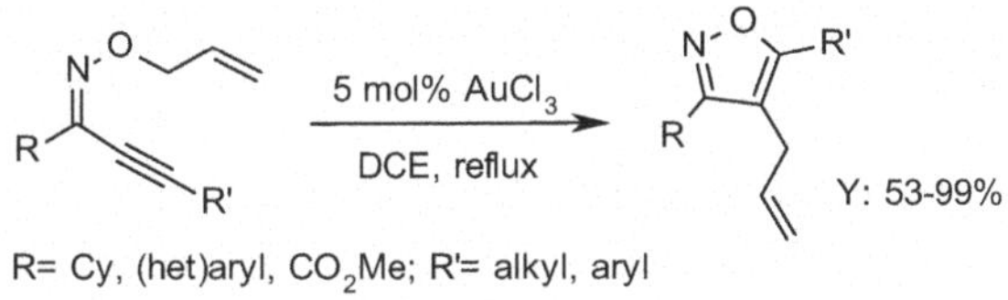

It has long been recognized that the transition metal-catalyzed carbon—heteroatom bond-forming reactions provide an easy access to heterocyclic compounds through direct ring closure. The copper-catalyzed synthesis of a collection of benzoxazole derivatives from *N*-acylated *ortho*-haloanilines has been reported by Batey [161]. The substitution of the aniline as well as the variation of the acyl moiety has been well tolerated except for some amino acid-derived amides.

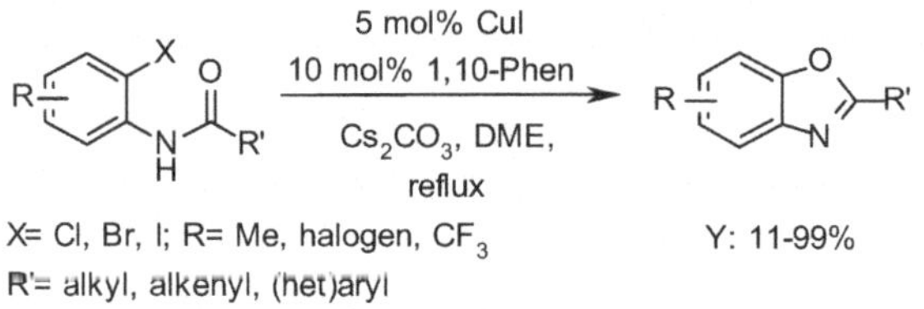

The copper-catalyzed cyclization of amides into benzoxazoles can also be achieved through C–H activation as reported by Nagasawa [162]. Depending on the substitution pattern of the aniline derivative, the authors usually observed a regioselective ring formation. Although steric factors favor the formation of 5- and 6-substituted benzoxazoles, for compounds that bear a functional group capable of coordinating the copper(II) ion, the selective formation of the 7-substituted benzoxazole was observed. Through the variation of the acyl group, a wide range of substituents were introduced into the 2-position.

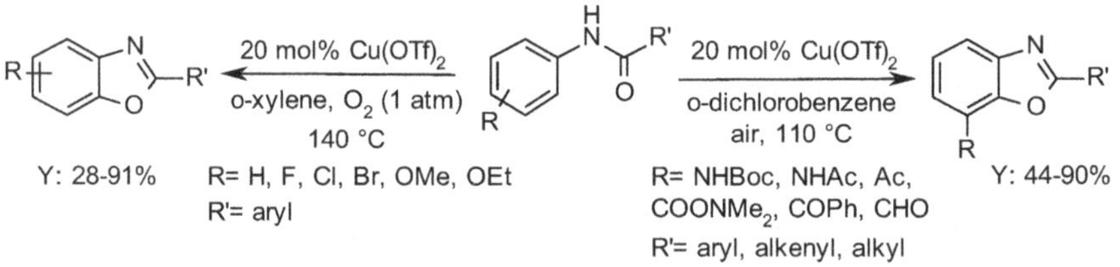

Tois reported the synthesis of 1,2-benzisoxazoles exploiting the same chemical transformation [163]. *ortho*-Bromoacetophenone oximes underwent copper-catalyzed cyclization at room temperature in the presence of a simple diamine ligand. According to the authors the key to the success is the use of *Z*-oximes to which they also devised a synthetic route.

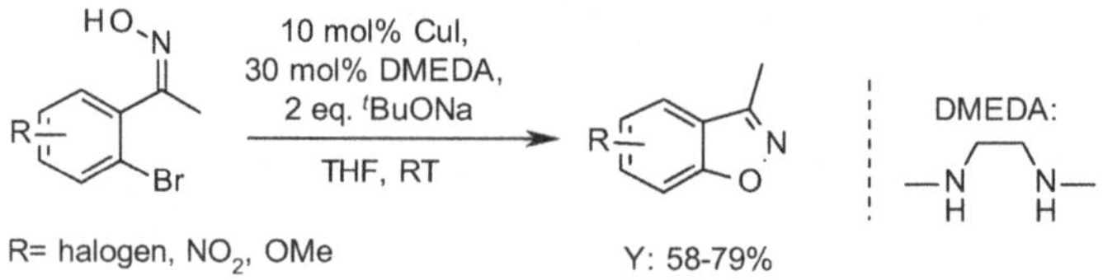

3.1.2 Electrophile-Induced Ring Closure

Besides the transition metal-catalyzed rearrangement of alkynyl oxime ethers to isoxazoles that were described in the previous chapter, it is worth mentioning that the same compounds can also undergo electrophile-induced ring closure as reported by Larock [164]. The added value of this transformation is the incorporation of an iodine into the 4-position of the isoxazole ring that might serve as a starting point for a myriad of follow-up reactions. The ring closure works equally efficiently with a wide selection of substituents thus introducing a diverse set of 3- and 5-substituents onto the isoxazole ring.

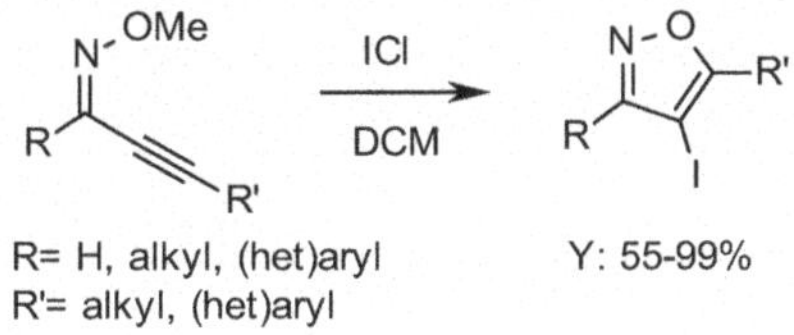

3.1.3 Simultaneous Formation of Two Ring Bonds

The gold(I)-catalyzed synthesis of 2,4-disubstituted oxazoles in the [3 + 2] annulation of terminal alkynes and carboxamides in the presence of an oxidizing agent has been reported by Zhang [165]. The reaction whose postulated key intermediate is an α-oxo gold carbine runs under mild conditions and the use of a bidentate *P,N*-ligand was crucial for its efficiency. The transformation worked equally well with (het)aryl and alkenyl carboxamides, while aliphatic carboxamides gave only poor yields.

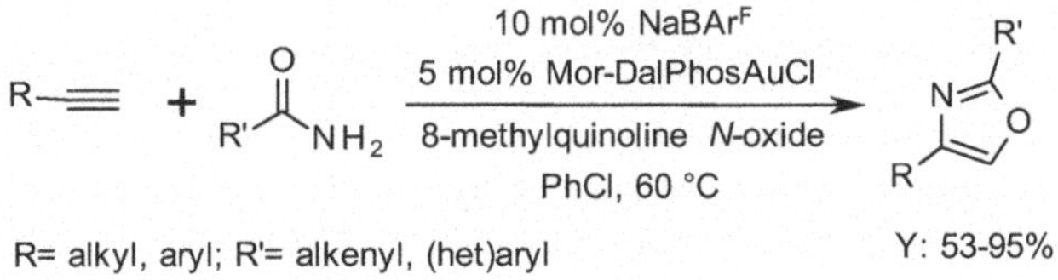

The gold(III)-catalyzed intermolecular [3 + 2] cycloaddition of conjugated *N*-ylides and *N*-alkynyl sulfonamides or cyclic carbamates were described by Davies [166]. The transformation gave the 2,5-disubstituted 4-aminooxazole derivatives with high selectivity. The reaction is not limited to ynamides since ethyl phenylethynyl ether gave the appropriate 4-ethoxyoxazole derivative albeit in a moderate yield.

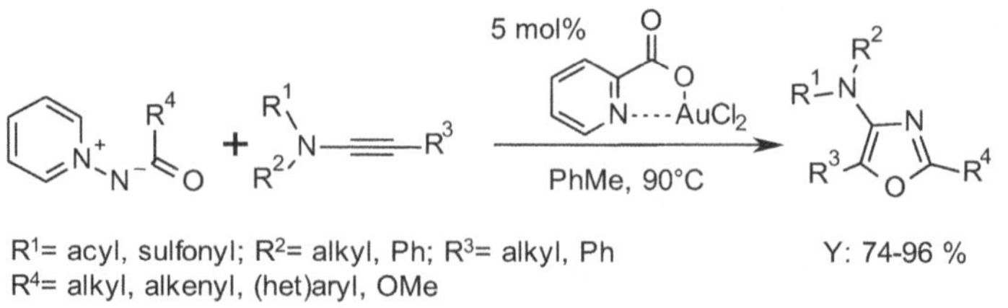

The reaction of copper(I) acetylides and nitrile oxides has been shown by Fokin [167] to yield 3,5-disubstituted isoxazoles. A simple and efficient one-pot variant of the same transformation starting from an aromatic aldehyde, a terminal alkyl or aryl acetylene, hydroxylamine, and NCS in the presence of a clay supported copper(II)/sodium azide catalyst precursor system has been reported by Vishwakarma [168].

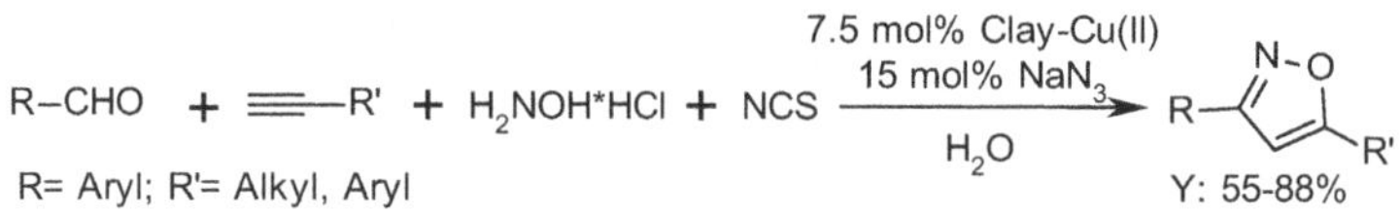

An opposite regioselectivity has been achieved by changing the catalyst to ruthenium. In the reaction of terminal aryl and alkyl acetylenes and differently substituted chlorooximes, the predominant formation of 3,4-disubstituted isoxazoles has been observed by Fokin [169]. Using internal alkynes the cyclization has also shown good selectivity: the more electronegative carbon center of the alkyne became C4 of the isoxazole ring unless a hydrogen bond donor was present that always ended up at C4.

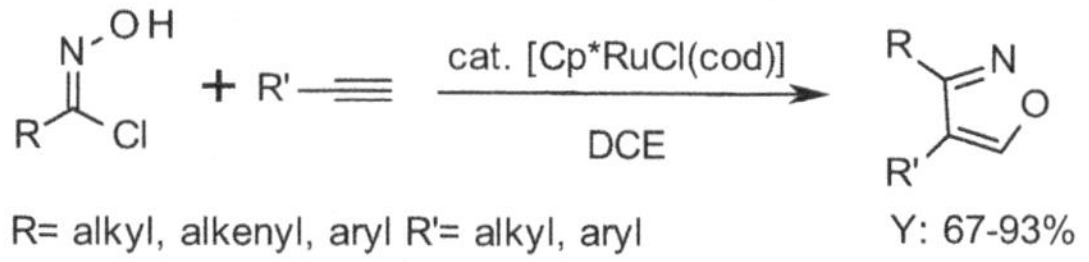

The room temperature palladium-catalyzed carbonylative coupling of terminal alkynes and aryl iodides in the presence of hydroxylamine led to the formation of 3,5-disubstituted isoxazoles as reported by Mori [170]. The transformation ran under 1 atm of carbon monoxide and the use of an excess of hydroxylamine hydrochloride and aqueous ammonia was found to be beneficial in avoiding the main side reaction (Sonogashira coupling).

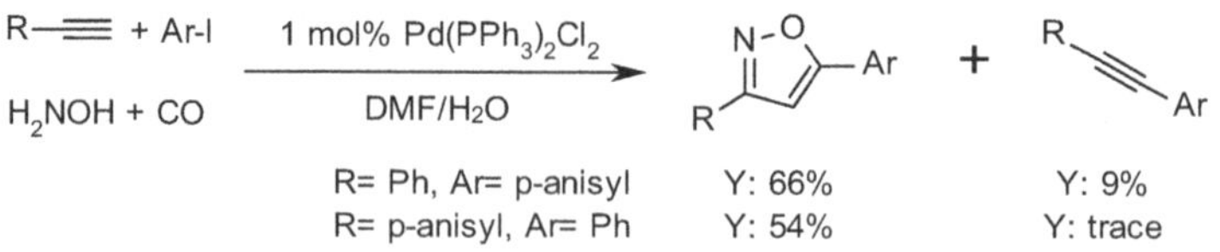

A mild and general protocol has been reported recently by Jiang [171] for the palladium-catalyzed oxidative cyclization of *ortho*-aminophenols and isocyanides to 2-aminobenzoxazole derivatives. The transformation tolerated a broad range of substituents on the benzene ring, and alkyl and aryl isocyanides were coupled successfully. Interestingly the direction of the ring closure could be altered by the change of the catalyst and solvent and elevation of the temperature giving rise to 3-aminobenzoxazines with equally good selectivity.

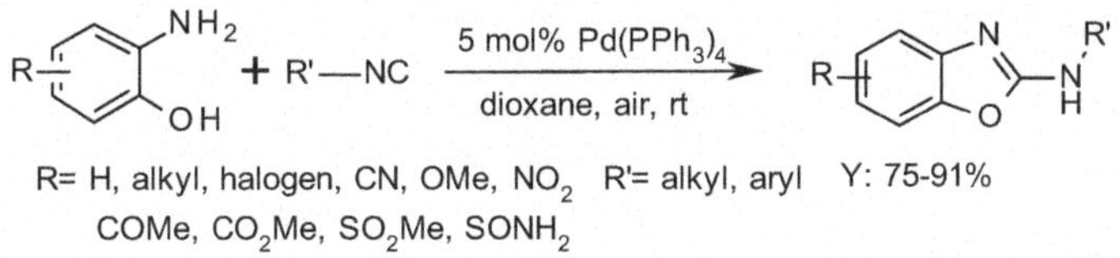

The synthesis of 2-arylbenzoxazoles has been achieved by Lang [172] in the reaction of *ortho*-aminophenols, *tert*-butyl isocyanide, and aryl halides. Although seemingly analogous to the above transformation, this reaction runs in the absence of oxidizing agent and uses a significant excess of the aminophenol.

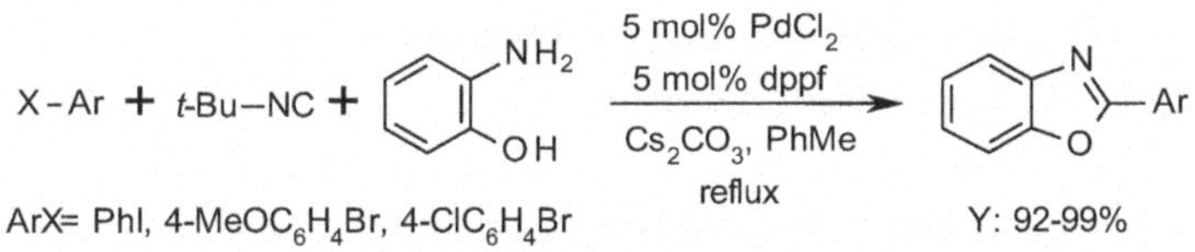

The copper(II) catalyzed oxidative rearrangement of bisaryloxime ethers to 2-arylbenzoxazoles was reported recently by Punniyamurthy [173]. The scope of the reaction is limited to *O*-aryl aldoximes and it uses molecular oxygen (1 atm) as the oxidant.

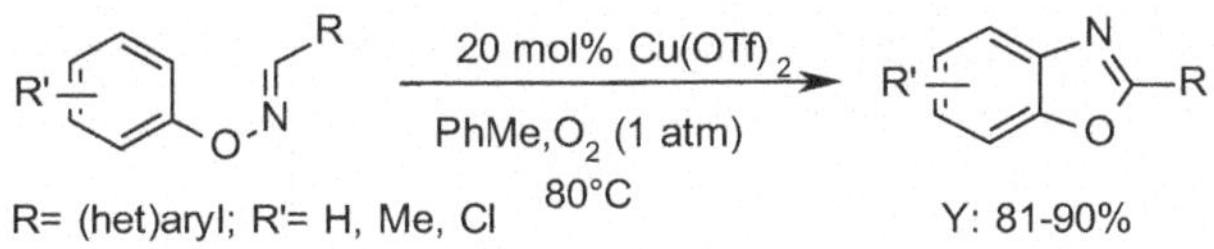

3.1.4 Miscellaneous

α-Diazocarbonyl compounds and nitriles underwent ruthenium-catalyzed ring closure to the appropriate oxazole derivatives in acceptable yield. The transformation reported by Lacour [174] utilizes the nitrile as the solvent.

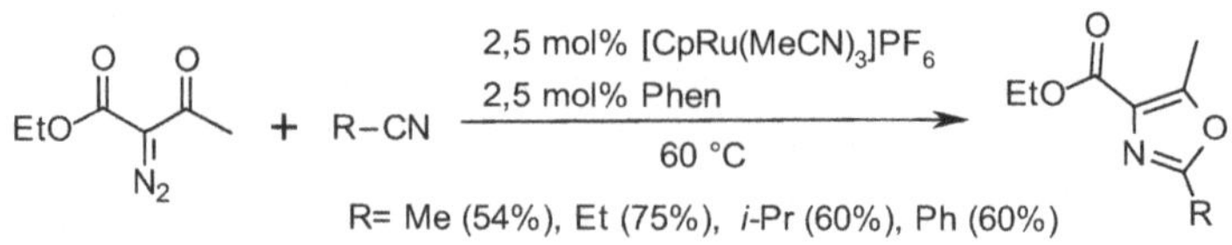

Iron(II) bromide was reported to catalyze the transformation of *ortho*-acyl arylazides into 2,1-benzisoxazoles by Driver [175]. The transformation tolerates a variety of different substituents on the benzene ring, while the carbonyl group might carry an aliphatic or aromatic substituent but not hydrogen.

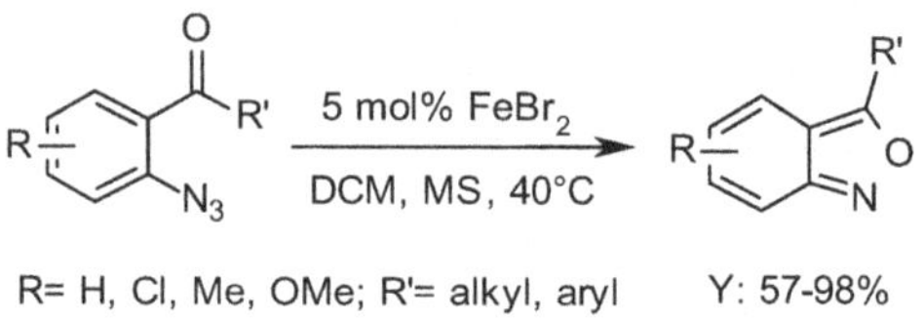

3.2 *Functionalization of Oxygen Heterocycles with Additional Heteroatom*

The transition metal-catalyzed functionalization of oxazoles, isoxazoles, and oxadiazoles, as well as their condensed derivatives has long been being explored due both to the inherent reactivity of these systems towards electrophilic reagents and the easy access to the appropriate halogenated azoles. As the following selection of examples demonstrates, a wide variety of substrates might be introduced onto these ring systems. This and the multitude of available transformations allow today's chemists to easily access diversely substituted ring systems in a highly selective manner. The presented reactions have been classified according to the nature of the newly formed bond.

3.2.1 Carbon–Carbon Bond Formation

The interest for the introduction of carbon-based substituents onto the five-membered heterocyclic core in a transition metal-catalyzed process has been widely explored because of the abundance of these cores in medicinal chemistry. Besides the well-established cross-coupling reactions, of which we only present a limited selection, in the recent years a great deal of attention has been given to the functionalization of azole C–H bonds (a.k.a. C–H activation). The inherent reactivity of the five-membered heterocycles towards electron-deficient reagents as well as the selectivity of the ensuing reactions made these transformations very popular as will be demonstrated by the diversity of examples.

C–X Functionalization

The uneven distribution of the electrons in the five-membered ring make the reactivity of the two halogens in 2,4-diiodooxazole different. Strotman and Chobanian [176] have screened a series of different palladium-based catalyst systems and found that by the proper selection of the used ligand a regioselective Suzuki coupling can be achieved in either position. When using Xantphos the traditionally more reactive 2-position was functionalized while in the presence of the phosphaadamantane-based ligand the transformation took place in the 4-position. Another advantage of these ligands is the limited formation of bis-arylated products, which allows for the sequential introduction of different substituents into the 2- and 4-position in subsequent Suzuki couplings. The conditions are applicable to the coupling of a wide variety of aryl-, heteroaryl-, cyclopropyl-, and vinylboronic acids.

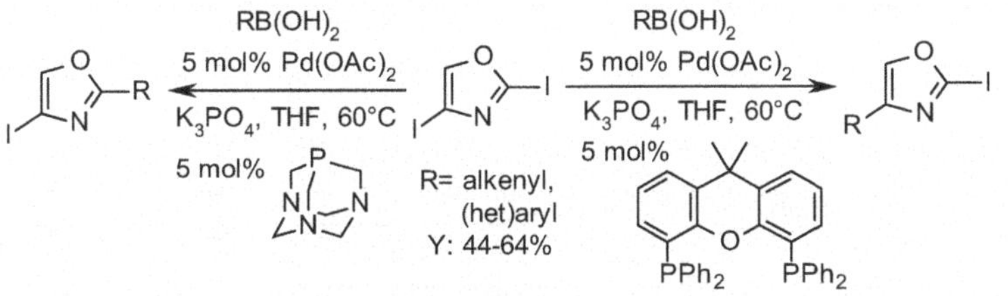

The use of methylthio derivatives in place of aryl halides in cross-coupling reactions in the presence of a copper(I) additive is a well-established protocol [177]. Stambuli reported [178] that 2-methylthiooxazoles undergo cross-coupling with a wide range of organozinc reagents in the presence of a palladium- or nickel-based catalyst system. Interestingly the addition of a copper(I) salt has blocked the coupling completely. The reaction was also extended to 2,5-bis(methylthio)-oxazole. By using a moderately active palladium-based catalyst and one equivalent of an organozinc chloride, the reaction could be stopped after the first coupling that took place in the 2-position. The change of the catalyst system to a more active nickel phosphine and the addition of another organozinc reagent initiated the cross-coupling in the 5-position and gave the desired 2,5-disubtituted oxazole derivative.

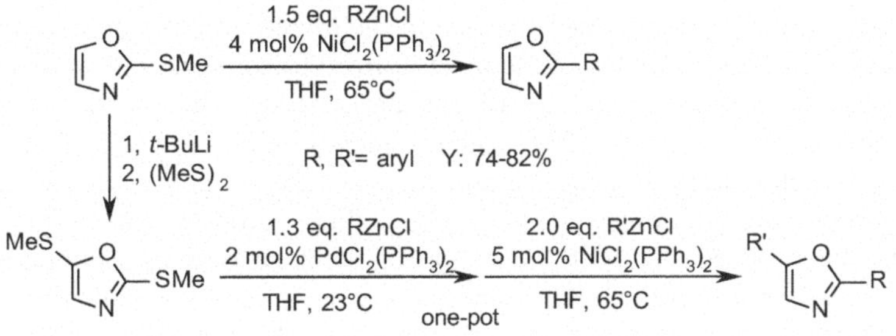

In a unique transformation reported by Tatibouet [179] a sugar-coupled oxazolinethione was alkynylated. The key to the process is the copper-mediated activation of the C–S bond of oxazole through its thiol tautomeric form. The process worked equally well with a series of acetylene derivatives and could also be extended to oxazolidinethiones.

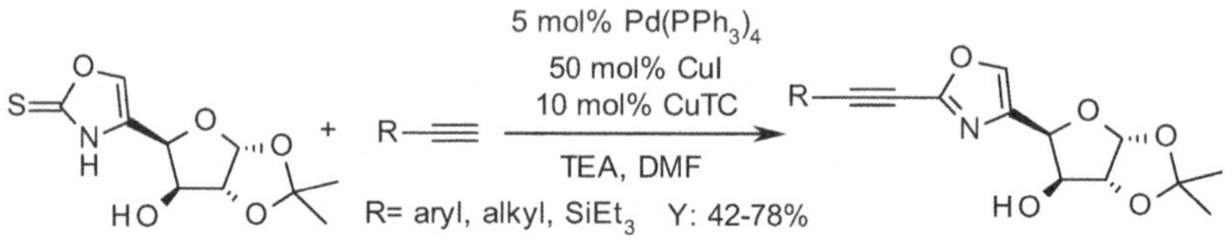

The decarboxylative cross-coupling reaction of aryl halides and aryl carboxylic acids has attracted significant attention since the groundbreaking work of Goossen [180]. The palladium-catalyzed cross-coupling of oxazole 5-carboxylic acids and a variety of aryl and hetaryl halides was reported by Greaney [181]. The success of the reaction requires the addition of an equivalent amount of silver carbonate.

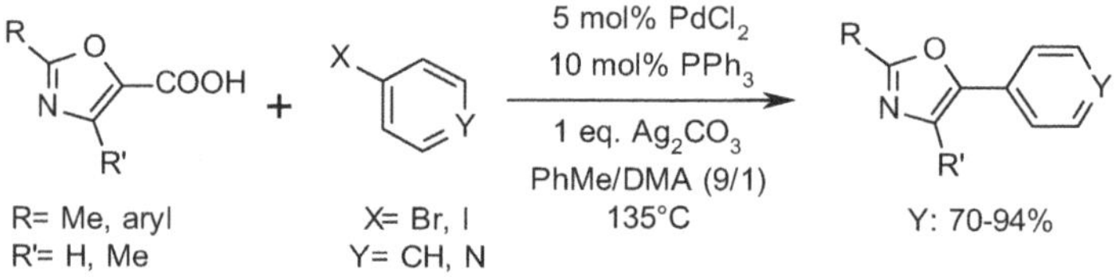

C–H Activation

Traditionally transition metal-catalyzed cross-coupling reactions utilize an aryl halides and an organometallic reagent as coupling partners. With the better understanding of their mechanism and the drive to develop transformations that have a decreased ecological footprint, the search intensified for such cross-coupling reactions where one or both reacting bonds are replaced by a C–H bond (a.k.a. C–H activation). The five-membered heteroaromatic systems proved to be an excellent reagent class for such transformations due to the increased electron density of the heteroaromatic ring [182]. One of the most utilized transformations of this class is the palladium-catalyzed coupling of an aryl halide with a five-membered heteroaromatic ring, an example of which is shown below. The commonly implicated mechanism features the electrophilic attack of the arylpalladium halide on the heteroaromatic ring giving rise to a bis-arylpalladium complex. An alternate mechanism has been proposed for the arylation of benzoxazoles by Zhuravlev [183] where the opening step is the base-promoted ring opening of the azole ring that in a ligand substitution-ring closure sequence leads to the formation of the shown bis-arylpalladate complex. The concluding step in each case is the reductive elimination of the product and regeneration of the active catalyst.

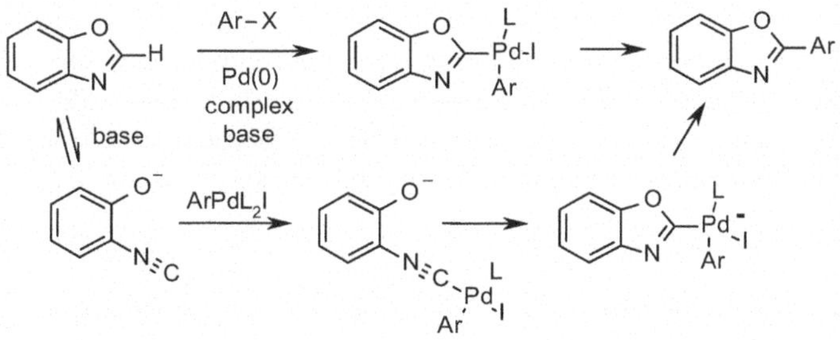

Studying the palladium-catalyzed coupling of aryl halides and oxazole under different conditions Strotman and Chobanian [184] established complementary conditions for the direct arylation of oxazole with very high regioselectivity both at the 2- and 5-position. The scope of reagents covered a wide range of aryl and heteroaryl bromides, chlorides, iodides, and triflates. Using polar solvents and *n*-butyl diadamantylphosphine as ligand, C-5 arylation is preferred, whereas switching to nonpolar solvents and RuPhos C-2 arylation became preferred. The authors reason that in C-5 selective reactions, the second palladium-carbon bond is formed in a concerted metalation-deprotonation reaction proposed by Fagnou [185], while in C-2 selective reactions the deprotonated form of the oxazole or its ring-opened analogue might react with the monoarylpalladium complex.

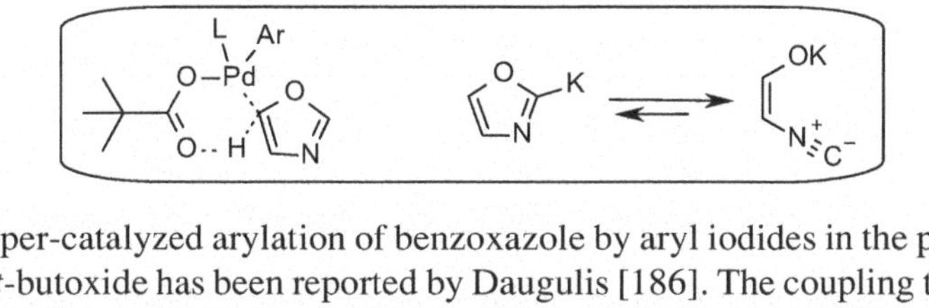

The copper-catalyzed arylation of benzoxazole by aryl iodides in the presence of lithium *tert*-butoxide has been reported by Daugulis [186]. The coupling takes place selectively in the 2-position and works well for sterically hindered aromatic moieties such as the mesitylene group. Preliminary mechanistic studies using deuterated substrates ruled out the presence of a benzyne-like mechanism. A rhodium-catalyzed variant of this reaction has been reported by Bergman and Ellman [187] in 2004, while Santelli [188] described the similar palladium-catalyzed arylation of 3,5-dimethylisoxazole in the 4-position using a wide variety of aryl bromides.

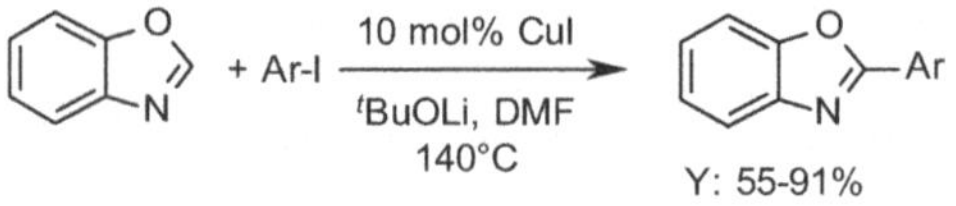

Analogously to other cross-coupling reactions conducting the arylation of benzoxazole with aryl iodides in the presence of carbon monoxide, Beller [189] achieved the synthesis of 2-aroyl-benzoxazoles. The transformation required the use of a palladium catalyst system and copper(I) iodide as an additive.

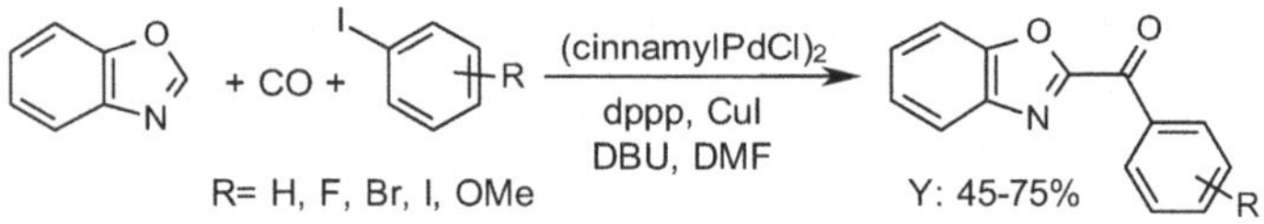

The apparent ease of arylation of oxazoles in the 2-position through C–H activation has led to a widespread effort to extend the scope of arylating agents. Using moisture stable aryl sulfamates Ackermann [190] established an efficient palladium-catalyzed protocol for the 2-arylation of benzoxazoles. The same catalyst system worked equally well with cycloalkenyl and benzyl phosphates expanding the scope of substituents to the benzyl moiety.

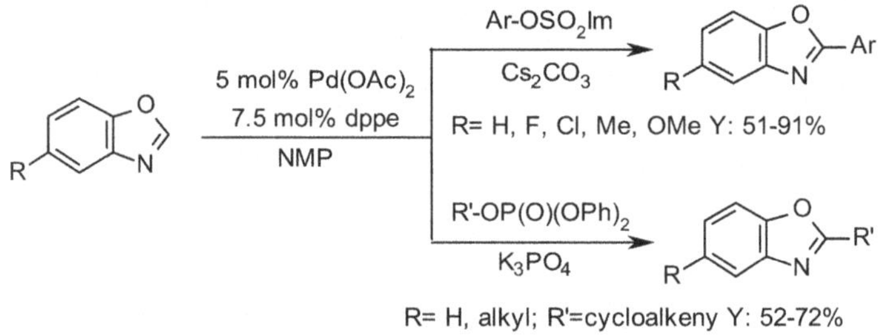

Another alternate to the use of aryl halides is the use of carboxylic acid derivatives. Greaney [191] has successfully coupled azolyl carboxylic acids with oxazole derivatives in the presence of a palladium catalyst. The use of an electron rich and bulky bidentate ligand was crucial to the selective coupling. The reaction uses an excess of copper(II) carbonate as base and oxidant and probably proceeds through an azolylcopper complex. The use of silver carbonate has led to a similar outcome.

Switching the catalyst from palladium to nickel, Yamaguchi and Itami [192] has successfully extended the scope of the coupling to aryl carboxylic acid phenyl esters. Under these conditions the copper(II) salt could be replaced by potassium phosphate. The coupling worked equally well with 5- and 6-membered heteroaromatic esters as well as carbocyclic analogues.

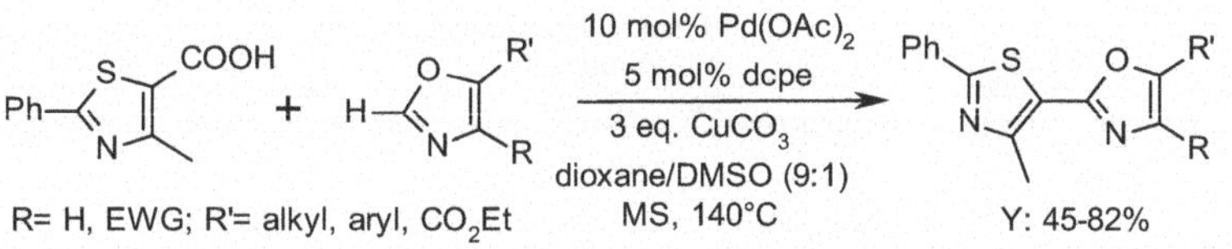

A highly chemo- and regioselective deamidative arylation of oxazoles, oxadiazoles, and benzoxazoles with arylamides has been reported by Wang [193]. The palladium-catalyzed reaction proceeds smoothly to generate the corresponding products in good yields via a tandem decarbonylation–C–H activation.

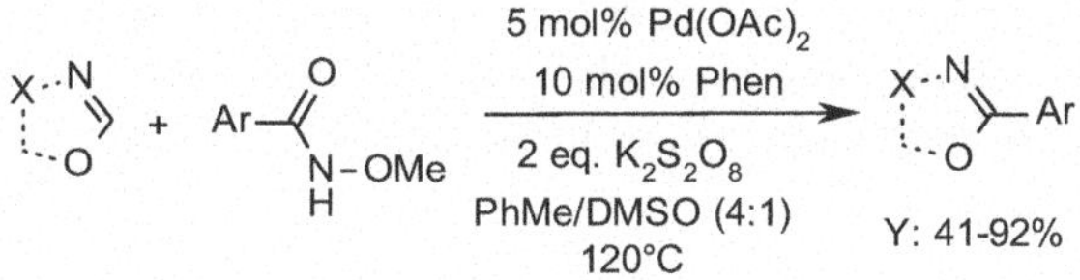

The use of arylboronic acids in the palladium-/copper-catalyzed arylation of benzoxazole has been reported by Liu [194]. The oxidative couplings were run under air and enabled the introduction of benzene derivatives and pyridine into the 2-position. A subsequent communication by Xu, Yu, and Wang [195] describes the analogous coupling of arylboronic esters, using copper as catalyst and molecular oxygen as the oxidant. Besides a similar substrate scope for the arylation of benzoxazoles, the authors were also successful in arylating oxazole and its 5-aryl derivatives.

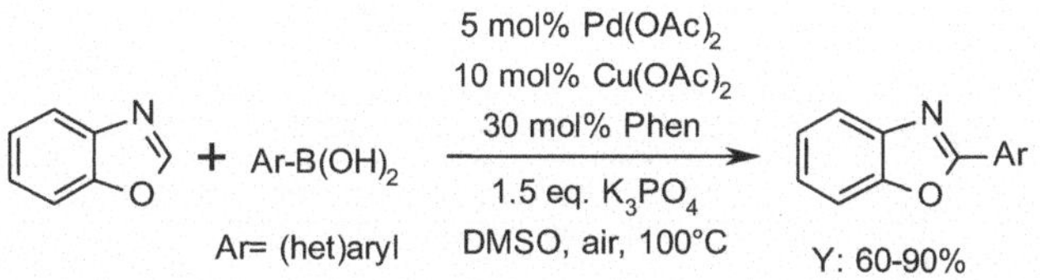

A further extension of the scope of arylating agent has been reported by Miura [196] who coupled aryl and alkenyl triethoxysilanes with oxazole and benzoxazole derivatives in the presence of a nickel-based catalyst system and stoichiometric amounts of cesium fluoride and copper(II) fluoride. In a similar experiment Ofial [197] coupled successfully oxazole and benzoxazoles with trialkoxy(aryl)silanes using palladium(II) acetate as catalyst and copper(II) acetate and silver fluoride as additives. The reaction was successfully extended to allyltriphenylstannane that gave the 2-arylazoles in good (65–76%) yield.

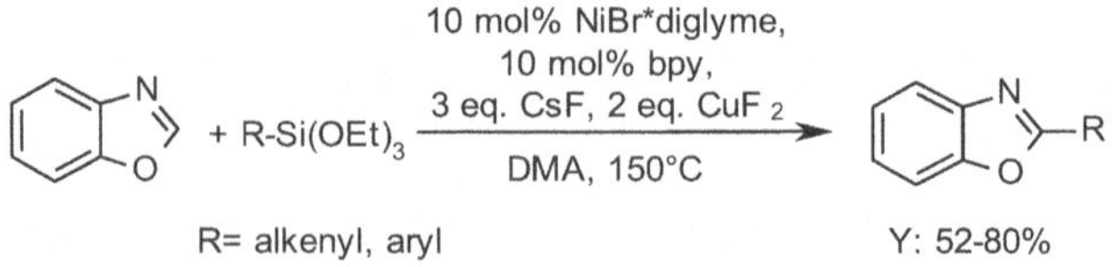

The palladium-catalyzed direct desulfitative C-arylation of the 2-position of benzoxazole with arenesulfonyl chlorides has been reported by Chen [198]. The procedure tolerates a wide range of functional groups on the arylsulfonyl chlorides. Copper salt and air were used as oxidant. In an analogous coupling You [199] used the easily available, air-stable, and easy to handle sodium sulfinates as the coupling partner. Oxazole, oxadiazole, and benzoxazole were arylated with similar yields. Using the stable and easily accessible arylsulfonyl hydrazides with a palladium catalyst and copper(II) acetate as oxidant, Li and Wan [200] arylated differently substituted oxazoles and benzoxazoles in the 2-position in a similar fashion.

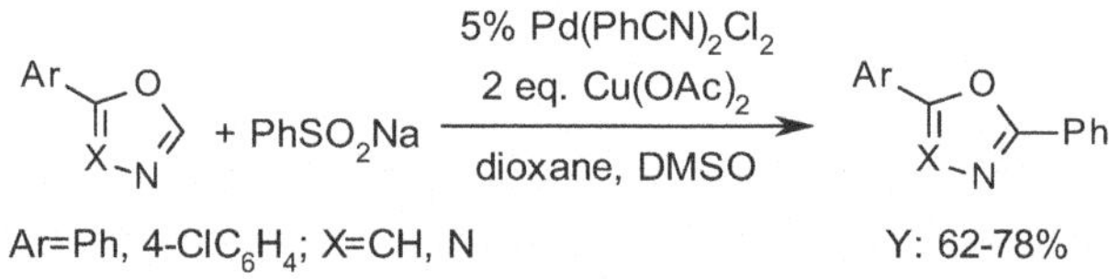

The direct arylation of benzoxazoles and 2-phenyloxadiazole in cross-coupling with tautomerizable heterocycles has been reported by van der Eycken [201]. The reaction proceeds through a PyBroP-mediated and Pd/Cu-catalyzed sequential C–O/C–H activation. The methodology worked equally well for 1,2-, 1,3-, and 1,4-diazines.

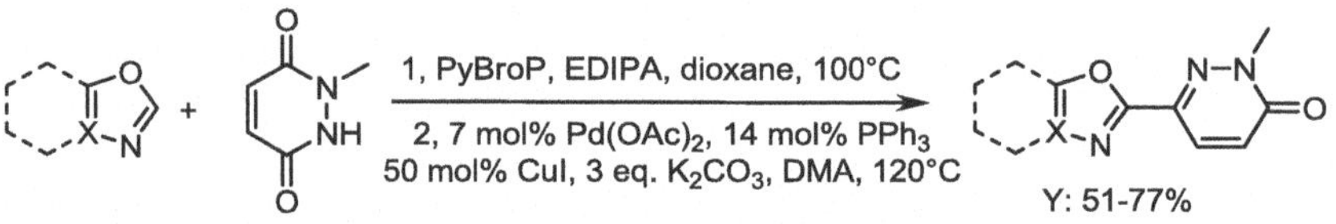

The functionalization of azole derivatives through C–H activation is not limited to arylation reactions. Piguel [202] has reported the copper-catalyzed alkenylation of oxazoles and benzoxazole with bromoalkenes in a regio- and stereoselective process.

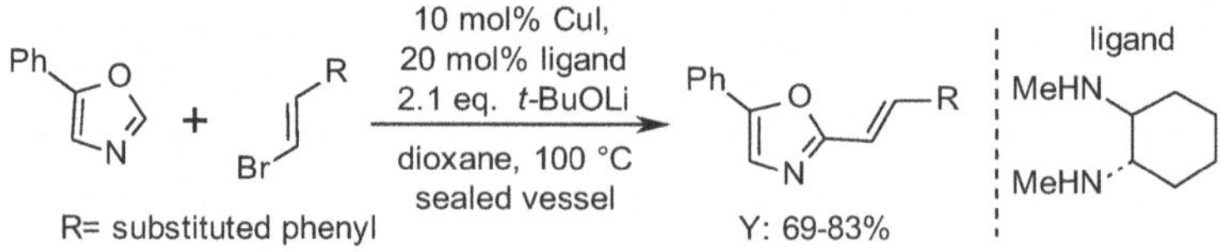

The alkynylation of oxazole and benzoxazole using bromoacetylene derivatives has been reported by Miura. The nickel-catalyzed coupling worked well for several substrates, but in certain cases the addition of copper(I) iodide as co-catalyst was required to achieve good yield [203]. The analogous copper(I)-catalyzed alkynylation of oxazole, benzoxazole, and oxadiazole derivatives with bromoalkynes was reported concomitantly by Piguel [204].

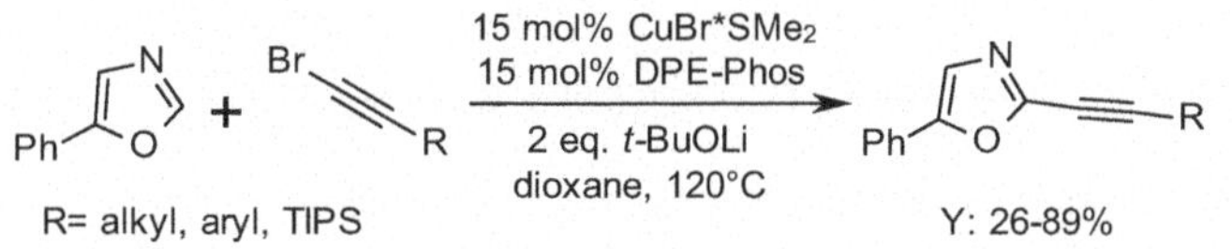

The copper-catalyzed direct alkynylation of azoles can also be achieved using 1,1-dibromo-1-alkenes as electrophiles as described by Piguel [205]. The reactive bromoacetylene is presumably generated by the added strong base in the opening step of the process. The scope of the process was similar to that of the above-described copper-catalyzed alkynylation, which might not be too surprising since the two transformations share the same catalyst system and only the amount of added base differs significantly. Ackermann [206] described an analogous palladium-catalyzed transformation using 1,1-dichloroalkenes as reagent.

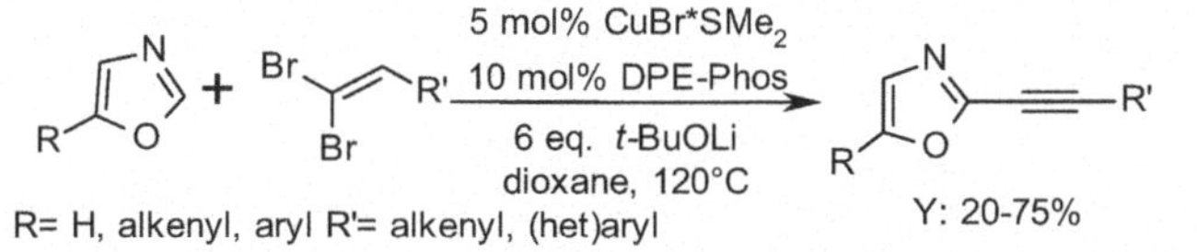

The alkylation of azoles through C–H activation has also been described. Miura and Hirano [207] reported an efficient palladium and nickel catalyst systems for the direct alkylation of oxazoles and benzoxazoles with alkyl bromides and chlorides. Interestingly for more complex reagents (e.g., 6-bromohexene), the outcome of the reaction depended on the catalyst used. Miura [208] has also described the direct benzylation of azoles with benzyl carbonates in the presence of a palladium catalyst system that affords the corresponding diarylmethanes in good yield. In addition, the same palladium catalyst enables the benzylation of the product on its C(sp^3)-H bond with a second benzyl carbonate without employing any external base.

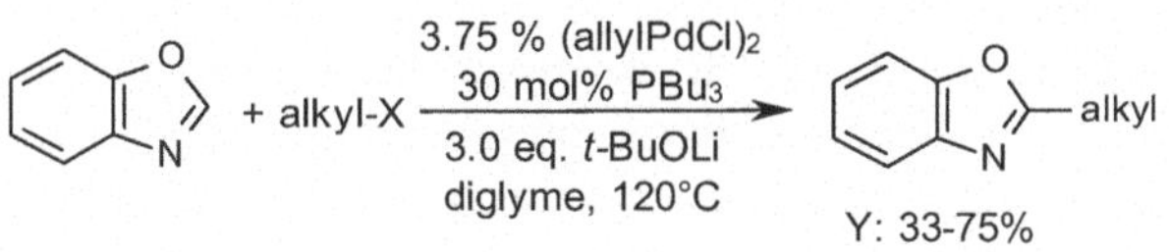

The use of tosylhydrazones of acetophenon and cyclohexenone derivatives as alkylating agent in a copper-catalyzed and base-mediated process has been reported by Wang [209]. The process leads to the direct benzylation or allylation of oxazole and benzoxazoles. An extension of this process reported by Miura and Hirano [210]

utilizes a nickel-based catalyst and works equally well with tosylhydrazones bearing non-activated alkyl groups.

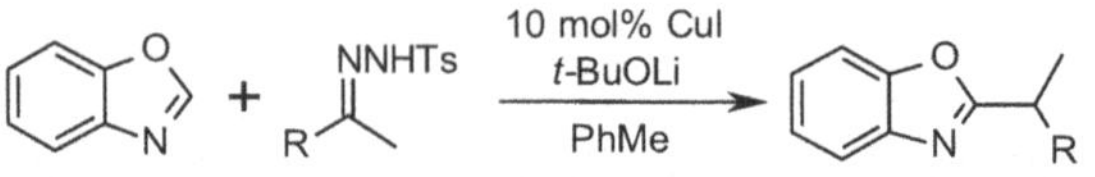

R= Ph (80%), 4-MeOC6H4 (58%), 4-CF3C6H4 (76%)

A unique example of the functionalization of azoles through C–H activation is the gold-catalyzed carboxylation of azoles with carbon dioxide. The procedure described by Nolan [211] employs gold(I)-carbene complexes as catalyst, a slight overpressure (1.5 bar) of carbon dioxide, and works equally well on oxazole, isoxazole (activation of 5-position), and benzoxazole. A similar copper-carbene complex-catalyzed transformation has been reported by Hou [212] who converted the carboxylate intermediates without isolation to the corresponding esters using different alkylating agents.

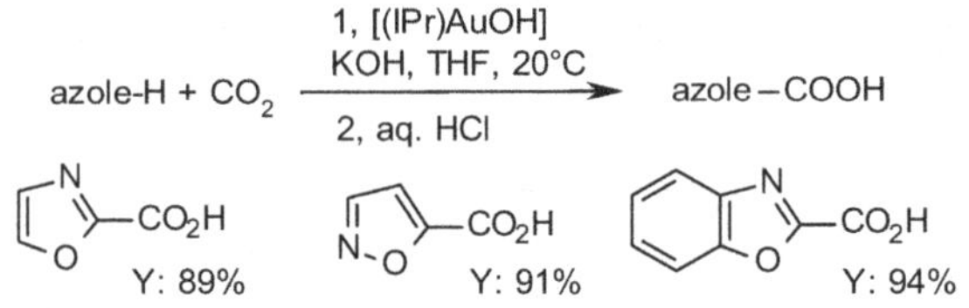

Double C–H Activation

In theory under oxidative conditions azoles and other reagents could be coupled in a double C–H activation, resulting in the formation of water as by-product. Although such a transformation would be quite appealing from the atom economy and environmental point of view its widespread application is limited mostly by the difficulty of fine tuning the conditions to achieve the selective activation of the desired C–H bonds.

These difficulties are not present in the oxidative homocoupling of azoles that was reported by Qian and Bao [213]. Benzoxazoles and oxadiazole were found to dimerize through their 2-position in the presence of a copper catalyst and air as oxidant to give the respective bisazoles in good yield. Attempts to extend this catalyst system to the heterocoupling of azoles led to mixtures with little deviation from statistic product distribution.

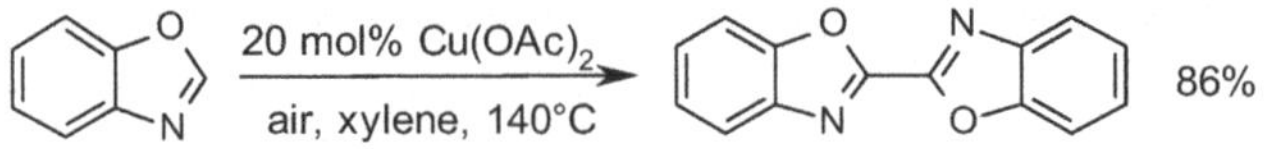

In a recent publication Xu, Yu, and Wang [214] claimed that stoichiometric copper(II) acetate in dimethyl sulfoxide under argon mediates the selective cross-coupling of different azoles in high yield and with excellent chemoselectivity. The transformation was sensitive to the applied reaction conditions and the presence of the acetate ions was also crucial for the successful coupling. They reported that the process worked well with a broad scope of azoles.

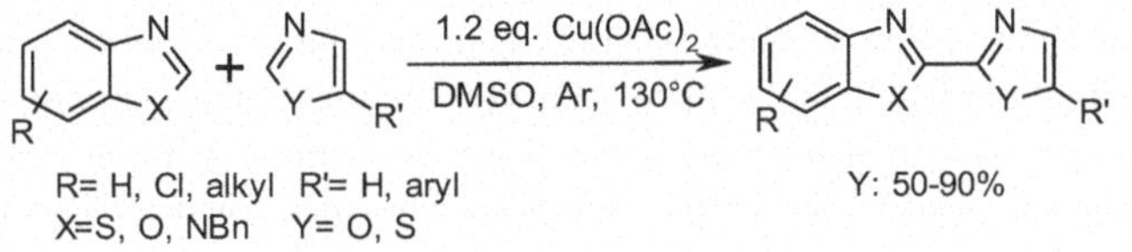

Similarly, outstanding selectivity was observed by You [215] in the coupling of oxazole derivatives with thiazoles and imidazoles. The reaction required the presence of a palladium catalyst, a copper co-catalyst, and analogously to the above-reported results, stoichiometric amounts of copper(II) acetate. In this paper the authors have also identified and quantified the formation of the homocoupled by-products.

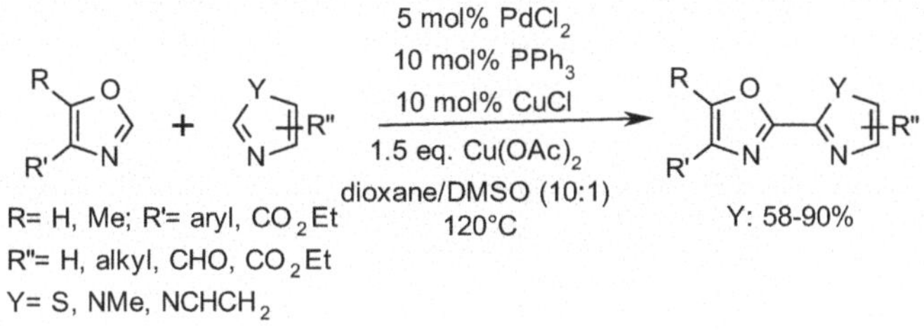

The palladium-catalyzed oxidative olefination of oxazoles at the 4-position through C–H bond activation has been reported by Antilla [216]. The transformation requires the presence of copper(II) acetate as oxidizing agent and the broad scope of applicable terminal olefins includes vinylstannanes, vinylsilanes, acrylamides, and butadiene derivatives alike. The formed products were also transformed into functionalized amino alcohol and homophenylalanine derivatives.

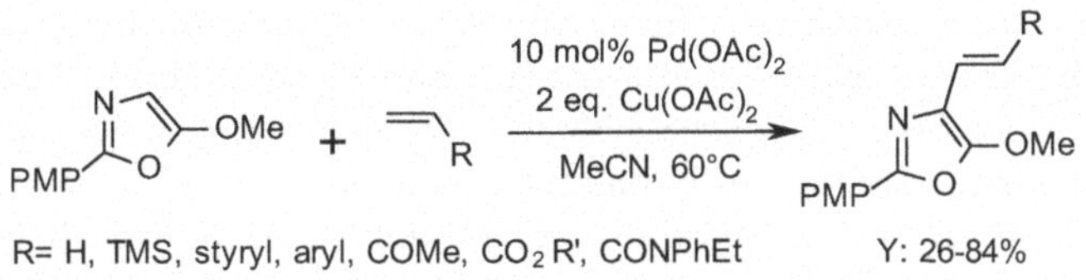

The alkenylation of the 2-position of oxazoles, oxadiazoles, and benzoxazoles can also be achieved in the cobalt-catalyzed addition of these azoles to internal alkynes as reported by Yoshikai [217]. The use of a bidentate phosphine ligand at least half equivalent of a Grignard reagent were necessary to achieve good yields.

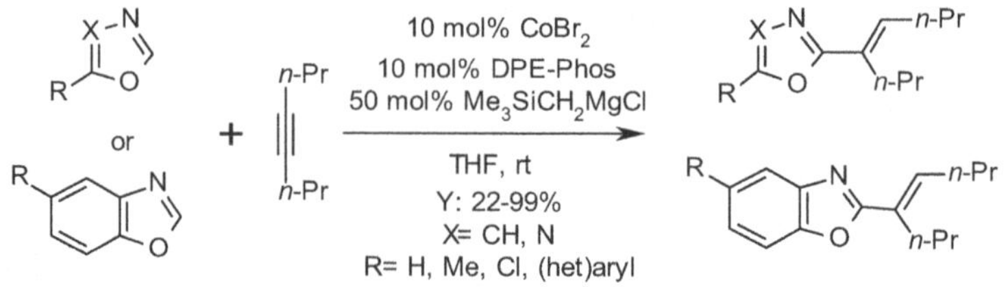

The oxidative coupling of benzoxazoles and terminal acetylene derivatives was reported by Chang [218]. The coupling required the use of the conventional tetrakis (triphenylphosphino)palladium catalyst and air as oxidant. It is interesting to note that no copper salt was added to the reaction. The reaction was also extended to 4-aryloxazoles.

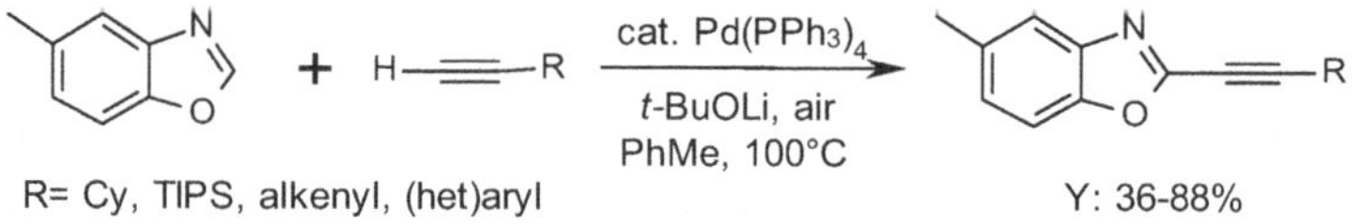

In a recent paper describing the nickel-catalyzed hydroheteroarylation of vinylarenes, Nakao and Hiyama [219] reported the phenethylation of oxazole and benzoxazole with styrene. The reaction that ran at elevated temperature gave the respective products in 41 and 97% yield.

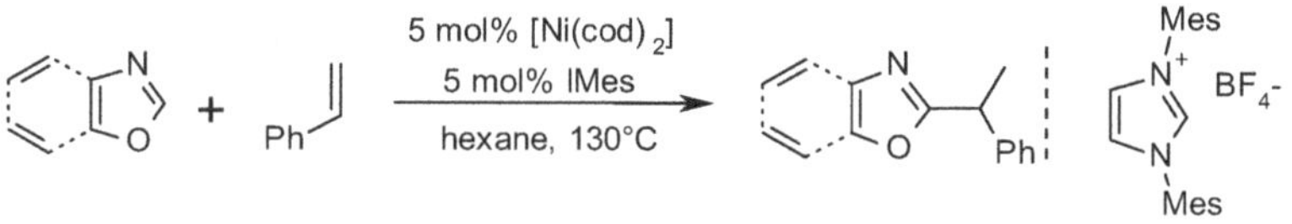

3.2.2 Carbon–Heteroatom Bond Formation

Heterocyclic compounds bearing heteroatom-based substituents have always been a primary target for organic synthesis. In case of 2-substituted oxazoles and benzoxazoles, their synthesis is frequently achieved through direct ring closure reactions. The marked reactivity of azoles towards electrophiles makes their nitro and halogen derivatives easily accessible too. The chemical pathways to transform these intermediates into the desired amine-, alcohol-, or thiol-substituted derivatives without the use of a transition metal catalyst, apart from heterogeneous hydrogenation reactions that is not discussed here, are also known. The following chapter collects representative examples for the transition metal-catalyzed introduction of heteroatom-based substituents onto the five-membered heterocyclic ring.

C–X Functionalization

The palladium- and copper-catalyzed conversion of haloarenes to their amino- or alkoxy-substituted derivatives is well established. Contrary to its widespread use for benzene derivatives and azines, the number of relevant examples for azoles is limited. 5-Bromooxazoles were reported by Doi and Takahashi [220] to undergo Buchwald-Hartwig coupling with different amines and a phenol. The expected products were isolated in acceptable yield. In case of octylamine the amidation of the pendant ester group was also observed (not shown for clarity). Spencer [221] published a similar transformation that was used in the generation of a small library of oxazole derivatives.

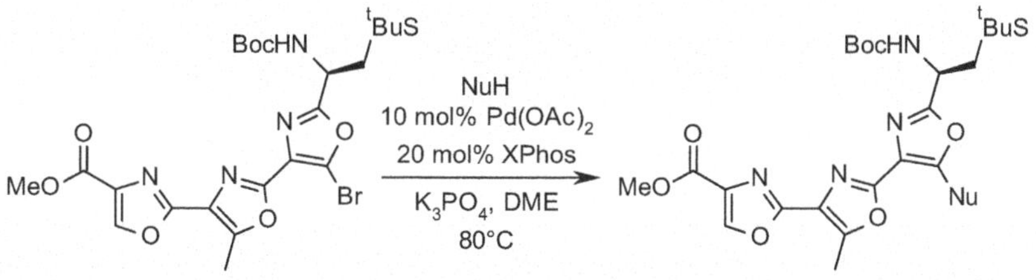

The literature of the transition metal-catalyzed amination of 4-halooxazoles is even scarcer. Boger [222] reported an example of the copper-catalyzed conversion of a 4-iodooxazole derivative to the methoxy analogue, while a recent patent application disclosed [223] the palladium-catalyzed sulfonamidation of 4-iodooxazoles.

The palladium-catalyzed amination of 6-chlorobenzoxazoles was reported by Buchwald [224]. The reaction proceeded smoothly both with anilines and piperidine in the presence of XPhos as ligand. For the efficient coupling of 3-cyanoaniline, the authors had to change the applied base from sodium *tert*-butoxide to potassium phosphate.

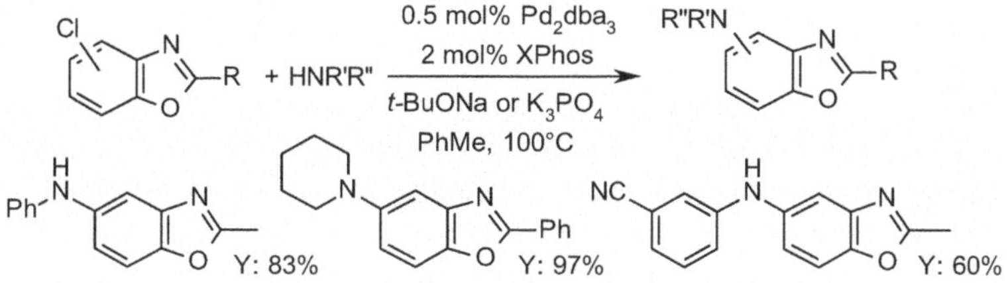

C–H Activation

The direct functionalization of azoles with nucleophiles in a C–H, Nu–H activation is of great synthetic interest. Since the keen participation of these heteroaromatic rings in coupling reactions including a C–H activation is well documented, it is not surprising that the analogous transition metal-catalyzed introduction of nucleophiles has also been studied recently by several groups. The copper(II)-catalyzed amination of benzoxazole reported by Mori [225] proceeds readily at elevated temperature under ambient pressure of oxygen. The introduction of diphenylamine, *N*-methylaniline, and piperidine was equally efficient (66–72% yields), while the yield using diethylamine was slightly inferior (47%).

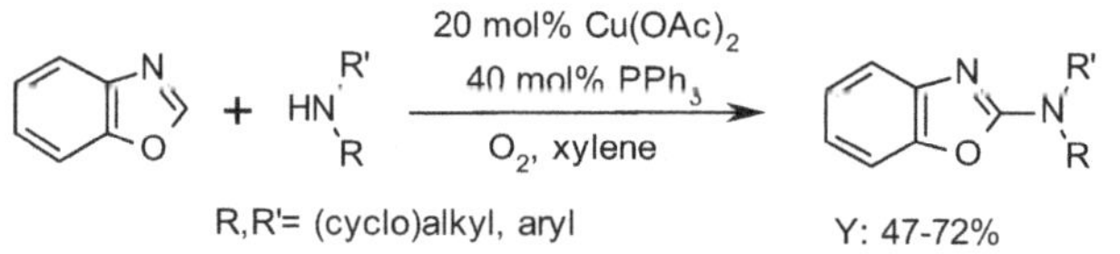

The copper(II) catalyst system is also able to catalyze the acylamination of oxazole as reported by Schreiber [226]. Pyrrolidinone reacted readily with a substituted oxazole derivative in the presence of oxygen. It is interesting to note that the analogous transformation of benzoxazole with *o*-trifluoromethylbenzamide required the use of a stoichiometric amount of copper(II) acetate. Although mechanistically distinct we should also mention the silver(I)-mediated amination of benzoxazoles in the 2-position using either amines or their formamide equivalents as reagent. The reaction was reported by Chang [227] to work well with a broad scope of amines and formamides but not on similar ring systems such as oxazole.

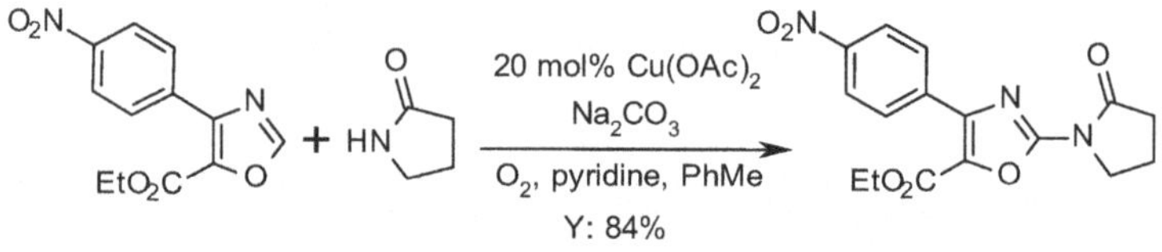

An interesting variant of the amination of benzoxazoles in the 2-position through C–H activation was reported by Yotphan [228] who used *O*-benzoyl hydroxylamine derivatives as amine source in a copper(I)-catalyzed reaction. A variety of secondary amines were introduced onto the benzoxazole core efficiently, while the use of *N*-monoalkyl derivatives proceeded only with poor yield. An example of the amination of 2-phenyl-oxadiazole was also reported.

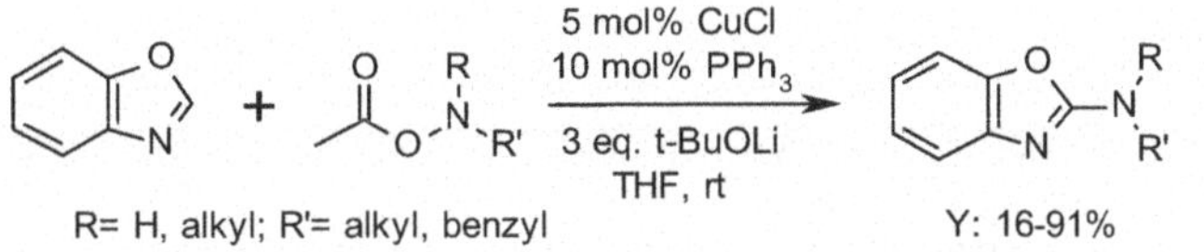

The cross-coupling reaction of benzoxazole with aryl thiols and diaryl disulfides was reported by Fukuzawa [229]. In the presence of a copper-bipyridine catalyst and oxygen, the corresponding aryl benzoxazolyl sulfides were isolated in moderate to good yields. The coupling reaction proceeded equally well in the presence of selected copper(I) and copper(II) salts indicating the rapid oxidation of the former under the applied conditions. The oxidation of the aryl thiols to the diaryl disulfides under the applied conditions is also likely.

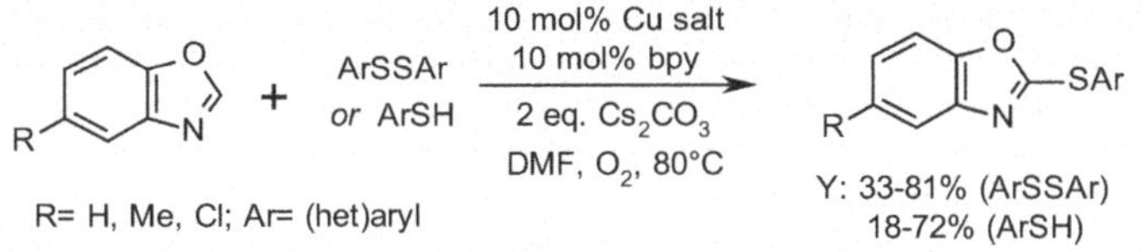

In the presence of a Lewis acid (typically silver trifluoroacetate) and a stoichiometric amount of copper(II) acetate oxazole, oxadiazole and benzoxazole derivatives reacted with butyl mercaptan to give the 2-butylthio derivatives in good yield [230]. Switching the silver catalyst to silver fluoride and the solvent to dimethyl sulfoxide, the methylthiolation was observed implying the dual role of the solvent.

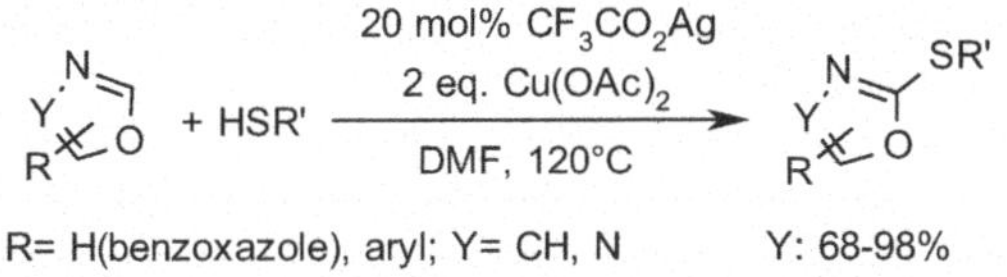

Summary, Conclusions, Outlook

In the previous chapters we collected a bouquet of transition metal-catalyzed transformations that are useful for the synthesis and derivatization of five-membered oxygen-containing heterocycles. The richness and diversity of this collection, and the newly developed methodologies in particular, mean that this area of chemistry will remain a burgeoning field in the near future. We firmly believe that the vast capabilities of the outlined chemical transformations will continue to drive the use of these ring systems in applied science, in medicinal chemistry in particular, while their increasing presence in applied science in turn will further fuel basic research in organic synthesis.

References

1. Dean FM (1982) In: Katritzky AR (ed) Advances in heterocyclic chemistry, vol 30. Academic, New York, pp 167–238
2. Dean FM, Sargent MV (1984) In: Bird CW, Cheeseman GWH (eds) Comprehensive heterocyclic chemistry, part 3, vol 4. Pergamon Press, New York, pp 531–598
3. Carril M, Correa A, Bolm C (2008) Iron-catalyzed Sonogashira reactions. Angew Chem Int Ed 47:4862
4. Gonda Z, Tolnai GL, Novák Z (2010) Dramatic impact of ppb levels of palladium on the "copper-catalyzed" Sonogashira coupling. Chem Eur J 16:11822
5. Wang R, Mo S, Lu Y, Shen Z (2011) Domino Sonogashira coupling/cyclization reaction catalyzed by copper and ppb levels of palladium: a concise route to indoles and benzo[b]furans. Adv Synth Catal 353:713
6. Jaseer EA, Prasad DJC, Sekar G (2010) Domino synthesis of 2-arylbenzo[b]furans by copper (II)-catalyzed coupling of o-iodophenols and aryl acetylenes. Tetrahedron 66:2077
7. Cano R, Yus M, Ramon DJ (2012) Impregnated copper or palladium copper on magnetite as catalysts for the domino and stepwise Sonogashira-cyclization processes: a straightforward synthesis of benzo[b]furans and indoles. Tetrahedron 68:1393
8. Saha D, Dey R, Ranu BC (2010) A simple and efficient one-pot synthesis of substituted benzo [b]furans by Sonogashira coupling–5-endo-dig cyclization catalyzed by palladium nanoparticles in water under ligand- and copper-free aerobic conditions. Eur J Org Chem 6067
9. Schumacher RF, Honraedt A, Bolm C (2012) Synthesis of N-methyl-2-indolyl- and N-methyl-2-benzo[b]furyl-substituted sulfoximines by Pd/Cu Co-catalyzed domino cross-coupling/cyclization reactions. Eur J Org Chem 3737
10. Wang JR, Manabe K (2010) Mechanistic studies and improvement of coinage metal-catalyzed transformation of alkynyloxiranes to furans: an alcohol addition-cyclization-elimination cascade. J Org Chem 75:5342
11. Yamaguch M, Katsumata H, Manabe K (2013) One-pot synthesis of substituted benzo[b] furans from mono- and dichlorophenols using palladium catalysts bearing dihydroxyterphenylphosphine. J Org Chem 78:9270
12. Arcadi A, Blesi F, Cacchi S, Fabrizi G, Goggiamani A, Marinelli F (2013) Multisubstituted benzo[b]furans through a copper- and/or palladium-catalyzed assembly and functionalization process. Tetrahedron 69:1857
13. Markina NA, Chen Y, Larock RC (2013) Efficient microwave-assisted one-pot three-component synthesis of 2,3-disubstituted benzofurans under Sonogashira conditions. Tetrahedron 69:2701
14. Russo O, Messaudi S, Hamze A, Olivi N, Peyrat JF, Brion JD, Sicsic S, Berque-Bestel I, Alami M (2007) Three-component one-pot process to propargylic amines and related amide and sulfonamide compounds: application to the construction of 2-(aminomethyl)benzofurans and indoles. Tetrahedron 63:10671
15. Csékei M, Novák Z, Kotschy A (2008) Development of a one-pot sequential Sonogashira coupling for the synthesis of benzofurans. Tetrahedron 64:8992
16. Álvarez R, Martínez C, Madich Y, Denis JG, Aurrecoechea JM, de Lera ÁR (2010) A general synthesis of alkenyl-substituted benzofurans, indoles, and isoquinolones by cascade palladium-catalyzed heterocyclization/oxidative Heck coupling. Chemistry 16:12746
17. Hu Y, Nawoschik KJ, Liao Y, Ma J, Fathi R, Yang Z (2004) Synthesis of conformationally restricted 2,3-diarylbenzo[b]furan by the Pd-catalyzed annulation of o-alkynylphenols: exploring a combinatorial approach. J Org Chem 69:2235
18. Zhou L, Shi Y, Xiao Q, Liu Y, Ye F, Zhang Y, Wang J (2011) CuBr-catalyzed coupling of N-tosylhydrazones and terminal alkynes: synthesis of benzofurans and indoles. Org Lett 13:968

19. Xiao T, Dong X, Zhou L (2013) Benzofuran and indole synthesis via Cu(I)-catalyzed coupling of N-tosylhydrazone and o-hydroxy or o-amino phenylacetylene. Org Biomol Chem 11:1490
20. Liao Y, Smith J, Fathi R, Yang Z (2005) Novel PdII-Mediated Cascade Carboxylative Annulation to Construct Benzo[b]furan-3-carboxylic Acids. Org Lett 7:2707
21. Nakamura M, Ilies L, Otsubo S, Nakamura E (2006) 2,3-disubstituted benzofuran and indole by copper-mediated C-C bond extension reaction of 3-zinciobenzoheterole. Org Lett 8:2803
22. Martinez C, Alvarez R, Aurrecoechea JM (2009) Palladium-catalyzed sequential oxidative cyclization/coupling of 2-alkynylphenols and alkenes: a direct entry into 3-alkenylbenzofurans. Org Lett 11:1083
23. Isono N, Lautens M (2009) Rhodium(I)-catalyzed cyclization reaction of o-alkynyl phenols and anilines. Domino approach to 2,3-disubstituted benzofurans and indoles. Org Lett 11:1329
24. Boyer A, Isono N, Lackner S, Lautens M (2010) Domino rhodium(I)-catalysed reactions for the efficient synthesis of substituted benzofurans and indoles. Tetrahedron 66:6468
25. Wang H, Han X, Lu X (2011) Cationic palladium(II)-catalyzed synthesis of 2-substituted 3-hydroxymethylbenzo[b]furans. Synlett 2590
26. Nakamura I, Mizushima Y, Yamamoto Y (2005) Synthesis of 2,3-disubstituted benzofurans by platinum-olefin-catalyzed carboalkoxylation of o-alkynylphenyl acetals. J Am Chem Soc 127:15022
27. Fürstner A, Davies PW (2005) Heterocycles by PtCl2-catalyzed intramolecular carboalkoxylation or carboamination of alkynes. J Am Chem Soc 127:15024
28. Du HA, Zhang XG, Tang RY, Li JH (2009) PdCl2-promoted electrophilic annulation of 2-alkynylphenol derivatives with disulfides or diselenides in the presence of iodine. J Org Chem 74:7844
29. Swamy NK, Yazici A, Pyne SG (2010) Copper-mediated cyclization-halogenation and cyclization-cyanation reactions of β-hydroxyalkynes and o-alkynylphenols and anilines. J Org Chem 75:3412
30. Gay RM, Manarin F, Schneider CC, Barancelli DA, Costa MD, Zeni G (2010) FeCl3-diorganyl dichalcogenides promoted cyclization of 2-alkynylanisoles to 3-chalcogen benzo [b]furans. J Org Chem 75:5701
31. Tsuchikama K, Hashimoto Y, Endo K, Shibata T (2009) Iridium-catalyzed selective synthesis of 4-substituted benzofurans and indoles via directed cyclodehydration. Adv Synth Catal 351:2850
32. Shibata T, Hashimoto Y, Otsuka M, Tsuchikama K, Endo K (2011) Ir(III)-catalyzed room-temperature synthesis of multisubstituted benzofurans initiated by C–H activation of a-aryloxy ketones. Synlett 2075
33. Youn SW, Eom JI (2005) Facile construction of the benzofuran and chromene ring systems via PdII-catalyzed oxidative cyclization. Org Lett 7:3355
34. Gabriele B, Mancuso R, Lupinacci E, Salerno G, Veltri L (2010) Tandem catalysis in ionic liquids: a recyclable catalytic synthesis of benzofuran derivatives. Tetrahedron 66:6156
35. Gabriele B, Mancuso R, Salerno G, Costa M (2007) Cascade reactions: a new synthesis of 2-benzofuran-2-ylacetamides by sequential Pd(0)-catalyzed deallylation-Pd(II)-catalyzed aminocarbonylative heterocyclization of 1-(2-allyloxyaryl)-2-yn-1-ols. J Org Chem 72:9278
36. Liu J, Chen W, Wang L (2013) Synthesis of 2-selenyl(sulfenyl)benzofurans via Cu catalyzed tandem reactions of 2-(gemdibromovinyl) phenols with diorganyl diselenides(disulfides). RSC Adv 3:4723
37. Nagamochi M, Fang Y-Q, Lautens M (2007) A general and practical method of alkynyl indole and benzofuran synthesis via tandem Cu- and Pd-catalyzed cross-couplings. Org Lett 9:2955
38. Newman SG, Aureggi V, Bryan CS, Lautens M (2009) Intramolecular cross-coupling of gem-dibromoolefins: a mild approach to 2-bromo benzofused heterocycles. Chem Commun 5236

39. Ye S, Liu G, Pu S, Wu J (2012) Synthesis of 2-(polyfluoroaryl)benzofurans via a copper(I)-catalyzed reaction of 2-(2,2-dibromovinyl)phenol with polyfluoroarene. Org Lett 14:70
40. Li H, Liu J, Yan B, Li Y (2009) New domino approach for the synthesis of 2,3-disubstituted benzo[b]furans via copper-catalyzed multi-component coupling reactions followed by cyclization. Tetrahedron Lett 50:2353
41. Carril M, SanMartin R, Tellitu I, Domínguez E (2006) On-water chemistry: copper-catalyzed straightforward synthesis of benzo[b]furan derivatives in neat water. Org Lett 8:1467
42. Bonnamour J, Piedrafita M, Bolm C (2010) Iron and copper salts in the synthesis of benzo[b] furans. Adv Synth Catal 352:1577
43. Faragó J, Kotschy A (2009) Synthesis of benzo[b]furans by palladium–NHC catalyzed ring closure of o-bromobenzyl ketones. Synthesis 85
44. Zeng W, Wu W, Jiang H, Huang L, Sun Y, Chen Z, Li X (2013) Facile synthesis of benzofurans via copper-catalyzed aerobic oxidative cyclization of phenols and alkynes. Chem Commun 49:6611
45. Zhu R, Wei J, Shi Z (2013) Benzofuran synthesis via copper-mediated oxidative annulation of phenols and unactivated internal alkynes. Chem Sci 4:3706
46. Wang S, Li P, Yu L, Wang L (2011) Sequential and one-pot reactions of phenols with bromoalkynes for the synthesis of (Z)-2-bromovinyl phenyl ethers and benzo[b]furans. Org Lett 13:5968
47. Yuan F-G, Han F-S (2013) Iron-catalyzed direct synthesis of densely substituted benzofurans and naphthopyrans from phenolic compounds and propargylic alcohols. Adv Synth Catal 355:537
48. Guo X, Yu R, Li H, Li Z (2009) Iron-catalyzed tandem oxidative coupling and annulation: an efficient approach to construct polysubstituted benzofurans. J Am Chem Soc 131:17387
49. Liang Z, Hou W, Du Y, Zhang Y, Pan Y, Mao D, Zhao K (2009) Oxidative aromatic C-O bond formation: synthesis of 3-functionalized benzo[b]furans by FeCl3-mediated ring closure of r-aryl ketones. Org Lett 11:4978
50. Huang H, Jiang H, Cao H, Zhao J, Shi D (2012) Palladium-catalyzed one-pot synthesis of polysubstituted furans from alkynoates and 2-yn-1-ols. Tetrahedron 68:3135
51. Hua C, Jiang H, Huang H (2011) Transition-metal-catalyzed domino reactions: efficient one-pot regiospecific synthesis of highly functionalized polysubstituted furans from electron-deficient alkynes and 2-Yn-1-ols. Synthesis 1019
52. Cao H, Jiang H, Yao W, Liu X (2009) Copper-catalyzed domino rearrangement/dehydrogenation oxidation/carbene oxidation for one-pot regiospecific synthesis of highly functionalized polysubstituted furans. Org Lett 11:1931
53. Cao H, Jiang H, Mai R, Zhu S, Qi C (2010) Silver-catalyzed one-pot cyclization reaction of electron- deficient alkynes and 2-Yn-1-ols: an efficient domino process to polysubstituted furans. Adv Synth Catal 352:143
54. Suhre MH, Reif M, Kirsch SF (2005) Gold(I)-catalyzed synthesis of highly substituted furans. Org Lett 7:3925
55. Jiang H, Yao W, Cao H, Huang H, Cao D (2010) Iron-catalyzed domino process for the synthesis of r-carbonyl furan derivatives via one-pot cyclization reaction. J Org Chem 75:5347
56. Zheng M, Huang L, Wu W, Jiang H (2013) Pd(II)-catalyzed sequential C_C/C_O bond formations: a new strategy to construct trisubstituted furans. Org Lett 15:1838
57. Zhang X, Lu Z, Fu C, Ma S (2010) Synthesis of polysubstituted furans based on a stepwise Sonogashira coupling of (Z)-3-iodoalk-2-en-1-ols with terminal propargylic alcohols and subsequent Au(I)- or Pd(II)-catalyzed cyclization-aromatization via elimination of H_2O. J Org Chem 75:2589
58. Liu Y, Song F, Song Z, Liu M, Yan B (2005) Gold-catalyzed cyclization of (Z)-2-En-4-yn-1-ols: highly efficient synthesis of fully substituted dihydrofurans and furans. Org Lett 7:5409

59. Du X, Song F, Lu Y, Chen H, Liu Y (2009) A general and efficient synthesis of substituted furans and dihydrofurans via gold-catalyzed cyclization of (Z)-2-en-4-yn-1-ols. Tetrahedron 65:1839
60. Du X, Chen H, Chen Y, Chen J, Liu Y (2011) Highly efficient synthesis of multisubstituted 2-acyl furans via PIFA/I2- mediated oxidative cycloisomerization of cis-2-En-4-yn-1-ols. Synlett 1010
61. Li E, Cheng X, Wang C, Shao Y, Li Y (2012) Palladium-catalyzed synthesis of 2,3,4-trisubstituted furans via cascade reactions of aryloxy-enynes with aryl halides. J Org Chem 7744
62. Li E, Yao W, Xie X, Wang C, Shao Y, Li Y (2012) Gold-catalyzed efficient synthesis of 2,4-disubstituted furans from aryloxyenynes. Org Biomol Chem 10:2960
63. Hayes SJ, Knight DW, Menzies MD, O'Halloran M, Tan WF (2007) An efficient furan synthesis using heterogeneous catalysis. Tetrahedron Lett 48:7709
64. Aponick A, Li C-Y, Malinge J, Marques EF (2009) An extremely facile synthesis of furans, pyrroles, and thiophenes by the dehydrative cyclization of propargyl alcohols. Org Lett 11:4624
65. Egi M, Azechi K, Akai S (2009) Cationic gold(I)-mediated intramolecular cyclization of 3-alkyne-1,2-diols and 1-amino-3-alkyn-2-ols: a practical route to furans and pyrroles. Org Lett 11:5002
66. Egi M, Azechi K, Akai S (2011) Reusable and durable immobilized-cationic gold(I) catalysts for environmentally benign bond-forming reactions. Adv Synth Catal 353:287
67. Blanc A, Tenbrink K, Weibel J-M, Pale P (2009) Silver(I)-catalyzed cascade: direct access to furans from alkynyloxiranes. J Org Chem 74:4360
68. Blanc A, Tenbrink K, Weibel J-M, Pale P (2009) Mechanistic studies and improvement of coinage metal-catalyzed transformation of alkynyloxiranes to furans: an alcohol addition-cyclization-elimination cascade. J Org Chem 74:5342
69. Aurrecoechea JM, Durana A, Pérez E (2008) Palladium-catalyzed cyclization/Heck- and cyclization/conjugate-addition-type sequences in the preparation of polysubstituted furans. J Org Chem 73:3650
70. Li Y, Wheeler KA, Dembinski R (2010) Gold(I)-catalyzed cycloisomerization of 2-fluoroalk-3-yn-1-ones: synthesis of 2,5-substituted 3-fluorofurans. Adv Synth Catal 352:2761
71. Li Y, Wheeler KA, Dembinski R (2012) Room temperature syntheses of entirely diverse substituted β-fluorofurans. Org Biomol Chem 10:2395
72. Liu R, Zhang J (2009) Tetrasubstituted furans by PdII-catalyzed three-component domino reactions of 2-(1-alkynyl)-2-alken-1-ones with nucleophiles and vinyl ketones or acrolein. Chemistry 15:9303
73. Xiao Y, Zhang J (2008) Tetrasubstituted furans by a PdII-catalyzed three-component Michael addition/cyclization/cross-coupling reaction. Angew Chem Int Ed 47:1903
74. Li W, Zhang J (2010) Tetrasubstituted furans by PdII/CuI-cocatalyzed three-component domino reactions of 2-(1-alkynyl)-2-alken-1-ones, nucleophiles and diaryliodonium salts. Chem Commun 46:8839
75. Li J, Liu L, Ding D, Sun J, Ji Y, Dong J (2013) Gold(III)-catalyzed three-component coupling reaction (TCC) selective toward furans. Org Lett 15:2884
76. Jiang Y, Zhong V, Khong Y, Lourdusamy E, Park C-M (2012) Synthesis of 2-aminofurans and 2-unsubstituted furans via carbenoid-mediated [3 + 2] cycloaddition. Chem Commun 48:3133
77. He C, Guo S, Ke J, Hao J, Xu H, Chen H, Lei A (2012) Silver-mediated oxidative C – H/C – H functionalization: a strategy to construct polysubstituted furans. J Am Chem Soc 134:5766
78. Lednden P, Entwistle DA, Willis MC (2011) An alkyne hydroacylation route to highly substituted furans. Angew Chem Int Ed 50:1065

79. Feng X, Tan Z, Chen D, Shen Y, Guo C-C, Xiang J, Zhu C (2008) Synthesis of tetrasubstituted furans via In-catalyzed propargylation of 1,3-dicarbonyl compounds-cyclization tandem process. Tetrahedron Lett 49:4110
80. Pan Y-M, Zhao S-Y, Ji W-H, Zhan Z-P (2009) One-pot synthesis of substituted furans using Cu(OTf)2-catalyzed propargylation/cycloisomerization tandem reaction. J Comb Chem 11:103
81. Zhan Z-P, Wang S-P, Cai X-B, Liu H-J, Yu J-L, Cui Y-Y (2007) Copper(II) triflate-catalyzed nucleophilic substitution of propargylic acetates with enoxysilanes. A straightforward synthetic route to polysubstituted furans. Adv Synth Catal 349:2097
82. Karpov AS, Merkul E, Oeser T, Müller TJJ (2006) One-pot three-component synthesis of 3-halofurans and 3-chloro-4-iodofurans. Eur J Org Chem 2991
83. Gabriele B, Mancuso R, Maltese V, Veltri L, Salerno G (2012) Synthesis of furan-3-carboxylic and 4-methylene-4,5-dihydrofuran- 3-carboxylic esters by direct palladium iodide catalyzed oxidative carbonylation of 3-Yne-1,2-diol derivatives. J Org Chem 77:8657
84. Wu X-F, Zhang M, Jiao H, Neumann H, Beller M (2013) Palladium-catalyzed synthesis of furans from double carbonylative coupling of aryl halides with terminal alkynes. Asian J Org Chem 2:135
85. Jiang H, Zeng W, Li Y, Wu W, Huang L, Fu W (2012) Copper(I)-catalyzed synthesis of 2,5-disubstituted furans and thiophenes from haloalkynes or 1,3-diynes. J Org Chem 77:5179
86. Nun P, Dupuy S, Gaillard S, Poater A, Cavallo L, Nolan SP (2011) Gold(I)-catalyzed synthesis of furans and pyrroles via alkyne hydration. Catal Sci Technol 1:58
87. Kramer S, Madsen JLH, Rottlander M, Skrydstrup T (2010) Access to 2,5-diamidopyrroles and 2,5-diamidofurans by Au(I)-catalyzed double hydroamination or hydration of 1,3-diynes. Org Lett 12:2758
88. Yang J, Liu S, Zheng J-F, Zhou J (2012) Room-temperature Suzuki–Miyaura coupling of heteroaryl chlorides and tosylates. Eur J Org Chem 6248
89. Salman GA, Nisa RU, Iaroshenko VO, Iqbal J, Ayub K, Langer P (2012) Pyrrole versus quinoline formation in the palladium catalyzed reaction of 2-alkynyl-3-bromothiophenes and 2-alkynyl-3-bromofurans with anilines. A combined experimental and computational study. Org Biomol Chem 10:9464
90. Salman GA, Ali A, Hussain M, Khera RA, Langer P (2011) Synthesis of functionalized benzofurans by a double Heck reaction of 2,3-dibromofurans and subsequent 6p-electrocyclization/dehydrogenation. Synthesis 2208
91. Molander GA, Brown AR (2006) Suzuki-Miyaura cross-coupling reactions of potassium vinyltrifluoroborate with aryl and heteroaryl electrophiles. J Org Chem 71:9681
92. Bárta J, Slavetinská L, Klapetárová B, Hocek M (2010) Modular synthesis of 5-substituted furan-2-yl C-2-deoxyribonucleosides and biaryl covalent base-pair analogues. Eur J Org Chem 5432
93. Fu HY, Zhao L, Bruneau C, Doucet H (2012) Palladium-catalysed direct heteroarylations of heteroaromatics using esters as blocking groups at C2 of bromofuran and bromothiophene derivatives: a One-step access to biheteroaryls. Synlett 2077
94. Chen Y-H, Knochel P (2008) Preparation of aryl and heteroaryl indium(III) reagents by the direct insertion of indium in the presence of LiCl. Angew Chem Int Ed 47:7648
95. Kinzel T, Zhang Y, Buchwald SL (2010) A new palladium precatalyst allows for the fast Suzuki-Miyaura coupling reactions of unstable polyfluorophenyl and 2-heteroaryl boronic acids. J Am Chem Soc 132:14073
96. Li J-H, Zhu Q-M, Xie Y-X (2006) Pd(OAc)2/DABCO-catalyzed Suzuki–Miyaura cross-coupling reaction in DMF. Tetrahedron 62:10888
97. Jiang S, Tala SR, Lu H, Abo-Dya NE, Avan I, Gyanda K, Lu L, Katritzky DAK (2011) Design, synthesis, and biological activity of novel 5-((arylfuran/1H-pyrrol-2-yl)methylene)-2- thioxo-3-(3-(trifluoromethyl)phenyl)thiazolidin-4-ones as HIV-1 fusion inhibitors targeting gp41. J Med Chem 54:572

98. Dubbaka SR, Vogel P (2004) Palladium-catalyzed Suzuki-Miyaura cross-couplings of sulfonyl chlorides and boronic acids. Org Lett 6:95
99. Su W, Urgaonkar S, McLaughlin PA, Verkade JG (2004) Highly active palladium catalysts supported by bulky proazaphosphatrane ligands for stille cross-coupling: coupling of aryl and vinyl chlorides, room temperature coupling of aryl bromides, coupling of aryl triflates, and synthesis of sterically hindered biaryls. J Am Chem Soc 126:16433
100. Denmark SE, Baird JD, Regens CS (2008) Palladium-catalyzed cross-coupling of five-membered heterocyclic silanolates. J Org Chem 73:1440
101. Denmark SE, Baird JD (2006) Palladium-catalyzed cross-coupling reactions of heterocyclic silanolates with substituted aryl iodides and bromides. Org Lett 8:793
102. Gavryushin A, Kofink C, Manolikakes G, Knochel P (2006) An efficient Negishi cross-coupling reaction catalyzed by nickel(II) and diethyl phosphite. Tetrahedron 62:7521
103. Groll K, Blümke TD, Unsinn A, Haas D, Knochel P (2012) Direct Pd-catalyzed cross-coupling of functionalized organoaluminum reagents. Angew Chem Int Ed 51:11157
104. Sévigny S, Forgione P (2013) Efficient desulfinylative cross-coupling of thiophene and furan sulfinates with aryl bromides in aqueous media. N J Chem 37:589
105. Goossen L, Zimmermann B, Linder C, Rodrígez N, Lange P, Hartung J (2009) Synthesis of biaryls and aryl ketones via microwave-assisted decarboxylative cross-couplings. Adv Synth Catal 351:2667
106. Goossen L, Rodrígez N, Lange P, Linder C (2010) Decarboxylative cross-coupling of aryl tosylates with aromatic carboxylate salts. Angew Chem Int Ed 49:1111
107. Organ MG, Calimsiz S, Sayah M, Hoi KH, Lough AJ (2009) Pd-PEPPSI-IPent: an active, sterically demanding cross-coupling catalyst and its application in the synthesis of tetra-ortho- substituted biaryls. Angew Chem Int Ed 48:2383
108. Lu L, Yan H, Sun P, Zhu Y, Yang H, Liu D, Rong G, Mao J (2013) Synthesis of internal alkynes through the Pd-catalyzed coupling of heteroaryl halides with terminal alkynes. Eur J Org Chem 1644
109. Nguyen TTB, Lomberget T, Tran NC, Barret R (2013) Synthesis of (Z) isomers of benzoheterocyclic derivatives of combretastatin A-4: a comparative study of several methods. Tetrahedron 69:2336
110. Wang Z, Elokdah H, McFairlane G, Pan S, Antane M (2006) Regioselective Suzuki coupling of benzofuran or benzothiophene boronic acids and dibromo substituted naphthalenes: synthesis of a potent inhibitor of plasminogen activator inhibitor-1. Tetrahedron Lett 47:3365
111. Denmark SE, Smith RC, Chang W-TT, Muhuhi JM (2009) Cross-coupling reactions of aromatic and heteroaromatic silanolates with aromatic and heteroaromatic halides. J Am Chem Soc 131:3104
112. Matsuda S, Takahashi M, Monguchi D, Mori A (2009) C–H and C–Si functionalization of furan derivatives: palladium-catalyzed homocoupling and arylation reactions. Synlett 1941
113. Manarin F, Roehrs JA, Brandao R, Nogueira CW, Zeni G (2009) Synthesis of 3-alkynyl-2-(methylsulfanyl)benzo[b]furans via Sonogashira cross-coupling of 3-iodo-2-(methylsulfanyl) benzo[b]furans with terminal alkynes. Synthesis 4001
114. Okitsu T, Nakazawa D, Nakagawa K, Okano T, Wada A (2010) Synthesis and biological evaluation of 9Z-retinoic acid analogs having 2-substituted benzo[b]furan. Chem Pharm Bull 58:418
115. Darweesh AF, Shaaban MR, Farag AM, Metz P, Dawood KM (2010) Facile access to biaryls and 2-acetyl-5-arylbenzofurans via Suzuki coupling in water under thermal and microwave conditions. Synthesis 3163
116. Arcadi A, Blesi F, Cacchi S, Fabrizi G, Goggiamani A (2011) 2,5,7-trisubstituted benzo[b] furans through a copper- and/or palladium-catalyzed assembly and functionalization process. Tetrahedron Lett 52:5149
117. Ghosh D, Lee HM (2012) Efficient Pd-catalyzed direct arylations of heterocycles with unreactive and hindered aryl chlorides. Org Lett 14:5534

118. Nadres ET, Lazareva A, Daugulis O (2011) Palladium-catalyzed indole, pyrrole, and furan arylation by aryl chlorides. J Org Chem 76:471
119. Baghbandazeh M, Pilger C, Kappe CO (2011) Palladium-catalyzed direct arylation of heteroaromatic compounds: improved conditions utilizing controlled microwave heating. J Org Chem 76:8138
120. Chen L, Bruneau C, Dixneuf PH, Doucet H (2013) Palladium-acetate catalyst for regioselective direct arylation at C2 of 3-furanyl or 3-thiophenyl acrylates with inhibition of Heck type reaction. Tetrahedron 69:4381
121. Roy D, Mom S, Lucas D, Cattey H, Hierso J-C, Doucet H (2011) Direct arylation of heteroaromatic compounds with congested, functionalised aryl bromides at low palladium/triphosphane catalyst loading. Chemistry 17:6453
122. Roger J, Doucet H (2010) Palladium-catalysed direct 5-arylation of furfurylamine or 2-(aminoalkyl)-thiophene derivatives. Eur J Org Chem 4412
123. Liégault B, Lapointe D, Caron L, Vlassova A, Fagnou K (2009) Establishment of broadly applicable reaction conditions for the palladium-catalyzed direct arylation of heteroatom-containing aromatic compounds. J Org Chem 74:1826
124. Battace A, Lemhadri M, Zair T, Doucet H, Santelli M (2007) Palladium-catalyzed direct arylation of furans via C-H functionalization at low catalyst loadings. Organometallics 26:472
125. Laidaoui N, Roger J, Miloudi A, El Abed D, Doucet H (2011) Palladium-catalyzed direct arylations of five-membered heteroarenes bearing N-monoalkylcarboxamide substituents. Eur J Org Chem 4373
126. Chen L, Roger J, Bruneau C, Dixneuf PH (2011) Phosphine-free palladium catalytic system for the selective direct arylation of furans or thiophenes bearing alkenes and inhibition of Heck-type reaction. Adv Synth Catal 353:2749
127. Della Ca N, Maestri G, Catellani M (2009) Palladium/norbornene-catalyzed synthesis of heteroatom-containing o-teraryls from aryl iodides and heteroarenes through double C_H activation in sequence. Chemistry 15:7850
128. Yanagisawa S, Sudo T, Noyori R, Itami K (2006) Direct C-H arylation of (hetero)arenes with aryl iodides via rhodium catalysis. J Am Chem Soc 128:11748
129. René O, Fagnou K (2010) Evaluation of electron-deficient phosphine ligands for direct arylation of heterocycles. Adv Synth Catal 352:2116
130. Qian YY, Wong KL, Zhang MW, Kwok TY, To CT, Chan KS (2012) Catalytic C–H arylation of unactivated heteroaromatics with aryl halides by cobalt porphyrin. Tetrahedron Lett 53:1571
131. Join B, Yamamoto T, Itami K (2009) Iridium catalysis for C_H bond arylation of heteroarenes with iodoarenes. Angew Chem Int Ed 48:3644
132. Zhao J, Huang L, Cheng K, Zhang Y (2009) Palladium-catalyzed alkenation of thiophenes and furans by regioselective C–H bond functionalization. Tetrahedron Lett 50:2758
133. Maehara A, Satoh T, Miura M (2008) Palladium-catalyzed direct oxidative vinylation of thiophenes and furans under weakly basic conditions. Tetrahedron 64:5982
134. Aouf C, Thiery E, Le Bras J, Muzart J (2009) Palladium-catalyzed dehydrogenative coupling of furans with styrenes. Org Lett 11:4096
135. Vasseur A, Muzart J, Le Bras J (2011) Dehydrogenative Heck reaction of furans and thiophenes with styrenes under mild conditions and influence of the oxidizing agent on the reaction rate. Chemistry 17:12556
136. Zhang Y, Li Z, Liu Z-Q (2012) Pd-catalyzed olefination of furans and thiophenes with allyl esters. Org Lett 14:226
137. Jiang Z, Zhang L, Dong C, Cai Z, Tang W, Li H, Xu L, Xiao J (2012) Palladium-catalyzed highly regioselective arylation of allylamines with thiophenes and furans. Adv Synth Catal 354:3225
138. Juwaini NAB, Ng JKP, Seayad J (2012) Catalytic regioselective oxidative coupling of furan-2-carbonyls with simple arenes. ACS Catal 2:1787

139. Xi P, Yang F, Qin S, Zhao D, Lan J, Gao G, Hu C, You J (2010) Palladium(II)-catalyzed oxidative C-H/C-H cross-coupling of heteroarenes. J Am Chem Soc 132:1822
140. He C-Y, Fan S, Zhang X (2010) Pd-catalyzed oxidative cross-coupling of perfluoroarenes with aromatic heterocycles. J Am Chem Soc 132:12850
141. Kuhl N, Hopkinson MN, Glorius F (2012) Selective rhodium(III)-catalyzed cross-dehydrogenative coupling of furan and thiophene derivatives. Angew Chem Int Ed Engl 51:8230
142. Gottumukkala A, Doucet H (2008) Palladium-catalyzed direct C-4 arylation of 2,5-disubstituted furans with aryl bromides. Adv Synth Catal 350:2183
143. Cao H, Shen D, Zhan H, Yang L (2011) Palladium-catalyzed direct arylation reaction of 2,3,5-trisubstituted furans with aryl iodides or aryl bromides. Synlett 1472
144. Li P, Gu J-W, Ying Y, He Y-M, Zhang H-F, Zhao G, Zhu S-Z (2010) Palladium-catalyzed alkenylation of fluoro-substituted furans via CeH activation to form tetrasubstituted furans. Tetrahedron 66:8387
145. Iitsuka T, Schaal P, Hirano K, Satoh T, Bolm C, Miura M (2013) Rhodium-catalyzed C3-selective alkenylation of substituted thiophene-2-carboxylic acids and related compounds. J Org Chem 78:7216
146. Roy D, Mom S, Lucas D, Hierso J-C, Doucet H (2012) Palladium-catalyzed direct arylation of heteroaromatics with activated aryl chlorides using a sterically relieved ferrocenyl-diphosphane. ACS Catal 2:1033
147. Tamba S, Okubo Y, Tanaka S, Monguchi D, Mori A (2010) Palladium-catalyzed C-H functionalization of heteroarenes with aryl bromides and chlorides. J Org Chem 75:6998
148. Yanagisawa S, Itami K (2011) Palladium/2,20-bipyridyl/Ag2CO3 catalyst for CeH bond arylation of heteroareneswith haloarenes. Tetrahedron 67:4425
149. Yamaguchi K, Yamaguchi J, Studer A, Itami K (2012) Hindered biaryls by C–H coupling: bisoxazoline-Pd catalysis leading to enantioselective C–H coupling. Chem Sci 3:2165
150. Biajoli AFP, da Penha ET, Correira CRD (2012) Palladium catalysed regioselective arylation of indoles, benzofuran and benzothiophene with aryldiazonium salts. RSC Adv 2:11930
151. He C-Y, Min Q-Q, Zhang X (2012) Palladium-catalyzed aerobic dehydrogenative cross-coupling of polyfluoroarenes with thiophenes: facile access to polyfluoroarene – thiophene structure. Organometallics 31:1335
152. Liégault B, Petrov I, Gorelsky SI, Fagnou K (2010) Modulating reactivity and diverting selectivity in palladium-catalyzed heteroaromatic direct arylation through the Use of a chloride activating/blocking group. J Org Chem 75:1047
153. Carrer A, Brinet D, Florent J-C, Rouselle P, Bertounesque E (2012) Palladium-catalyzed direct arylation of polysubstituted benzofurans. J Org Chem 77:1316
154. Ueyama T, Mochida S, Fukutani T, Hirano K, Satoh T, Miura M (2011) Ruthenium-catalyzed oxidative vinylation of heteroarene carboxylic acids with alkenes via regioselective C-H bond cleavage. Org Lett 13:706
155. Hashmi ASK, Weyrauch JP, Frey W, Bats JW (2004) Gold catalysis: mild conditions for the synthesis of oxazoles from N-propargylcarboxamides and mechanistic aspects. Org Lett 6:4391
156. Milton MD, Inada Y, Nishibayashi Y, Uemura S (2004) Ruthenium- and gold-catalysed sequential reactions: a straightforward synthesis of substituted oxazoles from propargylic alcohols and amides. Chem Commun 2712
157. Wachenfeldt H, Rçse P, Paulsen F, Loganathan N, Strand D (2013) Catalytic three-component domino reaction for the preparation of trisubstituted oxazoles. Chemistry 19:7982
158. Wendlandt AE, Stahl SS (2012) Copper(II)-mediated oxidative cyclization of enamides to oxazoles. Org Biomol Chem 10:3866
159. Ueda M, Ikeda Y, Sato A, Ito Y, Kakiuchi M, Shono H, Miyoshi T, Naito T, Miyata O (2011) Silver-catalyzed synthesis of disubstituted isoxazoles by cyclization of alkynyl oxime ethers. Tetrahedron 67:4612

160. Ueda M, Sato A, Ikeda Y, Miyoshi T, Naito T, Miyata O (2010) Direct synthesis of trisubstituted isoxazoles through gold-catalyzed domino reaction of alkynyl oxime ethers. Org Lett 12:2594
161. Evindar G, Batey RA (2006) Parallel synthesis of a library of benzoxazoles and benzothiazoles using ligand-accelerated copper-catalyzed cyclizations of ortho-halobenzanilides. J Org Chem 71:1802
162. Ueda S, Nagasawa H (2009) Copper-catalyzed synthesis of benzoxazoles via a regioselective C-H functionalization/C-O bond formation under an air atmosphere. J Org Chem 74:4272
163. Udd S, Jokela R, Franzén R, Tois J (2010) Copper-catalyzed cyclization of *Z*-oximes into 3-methyl-1,2-benzisoxazoles. Tetrahedron Lett 51:1030
164. Waldo JP, Larock RC (2007) The synthesis of highly substituted isoxazoles by electrophilic cyclization: an efficient synthesis of valdecoxib. J Org Chem 72:9643
165. Luo Y, Ji K, Li Y, Zhang L (2012) Tempering the reactivities of postulated α-Oxo gold carbenes using bidentate ligands: implication of tricoordinated gold intermediates and the development of an expedient bimolecular assembly of 2,4-disubstituted oxazoles. J Am Chem Soc 134:17412
166. Davies PW, Cremonesi A, Dumitrescu L (2011) Intermolecular and selective synthesis of 2,4,5-trisubstituted oxazoles by a gold-catalyzed formal [3 + 2] cycloaddition. Angew Chem Int Ed 50:8931
167. Himo F, Lovell T, Hilgraf R, Rostovtsev VV, Noodleman L, Sharpless KB, Fokin VV (2005) Copper(I)-catalyzed synthesis of azoles. DFT study predicts unprecedented reactivity and intermediates. J Am Chem Soc 127:210
168. Bharate SB, Padala AK, Dar BA, Yadav RR, Singh B, Vishwakarma RA (2013) Montmorillonite clay Cu(II) catalyzed domino one-pot multicomponent synthesis of 3,5-disubstituted isoxazoles. Tetrahedron Lett 54:3558
169. Grecian S, Fokin VV (2008) Ruthenium-catalyzed cycloaddition of nitrile oxides and alkynes: practical synthesis of isoxazoles. Angew Chem Int Ed 47:8285
170. Ahmed MSM, Kobayashi K, Mori M (2005) One-pot construction of pyrazoles and isoxazoles with palladium-catalyzed four-component coupling. Org Lett 7:4487
171. Liu B, Yin M, Gao H, Wu W, Jiang H (2013) Synthesis of 2-aminobenzoxazoles and 3-aminobenzoxazines via palladium-catalyzed aerobic oxidation of o-aminophenols with isocyanides. J Org Chem 78:3009
172. Bochatay VN, Boissarie PJ, Murphy JA, Suckling CJ, Lang S (2013) Mechanistic exploration of the palladium-catalyzed process for the synthesis of benzoxazoles and benzothiazoles. J Org Chem 78:1471
173. Guru MM, Ali MA, Punniyamurthy T (2011) Copper(II)-catalyzed conversion of bisaryloxime ethers to 2-arylbenzoxazoles via C-H functionalization/C-N/C-O bonds formation. Org Lett 13:1194
174. Austeri M, Rix D, Zeghida W, Lacour J (2011) CpRu-catalyzed O-H insertion and condensation reactions of α-diazocarbonyl compounds. Org Lett 13:1394
175. Stokes BJ, Vogel CV, Urnezis LK, Pan M, Driver TG (2010) Intramolecular Fe(II)-catalyzed N-O or N-N bond formation from aryl azides. Org Lett 12:2884
176. Strotman NA, Chobanian HR, He J, Guo Y, Dormer PG, Jones CM, Steves JE (2010) Catalyst-controlled regioselective Suzuki couplings at both positions of dihaloimidazoles, dihalooxazoles, and dihalothiazoles. J Org Chem 75:1733
177. Liebeskind LS, Srogl J (2002) Heteroaromatic thioether-boronic acid cross-coupling under neutral reaction conditions. Org Lett 4:979
178. Lee K, Counceller CM, Stambuli JP (2009) Nickel-catalyzed synthesis of oxazoles via C-S activation. Org Lett 11:1457
179. Silva S, Sylla B, Suzenet F, Tatibouet A, Rauter AP, Rollin P (2008) Oxazolinethiones and oxazolidinethiones for the first copper-catalyzed desulfurative cross-coupling reaction and first Sonogashira applications. Org Lett 10:853

180. Goossen LJ, Deng G, Levy LM (2006) Synthesis of biaryls via catalytic decarboxylative coupling. Science 313:662
181. Zhang F, Greaney MF (2010) Decarboxylative cross-coupling of azoyl carboxylic acids with aryl halides. Org Lett 12:4745
182. Pivsa-Art S, Satoh T, Kawamura Y, Miura M, Nomura M (1998) Palladium-catalyzed arylation of azole compounds with aryl halides in the presence of alkali metal carbonates and the use of copper iodide in the reaction. Bull Chem Soc Jpn 71:467
183. Sanchez RF, Zhuravlev FA (2007) Mechanistic evidence for a ring-opening pathway in the Pd-catalyzed direct arylation of benzoxazoles. J Am Chem Soc 129:5824
184. Strotman NA, Chobanian HR, Guo Y, He J, Wilson JE (2010) Highly regioselective palladium-catalyzed direct arylation of oxazole at C-2 or C-5 with aryl bromides, chlorides, and triflates. Org Lett 12:3578
185. Gorelsky SI, Lapointe D, Fagnou K (2008) Analysis of the concerted metalation-deprotonation mechanism in palladium-catalyzed direct arylation across a broad range of aromatic substrates. J Am Chem Soc 130:10848
186. Do H-Q, Daugulis O (2007) Copper-catalyzed arylation of heterocycle C-H bonds. J Am Chem Soc 129:12404
187. Lewis JC, Wiedemann SH, Bergman RG, Ellman JA (2004) Arylation of heterocycles via rhodium-catalyzed C-H bond functionalization. Org Lett 6:35
188. Fall Y, Reynaud C, Doucet H, Santelli M (2009) Ligand-free-palladium-catalyzed direct 4-arylation of isoxazoles using aryl bromides. Eur J Org Chem 4041
189. Wu X-F, Anbarasan P, Neumann H, Beller M (2010) Palladium-catalyzed carbonylative C-H activation of heteroarenes. Angew Chem Int Ed 49:7316
190. Ackermann L, Barfüsser S, Pospech J (2010) Palladium-catalyzed direct arylations, alkenylations, and benzylations through C-H bond cleavages with sulfamates or phosphates as electrophiles. Org Lett 12:724
191. Zhang F, Greaney MF (2010) Decarboxylative C-H cross-coupling of azoles. Angew Chem Int Ed 49:2768
192. Amaike K, Muto K, Yamaguchi J, Itami K (2012) Decarbonylative C – H Coupling of Azoles and Aryl Esters: Unprecedented Nickel Catalysis and Application to the Synthesis of Muscoride A. J Am Chem Soc 134:13573
193. Li C, Li P, Yanga J, Wang L (2012) Palladium-catalyzed deamidative arylation of azoles with arylamides through a tandem decarbonylation–C–H functionalization. Chem Commun 48:4214
194. Ranjit S, Liu X (2011) Direct arylation of benzothiazoles and benzoxazoles with aryl boronic acids. Chemistry 17:1105
195. Yang F, Xu Z, Wang Z, Yu Z, Wang R (2011) Copper-catalyzed oxidative arylation of heteroarenes under mild conditions using dioxygen as the sole oxidant. Chemistry 17:6321
196. Hachiya H, Hirano K, Satoh T, Miura M (2010) Nickel-catalyzed direct C-H arylation and alkenylation of heteroarenes with organosilicon reagents. Angew Chem Int Ed 49:2202
197. Han W, Mayer P, Ofial AR (2011) Palladium-catalyzed direct arylations of azoles with aryl silicon and Tin reagents. Chemistry 17:6904
198. Zhang M, Zhang S, Liu M, Cheng J (2011) Palladium-catalyzed desulfitative C-arylation of a benzo[d]oxazole C–H bond with arene sulfonyl chlorides. Chem Commun 47:11522
199. Liu B, Guo Q, Cheng Y, Lan J, You J (2011) Palladium-catalyzed desulfitative C-H arylation of heteroarenes with sodium sulfinates. Chemistry 17:13415
200. Yu X, Li X, Wan B (2012) Palladium-catalyzed desulfitative arylation of azoles with arylsulfonyl hydrazides. Org Biomol Chem 10:7479
201. Sharma A, Vachhani D, van der Eycken E (2012) Direct heteroarylation of tautomerizable heterocycles into unsymmetrical and symmetrical biheterocycles via Pd/Cu-catalyzed phosphonium coupling. Org Lett 14:1854
202. Besselievre F, Piguel S, Mahuteau-Betzer F, Grierson DS (2008) Stereoselective direct copper-catalyzed alkenylation of oxazoles with bromoalkenes. Org Lett 10:4029

203. Matsuyama N, Hirano K, Satoh T, Miura M (2009) Nickel-catalyzed direct alkynylation of azoles with alkynyl bromides. Org Lett 11:4156
204. Besselievre F, Piguel S (2009) Copper as a powerful catalyst in the direct alkynylation of azoles. Angew Chem Int Ed 48:9553
205. Berciano BP, Lebrequier S, Besselievre F, Piguel S (2010) 1,1-dibromo-1-alkenes as valuable partners in the copper-catalyzed direct alkynylation of azoles. Org Lett 12:4038
206. Ackermann L, Kornhaass C, Zhu Y (2012) Palladium-catalyzed direct C-H bond alkynylations of heteroarenes using gem-dichloroalkenes. Org Lett 14:1824
207. Yao T, Hirano K, Satoh T, Miura M (2010) Palladium- and nickel-catalyzed direct alkylation of azoles with unactivated alkyl bromides and chlorides. Chemistry 16:12307
208. Mukai T, Hirano K, Satoh T, Miura M (2010) Palladium-catalyzed direct benzylation of azoles with benzyl carbonates. Org Lett 12:1360
209. Zhao X, Wu G, Zhang Y, Wang J (2011) Copper-catalyzed direct benzylation or allylation of 1,3-azoles with N-tosylhydrazones. J Am Chem Soc 133:3296
210. Yao T, Hirano K, Satoh T, Miura M (2012) Nickel- and cobalt-catalyzed direct alkylation of azoles with N-tosylhydrazones bearing unactivated alkyl groups. Angew Chem Int Ed 51:775
211. Boogaerts IIF, Nolan SP (2010) Carboxylation of C-H bonds using N-heterocyclic carbene gold(I) complexes. J Am Chem Soc 132:8858
212. Zhang L, Cheng L, Ohishi T, Hou Z (2010) Copper catalysed analogous synthesis of esters: copper-catalyzed direct carboxylation of C-H bonds with carbon dioxide. Angew Chem Int Ed 49:8670
213. Li Y, Jin J, Qian W, Bao W (2010) An efficient and convenient Cu(OAc)2/air mediated oxidative coupling of azoles via C–H activation. Org Biomol Chem 8:326
214. Mao Z, Wang Z, Xu Z, Huang F, Yu Z, Wang R (2012) Copper(II)-mediated dehydrogenative cross-coupling of heteroarenes. Org Lett 14:3854
215. Dong J, Huang Y, Qin X, Cheng Y, Hao J, Wan D, Li W, Liu X, You J (2012) Palladium(II)-catalyzed oxidative C-H/C-H cross-coupling between two structurally similar azoles. Chemistry 18:6158
216. Cui S, Wojtas L, Antilla JC (2011) Pd-catalyzed C4-olefination of oxazoles via CH bond activation: divergent synthesis of functionalized amino alcohol and amino acid derivatives. Org Lett 13:5040
217. Ding Z, Yoshikai N (2010) Cobalt-catalyzed addition of azoles to alkynes. Org Lett 12:4180
218. Kim SH, Yoon J, Chang S (2011) Palladium-catalyzed oxidative alkynylation of heterocycles with terminal alkynes under Air conditions. Org Lett 13:1474
219. Nakao Y, Kashihara N, Kanyiva KS, Hiyama T (2010) Nickel-catalyzed hydroheteroarylation of vinylarenes. Angew Chem Int Ed 49:4451
220. Shibata K, Yoshida M, Doi T, Takahashi T (2010) Derivatization of a tris-oxazole using Pd-catalyzed coupling reactions of a 5-bromooxazole moiety. Tetrahedron Lett 51:1674
221. Spencer J, Patel H, Amin J, Callear SK, Coles SJ, Deadman JJ, Furman C, Mansouri R, Chavatte P, Millet R (2012) Microwave-mediated synthesis and manipulation of a 2-substituted-5-aminooxazole-4-carbonitrile library. Tetrahedron Lett 53:1656
222. DeMartino JK, Garfunkle J, Hochstatter DG, Cravatt BF, Boger DL (2008) Exploration of a fundamental substituent effect of a-ketoheterocycle enzyme inhibitors: Potent and selective inhibitors of fatty acid amide hydrolase. Bioorg Med Chem Lett 18:5842
223. Cullis CA, Granger KE, Guo J, Hirose M, Li G, Mizutani M, Vos TJ (2012) Heteroaryls and uses thereof. PCT application: WO 2012/021615
224. Charles MD, Schultz P, Buchwald SL (2005) Efficient Pd-catalyzed amination of heteroaryl halides. Org Lett 7:3965
225. Monguchi D, Fujiwara T, Furukawa H, Mori A (2009) Direct amination of azoles via catalytic C-H, N-H coupling. Org Lett 11:1607
226. Wang Q, Schreiber SL (2009) Copper-mediated amidation of heterocyclic and aromatic C-H bonds. Org Lett 11:5178

227. Cho SH, Kim JY, Lee SY, Chang S (2009) Silver-mediated direct amination of benzoxazoles: tuning the amino group source from formamides to parent amines. Angew Chem Int Ed 48:9127
228. Yotphan S, Beukeaw D, Reutrakul V (2013) Synthesis of 2-aminobenzoxazoles via copper-catalyzed electrophilic amination of benzoxazoles with O-benzoyl hydroxylamines. Tetrahedron 69:6627
229. Fukuzawa S-I, Shimizu E, Atsuumi Y, Haga M, Ogata K (2009) Copper-catalyzed direct thiolation of benzoxazole with diaryl disulfides and aryl thiols. Tetrahedron Lett 50:2374
230. Dai C, Xu Z, Huang F, Yu Z, Gao YF (2012) Lewis acid-catalyzed, copper(II)-mediated synthesis of heteroaryl thioethers under base-free conditions. J Org Chem 77:4414

Top Heterocycl Chem (2016) 45: 305–358
DOI: 10.1007/7081_2014_137

Published online: 6 November 2014

Transition Metal-Catalyzed Coupling Reactions in Library Synthesis

János Gerencsér, Árpád Balázs, and György Dormán

Abstract It has been widely recognized that the global pharmaceutical and agrochemical industries are currently experiencing a dynamic change. To increase productivity and efficiency, agro and drug companies have been striving to make improvements in every aspect of the R&D. Consequently, new technologies providing novel chemical cores currently are of great demand at all major companies.

Furthermore, high-throughput screening requires the synthesis of larger and more diverse sets of compounds. High-throughput synthesis of heterocyclic compound libraries utilizes various combinatorial strategies including direct scaffold decoration, linear convergent, and divergent approaches.

Biphenyl frameworks (including heteroaromatic moieties) are found in many biologically active products and thus can be considered as a privileged motif in medicinal chemistry.

The emergence of transition metal-catalyzed coupling reactions readily facilitated library/scaffold diversification with the easy incorporation of aromatic/heteroaromatic rings into various chemotypes that are preferentially directed towards hydrophobic cavities of protein targets. Typically, carbon–halogen bonds are utilized for a variety of transition metal (mostly palladium)-catalyzed reactions, including Suzuki–Miyaura, Sonogashira, Buchwald–Hartwig, Stille, Negishi, and Heck couplings. Most recently, direct C–H activation/arylation protocols were also reported for direct derivatization of heterocyclic cores.

On the other hand, cross-coupling reactions were also implemented into domino and cascade reactions to enable multiple bond-forming and bond-cleaving events in a single synthetic operation yielding efficiently novel ring systems.

J. Gerencsér and Á. Balázs
ComInnex Inc., Záhony u. 7, 1031, Budapest, Hungary

G. Dormán (✉)
ThalesNano Inc., Záhony u. 7, 1031, Budapest, Hungary
e-mail: gyorgy.dorman@thalesnano.com

Keywords Catalysis • Cross-coupling • Library synthesis • Palladium metal

Contents

Abbreviations

BEMP	2-*Tert*-Butylimino-2-diethylamino-1,3-dimethylperhydro-1,3,2-diazaphosphorine
DavePhos	2-Dicyclohexylphosphino-2′-(*N*,*N*-dimethylamino)biphenyl
dba	Dibenzylideneacetone
DBU	1,8-Diazabicycloundec-7-ene
DCM	Dichloromethane, methylene chloride
DDQ	2,3-Dichloro-5,6-Dicyanobenzoquinone
DIC	*N*,*N*′-Diisopropylcarbodiimide
DIPEA	*N*,*N*-Diisopropylethylamine
DME	Dimethoxyethane
DMF	*N*,*N*-Dimethylformamide
dppf	1,1′- Bis(diphenylphosphino) ferrocene
DVB	Divinylbenzene
Fmoc	Fluorenylmethyloxycarbonyl
HMBA	10-[(3-Hydroxy-4-methoxybenzylidene)]-9(10*H*)-anthracenome
HOBt	Hydroxybenzotriazole
mCPBA	*Meta*-chloroperbenzoic acid
MeCN	Acetonitrile
MIDA	*N*-Methyliminodiacetic acid
MTBE	Methyl *tert*-butyl ether
MWI	Microwave irradiation
NBS	*N*-Bromosuccinic imide
NEM	*N*-Ethylmorpholine
NIS	*N*-Iodosuccinic imide

NMP	*N*-Methylpyrrolidone
POPd	Dihydrogendichlorobis(di-*tert*-butylphosphinito-*k*P)palladate(−2)
PS	Polystyrene
PTSA	4-Toluenesulfonic acid
rt	Room temperature
SPhos	2-Dicyclohexylphosphino-2′,6′-dimethoxybiphenyl
TBAB	Tetrabutylammonium bromide
TBAC	Tetrabutylammonium chloride
TBAI	Tetrabutylammonium iodide
TBS	*Tert*-butyl dimethyl
TBTU	*N*,*N*,*N*′,*N*′-Tetramethyl-*O*-(benzotriazol-1-yl)uronium tetrafluoroborate
TEA	Triethylamine
TFA	Trifluoroacetic acid
THF	Tetrahydrofuran
TosMIC	Toluenesulfonylmethyl isocyanide
TPP	Triphenylphosphine
Ts	Tosyl, 4-methylsulfonyl
Xantphos	4,5-Bis(diphenylphosphino)-9,9-dimethylxanthene
XPhos	2-Dicyclohexylphosphino-2′,4′,6′-triisopropylbiphenyl

1 Introduction

This chapter is dedicated to demonstrate how palladium-catalyzed cross-coupling reactions paved the way for efficient library synthesis and became a privileged tool for today's chemists in designing novel scaffolds (including highly conjugated and biaromatic systems) and building diversity.

Before palladium, copper was the choice of metal for carbon–carbon bond-forming reactions albeit palladium became more popular in the course of their development. Since the early discoveries of palladium-catalyzed carbon–carbon bond-forming reactions like the Suzuki–Miyaura, Heck, Negishi, Sonogashira, and Stille coupling, substantial effort has been invested into this field to develop a wide range of versatile and useful chemistries in order to gain access to valuable fine chemicals, intermediates, and drug candidates. Undoubtedly, the highest recognition of this endeavor was the Nobel Prize in Chemistry 2010 awarded jointly to Richard F. Heck, Ei-ichi Negishi, and Akira Suzuki "for palladium-catalyzed cross-couplings in organic synthesis" [1].

As a result of decades of fundamental research and the deeper understanding of this type of chemistry, nowadays, palladium-catalyzed cross-couplings became suitable not only for scaffold and target (overall library) synthesis but also to be incorporated into multicomponent [2, 3] and domino [4, 5] processes, as well, that greatly increased diversity and step economy. Today, classical palladium-mediated C–C bond-forming "name" reactions are standard instruments in the toolkit of

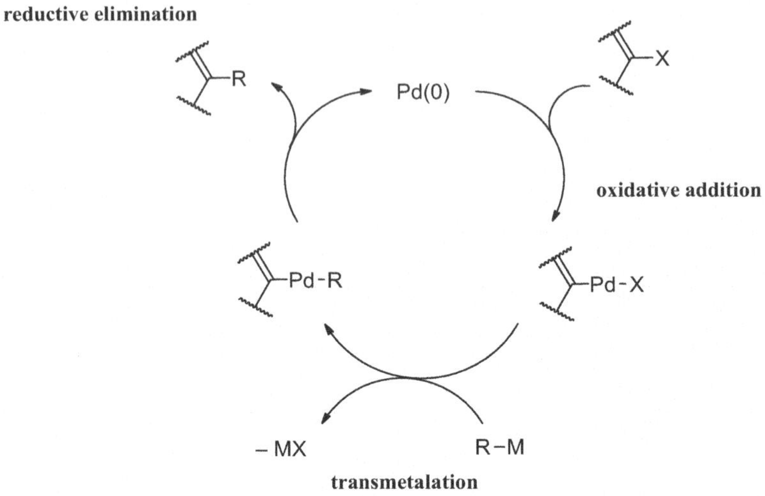

Scheme 1 General catalytic cycle for palladium catalyzed C–C bond formation (stereochemistry and ligands omitted)

bench chemists all around the globe, and the well-known catalytic cycle [6] (Scheme 1) of these types of reactions is indispensable from every chemistry textbook.

It is important to note that these chemical transformations are still under continuous development. As outstanding examples, they proved to be particularly suitable for microwave-assisted transformations [7] and most recently in continuous-flow methodologies [8] development, as well.

Contiguous research is ongoing towards the synthesis of new phosphine ligands [9] and circumstances also to improve catalyst activity, thus improving selectivity and yields while decreasing catalyst loading, reaction temperature, reaction time, and overall cost.

There is no doubt that research in this field will continue on, as state-of-the-art methodologies are currently under close investigation in order to meet the requirements of tomorrow's research regulations and environmental criteria. These innovative fields involve reactions run in/on water [10–12], the application of polymer-supported reagents and heterogeneous palladium catalysts [13, 14], and even the use of preformed "ready-to-use" catalyst-incorporated tablets [15]. Besides technology expansion, chemical development of palladium-mediated carbon–carbon bond formation is still an area of high interest nowadays. Direct C–H activation [16, 17] and oxidative palladium-catalyzed C–C bond formation [18, 19] are leading examples of this exciting field without being exhaustive.

Assorted examples that display best the utilization of palladium-catalyzed cross-coupling reactions in library design and synthesis with a focus on carbon–carbon bond-forming reactions from broadly the past decade are presented within this chapter.

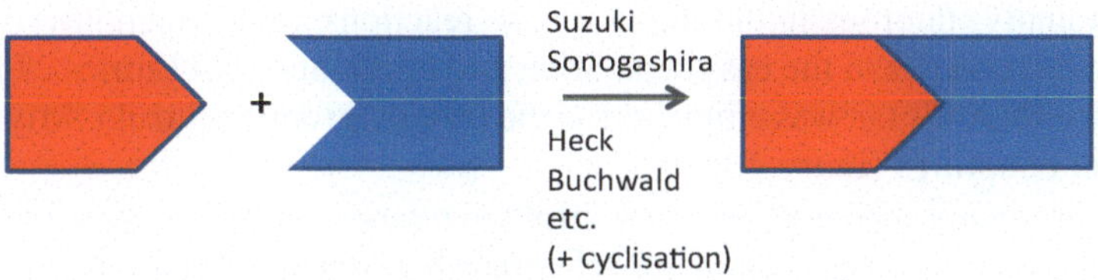

Scheme 2 Scaffold synthesis with the involvement of cross-coupling reactions

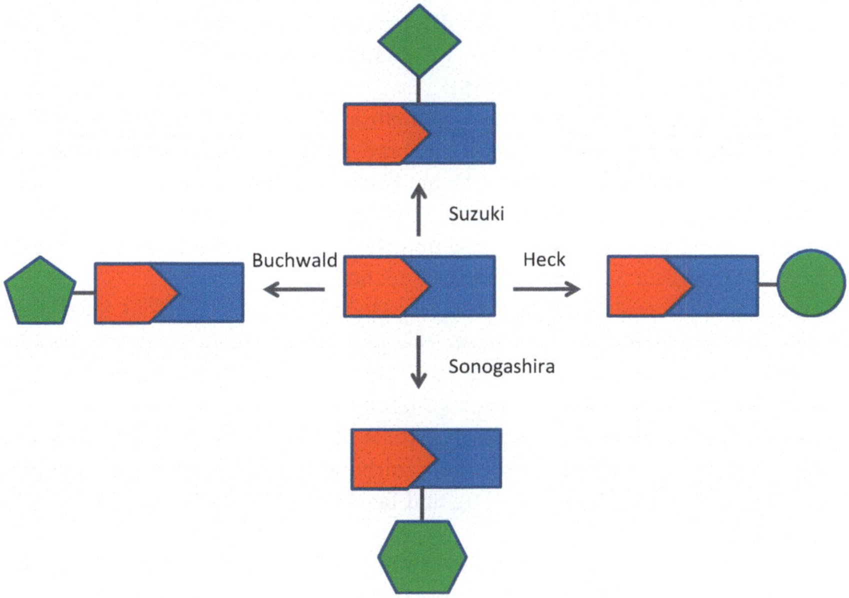

Scheme 3 Scaffold decoration and library diversification with cross-coupling reactions

2 General Synthetic Strategies

There are two main different types of synthetic strategies:

a. Scaffold synthesis with the involvement of cross-coupling reactions (Scheme 2)
b. Scaffold decoration and library diversification with cross-coupling reactions (Scheme 3)

Technically, both scaffold synthesis and decoration could be done in solid and solution phases. In solution phase solid-supported reagents and catalysts could also be used. Numerous solid-phase syntheses are reported for Suzuki–Miyaura (Entries 1 [20], 2 [21], 3 [22], 4 [23], 5 [24]) and for Sonogashira couplings (Entry 39 [25]). Stille coupling-induced ring closure on solid support was described (Entry 43 [26]) together with solution phase scaffold decoration.

Additionally, fluorous liquid–liquid phase reactions are also employed making use of the advantages of the easy separation and purification. In Entries 27 [27] and 28 [28], the fluorous tag was employed as the leaving group in Suzuki–Miyaura and in Negishi couplings (Entry 40 [29]).

There are many instances where scaffold synthesis, as well as scaffold decoration, employs cross-coupling reactions. Sonogashira reaction is particularly popular in scaffold synthesis since the inserted triple bond can be used in subsequent cyclization reactions in many ways. (Entries 58 [30], 59 [31], 60 [32], 62 [33], 63 [34]).

Suzuki–Miyaura coupling is generally performed on aromatic systems. However, there are two examples where only partially unsaturated rings are subjected to C–C coupling (Entries 22 [35], 32 [36]).

In a one-pot cross-coupling and amination reaction, aryl and unsaturated boronates were used in Entry 12 [37] at high temperatures (160°C) under MW heating. The ring halogen reacted with the boronates, while the aliphatic chloride was replaced with various amines; thus, two diversity elements were introduced selectively in one reaction. In this reaction the desired compounds were purified with scavengers using a "catch-to-release" technique.

Two identical aromatic groups can be introduced in double scaffold decoration if two identical halogens are present in the precursor structure (Entry 30 [38]). However, one-pot sequential double scaffold decoration was applied in Entry 24 [39] with different aromatic species making use of the different reactivity of the I and Br leaving groups or similarly triflate and Br (Entry 53 [40]). In another example ("sandwich sequence"), selective introduction of a bromine leaving group is followed by the first Suzuki coupling and facilitated the second Suzuki coupling (Entry 30 [38]).

Sequential Suzuki coupling was also reported leading to a chain of tiophenes (Entry 3 [22]) or thienylpyridyl garlands (Entry 16 [41]). Large majority of the synthetic targets are N- or N,O-/N,S/heterocyclic ring systems. The heterocyclic rings are either condensed ring systems or contain a chain of isolated rings (e.g., Entry: 16 [41]). Few examples lack any heteroatoms (Entries 12 [37], 13 [42], 58 [30]). Apart from some *O*-heterocycles (Entries 1 [20], 3 [22], 4 [23], 26 [43], 31 [44], 33 [45], 39 [25], 43 [26], 45 [46], 46 [47], 47 [48], 52 [49]), few S-heterocycles are reported (Entries 1 [20], 3 [22], 48 [50], 49 [51]).

Normally, the preformed scaffold contains the halogen or pseudohalogen (e.g., triflate) leaving group in the reaction. In one reported case a "reversed strategy" was used (Entry 9 [52]), namely, the triazole motif of the scaffold was converted to a boronate ester and reacted with various aryl bromides. The striking advantage of this strategy is that aryl halides are available on the market more readily and in higher numbers compared with boronates. Moreover, aryl halides are generally more reasonably priced.

In an interesting reaction sequence chromones were subjected to Suzuki coupling which was followed by subsequent ring opening and closure to pyrimidine ring (Entry: 57 [53]).

Multicomponent or tandem scaffold synthesis was reported in Entry 35 [54], which includes Heck reaction and tin-induced ring closure leading to dihydroindenoisoquinoline.

Only two examples were reported when Negishi coupling was employed in scaffold synthesis (Entries 40 [29], 41 [55]).

When a certain spatial distance is set between the aryl fragments, intramolecular direct CH activation is possible without using activating groups like boronates to induce coupling leading to fused ring systems (Entry 42 [56]).

The intermediate scaffolds having the proper leaving groups I, Br, or the pseudo halide triflates are suitable for conversion to decorated scaffolds with Suzuki–Miyaura, Negishi, Sonogashira, Heck, and Pd-catalyzed carbonylation reactions. There are several examples where many of these cross-coupling reactions were used for increasing the diversity around the scaffold or core structure (Entries 15 [57], 43 [26], 44 [58], 45 [46], 46 [47], 47 [48], 48 [50], 49 [51], 50 [59], 51 [60], 52 [49], 53 [40], 54 [61], 55 [62], 56 [63]).

Finally, there are so-called domino or tandem transformations that combine two cross-coupling reactions in one pot. In Entry 58 [30] intramolecular Heck reaction and Suzuki couplings lead to indenes. In this case, the scaffold synthesis and decoration take place in one single step and pot. Similarly, Sonogashira reaction followed by intramolecular ring closure leading to an indolizine library (Entry 59 [31]), pyrrolopyridazines (Entry 60 [32]), indoles (Entry 62 [33]), and quinolinones (Entry 63 [34]) has been reported. In Entry 61 [64] and Entry 34 [65], Heck reaction was followed by an aza-Michael-type ring closure.

Tandem Suzuki–Miyaura coupling–cyclization is reported in Entry 64 [66]. In Entry 65 [67] two scaffold fragments possessing diversity elements were merged in a Suzuki–Miyaura coupling, which was followed by intramolecular lactamization.

2.1 Components in the Suzuki Coupling

Since the Suzuki–Miyaura coupling is the most frequently used C–C bond-forming reaction in library synthesis, this subsection is primarily focused on the main factors of this type of cross-coupling reaction. There are four key components of the Suzuki coupling: boronates, Pd catalysts, organophosphine ligands, and inorganic salts or bases.

Boronates are typically used as boronic acids ($RB(OH)_2$), boronate esters ($RB(OR')_2$, and organoboranes (R_3B). Besides the common boronates, pinacol boronate is employed in Entry 10 [68]. Organotrifluoroborates are used in Entry 8 [69] that are tolerant to air and moisture and easy to handle and purify. An example of solid-supported boronate reagent (Fig. 1) is reported in Entry 6 [70]. This boronate is a variant of the MIDA boronate [71] (trivalent *N*-methyliminodiacetic acid (MIDA) ligand).

Regarding the Pd catalysts there are many variations employed (Fig. 2.): $Pd(PPh_3)_4$; $Pd(PPh_3)Cl_2$; $Pd_3(OAc)_6$ (Entry 13 [42]), the trimeric form of $Pd(OAc)_2$;

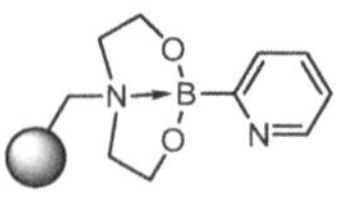

Fig. 1 Solid-supported boronate reagent for Suzuki coupling

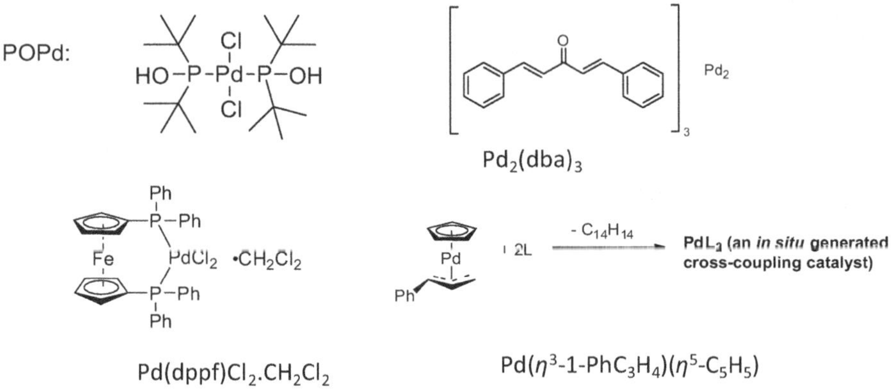

Fig. 2 Palladium catalysts

POPd (Entry 17 [72]); $Pd_2(dba)_3$, (tris(dibenzylideneacetone)dipalladium(0)] (Entry 14 [73]); and $Pd(dppf)Cl_2{\bullet}CH_2Cl_2$ [1,1′-Bis(diphenylphosphino)ferrocene] dichloro-palladium(II)] (Entries 18 [74], 27 [27], 28 [28], 65 [67]).

Heat- and air-stable compound $Pd(\eta^3\text{-}1\text{-}PhC_3H_4)(\eta^5\text{-}C_5H_5)$ [75] reacts rapidly with a wide variety of tertiary phosphines (L) to produce near-quantitative yields of the corresponding Pd(0) compounds PdL_2. This moiety is believed to be the active species in many often-used cross-coupling catalyst systems including $Pd(PPh_3)_4$, $Pd_2(dba)_3$, $PdCl_2$, and $Pd(OAc)_2$.

Polymer-supported PS-$Pd(PPh_3)_4$ is used in Entries 11 [76], 32 [36] under MW heating.

Organophosphine ligands increase the activity and stability of the Pd catalysts reducing its loading; therefore, many different forms were reported (Fig. 3): *DavePhos* (Suzuki, Entry 8 [69]; Buchwald, Entry 36 [77]); *SPhos* (Entries 13 [42], 55 [62], 64 [66]); *XPhos* (Buchwald: Entry 37 [78]); *Xantphos* (Buchwald: Entry 38 [79]); *dppf* (direct CH activation: Entry 42 [56]; palladium-catalyzed aminocarbonylation, Entry 47 [48]); *BINA*P (Buchwald: Entry 52 [49]); and *dimethoxy-triphenylphosphine* (Heck: Entry 53 [40]).

Inorganic salts are generally applied as bases in the Suzuki reaction to facilitate transmetallation [80] – CsF, KF, Cs, Na and K carbonates, bicarbonates, phosphates and hydroxides – while TEA is often used as an acid scavenging base (e.g., Entry 32 [36]). In the Sonogashira reaction, copper salts are typically used to activate the alkyne partner beside the Pd catalyst (CuI, e.g., in Entry 39 [25]).

Cross-coupling reactions often require high reaction temperature. Microwave-assisted rate acceleration is used in Entries 8 [69], 11 [76], 12 [37], 18 [74], 19 [81], 20 [82], 23 [83], 27 [27], 28 [28], 31 [44], and 55 [62] for Suzuki–Miyaura

Fig. 3 Organophosphine ligands

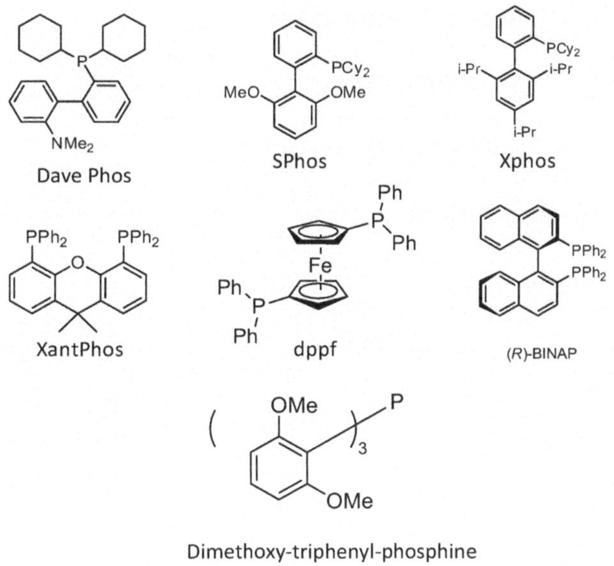

couplings, in Entry 33 [45] for Heck couplings, and in Entry 38 [79] for Buchwald–Hartwig couplings.

Nevertheless, the choice of solvent is an important factor in transition metal-catalyzed cross-couplings. While as an exception, the Suzuki reaction tolerates water as the reaction medium, other cross-coupling reactions often require the rigorous exclusion of even moisture. Typical reaction media for palladium-catalyzed coupling reactions are toluene, THF, dioxane, and DMF, although unique transformations can be run under solvent-free conditions.

3 Reaction Summaries

3.1 Suzuki Coupling

3.1.1 Solid-Phase Synthesis

One-Ring Systems

Entry 1

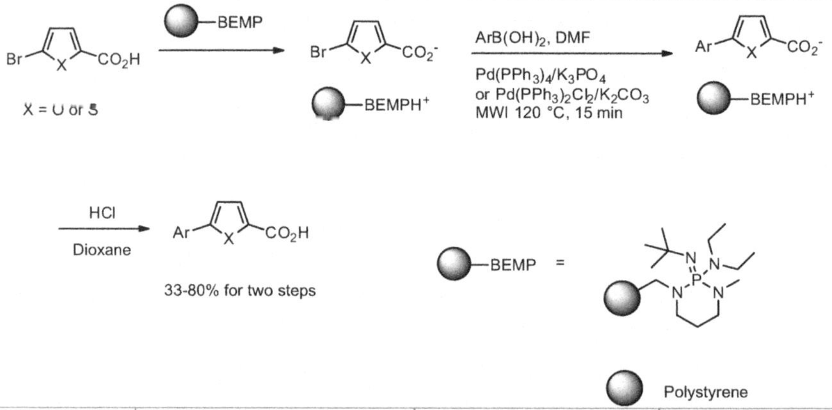

Core: Furan, thiophene	Library size: 33 examples (2 × 19, 6 unsuccessful)	Key step: scaffold synthesis and decoration	Method: solid phase – ionic immobilization
Yields: 33–80%	Biology: methionine aminopeptidase inhibitors	Comments:	[20]

Entry 2

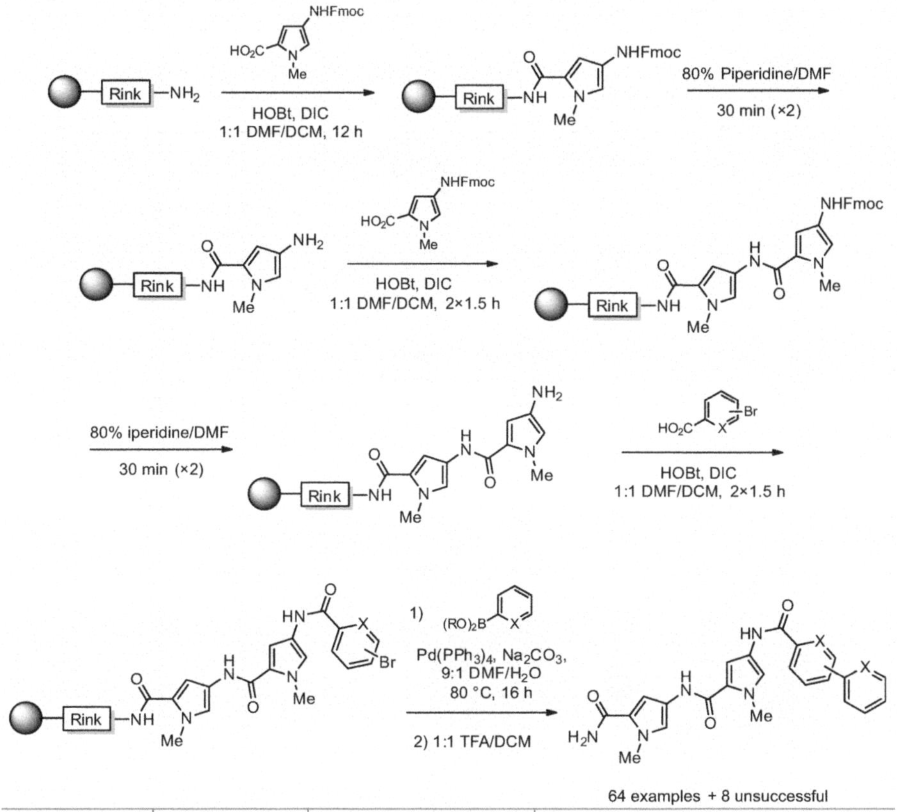

Core: distamycin analogs	Library size: 72 examples	Key step: scaffold decoration (last step)	Method: solid phase – Rink Amide Linker (RAM) SynPhase lantern
Yields: n/d	Biology: n/d	Comments:	[21]

Entry 3

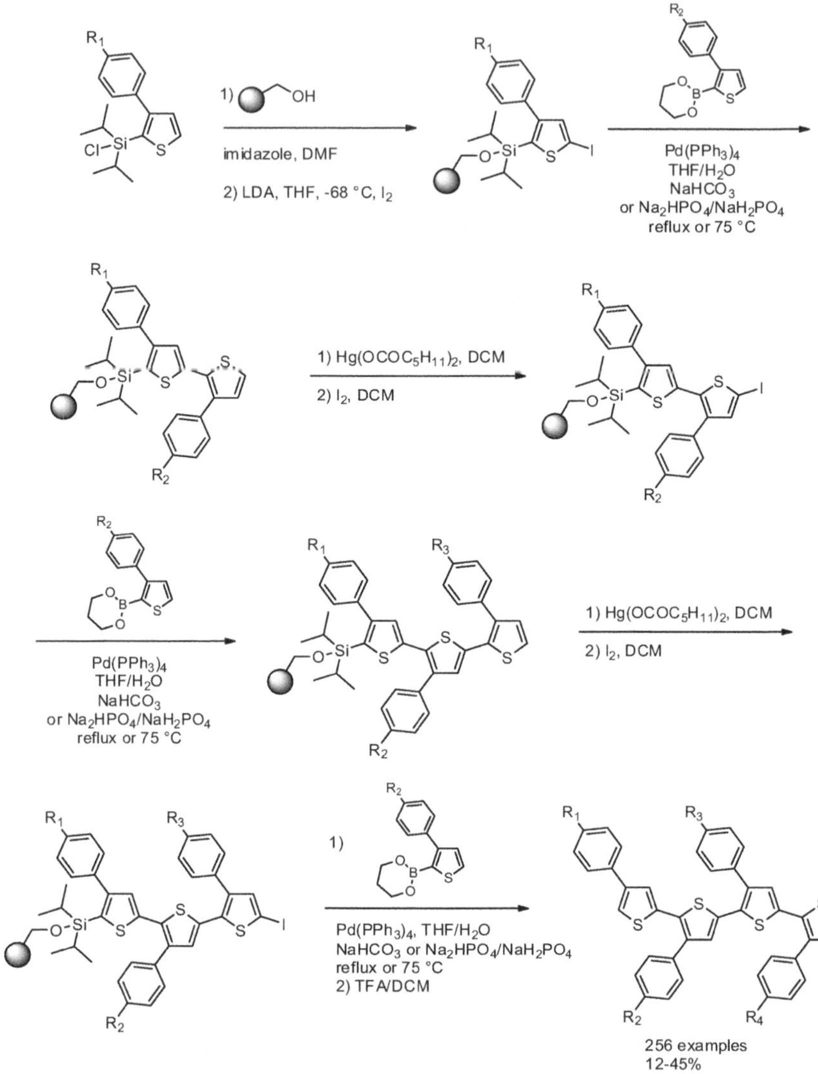

Core: quarter (3-arylthiophene)	Library size: 256 examples	Key step: scaffold decoration (last step)	Method: solid phase
Yields: 12–45%	Biology: n/d	Comments: sequential Suzuki coupling, organic electron transport materials	[22]

Two-Ring Systems

Entry 4

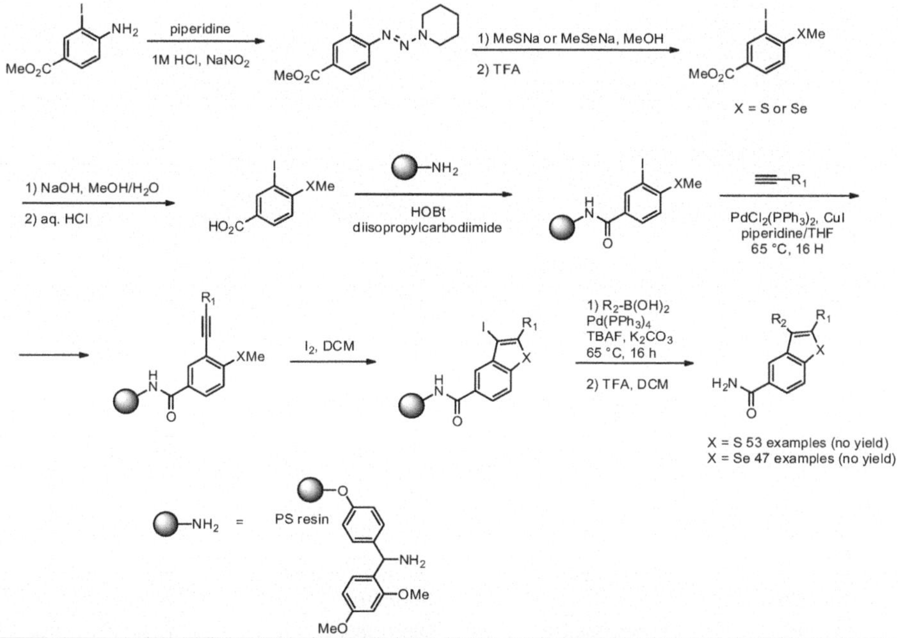

Core: benzo[*b*]thiophene, benzo[*b*]selenophene	Library size: 53 +47 examples	Key step: scaffold synthesis and decoration	Method: solid phase
Yields: n/d	Biology: n/d	Comments:	[23]

Three-Ring Systems

Entry 5

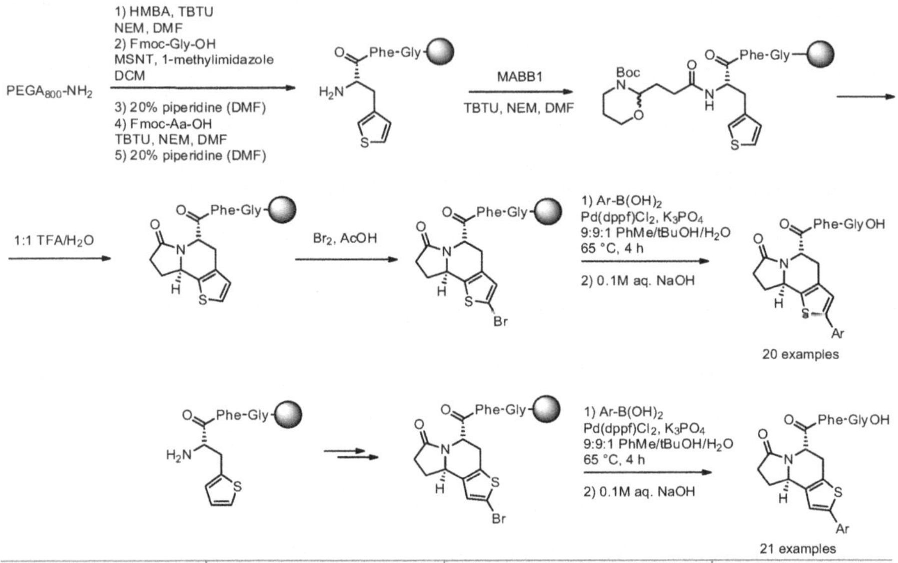

Core: thienoindolizine	Library size: 41 examples	Key step: scaffold decoration	Method: solid phase
Yields: n/d	Biology: n/d	Comments:	[24]

3.1.2 Solution Phase

One-Ring Systems

Entry 6

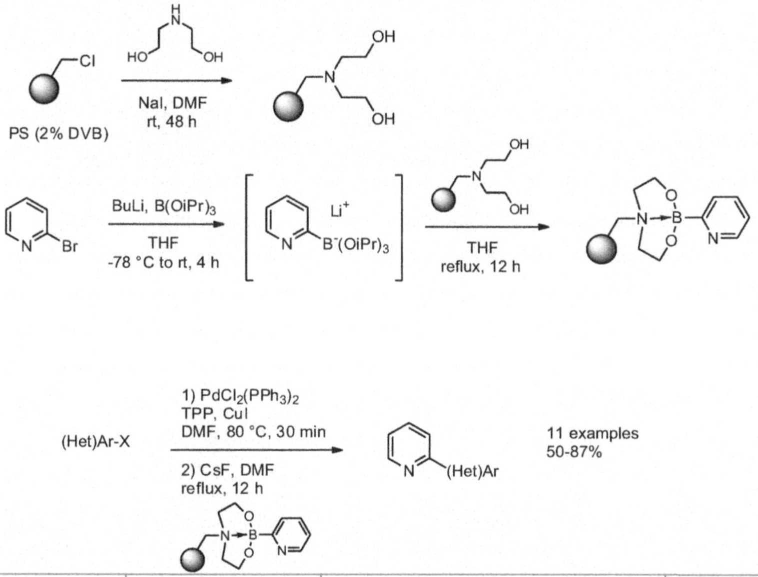

Core: pyridine	Library size: 11 examples	Key step: scaffold decoration	Method: solid-supported reagent
Yields: 50–87%	Biology: n/d	Comments: methodology for 2-pyridylboronates	[70]

Entry 7

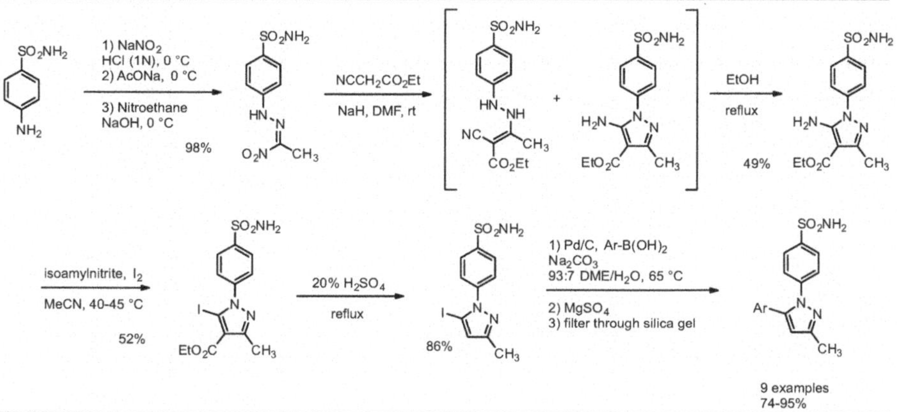

(continued)

Core: pyrazole	Library size: 9 examples	Key step: scaffold decoration	Method: solution phase
Yields: 74–95%	Biology: COX-2 inhibitor	Comments:	[84]

Entry 8

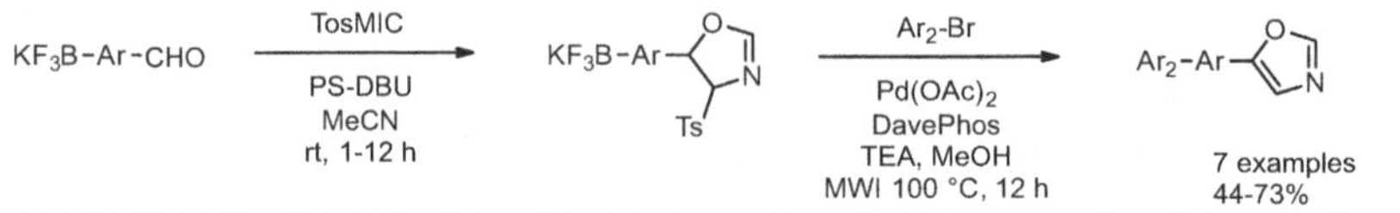

Core: oxazole	Library size: 7 examples	Key step: scaffold decoration	Method: solution phase
Yields: 44–73%	Biology: n/d	Comments:	[69]

Entry 9

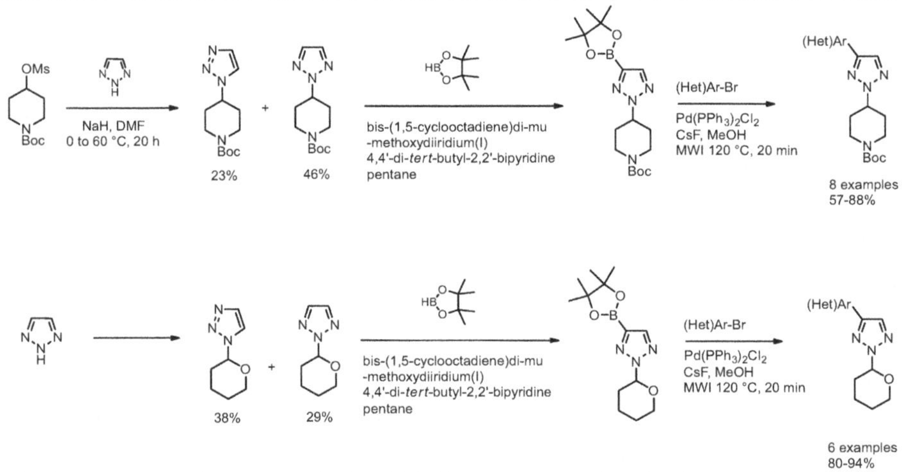

Core: triazole	Library size: 14 examples	Key step: scaffold decoration (last step)	Method: solution phase – reversed strategy!
Yields: 57–94%	Biology: n/d	Comments:	[52]

Entry 10

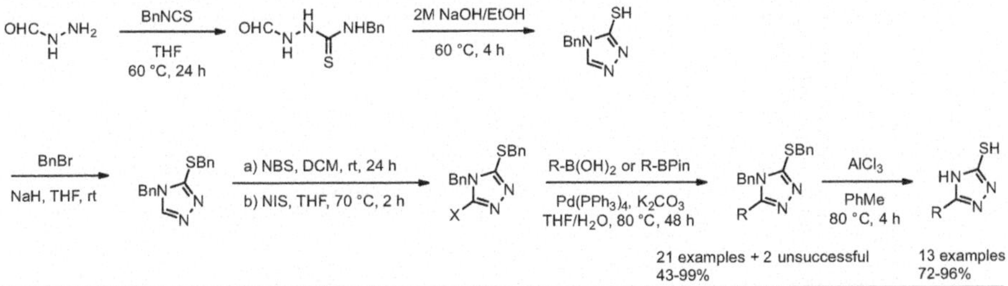

Core: triazole	Library size: 23 + 13	Key step: scaffold decoration (last step)	Method: solution phase
Yields: 43–99%, 72–96%	Biology: n/d	Comments:	[68]

Entry 11

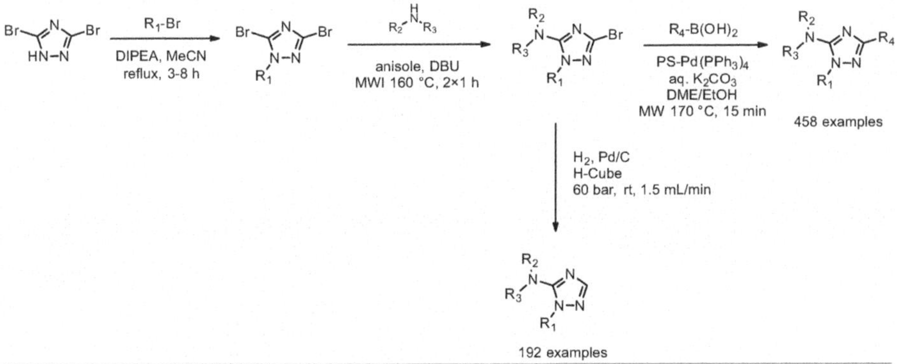

Core: triazole	Library size: 458 + 192 examples	Key step: scaffold decoration	Method: solution phase with solid-supported catalyst
Yields: n/d	Biology: n/d	Comments: flow hydrogenation	[76]

Entry 12

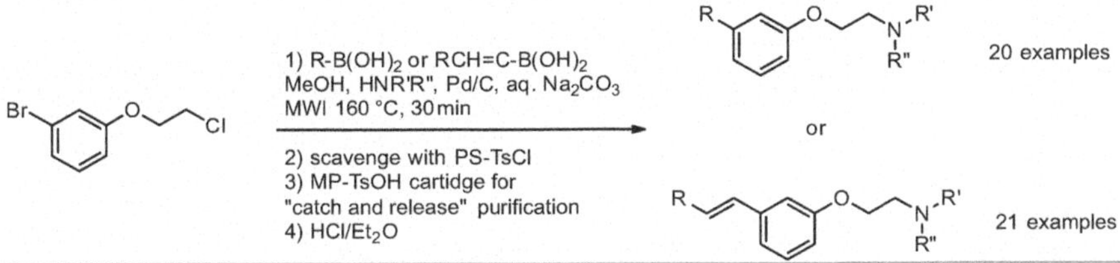

Core: biaryl and styrene	Library size: 41 examples	Key step: scaffold decoration	Method: solution phase
Yields: n/d	Biology: acetylcholine receptor agonists and antagonists nAChR	Comments: scavengers, catch and release	[37]

Entry 13

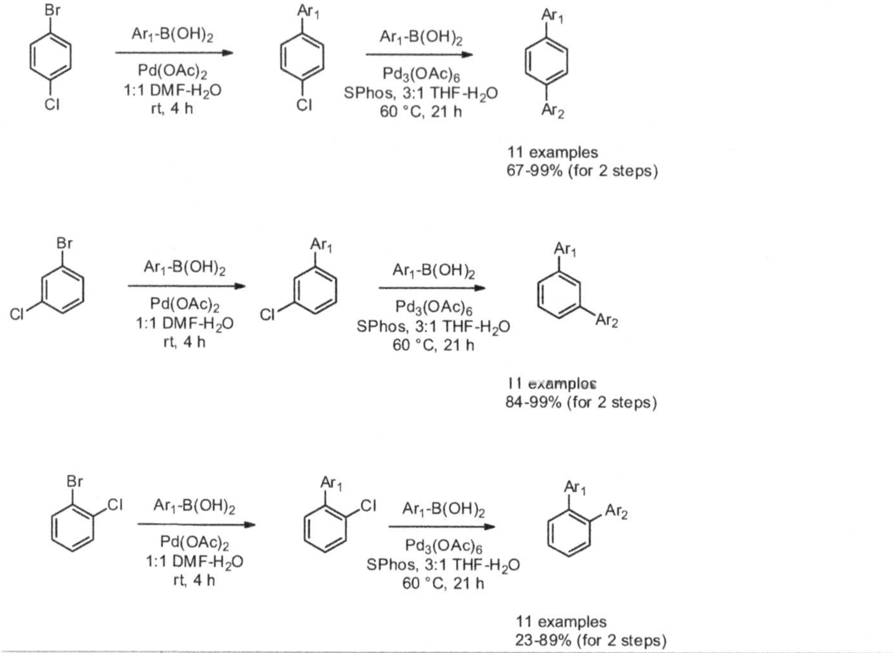

Core: *o-,m-, and p*-terphenyls	Library size: 33 examples	Key step: scaffold decoration (last step)	Method: solid phase
Yields: 23–99%	Biology: n/d	Comments: potential photochemical properties used in organic electroluminescent (OEL) devices	[42]

Entry 14

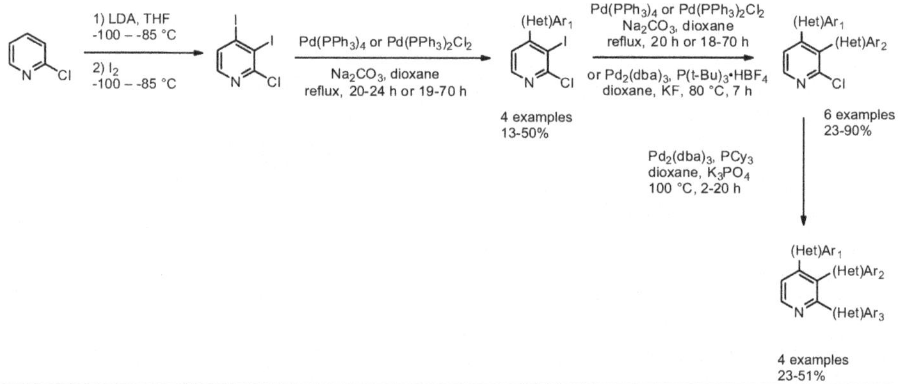

(continued)

Core: Pyridine	Library size: 14 examples	Key step: scaffold decoration	Method: solution phase
Yields: 13–90%	Biology: n/d	Comments: sequential diarylation	[73]

Entry 15

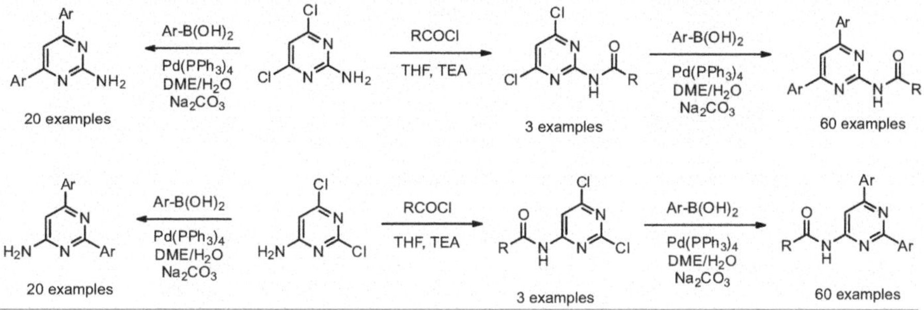

Core: pyrimidine	Library size: 160 examples	Key step: scaffold decoration	Method: solution phase
Yields: 59–94%	Biology: A3 adenosine receptor antagonist	Comments:	[57]

Entry 16

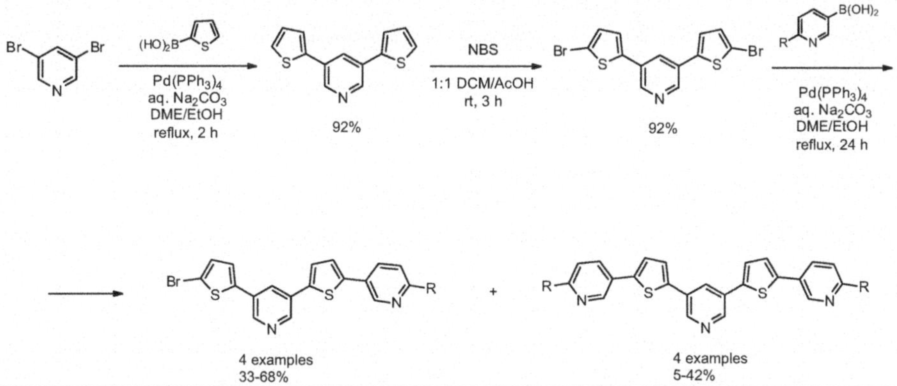

Core: thienylpyridyl garlands	Library size: 8 examples	Key step: scaffold decoration	Method: solution phase
Yields: 5–68%	Biology: non-peptidic alpha helix mimetics with potential PPI inhibitors	Comments: mono- and bisarylation	[41]

Entry 17

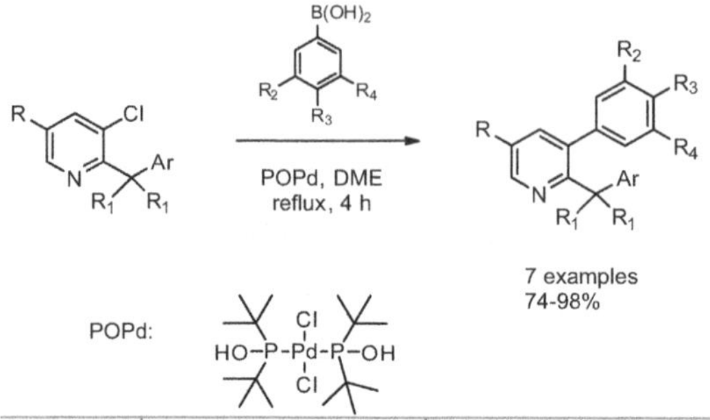

Core: pyridine	Library size: 7 examples	Key step: scaffold decoration	Method: solution phase
Yields: 74–98%	Biology: potential COX-2 inhibitors	Comments: methodology for sterically hindered pyridyl halides/air-stable catalyst POPd	[72]

Entry 18

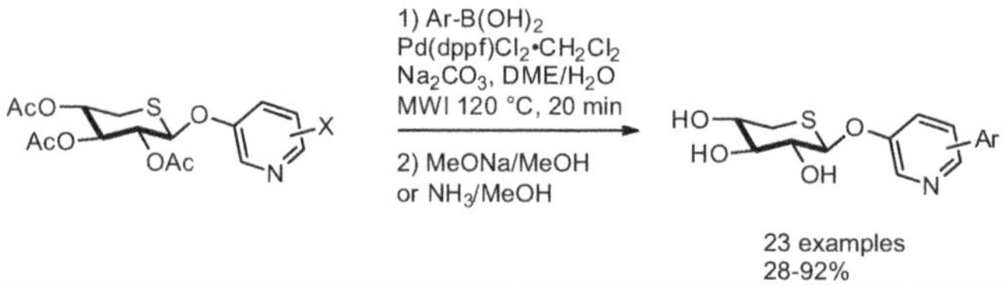

Core: 5-thio-xylopyranoside	Library size: 23 examples	Key step: scaffold decoration	Method: solution phase
Yields: 28–92%	Biology: antithrombotic effect	Comments:	[74]

Entry 19

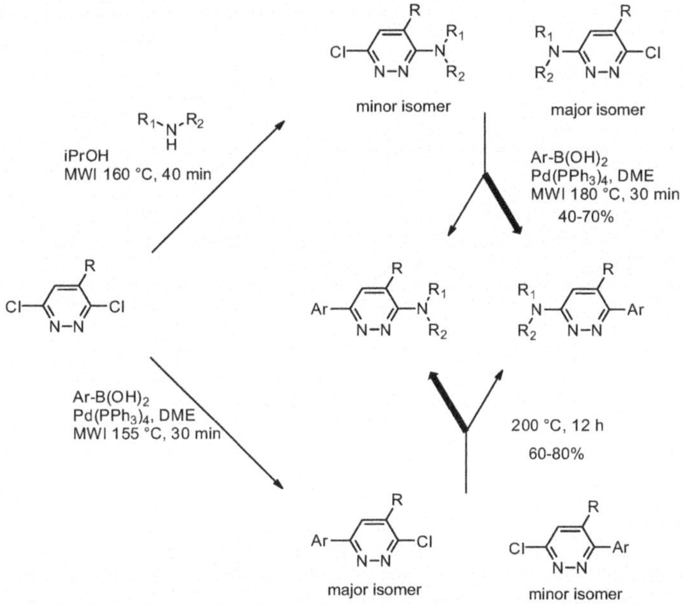

Core: pyridazine	Library size: two sublibraries as mixtures – $4 \times 5 \times 3 = 60$	Key step: scaffold decoration	Method: solution phase
Yields: 40–70%, 60–80%	Biology: n/d	Comments: sequential amination and Suzuki coupling (mixture synthesis)	[81]

Two-Ring Systems

Entry 20

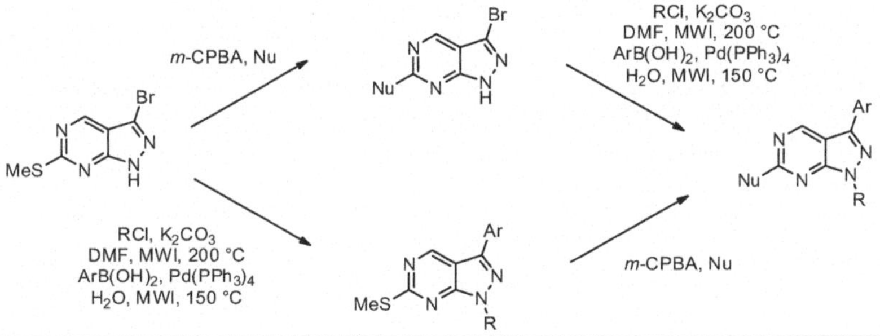

(continued)

Core: pyrazolopyrimidine	Library size: 29 examples	Key step: scaffold synthesis in three steps – 14 upper route + 15 lower route	Method: solution phase
Yields: 51–93%, (60–86%)	Biology: n/d	Comments:	[82]

Entry 21

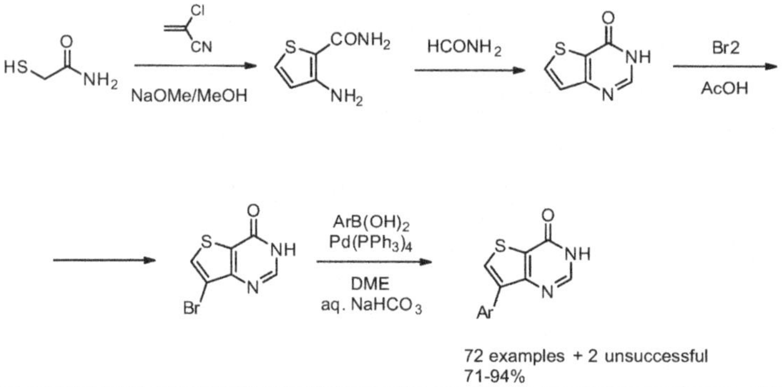

Core: thienopyrimidine	Library size: 72 examples + 2 unsuccessful	Key step: scaffold decoration (last step)	Method: solution phase
Yields: 71–94%	Biology: n/d	Comments:	[85]

Entry 22

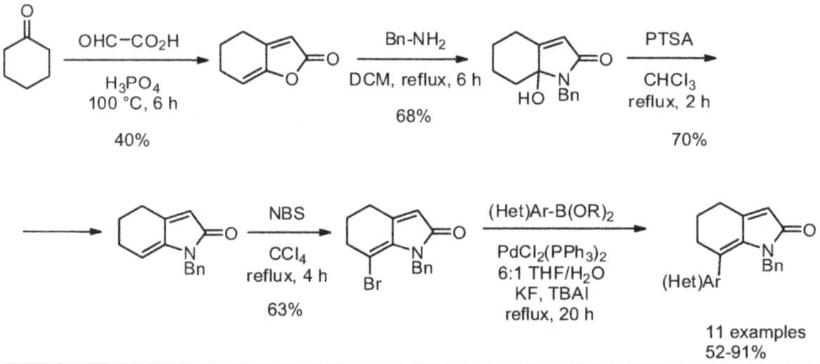

Core: 5,6-dihydroindol-2-one	Library size: 11 examples	Key step: scaffold decoration	Method: solution phase
Yields: 52–91%	Biology: n/d	Comments:	[35]

Entry 23

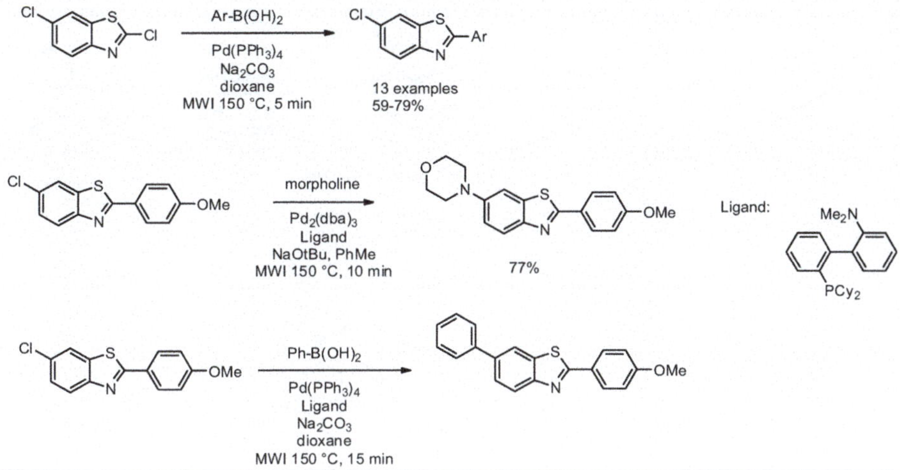

Core: benzothiazole	Library size: 13 examples	Key step: scaffold decoration	Method: solution phase
Yields: 59–79%	Biology: n/d	Comments:	[83]

Entry 24

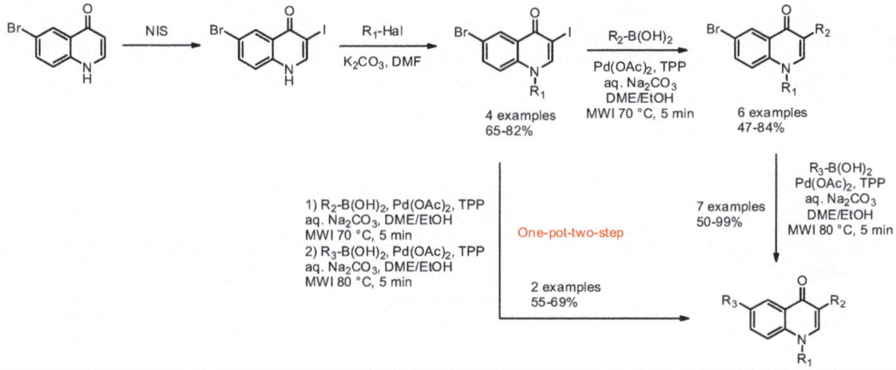

Core: quinolinone	Library size: 13 examples	Key step: scaffold decoration	Method solution phase
Yields: 47–99%	Biology: n/d	Comments: One-pot sequential double scaffold decoration	[39]

Entry 25

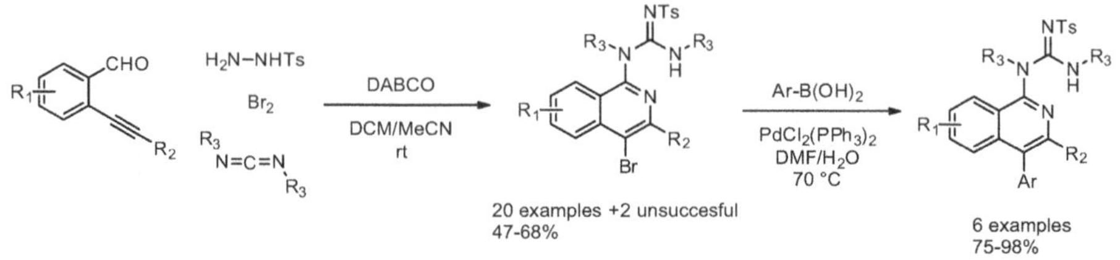

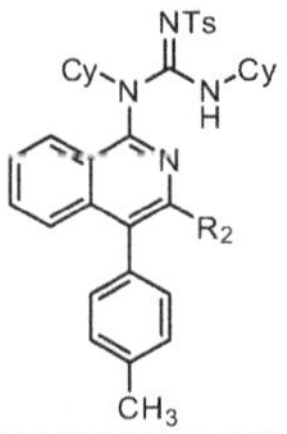

IC$_{50}$: 2.05 uM

Core: isoquinoline	Library size: 6 examples	Key step: scaffold decoration	Method: solution phase
Yields: 75–98%	Biology: protein tyrosine phosphatase 1B (PTP1B)	Comments:	[86]

Entry 26

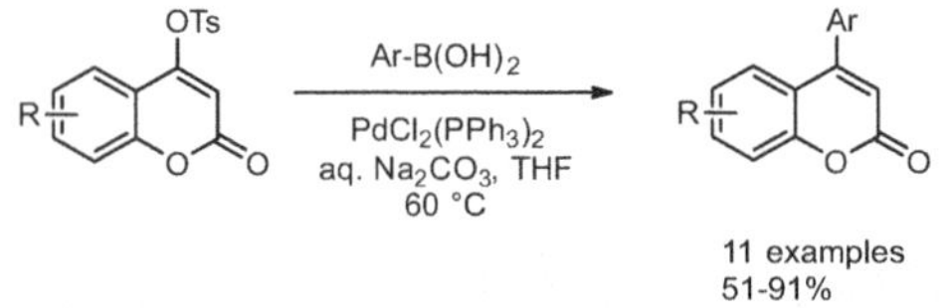

Core: cumarine	Library size: 11 examples	Key step: scaffold decoration	Method: solution phase
Yields: 51–91%	Biology: n/d	Comments:	[43]

Entry 27

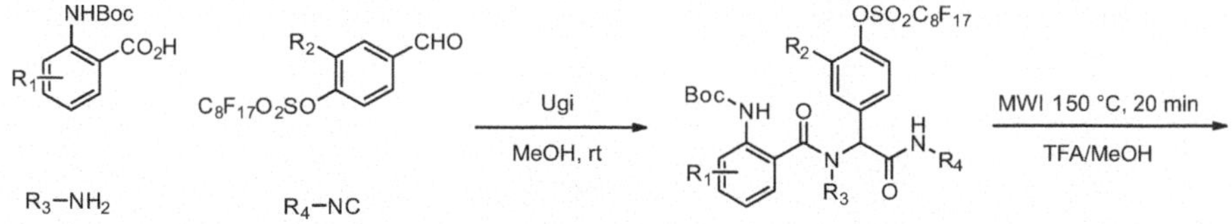

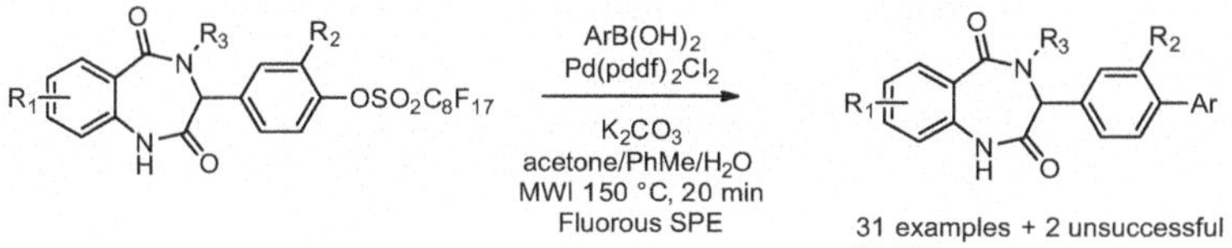

31 examples + 2 unsuccessful
20-67%

Core: benzodiazepinedione	Library size: 31 examples	Key step: scaffold decoration	Method: fluorous synthesis
Yields: 20–67%	Biology: n/d	Comments: cyclohexyl and methyl 2-isocyanoacetate as convertible isonitriles	[27]

Entry 28

NX = NHBoc or NO_2

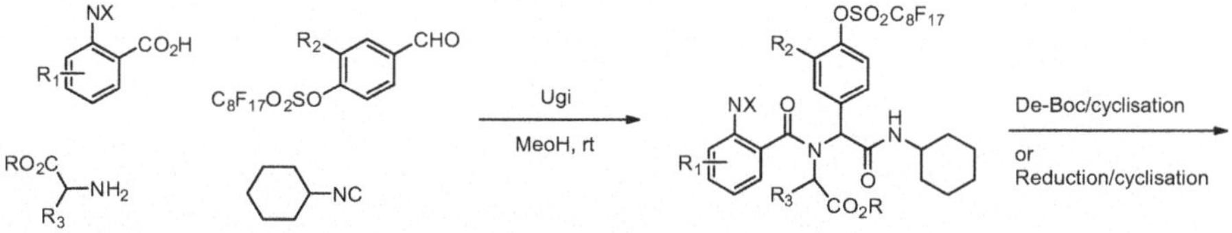

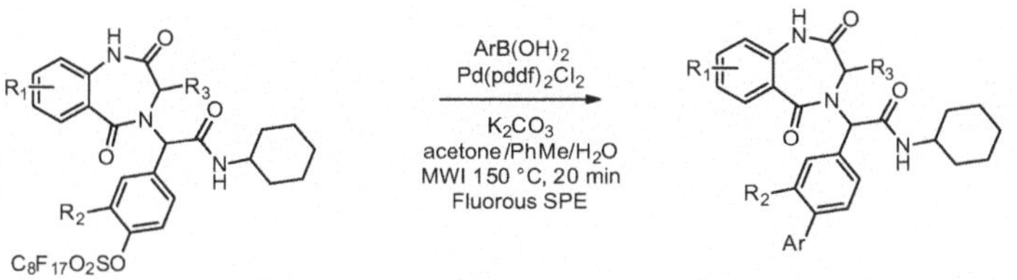

36 examples + 4 unsuccessful
10-76%

Core: benzodiazepinedione 2	Library size: 36 examples	Key step: scaffold decoration	Method: fluorous synthesis
Yields: 10–76% (generally 20–40%)	Biology: n/d	Comments:	[28]

Three-Ring Systems

Entry 29

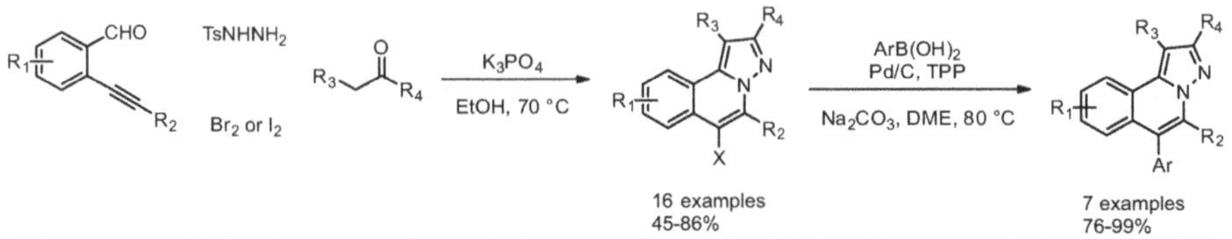

Core: pyrazole-isoquinoline	Library size: 7 examples	Key step: scaffold decoration	Method: solution phase
Yields: 76–99%	Biology: n/d	Comments:	[87]

Entry 30

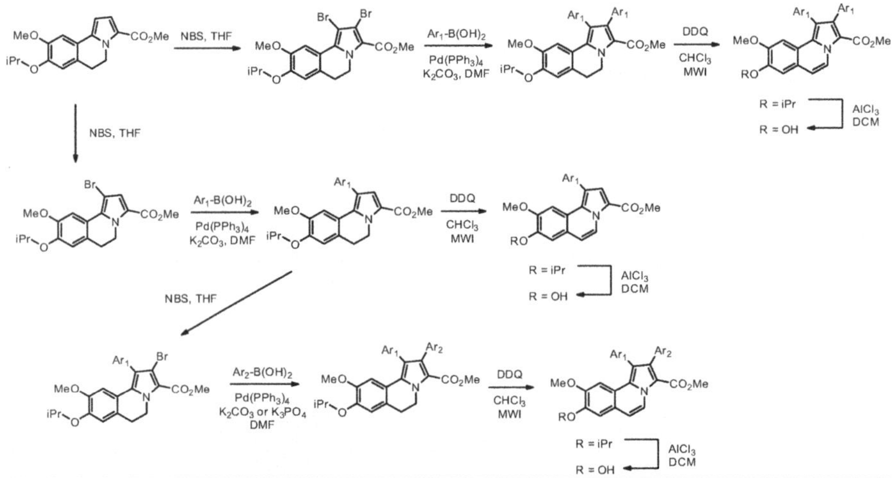

Core: pyrroloisoquinoline	Library size: 45 examples; 9 sublibraries	Key step: scaffold decoration	Method: solution phase
Yields: n/d	Biology: cytotoxicity data given	Comments: double and sequential scaffold decorations, Lamellarin D analogs	[38]

Entry 31

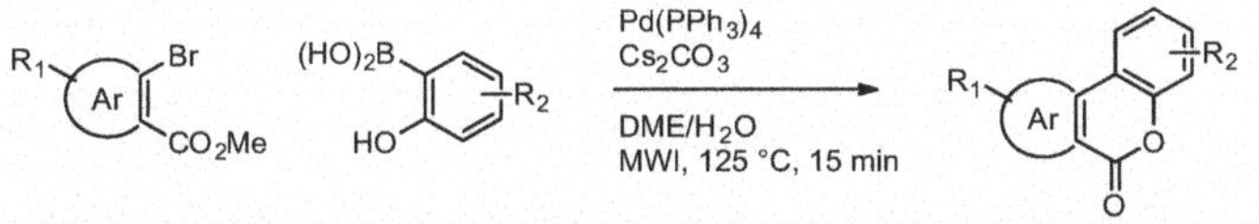

Core: dibenzopyranones	Library size: 32 examples	Key step: scaffold synthesis	Method: solution phase
Yields: 68–98%	Biology: n/d	Comments:	[44]

Entry 32

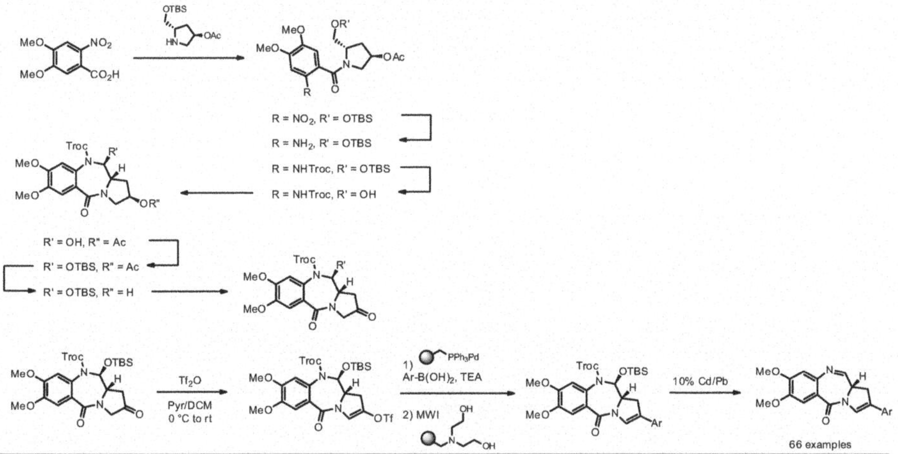

Core: pyrrolobenzodiazepine	Library size: 66 examples	Key step: scaffold decoration (last step)	Method: solution phase, solid-supported Pd
Yields: 4–57%	Biology: n/d	Comments:	[36]

3.2 *Heck Reaction*

Entry 33

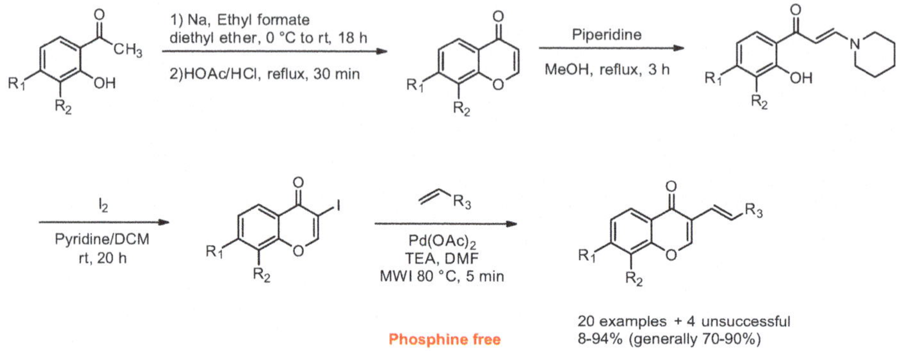

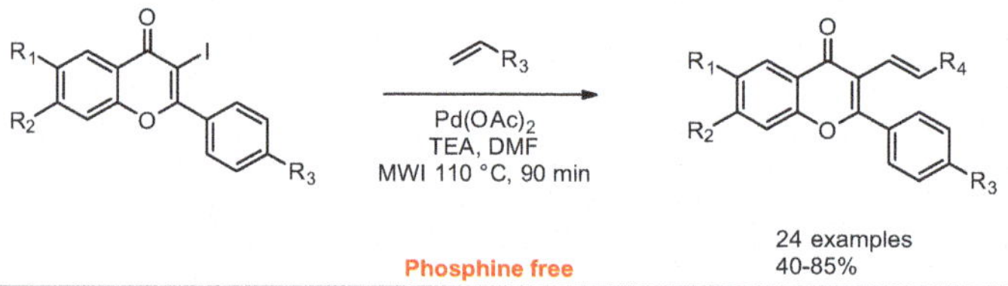

Core: chromanone, flavone	Library size: 24 + 24 examples	Key step: scaffold decoration	Method: solution phase
Yields: 8–94%, 40–85%	Biology: n/d	Comments: phosphine-free Heck	[45]

Entry 34

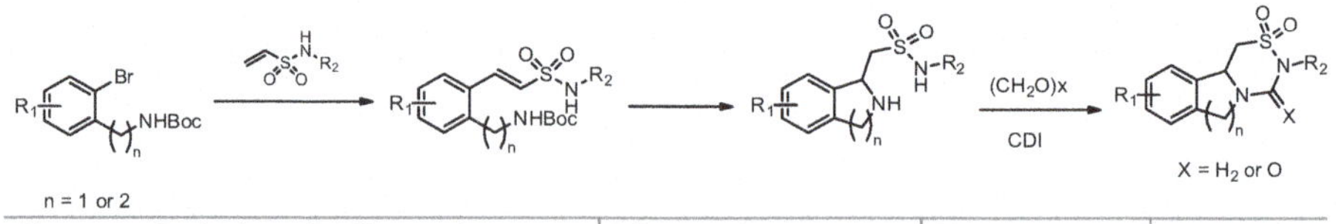

Core: tetrahydroisoquinoline and dihydroisoindoline containing tricyclic sultams	Library size: 120 members (40 + 40 + 40)	Key step: scaffold synthesis	Method: solution phase
Yields: 50–99%, 53–99%, 35–99%, 22–88%	Biology: n/d	Comments:	[65]

Entry 35

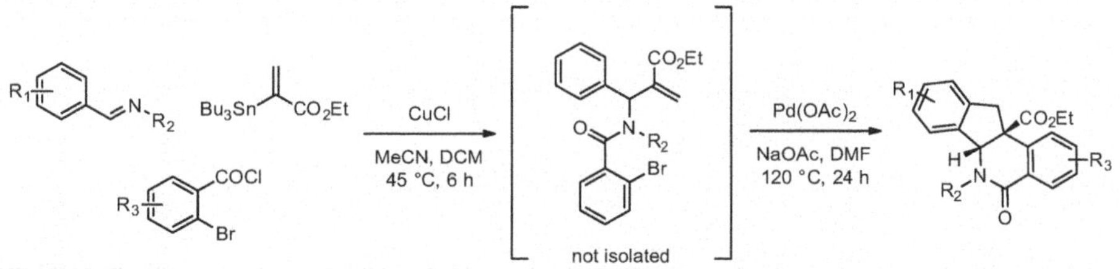

Core: dihydroindenoisoquinoline	Library size: 111 members	Key step: scaffold synthesis	Method: solution phase
Yields: 6–71% for two steps (generally 20–30%)	Biology: n/d	Comments: multicomponent or tandem scaffold synthesis (Heck, and tin-induced ring closure)	[54]

3.3 *Buchwald–Hartwig Amination*

Entry 36

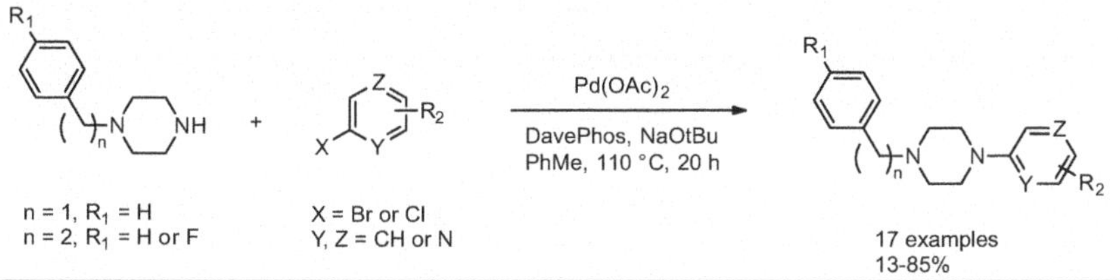

Core: piperazine	Library size: 17 examples	Key step: scaffold decoration	Method: solution phase
Yields: 13–85%	Biology: potential 5-HT2a	Comments: Ligand = DavePhos	[77]

Entry 37

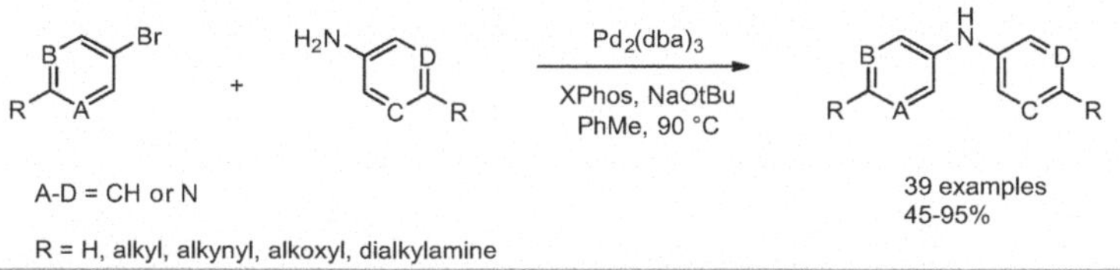

Core: bis-(het) arylamine	Library size: 39 examples	Key step: scaffold decoration (last step)	Method: solution phase
Yields: 45–95%	Biology: n/d	Comments: in situ generated Pd catalyst $Pd(\eta^3\text{-}1\text{-}PhC_3H_4)(\eta^5\text{-}C_5H_5)$	[78]

Entry 38

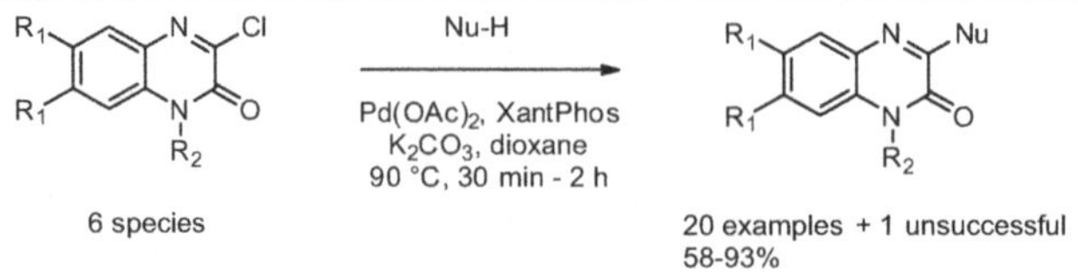

Nu-H: amides, lactames, carbamates, weak amine nucleophiles

Core: quinoxalinone	Library size: 21 examples	Key step: scaffold decoration	Method: solution phase
Yields: 58–93%	Biology: cannabinoid CB2 receptor agonist	Comments:	[79]

3.4 Sonogashira Reaction

Entry 39

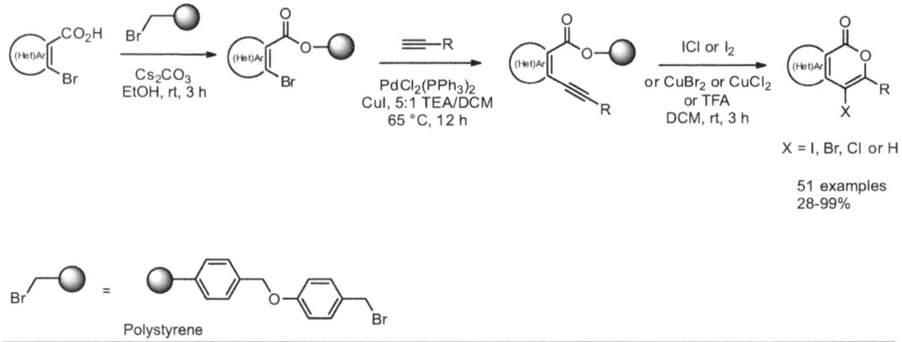

Core: isocoumarines and heterocyclic derivatives	Library size: 51 examples	Key step: scaffold synthesis	Method: solid phase
Yields: 28–99%	Biology: n/d	Comments:	[25]

3.5 Negishi Coupling

Entry 40

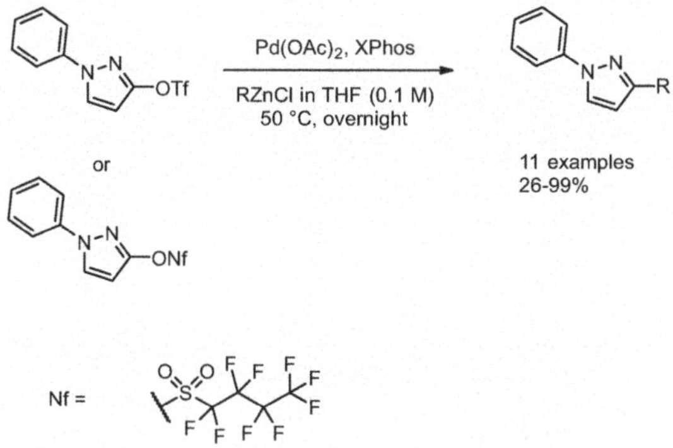

Core: pyrazole	Library size: 11 examples	Key step: scaffold decoration suitable for library synthesis	Method: fluorous synthesis
Yields: 26–99%	Biology: n/d	Comments: fluorous tag was used as a leaving group	[29]

Entry 41

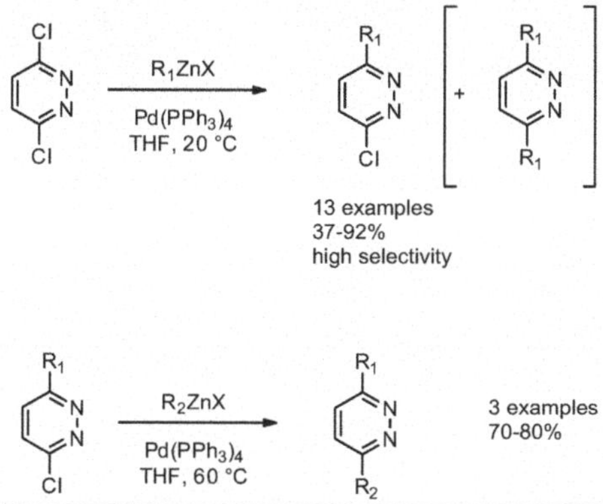

Core: pyridazine	Library size: 13 examples	Key step: scaffold synthesis suitable for library synthesis	Method: solid phase
Yields: 37–92%	Biology: n/d	Comments:	[55]

3.6 Direct CH Activation

Entry 42

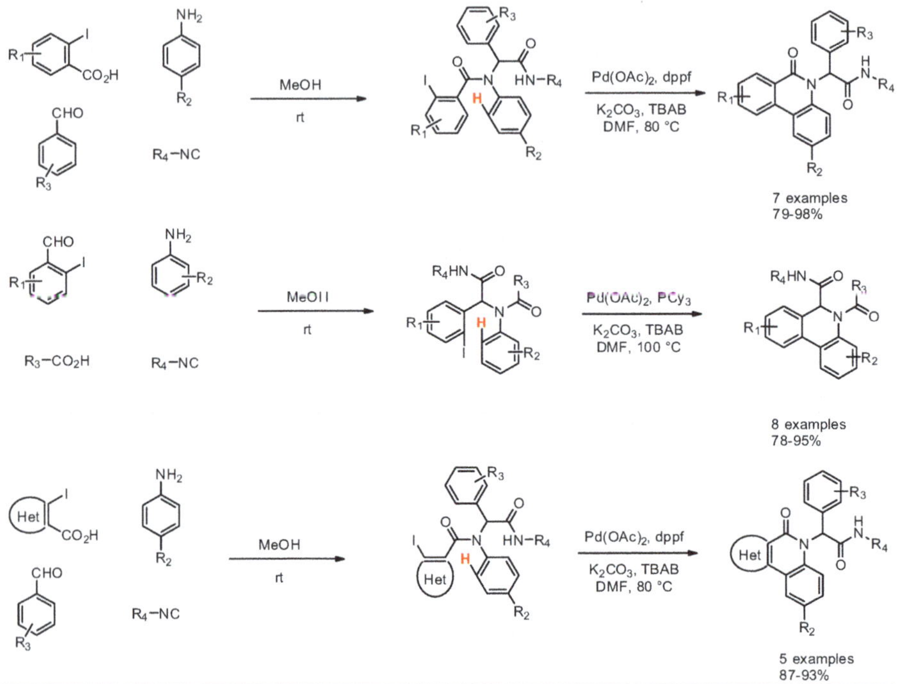

Core: quinolinone	Library size: 20 examples	Key step: scaffold decoration (last step)	Method: solution phase
Yields: 78–98%	Biology: n/d	Comments: Direct CH activation	[56]

3.7 *Multiple Cross-Couplings*

3.7.1 Solid Phase

Entry 43

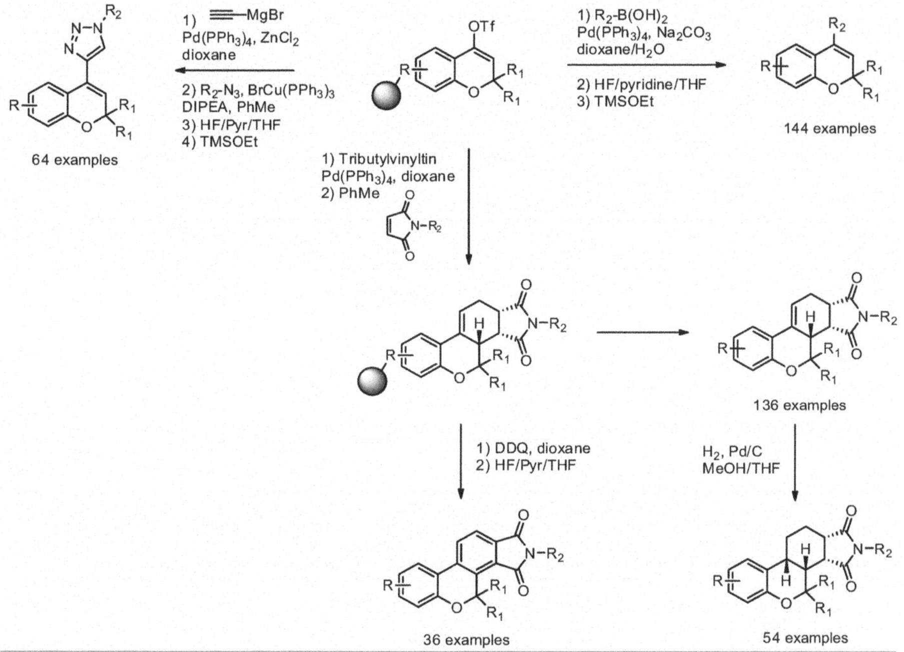

Core: benzopyrane	Library size: 424 examples	Key step: scaffold decoration	Method: solid phase
Yields: n/d	Biology: n/d	Comments: multiple cross-coupling Suzuki, Stille, Negishi	[26]

3.7.2 Solution Phase

One-Ring Systems

Entry 44

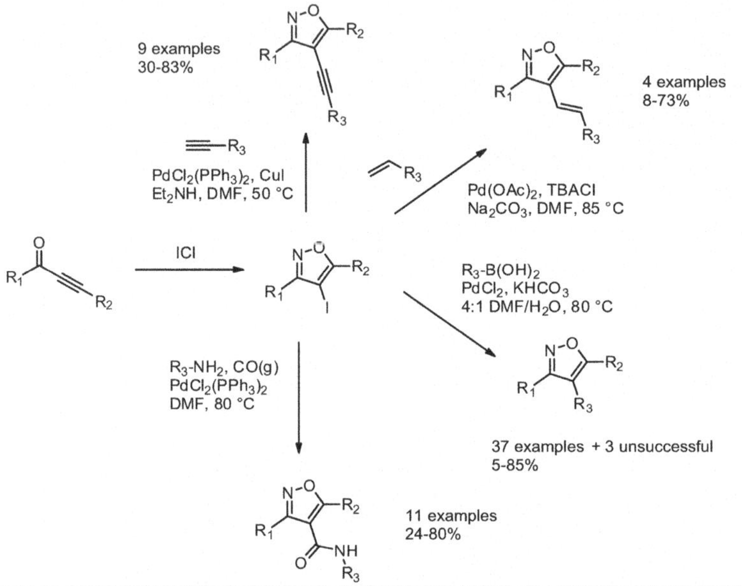

Core: isoxazole	Library size: 64 examples	Key step: scaffold decoration	Method: solution phase
Yields: 8–85%	Biology: n/d	Comments: multiple cross-coupling Suzuki, Sonogashira, Heck, Pd-catalyzed aminocarbonylation	[58]

Two-Ring Systems

Entry 45

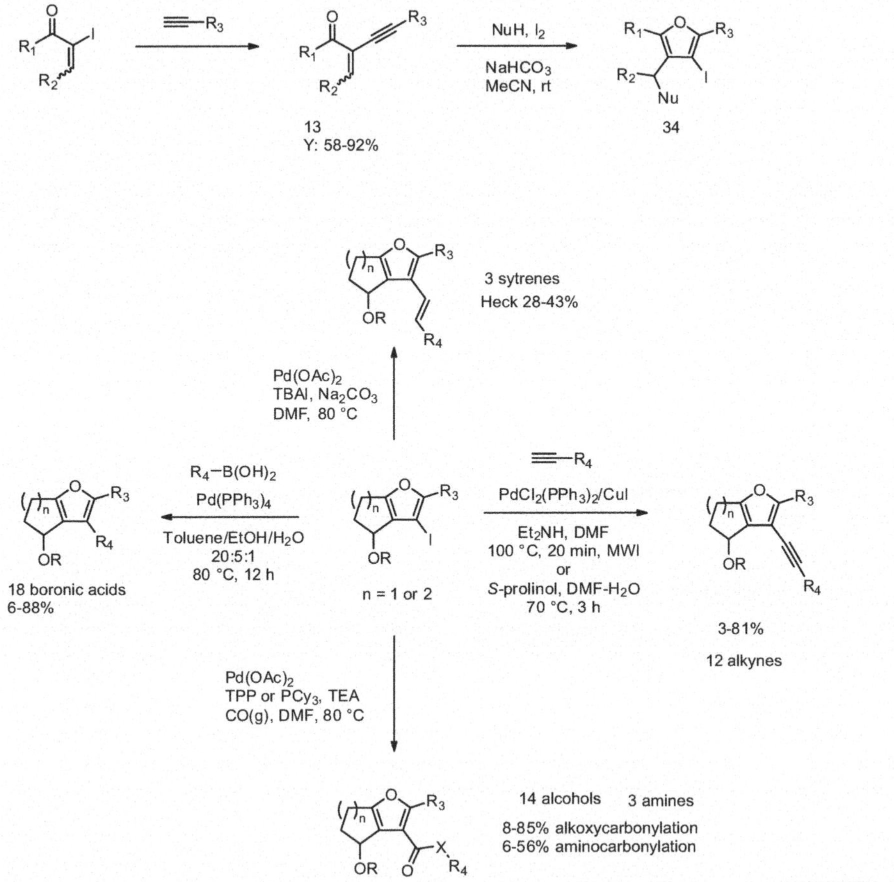

Core: furane	Library size: 163 examples, 46 (Suzuki) +53 (Sonogashira) +3 (Heck) +6 (amides) +55 (esters)	Key step: scaffold decoration	Method: solution phase
Yields: 6–88%	Biology: n/d	Comments: multiple type cross-couplings Suzuki, Sonogashira, Heck	[46]

Entry 46

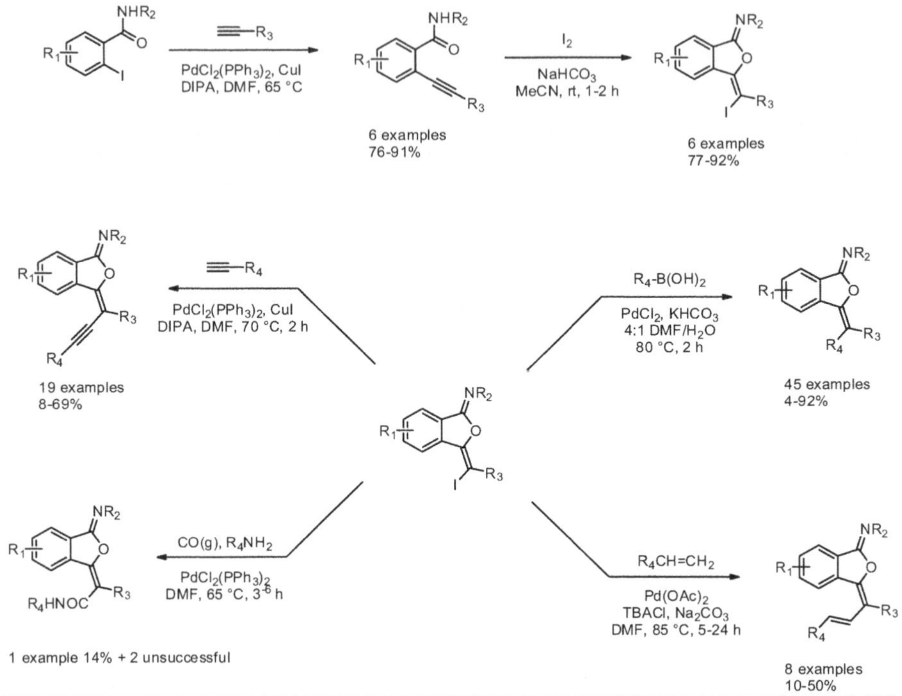

Core: cyclic imidate, isobenzofurane	Library size: 75 examples	Key step: scaffold decoration	Method: solution phase
Yields: 4–92%	Biology: n/d	Comments: multiple type cross-couplings Suzuki, Sonogashira, Heck, Pd-amino carbonylation	[47]

Entry 47

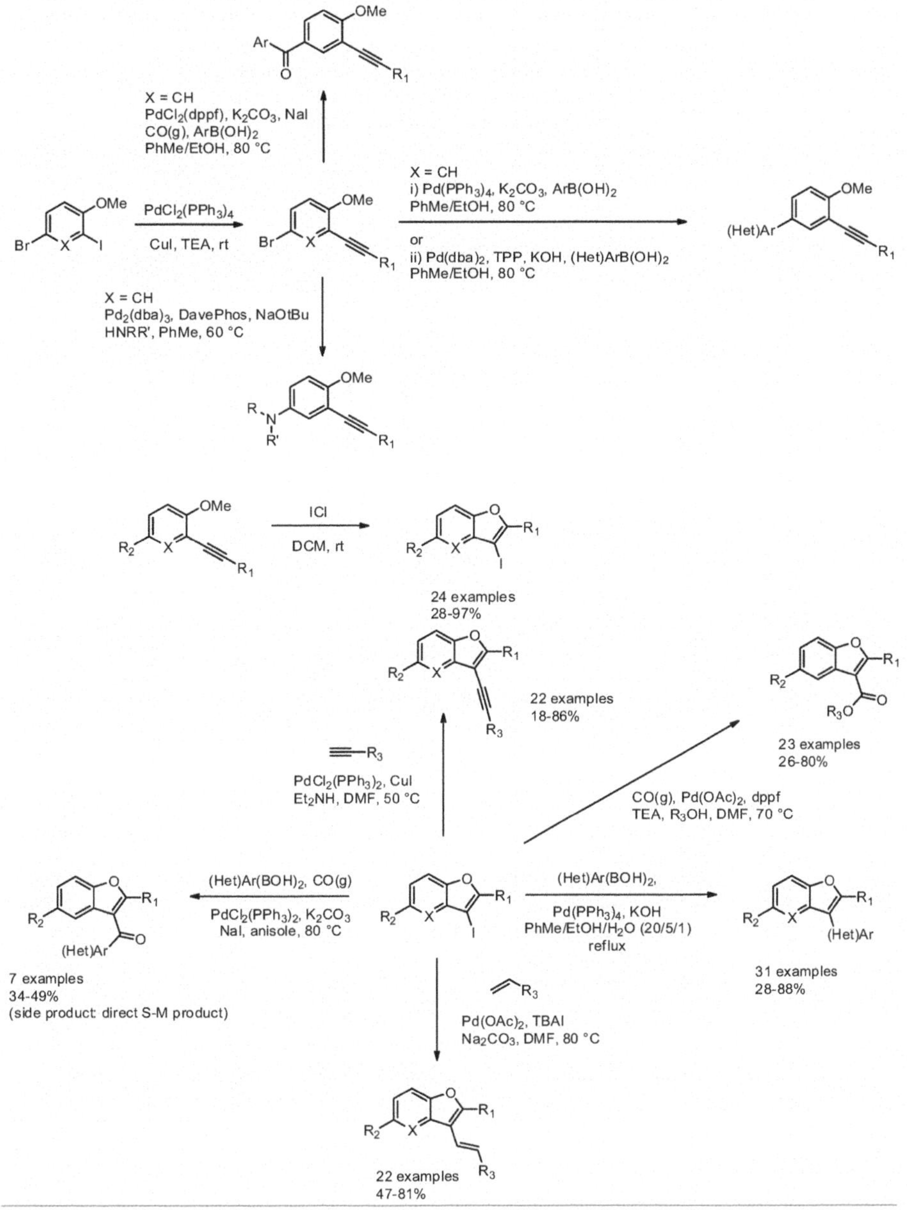

(continued)

Core: benzofurane	Library size: 121 examples	Key step: scaffold synthesis and decoration	Method: solution phase
Yields: 18–88%	Biology: n/d	Comments: scaffold synthesis (Sonogashira); multiple cross-coupling Suzuki, Suzuki carbonylation., Sonogashira, Heck, Pd-catalyzed alkoxycarbonylation	[48]

Entry 48

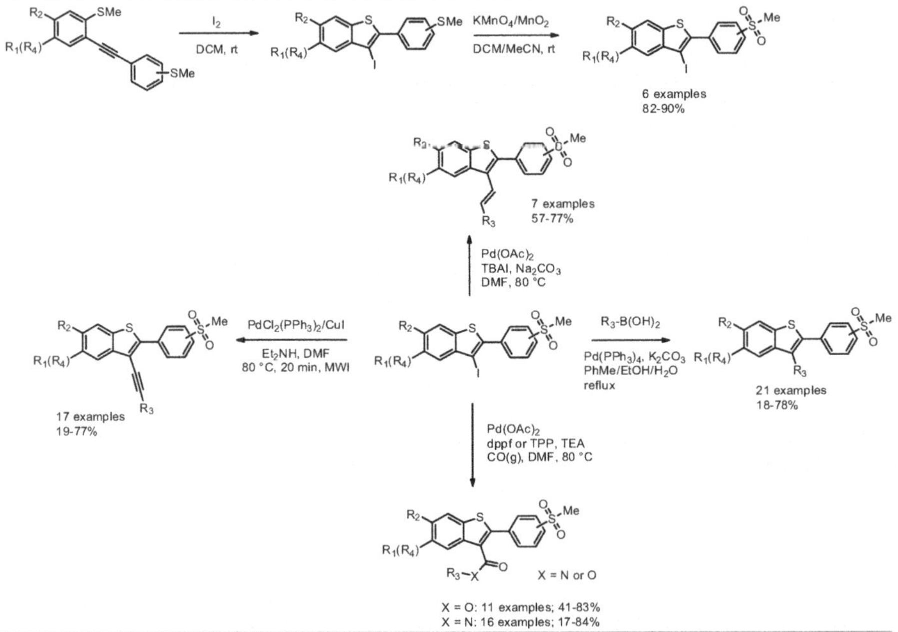

Core: benzothiophene	Library size: 72 examples – 21 (Suzuki) + 17 (Sonogashira) + 7 (Heck) + 11 (alkoxycarbonylation) + 16 (aminocarbonylation)	Key step: scaffold decoration	Method: solution phase
Yields: n/d	Biology: n/d	Comments: multiple cross-coupling Suzuki, Sonogashira, Heck, Pd-cat. amino- and alkoxycarbonylation	[50]

Entry 49

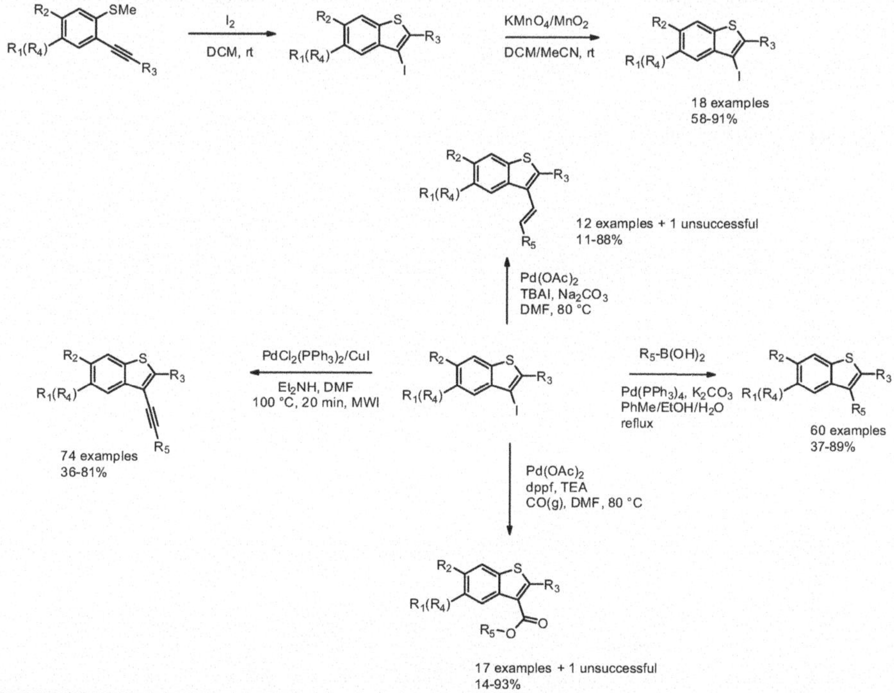

Core: benzothiophene	Library size: 165 examples	Key step: scaffold decoration	Method: solution phase
Yields: n/d	Biology: n/d	Comments: multiple cross-coupling Suzuki, Sonogashira, Heck, Pd-catalyzed alkoxycarbonylation	[51]

Entry 50

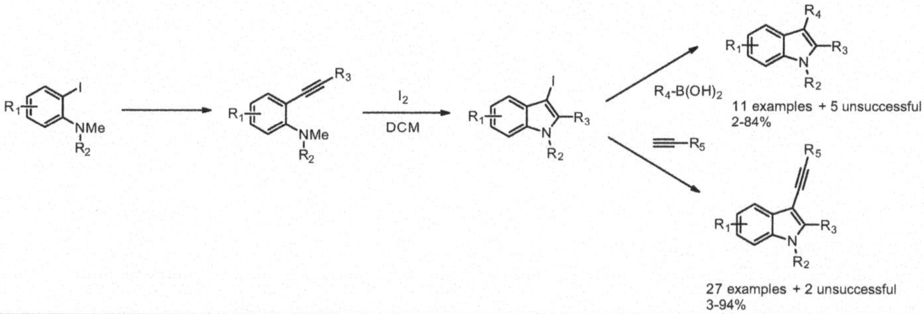

(continued)

Core: indole	Library size: 38 examples (11 Suzuki + 27 Sonogashira)	Key step: scaffold synthesis and decoration	Method: solution phase – few solid-phase examples also discussed
Yields: 2–84%, 3–94%	Biology: n/d	Comments: multiple cross-coupling Suzuki, Sonogashira	[59]

Entry 51

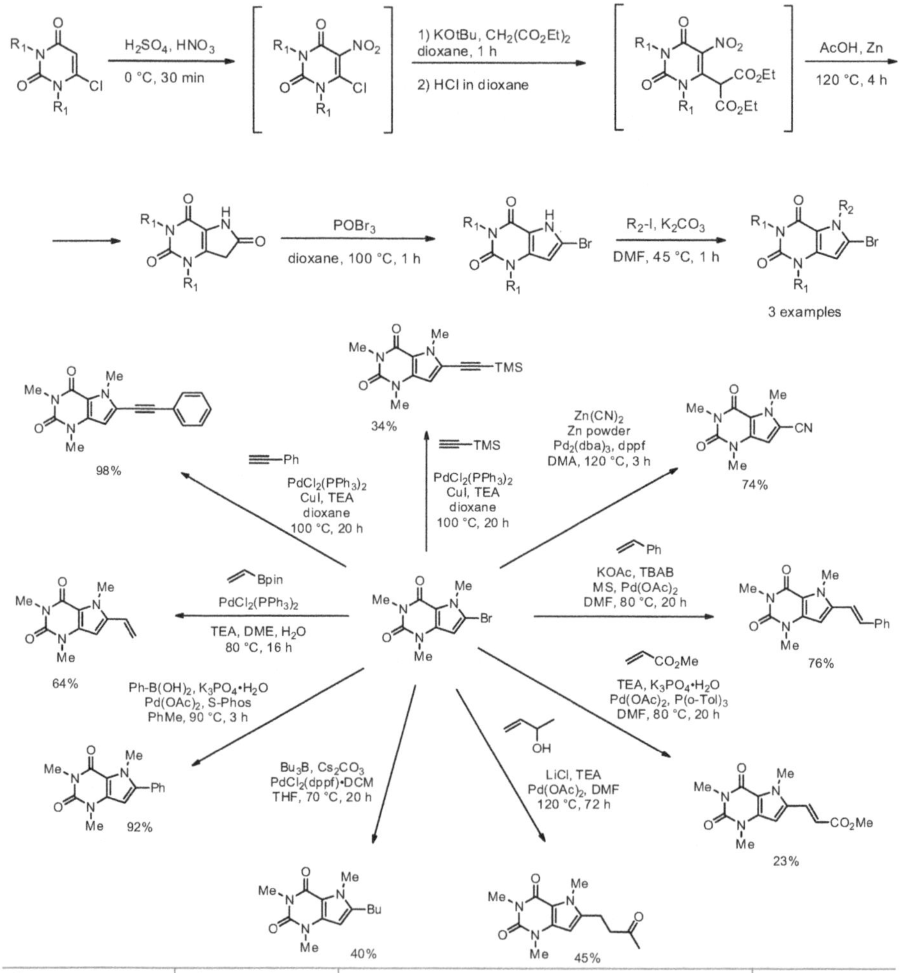

Core: 9-deazaxanthine	Library size: 9 examples	Key step: scaffold decoration	Method: solution phase
Yields: 23–98%	Biology: n/d	Comments: multiple cross-coupling Suzuki, Sonogashira, Heck, etc.	[60]

Entry 52

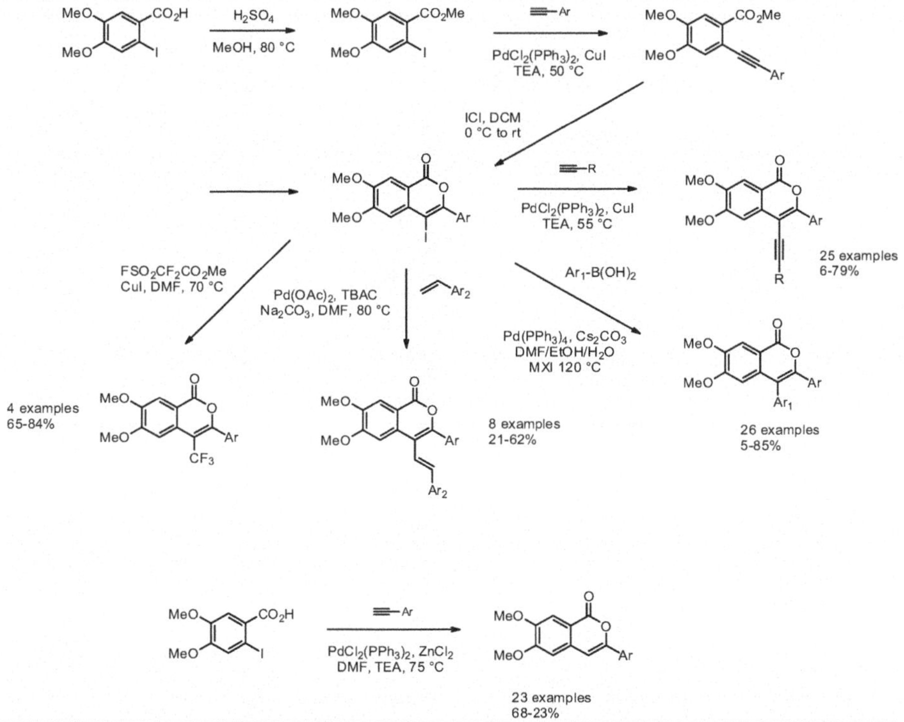

Core: isocoumarin	Library size: 82 examples – 26 (Suzuki) + 25 (Sonogashira) + 8 (Heck) + 23 (isocoumarin)	Key step: scaffold synthesis and decoration	Method: solution phase
Yields: n/d	Biology: n/d	Comments: scaffold synthesis (Sonogashira); multiple coupling Suzuki, Sonogashira, Heck	[49]

Entry 53

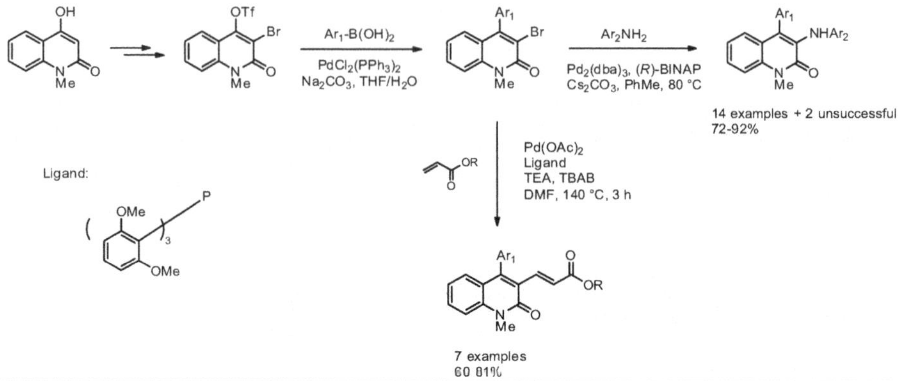

Core: quinoline	Library size: 23 examples	Key step: sequential double scaffold decoration	Method: solution phase
Yields: 60–81%, 72–92%	Biology: n/d	Comments: multiple cross-coupling Buchwald, Heck	[40]

Entry 54

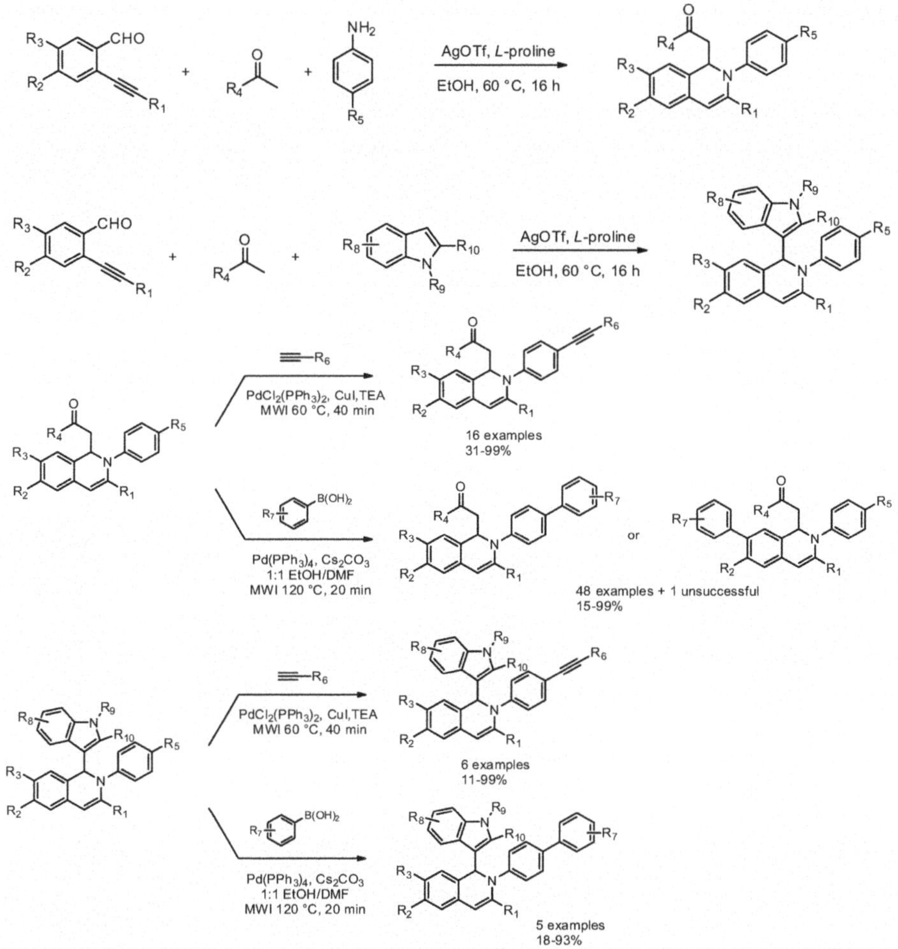

Core: dihydroisoquinoline	Library size: 75 examples	Key step: scaffold decoration	Method: solution phase
Yields: 11–99%	Biology: n/d	Comments: multiple cross-coupling: Suzuki, Sonogashira	[61]

Three-Ring Systems

Entry 55

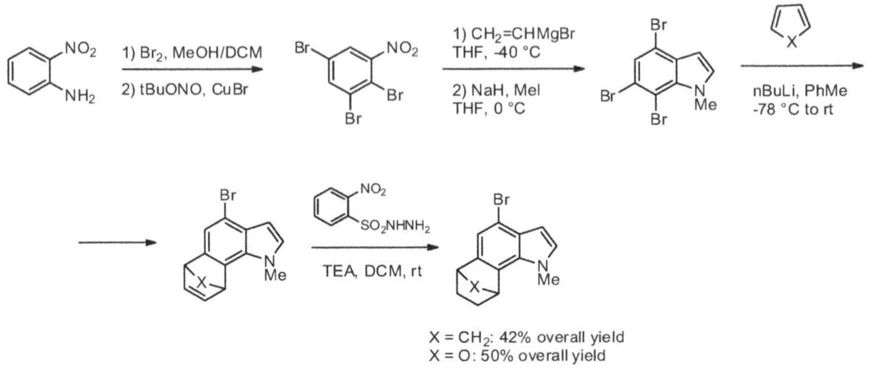

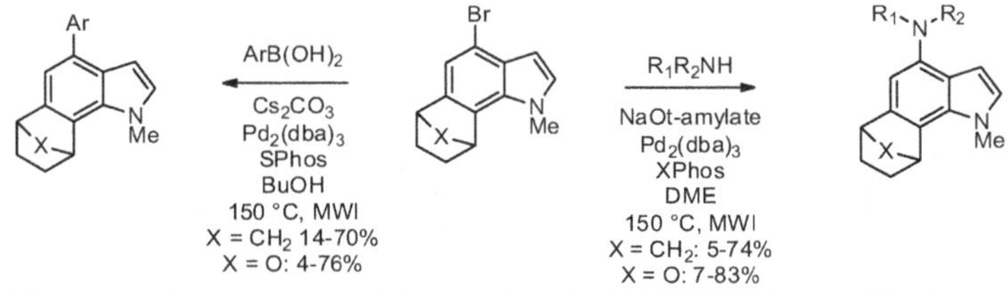

Core: indole-based NPL	Library size: 93 examples – 51 Suzuki, 42 Buchwald	Key step: scaffold decoration	Method: solution phase
Yields: 4–76%, 5–83%	Biology: n/d	Comments: multiple type cross-couplings – Suzuki, Buchwald–Hartwig	[62]

Entry 56

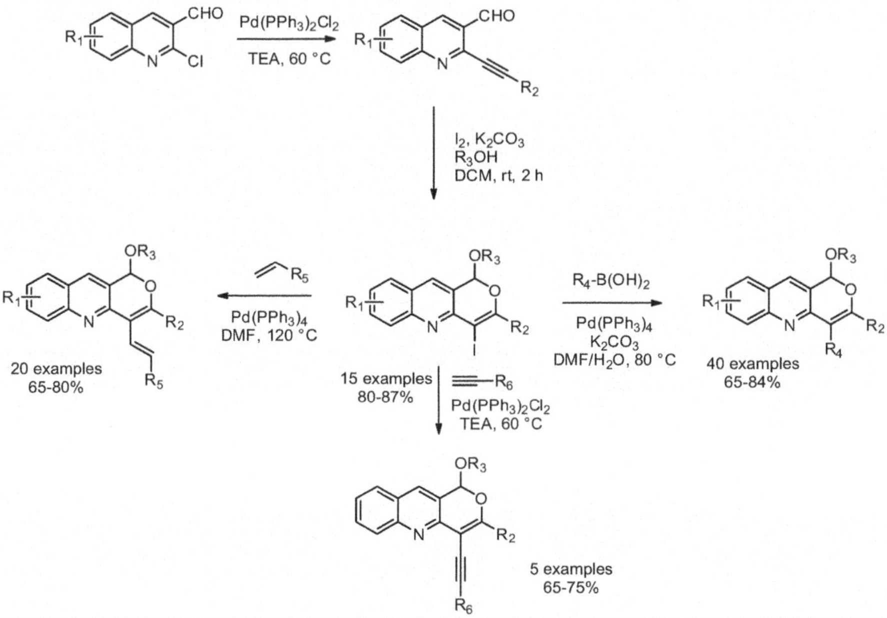

Core: pyranoquinolines	Library size: 65 examples – 40 Suzuki, 20 Heck, 5 Sonogashira	Key step: scaffold synthesis and decoration	Method: solution phase
Yields: 65–84%, 65–80%, 65–75%	Biology: n/d	Comments: scaffold synthesis (Sonogashira); multiple type cross-couplings – Suzuki, Heck, Sonogashira	[63]

3.8 *Domino–Tandem Cross-Coupling Reactions*

3.8.1 One-Ring Systems

Entry 57

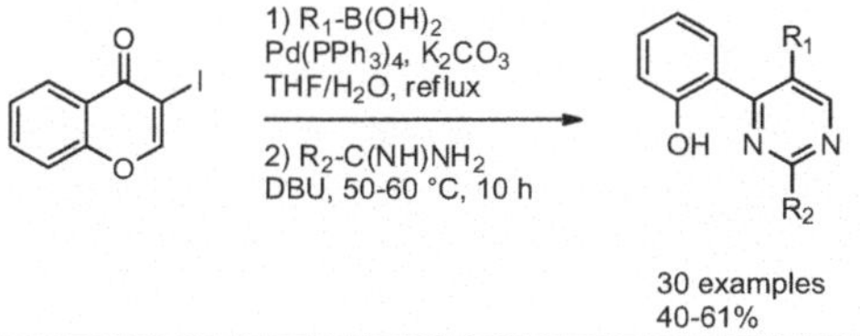

(continued)

Core: pyrimidine	Library size: 30 examples	Key step: scaffold synthesis	Method: solution phase
Yields: 40–61%	Biology: human hepatocellular carcinoma BEL-7402 cells	Comments: two-step one-pot Suzuki–cyclization	[53]

3.8.2 Two-Ring Systems

Entry 58

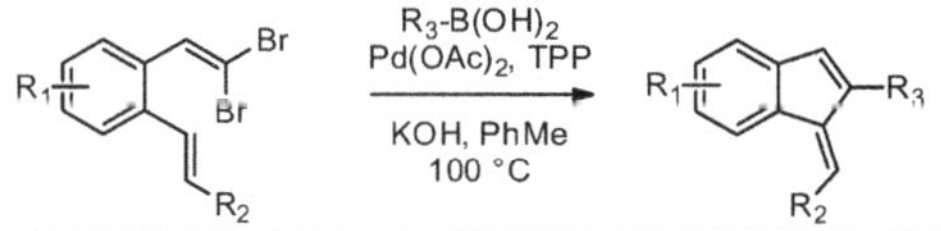

Core: indene	Library size: 20 examples	Key step: one-pot scaffold synthesis and decoration	Method: solution phase
Yields: 67–98%	Biology: n/d	Comments: tandem Suzuki–Heck	[30]

Entry 59

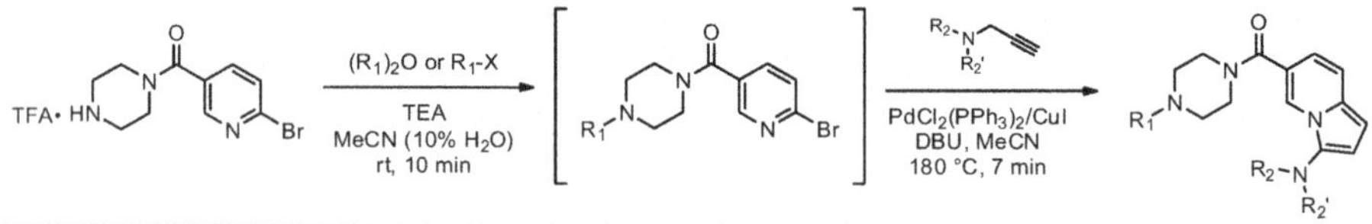

Core: Indolizine	Library size: 25 members (5 × 5)	Key step: one-pot scaffold synthesis and decoration	Method: solution phase
Yields: 28–78%	Biology: n/d	Comments: tandem Sonogashira–cycloisomerization, flow chemistry	[31]

Entry 60

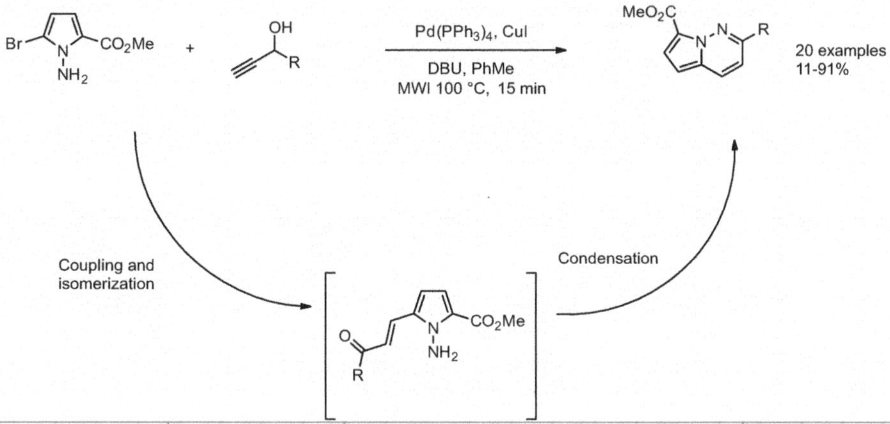

Core: pyrrolo-pyridazine	Library size: 20 examples	Key step: one-pot scaffold synthesis and decoration	Method: solution phase
Yields: 11–91%	Biology: n/d	Comments: Sonogashira coupling–isomerization–condensation	[32]

Entry 61

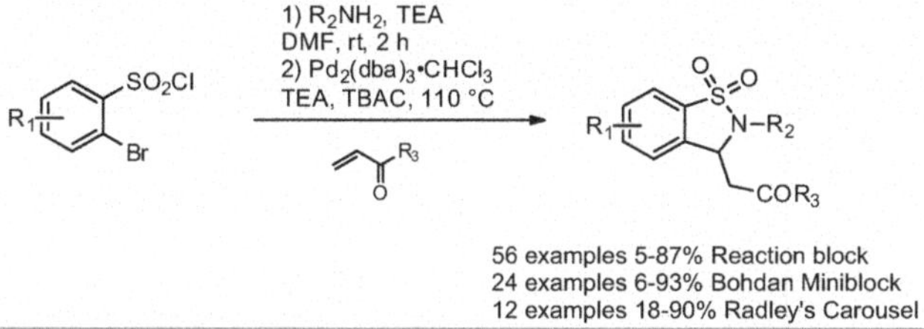

Core: sultam; 1,1-dioxido-1,2-benzisothiazoline	Library size: 92 examples	Key step: one-pot scaffold synthesis and decoration	Method: solution phase
Yields: 5–93%	Biology: n/d	Comments: tandem Heck–aza-Michael	[64]

Entry 62

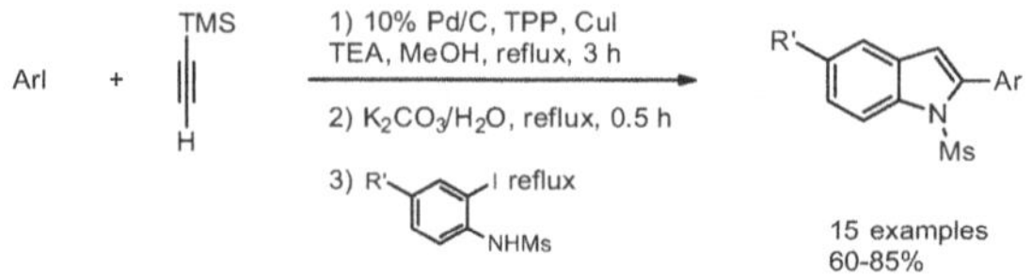

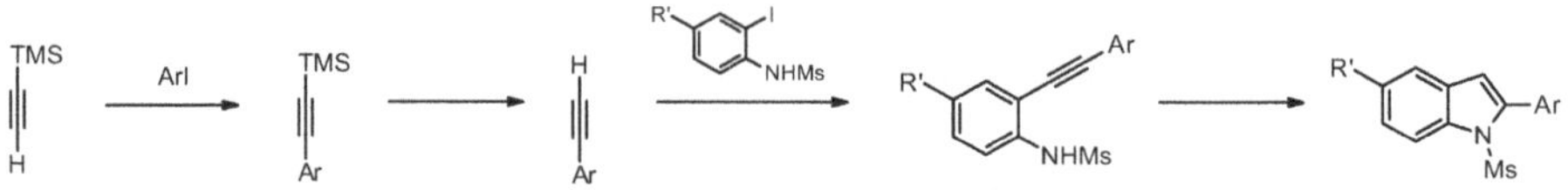

Core: indole	Library size: 15 examples	Key step: one-pot scaffold synthesis and decoration	Method: solution phase
Yields: 60–85%	Biology: SIRT1	Comments: Sonogashira–desilylation Sonogashira–cyclization	[33]

Entry 63

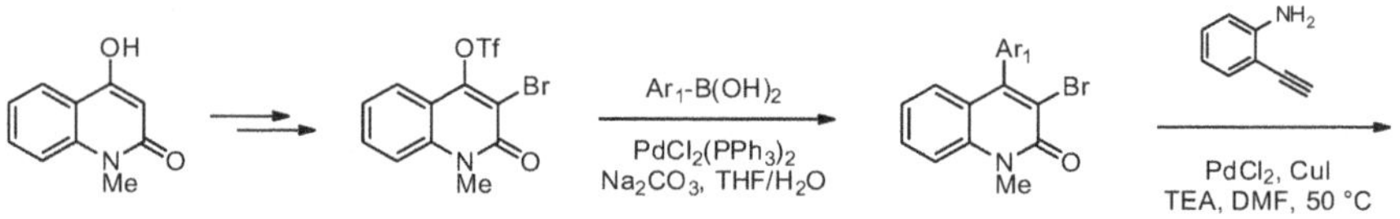

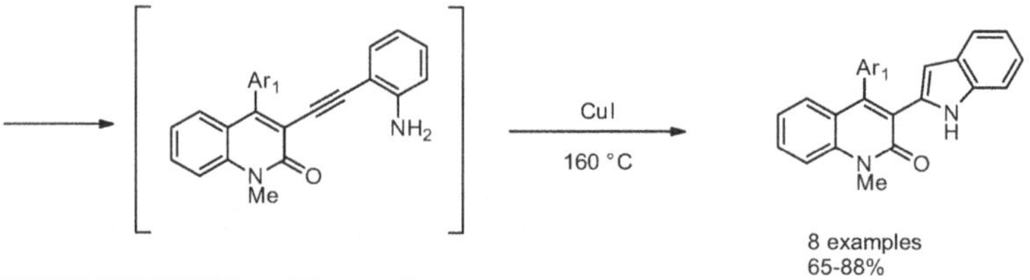

Core: quinolinone	Library size: 8 examples	Key step: one-pot scaffold synthesis and decoration	Method: solution phase
Yields: 65–88%	Biology: n/d	Comments: Suzuki, Sonogashira, cyclization	[34]

3.8.3 Three-Ring Systems

Entry 64

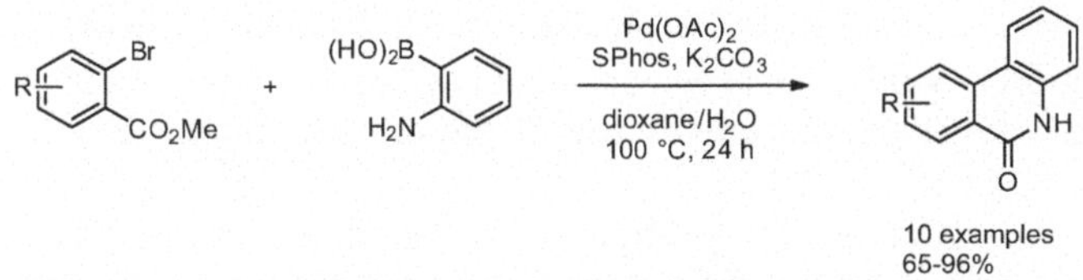

Core: phenanthridinone	Library size: 10 examples	Key step: one-pot scaffold synthesis and decoration	Method: solid phase
Yields: 65–96%	Biology: n/d	Comments: tandem Suzuki–cyclization	[66]

3.8.4 Four-Ring Systems

Entry 65

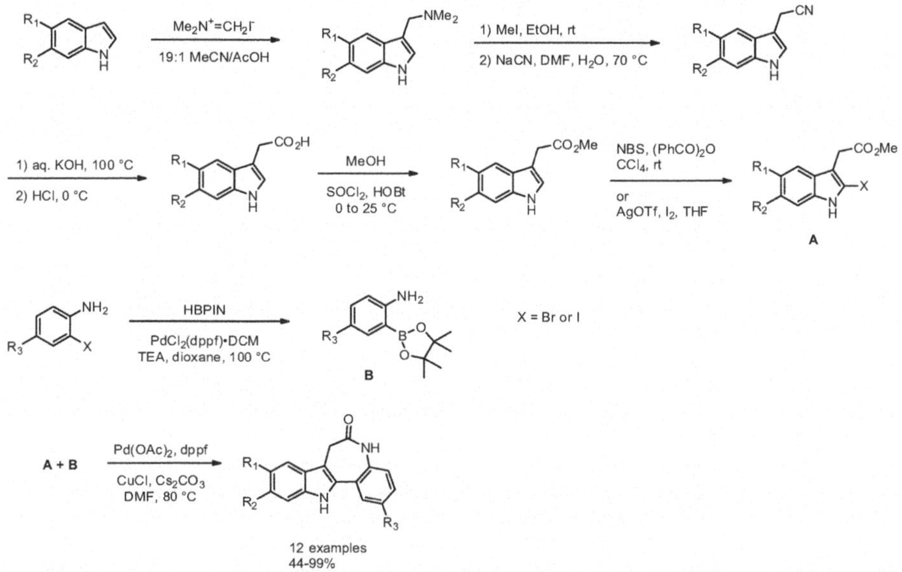

Core: dihydroindolo-benzazepinone (paullone)	Library size: 12 examples	Key step: one-pot scaffold synthesis, joining two diversity holding fragments	Method: solution phase
Yields: 44–99%	Biology: SIRT1 inhibitors	Comments: tandem Suzuki–intramolecular amide formation	[67]

Conclusion

In the last decade cross-coupling reactions became a standard laboratory technique in library synthesis. In the meantime biphenyl frameworks (including heteroaromatic ring systems) have extensively been reported in many biologically active products and have been considered as a privileged motif in medicinal chemistry [88]; thus, the importance of the cross-coupling reactions was further justified.

While there are some examples for solid-phase synthesis, mostly solution phase parallel synthesis is employed including the application of solid-supported reagents and fluorous two-phase reactions. Suzuki–Miyaura coupling is the most widely used C–C coupling reaction, while Sonogashira reaction is favorably used in scaffold synthesis followed by ring closure involving alkynes. Apart from the Suzuki and Sonogashira coupling, additional Pd-catalyzed reactions (Heck, Negishi, Buchwald–Hartwig amination, etc.) were used to increase the diversity around the core structure. Typically such diversity enhancement can be realized through C–C and C–N single bonds (Suzuki–Miyaura and Buchwald–Hartwig amination), double bonds (Heck), triple bonds (Sonogashira), etc. In summary, these techniques greatly increase the toolbox of the organic and medicinal chemists and contribute to afford novel chemotypes for the early phase of drug discovery.

References

1. Nobel Media AB (2013) The Nobel prize in chemistry 2010. www.Nobelprize.org
2. Balme G, Bouyssi D, Monteiro N (2006) Palladium-mediated cascade or multicomponent reactions: a new route to carbo- and heterocyclic compounds. Pure Appl Chem 78(2):231–239
3. Zhu J, Bienayme H (2006) Multicomponent reactions. Wiley-VCH, Weinheim
4. de Meijere A, von Zezschwitz P, Bräse S (2005) The virtue of palladium-catalyzed domino reactions – diverse oligocyclizations of acyclic 2-bromoenynes and 2-bromoenediynes. Acc Chem Res 38(5):413–422
5. Pellissier H (2013) Asymmetric domino reactions. RSC Publishing, London
6. Schröter S, Stock C, Bach T (2005) Regioselective cross-coupling reactions of multiple halogenated nitrogen-, oxygen-, and sulfur-containing heterocycles. Tetrahedron 61 (9):2245–2267
7. de la Hoz A, Loupy A (2012) Microwaves in organic synthesis. Wiley-VCH, Weinheim
8. Noël T, Buchwald SL (2011) Cross-coupling in flow. Chem Soc Rev 40(10):5010–5029
9. Martin R, Buchwald SL (2008) Palladium-catalyzed Suzuki–Miyaura cross-coupling reactions employing dialkylbiaryl phosphine ligands. Acc Chem Res 41(11):1461–1473
10. Li B, Dixneuf PH (2013) sp^2 C–H bond activation in water and catalytic cross-coupling reactions. Chem Soc Rev 42(13):5744–5767
11. Lipshutz BH, Taft BR, Abela AR, Ghorai S, Krasovskiy A, Duplais C (2012) Catalysis in the service of green chemistry: Nobel prize-winning palladium-catalysed cross-couplings, run in water at room temperature. Platinum Metals Rev 56(2):62–74
12. Dixneuf PH, Cadierno V (2013) Metal-catalyzed reactions in water. Wiley-VCH, Weinheim
13. Yin L, Liebsche J (2007) Carbon–carbon coupling reactions catalyzed by heterogeneous palladium catalysts. Chem Rev 107(1):133–173

14. Molnár Á (2013) Palladium-catalyzed coupling reactions. Wiley-VCH, Weinheim
15. Ruhland T, Nielsen SD, Holm P, Christensen CH (2007) Nanoporous magnesium aluminometasilicate tablets for precise, controlled, and continuous dosing of chemical reagents and catalysts: applications in parallel solution-phase synthesis. J Comb Chem 9(2):301–305
16. Lyons TW, Sanford MS (2010) Palladium-catalyzed ligand-directed C–H functionalization reactions. Chem Rev 110(2):1147–1169
17. Mousseau JJ, Charette AB (2013) Direct functionalization processes: a journey from palladium to copper to iron to nickel to metal-free coupling reactions. Acc Chem Res 46(2):412–424
18. Chen X, Engle KM, Wang D-H, Yu J-Q (2009) Palladium(II)-catalyzed C–H activation/C–C cross-coupling reactions: versatility and practicality. Angew Chem Int Ed 48(28):5094–5115
19. Wu Y, Wang J, Mao F, Kwong FY (2014) Palladium-catalyzed cross-dehydrogenative functionalization of C(sp2)-H bonds. Chem Asian J 9(1):26–47
20. Vedantham P, Guerra JM, Schoenen F, Huang M, Gor PJ, Georg GI, Wang JL, Neuenswander B, Lushington GH, Mitscher LA, Ye Q-Z, Hanson PR (2008) Ionic immobilization, diversification, and release: application to the generation of a library of methionine aminopeptidase inhibitors. J Comb Chem 10(2):185–194
21. Brucoli F, Howard PW, Thurston DE (2009) Efficient solid-phase synthesis of a library of distamycin analogs containing novel biaryl motifs on synphase lanterns. J Comb Chem 11 (4):576–586
22. Briehn CA, Bäuerle P (2002) Design and synthesis of a 256-membered ð-conjugated oligomer library of regioregular head-to-tail coupled quater(3-arylthiophene)s. J Comb Chem 4(5):457–469
23. Bui CT, Flynn BL (2006) Solid-phase synthesis of 2,3-disubstituted Benzo[b]thiophenes and benzo[b]selenophenes. J Comb Chem 8(2):163–167
24. Le Quement ST, Nielsen TE, Meldal M (2008) Solid-phase synthesis of aryl-substituted thienoindolizines: sequential pictet-spengler, bromination and Suzuki cross-coupling reactions of thiophenes. J Comb Chem 10(3):447–455
25. Peuchmaur M, Lisowski V, Gandreuil C, Maillard LT, Martinez J, Hernandez J-F (2009) Solid-phase synthesis of isocoumarins: a traceless halocyclization approach. J Org Chem 74 (11):4158–4165
26. Oh S, Jang HJ, Ko SK, Ko Y, Park SB (2010) Construction of a polyheterocyclic benzopyran library with diverse core skeletons through diversity-oriented synthesis pathway. J Comb Chem 12(4):548–558
27. Zhou H, Zhang W, Yan B (2010) Use of cyclohexylisocyanide and methyl 2-isocyanoacetate as convertible isocyanides for microwave-assisted fluorous synthesis of 1,4-benzodiazepine-2,5-dione library. J Comb Chem 12(1):206–214
28. Liu A, Zhou H, Su G, Zhang W, Yan B (2009) Microwave-assisted fluorous synthesis of a 1,4-benzodiazepine-2,5-dione library. J Comb Chem 11(6):1083–1093
29. Zhang T, Gao X, Wood HB (2011) Pd-catalyzed Negishi coupling of pyrazole triflates with alkyl zinc halides. Tetrahedron Lett 52(2):311–313
30. Ye S, Ren H, Wu J (2010) Efficient assembly of 1-methylene-1H-indenes via palladium-catalyzed tandem reaction of 1-(2,2-dibromovinyl)-2-alkenylbenzene with arylboronic acid. J Comb Chem 12(5):670–675
31. Lange PP, James K (2012) Rapid access to compound libraries through flow technology: fully automated synthesis of a 3-aminoindolizine library via orthogonal diversification. ACS Comb Sci 14(10):570–578
32. Wang M, Tan C, He Q, Xie Y, Yang C (2013) A novel convenient approach towards pyrrolo [1,2-b]pyridazines through a domino coupling–isomerization–condensation reaction. Org Biomol Chem 11:2574–2577
33. Rao RM, Reddy U, Alinakhi CH, Mulakayala N, Alvala M, Arunasree MK, Poondra RR, Javed J, Pal M (2011) Sequential coupling/desilylation–coupling/cyclization in a single pot under Pd/C–Cu catalysis: synthesis of 2-(hetero)aryl indoles. Org Biomol Chem 9:3808–3816

34. Wang Z, Wu J (2008) Synthesis of 1H-indol-2-yl-(4-aryl)-quinolin-2(1H)-ones via Pd-catalyzed regioselective cross-coupling reaction and cyclization. Tetrahedron 64 (8):1736–1742
35. Goh WK, StC Black D, Kumar N (2007) Synthesis of novel 7-substituted 5,6-dihydroindol-2-ones via a Suzuki–Miyaura cross-coupling strategy. Tetrahedron Lett 48(51):9008–9011
36. Antonow D, Cooper N, Howard PW, Thurston DE (2007) Parallel synthesis of a novel C2-aryl pyrrolo[2,1-c][1,4]benzodiazepine (PBD) library. J Comb Chem 9(3):437–445
37. Organ MG, Mayer S, Lepifre F, N'Zemba B, Khatri J (2003) Combining the use of solid-supported transition metal catalysis with microwave irradiation in solution-phase parallel library synthesis. Mol Divers 7:211–227
38. Pla D, Marchal A, Olsen CA, Francesch A, Cuevas C, Albericio F, Álvarez M (2006) Synthesis and structure–activity relationship study of potent cytotoxic analogues of the marine alkaloid lamellarin D. J Med Chem 49(11):3257–3268
39. Mugnaini C, Falciani C, De Rosa M, Brizzi A, Pasquini S, Corelli F (2011) Regioselective functionalization of quinolin-4(1H)-ones via sequential palladium-catalyzed reactions. Tetrahedron 67(32):5776–5783
40. Wang Z, Wang B, Wu J (2007) Diversity-oriented synthesis of functionalized quinolin-2(1H)-ones via Pd-catalyzed site-selective cross-coupling reactions. J Comb Chem 9(5):811–817
41. De Giorgi M, Voisin-Chiret AS, Sopková-de Oliveira Santos J, Corbo F, Franchini C, Rault S (2011) Design and synthesis of thienylpyridyl garlands as non-peptidic alpha helix mimetics and potential protein–protein interactions disruptors. Tetrahedron 67(34):6145–6154
42. Miguez JMA, Adrio LA, Sousa-Pedrares A, Vila JM, Hii KK (2007) A practical and general synthesis of unsymmetrical terphenyls. J Org Chem 72(20):7771–7774
43. Wu J, Wang L, Fathi R, Yang Z (2002) Palladium-catalyzed cross-coupling reactions of 4-tosylcoumarin and arylboronic acids: synthesis of 4-arylcoumarin compounds. Tetrahedron Lett 43(24):4395–4397
44. Vishnumurthy K, Makriyannis A (2010) Novel and efficient one-step parallel synthesis of dibenzopyranones via Suzuki–Miyaura cross coupling. J Comb Chem 12(5):664–669
45. Zhang Y, Lv Z, Zhong H, Zhang M, Zhang T, Zhang W, Li K (2012) Efficient Heck cross-coupling of 3-iodo-benzopyrones with olefins under microwave irradiation without phosphine. Tetrahedron 68(47):9777–9787
46. Cho C-H, Shi F, Jung D-I, Neuenswander B, Lushington GH, Larock RC (2012) Solution-phase synthesis of a highly substituted furan library. ACS Comb Sci 14(7):403–414
47. Mehta S, Waldo JP, Neuenswander B, Lushington GH, Larock RC (2013) Solution-phase parallel synthesis of a multisubstituted cyclic imidate library. ACS Comb Sci 15(5):247–254
48. Cho C-H, Neuenswander B, Lushington GH, Larock RC (2008) Parallel synthesis of a multi-substituted benzo[b]furan library. J Comb Chem 10(6):941–947
49. Roy S, Roy S, Neuenswander B, Hill D, Larock RC (2009) Solution-phase synthesis of a diverse isocoumarin library. J Comb Chem 11(6):1128–1135
50. Cho C-H, Neuenswander B, Larock RC (2010) Diverse methyl sulfone-containing benzo[b] thiophene library via iodocyclization and palladium-catalyzed coupling. J Comb Chem 12 (2):278–285
51. Cho C-H, Neuenswander B, Lushington GH, Larock RC (2009) Solution-phase parallel synthesis of a multi-substituted benzo[b]thiophene library. J Comb Chem 11(5):900–906
52. Davey PRJ, Delouvrié B, Dorison-Duval D, Germain H, Harris CS, Magnien F, Ouvry G, Tricotet T (2012) Facile preparation and Suzuki–Miyaura cross-coupling of N-2-alkylated 2H-1,2,3-triazole 4-boronates. Tetrahedron Lett 53(50):6849–6852
53. Xie F, Li S, Bai D, Lou L, Hu Y (2007) Three-component, one-pot synthesis of 2,4,5-substituted pyrimidines library for screening against human hepatocellular carcinoma BEL-7402 cells. J Comb Chem 9(1):12–13
54. Kumar S, Painter TO, Pal BK, Neuenswander B, Malinakova HC (2011) Application of sequential Cu(I)/Pd(0)-catalysis to solution-phase parallel synthesis of combinatorial libraries of dihydroindeno[1,2-c]isoquinolines. ACS Comb Sci 13(5):466–477

55. Chekmarev DS, Stepanov AE, Kasatkin AN (2005) Highly selective mono-substitution in Pd-catalyzed cross-coupling reactions of 3,6-dichloropyridazine with organozinc compounds. Tetrahedron Lett 46(8):1303–1305
56. Ma Z, Xiang Z, Luo T, Lu K, Xu Z, Chen J, Yang Z (2006) Synthesis of functionalized quinolines via Ugi and Pd-catalyzed intramolecular arylation reactions. J Comb Chem 8 (5):696–704
57. Yaziji V, Coelho A, El Maatougui A, Brea J, Loza MI, Garcia-Mera X, Sotelo E (2009) Divergent solution-phase synthesis of diarylpyrimidine libraries as selective A3 adenosine receptor antagonists. J Comb Chem 11(4):519–522
58. Waldo JP, Mehta S, Neuenswander B, Lushington GH, Larock RC (2008) Solution phase synthesis of a diverse library of highly substituted isoxazoles. J Comb Chem 10(5):658–663
59. Worlikar SA, Neuenswander B, Lushington GH, Larock RC (2009) Highly substituted indole library synthesis by palladium-catalyzed coupling reactions in solution and on a solid support. J Comb Chem 11(5):875–879
60. Bartoccini F, Piersanti G, Mor M, Tarzia G, Minetti P, Cabri W (2012) Divergent synthesis of novel 9-deazaxanthine derivatives via late-stage cross-coupling reactions. Org Biomol Chem 10:8860–8867
61. Markina NA, Mancuso R, Neuenswander B, Lushington GH, Larock RC (2011) Solution-phase parallel synthesis of a diverse library of 1,2-dihydroisoquinolines. ACS Comb Sci 13 (3):265–271
62. Thornton PD, Brown N, Hill D, Neuenswander B, Lushington GH, Santini C, Buszek KR (2011) Application of 6,7-indole aryne cycloaddition and Pd(0)-catalyzed Suzuki–Miyaura and Buchwald–Hartwig cross-coupling reactions for the preparation of annulated indole libraries. ACS Comb Sci 13(5):443–448
63. Aggarwal T, Imam M, Kaushik NK, Chauhan VS, Verma AK (2011) Pyrano[4,3-b]quinolines library generation via iodocyclization and palladium-catalyzed coupling reactions. ACS Comb Sci 13(5):530–536
64. Rolfe A, Young K, Volp K, Schoenen F, Neuenswander B, Lushington GH, Hanson PR (2009) One-pot, three-component, Domino Heck-aza-Michael approach to libraries of functionalized 1,1-dioxido-1,2-benzisothiazoline-3-acetic acids. J Comb Chem 11(4):732–738
65. Zang Q, Javed S, Porubsky P, Ullah F, Neuenswander B, Lushington GH, Basha FZ, Organ MG, Hanson PR (2012) Synthesis of a unique isoindoline/tetrahydroisoquinoline-based tricyclic sultam library utilizing a Heck-aza-Michael strategy. ACS Comb Sci 14(3):211–217
66. Tanimoto K, Nakagawa N, Takeda K, Kirihata M, Tanimori S (2013) A convenient one-pot access to phenanthridinones via Suzuki–Miyaura cross-coupling reaction. Tetrahedron Lett 54 (28):3712–3714
67. Soto S, Vaz E, Dell'Aversana C, Álvarez R, Altucci L, de Lera ÁR (2012) New synthetic approach to paullones and characterization of their SIRT1 inhibitory activity. Org Biomol Chem 10:2101–2112
68. Katkevica S, Salun P, Jirgensons A (2013) Synthesis of 5-substituted 3-mercapto-1,2,4-triazoles via Suzuki–Miyaura reaction. Tetrahedron Lett 54(34):4524–4525
69. Molander GA, Febo-Ayala W, Jean-Gérard L (2009) Condensation reactions to form oxazoline-substituted potassium organotrifluoroborates. Org Lett 11(17):3830–3833
70. Gros P, Doudouh A, Fort Y (2004) New polystyrene-supported stable source of 2-pyridylboron reagent for Suzuki couplings in combinatorial chemistry. Tetrahedron Lett 45(33):6239–6241
71. Lee SJ, Gray KC, Paek JS, Burke MD (2008) Simple, efficient, and modular syntheses of polyene natural products via iterative cross-coupling. J Am Chem Soc 130(2):466–468
72. Khanapure SP, Garvey DS (2004) Use of highly reactive, versatile and air-stable palladium–phosphinous acid complex $[(t\text{-Bu})_2P(OH)]_2PdCl_2$ (POPd) as a catalyst for the optimized Suzuki–Miyaura cross-coupling of less reactive heteroaryl chlorides and arylboronic acids. Tetrahedron Lett 45(27):5283–5286

73. Daykin LM, Siddle JS, Ankers AL, Batsanov AS, Bryce MR (2010) Iterative and regioselective cross-couplings of 2-chloro-3,4-diiodopyridine leading to 2,3,4-triheteroarylpyridines. Tetrahedron 66(3):668–675
74. Bondoux M, Mignon L, Ou K, Renaut P, Thomas D, Barberousse V (2009) Palladium-catalyzed C–C coupling: efficient preparation of new 5-thio-β-D-xylopyranosides as oral venous antithrombotic drugs. Tetrahedron Lett 50(27):3872–3876
75. Fraser AW, Besaw JE, Hull LE, Baird MC (2012) Pd(η_{3-1}-PhC_3H_4)(η_5-C_5H_5), an unusually effective catalyst precursor for Suzuki–Miyaura cross-coupling reactions catalyzed by bis-phosphine palladium(0) compounds. Organometallics 31(6):2470–2475
76. Szommer T, Lukács A, Szabó MJ, Hoffmann MG, Schmitt MH, Gerencsér J (2012) Parallel synthesis of 1,2,4-triazole derivatives using microwave and continuous-flow techniques. Mol Divers 16:81–90
77. Michalik D, Kumar K, Zapf A, Tillack A, Arlt M, Heinrich T, Beller M (2004) A short and efficient synthesis of N-aryl- and N-heteroaryl-N′-(arylalkyl)piperazines. Tetrahedron Lett 45 (10):2057–2061
78. Hanthorn JJ, Valgimigli L, Pratt DA (2012) Preparation of highly reactive pyridine- and pyrimidine-containing diarylamine antioxidants. J Org Chem 77(16):6908–6916
79. Brachet E, Peyrat J-F, Brion J-D, Messaoudi S, Alami M (2013) A palladium-catalyzed coupling of 3-chloroquinoxalinones with various nitrogen-containing nucleophiles. Org Biomol Chem 11:3808–3816
80. Amatore C, Jutand A, LeDuc G (2011) Kinetic data for the transmetalation/reductive elimination in palladium-catalyzed Suzuki–Miyaura reactions: unexpected triple role of hydroxide ions used as base. Chem Eur J 17(8):2492–2503
81. Schmitt M, de Araújo-Júnior JX, Oumouch S, Bourguignon J-J (2006) Use of 4-bromo pyridazine 3,6-dione for building 3-amino pyridazine libraries. Mol Divers 10:429–434
82. Liu J, Wang X (2011) Microwave-assisted, divergent solution-phase synthesis of 1,3,6-trisubstituted pyrazolo[3,4-d]pyrimidines. ACS Comb Sci 13(4):414–420
83. Heo Y, Song YS, Kim BT, Heo J-N (2006) A highly regioselective synthesis of 2-aryl-6-chlorobenzothiazoles employing microwave-promoted Suzuki–Miyaura coupling reaction. Tetrahedron Lett 47(18):3091–3094
84. Organ MG, Mayer S (2003) Synthesis of 4-(5-Iodo-3-methylpyrazolyl) phenylsulfonamide and its elaboration to a COX II inhibitor library by solution-phase Suzuki coupling using Pd/C as a solid-supported catalyst. J Comb Chem 5(2):118–124
85. Peng J, Lin W, Jiang D, Yuan S, Chen Y (2007) Preparation of a 7-arylthieno[3,2-d]pyrimidin-4-amine library. J Comb Chem 9(3):431–436
86. Ye C, Chen Z, Wang H, Wu J (2012) Generation of diverse 1-(isoquinolin-1-yl)guanidines via a sequential multi-component/cross-coupling reaction. Tetrahedron 68(26):5197–5202
87. Yu X, Pan X, Wu J (2011) An efficient route to diverse H-pyrazolo[5,1-a]isoquinolines via sequential multi-component/cross-coupling reactions. Tetrahedron 67(6):1145–1149
88. Welsch ME, Snyder SA, Stockwell BR (2010) Privileged scaffolds for library design and drug discovery. Curr Op Chem Biol 14:1–15

Index

The manufacturer's authorised representative in the EU is Springer Nature Customer Service Centre GmbH, Europaplatz 3, 69115 Heidelberg, Germany. If you have any concerns regarding our products, please contact ProductSafety@springernature.com

Printed and bound by CPI Group (UK) Ltd, Croydon, CR0 4YY
30/09/2026
02225652-0012